# BIOLOGY:

## THE STUDY OF LIVING ORGANISMS

A Complete Course
With 900 Questions and Answers

•

**GEORGE H. FRIED, Ph.D.**

*Former Professor of Biology*
*Brooklyn College*

## SCHAUM'S OUTLINE SERIES

### McGRAW-HILL, INC.

New York  St. Louis  San Francisco  Auckland  Bogotá  Caracas  Lisbon
London  Madrid  Mexico City  Milan  Montreal  New Delhi
San Juan  Singapore  Sydney  Tokyo  Toronto

GEORGE H. FRIED was Professor of Biology at Brooklyn College. He received his A.B. from Brooklyn College and his M.S. and Ph.D. from the University of Tennessee at Knoxville. His research interests have centered on metabolic aspects of comparative physiology and the enzymatic factors in genetic and experimental obesities. From 1983 to 1987 he served as chairperson of the Biology Department of Brooklyn College. He has taught courses in general biology and animal physiology and has developed a course in Biology and Society during a 30-year teaching career.

 This book is printed on recycled paper containing a minimum of 50% total recycled fiber with 10% postconsumer de-inked fiber. Soybean based inks are used on the cover and text.

A slightly different version of this book is published as *Schaum's Outline of Theory and Problems of Biology*, first edition, © 1990.

BIOLOGY: The Study of Living Organisms

1 2 3 4 5 6 7 8 9 10 11 12 13 14 15 16 17 18 19 20 BAW BAW 9 6 5

ISBN 0-07-022402-1

Sponsoring Editor, Jeanne Flagg
Production Supervisor, Leroy Young
Editing Supervisor, Patty Andrews

**Library of Congress Cataloging-in-Publication Data**

Fried, George, date
    Biology : the study of living organisms : a complete course with
900 questions and answers / George H. Fried.
        p.      cm.—(Schaum's outline series)
    Rev. ed. of: Schaum's outline of theory and problems of biology.
© 1990.
    Includes index.
    ISBN 0-07-022402-1
    1. Biology.   I. Fried, George, date. Schaum's outline of theory
and problems of biology.   II. Title.   III. Series.
QH308.2.F75   1995
574—dc20                                    94-44786
                                             CIP

*To my wife, Lillian, for her patience*

*and to Sylvia, Ellen, and Judy*

# Preface

Biology has undergone tremendous changes since the seminal contributions of Watson and Crick inaugurated the era of molecular biology (1953). The more descriptive aspects of the field, long associated with the older notion of biology as natural history, have been complemented by investigative insights that afford an understanding of life in terms of the precise characteristics of macromolecules such as DNA, RNA, and protein. To a marked extent, heredity, development, the control of cell function and even evolution have become better understood by applying the probes of molecular biology.

However, the drama of life in its evolving diversity and scope requires a perspective of time and an appreciation of meticulous descriptive detail to be truly appreciated. To cover these descriptive and historical aspects while elucidating with thoroughness the biochemical and molecular approaches imposes a daunting challenge to anyone undertaking a concise treatment of modern biology. Maintaining such a balance has been a continuing aim here.

All of the major themes expected in a one-year biology course at the introductory level are covered in the thirty-three chapters. Each chapter ends with a series of probing questions with answers which provides both greater depth and an opportunity to clarify material which may not have been completely understood.

Biology need not demand relevance. Our very aliveness underscores the intimacy of the connection between the discipline of biology and our everyday concerns. I trust that the material which has been gathered here will inspire understanding in its readers and promote an excitement and reverence for life as well as responsible conserving citizenship.

GEORGE H. FRIED

# Contents

CONTENTS

CONTENTS

CONTENTS

# Chapter 1

# The Basic Structure of Science

## 1.1 THE METHODS OF SCIENCE

*Science* is an organized system for the rigorous study of the natural world. It involves the application of the *scientific method* to problems formulated by trained minds in particular disciplines. Scientists may be interested in different aspects of nature, but they use a similar intellectual approach to guide their investigations.

Scientists must first formulate a problem to which they can then seek an answer. The answer generally involves an explanation relating to order or process in nature. The scientist is primarily interested in the mechanisms by which the natural world works rather than in questions of ultimate purpose.

Once a question has been raised, the scientist seeks answers by collecting data relevant to the problem. The data, which may consist of observations, measurements, counts, and a review of past records, are carefully sifted for regularity and relationships.

An educated guess, called a *hypothesis*, is then drawn up; this places the data into a conceptual framework. *Inductive logic* is used to formulate a hypothesis. This is a process of reasoning that usually begins with specific (or individual) pieces of data from which a general (or universal) premise is inferred.

**EXAMPLE 1**   A man takes up birdwatching and has occasion to observe mated pairs of many different kinds of birds. The man repeatedly sees only the drabber bird of any given pair lay eggs. From these observations, the man concludes that all male birds are colorful and all female birds are drab.

A hypothesis must be both logical and testable. Although the conclusion in Example 1 demonstrates the use of inductive logic, the conclusion cannot be tested and so, as stated, is useless as a scientific hypothesis. *Deductive logic*, in which the thought process is from the general to the specific, is used to state a hypothesis that can be tested. The "If . . . , then . . . " format is often used for this.

**EXAMPLE 2**   The conclusion in the previous example could be restated as: If birds of a particular *species* (i.e., birds capable of interbreeding to produce viable young) differ in color, then the more colorful ones are the males.*

After a workable hypothesis has been formulated, it is tested by constructing experiments and gathering new data, which in the end will either support or refute the hypothesis. *Note*: the application of the scientific method can be used to disprove a hypothesis, but *it can never prove anything absolutely.* Hence, a hypothesis that withstands the rigors of today's tests may have to be altered in the light of tomorrow's evidence.

An experiment must be so structured that the data gathered are free of bias and sampling error. Therefore, the validity of an experiment depends on a careful selection of organisms for the control and experimental groups, so that differences in age, genetic factors, previous treatment, etc., will not influence the results. Adequate numbers of individuals within each group are also crucial, since with small groups, individual peculiarities may be magnified. In addition, an experiment must be reproducible—i.e., other scientists must be able to repeat the experiment and get the same results.

**EXAMPLE 3**   A scientist wishes to know whether the addition of bone meal to the diet of cattle will improve their growth. On the basis of previous evidence of dietary benefits of bone meal to other animals, the scientist sets forth the hypothesis that the addition of bone meal to cattle feed will enhance growth in cattle. (*Note*: since all the cattle that have ever lived cannot be examined, this general statement can never be proved completely.)

---

* Although this is often the case, the reverse is true for some species of sexually dimorphic birds.

To test the hypothesis, the scientist sets up two comparable groups of cattle. The *experimental group* is given bone meal in addition to other requisites for growth, while the second group, the *control group*, receives identical treatment *except* no bone meal is given. In a properly constructed experiment, any differences that develop between the control and experimental groups will be due to the single factor being tested. The two groups in this case differ only in presence or absence of bone meal in their diet, so any differences in growth patterns must be attributed to this substance. If the experimental group demonstrates improved growth relative to the control group, the results would support the hypothesis. Should the experimental group fail to undergo improvement in growth in comparison with the control group, then the hypothesis would be refuted. A poorer growth performance by the experimental group would not merely refute the hypothesis, but would suggest a possible inhibitory effect of bone meal on cattle growth; such a finding would lead to a new hypothesis.

As seen in Example 3, once the experiments have been completed, the results must be weighed to see if the hypothesis should be accepted, modified, or rejected.

It should be noted that scientists only rarely follow a prescribed program in a rigid manner. Hypotheses may precede the actual accumulation of data, or the data may be accumulated and analyzed and the hypothesis developed simultaneously rather than in an orderly progression. Also, although scientists are very inquisitive and highly creative in their thought processes, their curiosity may be constrained by previous, long-accepted views. Revolutionary departures from established concepts are relatively rare.

## 1.2  BIOLOGY AS A SCIENCE

Biologists apply the methods of science to arrive at an understanding of living organisms. Within the context of biology, it is useful to regard life as complex matter that is susceptible to analysis by chemical and physical approaches. Although there are many phenomena within living systems that appear to lie beyond this *mechanistic* approach, biologists have been most successful at reaching an understanding of life by focusing on those processes involving transformations of matter and energy. *A living organism* may thus be defined as a complex unit of physicochemical materials that is capable of self-regulation, metabolism, and reproduction. Furthermore, a living organism demonstrates the ability to interact with its environment, grow, move, and adapt.

Biologists cannot study all of life in their own lifetimes. Therefore, they divide the vastness of the living world into many different kinds of organisms and may confine their investigations to a particular type of organism or, alternatively, may study particular aspects of different kinds of organisms and their interactions with one another.

**EXAMPLE 4**  *Entomologists*, specialists in insect biology, devote their efforts to understanding the various facets of insects but do not become involved with other kinds of organisms. On the other hand, *developmental biologists* investigate the characteristics of embryo development in many different kinds of organisms but do not venture into investigating other areas.

The boundaries that mark these different areas of investigation provide biology with its specific disciplines, but these boundaries are in a constant state of flux.

## 1.3  THE SIGNIFICANCE OF EVOLUTION

In pursuing their investigation of the living world, biologists are guided by theories that bring order to life's diversity. In science, a *theory* is a hypothesis that has withstood repeated testing over a long period of time (in contrast to the lay meaning of unproved supposition or fanciful idea). The single significant theme that unifies all branches of biology is the concept of *evolution*, the theory that all living organisms have arisen from ancestral forms by continual modification through time. Evolution conveys the notion of change and development. The patterns of these changes reflect upon major investigative trends in all disciplines of biology.

The acceptance of evolution as an explanation of present-day biological diversity is comparatively recent. Many respected biologists of the nineteenth and early twentieth centuries firmly believed in the fixity of species. Even Charles Darwin only reluctantly came to accept evolution as an explanation for the diversity of life. A vestige of this long history of undynamic explanations for *speciation* (differentiation into new species) is the current *creationist* movement.

Although not widely accepted until recently, the concept of evolution is not new; however, an understanding of the *mechanism* of evolutionary change is only a little more than a century old. In 1801, Jean Baptiste Lamarck proposed the first comprehensive explanation for the mechanism of evolution. Lamarck believed that an adult organism acquired new characteristics in direct response to survival needs and then passed these new characteristics on to its offspring. We now know that inheritance is determined by genes, so that *acquired* characteristics cannot be passed on to offspring. Today, the mechanism of evolution is believed to be *natural selection,* a concept outlined by Charles Darwin in his book *On the Origin of Species by Means of Natural Selection,* published in 1859. In the book, Darwin presented a cogent series of arguments for evolution being the pervading theme of life.

Darwin was influenced not only by his experiences as a naturalist (biologist) during his 5-year voyage aboard a surveying vessel, the *Beagle,* but also by the findings of geologists, economists, and even farmers of his community. The universality of science is aptly illustrated in Darwin's conceptual development.

Natural selection favors the survival of those individuals whose characteristics render them best-adapted to their environment. Slight variations occur among offspring of all species, making them slightly different from their parents. If a variation is not favorable for survival, then the individuals having that trait either do not survive to reproduce or survive but produce fewer offspring. As a result, the unfavorable trait eventually disappears from the population. If, however, a variation enhances survival in that particular environment, the individuals possessing it are more likely to reproduce successfully and thereby pass the trait on to their offspring. In the course of time, the trait favoring survival becomes part of the general population.

**EXAMPLE 5**  Gibbons are small apes that spend most of their time in the uppermost parts of trees; they rarely descend to the ground and travel instead by *brachiating* (swinging from branch to branch). They feed on the foliage and fruits found in the tops of trees in their native southeastern Asia and East Indies. Gibbons' hands are long and spindly, with very short, thin thumbs. This anatomy enables gibbons to grasp branches easily and to dangle from branches, as well as to pluck fruits and buds. They cannot, however, easily pick up objects off a flat surface (e.g., the ground) or be otherwise dextrous with their hands (in contrast to gorillas and chimpanzees). The gibbons' environment does not require the latter characteristics for survival.

Descended from a common ancestor of all apes, the gibbons are possessed of a hand anatomy that evolved by the chance occurrence of traits that were then acted on by natural selection pressures of their environment— the tops of trees, a place where the species encounters little competition for food and faces few dangers from predators.

## 1.4  ORGANIZATION OF LIFE

The study of evolution is particularly useful for classifying organisms into groups because it reveals how organisms are chronologically and morphologically (by form and structure) related to each other. The classification of organisms is known as *taxonomy.* Taxonomists utilize evolutionary relationships in creating their groupings. Although classification schemes are, of necessity, somewhat arbitrary, they probably do reflect the "family tree" of today's diverse living forms.

All organisms belong to one of five major kingdoms. A *kingdom* is the broadest taxonomic category. The five kingdoms are Monera, Protista, Fungi, Plantae, and Animalia. The Monera consists of unicellular organisms that lack a nucleus and many of the specialized cell parts, called *organelles.* Such organisms are said to be *prokaryotic* (*pro* = "before"; *karyotic* = "kernal," "nucleus") and consist of bacteria. All of the other kingdoms consist of *eukaryotic* (*eu* = "true") organisms, which have cells that contain a nucleus and a fuller repertory of organelles. Unicellular eukaryotes are placed in kingdom

Protista, which includes the protozoans and plant and funguslike protists. Multicellular organisms that manufacture their own food are grouped into kingdom Plantae; flowers, mosses, and trees are examples. Uni- and multicellular plantlike organisms that absorb food from their environment are placed in kingdom Fungi, which includes the yeasts and molds. Multicellular organisms that must capture their food and digest it internally are grouped into kingdom Animalia; snakes and humans are examples. Viruses are not cells and do not belong to one of the five kingdoms. They depend on invasion and metabolic takeover of cells from all the kingdoms.

# Solved Problems

**1.1**     What is science?

Science is the systematic study of particular aspects of the natural world. The scope of science is limited to those things that can be apprehended by the senses (sight, touch, hearing, etc.). Generally, science stresses an *objective approach* to the phenomena that are studied. Questions about nature addressed by scientists tend to emphasize *how* things occur rather than *why* they occur.

**1.2**     What is the scientific method?

In the broadest sense, the scientific method refers to the working habits of practising scientists as their curiosity guides them in discerning regularities and relationships among the phenomena they are studying. A rigorous application of common sense to the study and analysis of data also describes the methods of science. In a more formal sense, the scientific method refers to the model for research developed by Francis Bacon (1561–1626). This model involves the following sequence:

1.  Identifying the problem
2.  Collecting data within the problem area (by observations, measurements, etc.)
3.  Sifting the data for correlations, meaningful connections, and regularities
4.  Formulating a hypothesis (a generalization), which is an educated guess that explains the existing data and suggests further avenues of investigation
5.  Testing the hypothesis rigorously by gathering new data
6.  Confirming, modifying, or rejecting the hypothesis in light of the new findings

**1.3**     What is the significance of the hypothesis in the scientific method?

The hypothesis makes up the lattice-work upon which scientific understanding is structured. Often called an "educated guess," the hypothesis constitutes a generalization that describes the state of affairs within an area of investigation. The formulation of fruitful hypotheses is the hallmark of the creative scientific imagination.

**1.4**     What is the inductive method in science?

In logic, *induction* usually refers to a movement from the particular to the general. Thus, the creation of a hypothesis (a generalization) from the particulars (specifics) of the data constitutes an inductive leap within the scientific method. Since the scientific method involves such an inductive process at its very core, it is often described as the *inductive method*.

It is of considerable historic interest that Bacon, who first developed what we now call the *scientific method*, was extremely suspicious of the inductive step for the development of hypotheses. He thought that with the garnering of sufficient data and the establishment of a large network of museums, the hidden truths of nature would be apparent without invoking induction.

**1.5**    Are hypotheses always designed to be true assumptions of an actual state of affairs?

Hypotheses are not designed to be true for all time. In fashioning a hypothesis, the scientist is aiming for *operational* truth, a "truth" that works as an explanation of the data but may be replaced as new data are found, rather like a mountain climber who clambers from one handhold to another in scaling a mountain. A hypothesis must be consistent with all data available and must provide a logical explanation of such data. However, many hypotheses do just that but appear to contradict a commonsense notion of truth. For example, light was found to exhibit the properties of a wave. Later, it was discovered to act also as a discrete particle. Which is correct? A hypothesis called *quantum theory* maintains that light is both a wave and a particle. Although this may offend our common sense and even challenge our capacity to construct a model of such a contradictory phenomenon, quantum theory is consistent with the data, explains it, and is readily accepted by physicists.

**1.6**    What are the characteristics of a good hypothesis?

1.    A hypothesis must be consistent with and explain the data already obtained.
2.    A hypothesis must be falsifiable through its predictions; that is, results must be obtainable that can clearly demonstrate whether the hypothesis is untrue.

If several competing hypotheses are contending for acceptance, the one that is simplest and clearest will probably be accepted within the scientific community. In its evaluation of hypotheses, science is also concerned with fruitfulness as well as with truthfulness.

**1.7**    What is the fate of hypotheses after they have been formulated?

A hypothesis undergoes rigorous testing and may be confirmed by experimental testing of its predictions. Repeated confirmations elevate the hypothesis to the status of a theory. Occasionally, the major tenets of a hypothesis are confirmed, but some modification of the hypothesis may occur in light of new evidence. When hypotheses have been repeatedly confirmed over long periods of time, they are sometimes designated as *laws*, although some philosophers of science disagree with the use of the term "scientific law." When hypotheses are substantially contradicted by new findings, they are rejected to make way for new hypotheses.

**1.8**    What factors might lead to the formulation of a hypothesis that does not stand up to further evidence?

Hypotheses are designed to explain what is currently known. New developments may lead to a broader view of reality that exposes inadequacies in a hypothesis formulated at an earlier time. More often, an investigator uncovers a group of facts not truly representative of the total and bases a hypotheses on this small or unrepresentative sample. Such *sampling error* can be minimized by using statistical techniques. Also, while science deservedly prides itself on its objectivity and basic absence of prejudgement, a *subjective bias* may intrude during the collection of data or in the framing of a hypotheses and thereby lead an investigator to ignore evidence that does not support a preconceived notion. Bias may also be involved in the tendency to assume the well-accepted ideas of established authorities.

**1.9**    Does adherence to the scientific method completely explain the development of modern science?

Present technology can be attributed in large part to trained professionals following the classical methods of science. However, such recent philosophers of science as Thomas Kühne have pointed out the role that subjective factors such as intuition and cultural influences have played in the world of science. In their view, attitudes and the potential for breakthroughs are influenced by a particular collective mindset, with science moving from one set of *paradigms* (intellectual models) to another in fits and starts. The generally accepted view that scientists are automatons proceeding from one phase of the scientific method to the next is simply not true. The personality and humanity of scientists will be coming under increasing scrutiny as society recognizes that financial backing, problems of ethics, and questions of survival are all part of the equation of modern science.

**1.10** Can science be expected to solve all the world's mysteries and problems?

No. Science can successfully explain the forces that determine natural phenomena and has given us the power to control our environment to a considerable extent. However, science does not deal with the origin of the natural world but accepts its existence as a given. Nor can science answer questions of why the world exists as it does. Since hypotheses are judged in terms of their operational effectiveness, science cannot guide us in terms of morality, of the rightness or wrongness of particular courses of action. Science may best be viewed as an instrument for understanding rather than as a guide for social action.

**1.11** What is a living organism?

A living organism is primarily physicochemical material that demonstrates a high degree of complexity, is capable of self-regulation, possesses a metabolism, and perpetuates itself through time. To many biologists, life is an arbitrary stage in the growing complexity of matter, with no sharp dividing line between the living and non-living worlds.

Living substance is composed of a highly structured array of macromolecules, such as proteins, lipids, nucleic acids, and polysaccharides, as well as smaller organic and inorganic molecules. A living organism has built-in regulatory mechanisms and interacts with the environment to sustain its structural and functional integrity. All reactions occurring *within* an individual living unit are called its *metabolism*. Specific molecules containing information in their structure are utilized both in the regulation of internal reactions and in the production of new living units.

**1.12** What are the attributes of living organisms?

Living organisms generally demonstrate:

1. *Movement*: the motions within the organisms or movement of the organisms from one place to another (locomotion)
2. *Irritability*: the capacity of organisms to respond in a characteristic manner to changes—known as *stimuli*—in the internal and external environments
3. *Growth*: the ability of organisms to increase their mass of living material by assimilating new materials from the environment
4. *Adaptation*: the tendency of organisms to undergo or institute changes in their structure, function, or behavior that improve their capacity to survive in a particular environment
5. *Reproduction*: the ability of organisms to produce new individuals like themselves

**1.13** How do biologists study living organisms?

The vast panorama of life is much too complicated to be studied in its entirety by any single investigator. The world of living things may be studied more readily by (1) dividing organisms into various kinds and studying one type intensively or (2) separating the investigative approaches and specializing in one or another of them.

Systems of classification of living organisms that permit the relative isolation of one or another type of organism for organized investigation have been constructed within biology. At one time all living organisms were subdivided into two fundamental groups, or *kingdoms*: the plants, the subject matter of *botany*, and the animals, the subject matter of *zoology*. At present, there are grounds for classifying all of life into five kingdoms. These kingdoms are further subdivided into smaller categories that give particular disciplines their subject material. Thus, biologists who study hairy, four-legged creatures that nurse their young (mammals) are called *mammologists*. Those who investigate soft-bodied, shelled animals are *malacologists*. The study of simple plants such as the mosses is carried out by *bryologists*.

Biological disciplines may also be differentiated according to *how* living organisms are studied. For example, *morphologists* concentrate on structure, while *physiologists* consider function. *Taxonomists* devote themselves to the science of classification, and *cytologists* study the cells, which are the basic units for all life. *Ecologists* deal with the interaction of organisms with each other and with their external environment. A relatively new but extremely exciting and fruitful branch of biology is *molecular biology*, which is the study of life in terms of the behavior of such macromolecules as proteins and nucleic acids. It is this branch of biology that has enabled us to understand life at the molecular level and even to change the hereditary characteristics of certain organisms in order to serve the needs of society.

**1.14    Why does evolutionary theory occupy a central position in biology?**

The variety and complexity of life require organizing principles to help understand so diverse a subject area. Evolution is a concept that provides coherence for understanding life in its totality. It presents a narrative that places living things in a historical perspective and explains the diversity of living organisms in the present. It also illuminates the nature of the interaction of organisms with each other and with the external environment. Classification today is almost entirely based on evolutionary relationships. Even the findings of molecular biology have been focused on the nature of evolutionary changes. Evolution is the key to understanding the dynamic nature of an unfolding world of living organisms.

**1.15    What is evolution?**

Evolution is a continuously substantiated theory that all living things have descended with modification from ancestral organisms in a long process of adaptive change. These changes have produced the organisms that have become extinct as well as the diverse forms of life that exist today. Although the pace of evolutionary changes in the structure, function, and behavior of groups of organisms is generally thought to be constant when viewed over very long periods of time, lively debate has ensued about the tempo of change when examined over shorter periods. The rate of change may not always be even but may occur in rapid bursts, and such abrupt changes have, in fact, been observed in some organisms.

**1.16    Are there alternatives to the theory of evolution?**

Although almost every practicing biologist strongly supports the theory of evolution, some nonbiologists believe that all living forms were individually created by a supernatural being and do not change in time. This view, known as *special creation*, is consistent with the biblical account of the origin and development of life. More recently, certain scientific facts have been incorporated into a more cohesive theory of *scientific creationism*, which attempts to meld the scientific with the biblical explanations by stating that life has indeed had a longer history than biblical accounts would support, but that living organisms show only limited changes from their initial creations. Although scientific creationists have sought to downplay the religious aspects of their theory and have demanded an opportunity to have their views represented in biology texts, most biologists do not accept these concepts as being valid scientifically. Thus far, the courts of the United States have interpreted scientific creationism to be an intrusion of religion into the secular realm of education.

**1.17    What is the difference between evolution and natural selection?**

Evolution is a scientifically accepted theory of the origin of present organisms from ancestors of the past, through a process of gradual modification. Natural selection is an explanation of how such changes might have occurred, i.e., the mechanism of evolution.

The concept of evolution existed among the Greeks of Athens. In the eighteenth century, the French naturalist Comte Georges de Buffon suggested that species may undergo change and that this may have contributed to the diversity of plant and animal forms. Erasmus Darwin, grandfather of Charles, also subscribed to the concept of changes in the lineage of most species, although his ideas do not seem to have played a role in the development of Charles Darwin's concept of evolutionary change.

The first comprehensive theory of a mechanism of evolution was advanced by Lamarck in 1801. Like Charles Darwin, Lamarck was profoundly influenced by new findings in geology, which suggested that the earth was extremely old and that present-day geological processes operated during past millennia.

**1.18    What are the basic concepts of Lamarck's theory of the mechanism of evolution?**

Lamarck believed that changes occur in an organism during its lifetime as a consequence of adapting to a particular environment. Those parts that are used tend to become prominent, while those that are not tend to degenerate (*use-disuse concept*). Further, the changes that occur in an organism during its lifetime are then passed onto its offspring; i.e., the offspring inherit these acquired characteristics. Integral to Lamarck's theory was the concept of a deep-seated impulse toward higher levels of complexity within the organism, as if each creature were endowed with the will to seek a higher station in life.

The chief defect in Lamarck's theory is the view that acquired characteristics are inherited. With our present understanding of the control of inheritance by the genetic apparatus, we realize that only changes in the makeup of genes could lead to permanent alterations in the offspring. However, at the time of Lamarck's formulation, little was known of the mechanism of genetics. Even Darwin incorporated some of the Lamarckian views of the inheritance of acquired characteristics into his own thinking.

Lamarck's theory of evolution should not be regarded as being merely a conceptual error. Rather, it should be viewed as a necessary step in a continuing development of greater exactness in the description of a natural process. Science moves in slow, tentative steps to arrive at greater certainty. The truths of today's science are dependent on the intellectual forays of earlier investigators. They provide the shoulders on which others may stand to reach for more fruitful explanations.

**1.19**   How does the theory of natural selection explain the process of evolution?

The Darwinian theory of the mechanism of evolution accounts for change in organisms as follows:

1.   In each generation many more young are produced than can possibly survive, given the limited resources of a habitat, the presence of predators, the physical dangers of the environment, etc.
2.   As a result, a competition for survival ensues within each species.
3.   The original entrants in the competition are not exactly alike but, rather, tend to vary to a greater or lesser degree.
4.   In this contest, those organisms that are better adapted to the environment tend to survive. Those that are less fit tend to die out. The natural environment is the delineating force in this process.
5.   The variants that survive and reproduce will pass their traits on to the next generation.
6.   Over the course of many generations, the species will tend to reflect the characteristics of those who have been most successful at surviving, while the traits of those less well adapted will tend to die out.

Darwin was not certain about the source of variation in offspring, but he was aware of the existence of heritable variations within a species. We would now attribute these variations to the shuffling of genes associated with sexual reproduction (Chap. 8) and to the changes, known as *mutations*, in the structure of genes.

**1.20**   What does "survival of the fittest" mean?

The selection process arises from the fact that the best-adapted organisms tend to survive, almost as if nature had handpicked a fortunate few for perpetuation. At its heart, fitness has little to do with which individuals survive the longest or are the strongest; rather, it is determined by which ones pass on their genes to the next generation. It is true, though, that the longest-lived individual may have more time to produce offspring and the strongest may have more opportunity to mate. In both cases, therefore, *reproductive* success is the key. Present organisms can trace their lineage through a long series of past reproductive winners in the battle for survival.

It should be realized that reproductive success is not just a matter of active combat for resources and mates, but may involve cooperative and altruistic features by which individual success may be enhanced. Nor is the competition an all-or-none affair in which there is a single winner and many losers. Rather, one might view the struggle for existence and the survival of the fittest as a mechanism for *differential* reproduction—those with better adaptations outproduce those of lesser "fitness." Over long periods of time, the species tends to hoard those genes that are passed on by the better-adapted individuals.

**1.21**   If evolution results in increasing fitness within each species, will we eventually reach a point of perfect fitness and end the possibility of further change?

No. This will not occur, because the environment is constantly changing and today's adapted group becomes tomorrow's anachronism. Thus, the process is never-ending. More than 95 percent of all the species that have evolved in time have become extinct, probably because of the changing features of the earth. *Fossils*, which are preserved remnants of once-living organisms, attest to the broad range of species that have perished in the continued quest for an adaptiveness that can produce only temporary success. The continual changes in lifestyle of all organisms are inextricably linked to the continuity of change upon the surface of the earth itself.

It should be noted that much of the success of human beings in populating the world has resulted from our ability to alter the environment to suit our needs, rather than having evolved into a form that is perfectly adapted to an ever-changing environment.

**1.22**   How can natural selection, a single mechanism for change, produce such diversity in living forms?

Mutation and shuffling of genes through sexual reproduction and chromosomal rearrangement produce tremendous variation, even among individuals of the same species. This variation provides the potential for many possible adaptive responses to selection pressure. The imperative to adapt or die operates in a similar fashion everywhere, but the interplay between the myriad environmental pressures existing on earth and the genetic variability available to meet these stresses has resulted in the vast diversity of life forms, each with its unique solution for survival.

Organisms are not required to follow a set path in their assembling of traits during their evolutionary development. The final result of evolution is not an ideal living type, but rather a set of features that works (much like a hypothesis). The sometimes strange assortment of creatures found on this planet is itself a form of evidence for evolutionary development as opposed to special creation, in which greater perfection and elegance of body plan might be expected.

**1.23**   Can order be imposed upon the diversity of life?

For purposes of clarity and convenience, all organisms are arranged into categories. These categories, or *taxa*, start with the broadest division: the *kingdom*. Kingdoms are subdivided into *phyla*. Phyla are further divided into *classes*, classes into *orders*, orders into *families*, families into *genera*, and genera into *species*. The species is the smallest and best-defined classification unit. A *species* is a group of similar organisms that share a common pool of genes; upon mating, they can produce fertile offspring. The assignment of an organism to a particular set of taxonomic categories is based on the presumed evolutionary relationships of the individual to other members of the taxonomic group. Thus monkeys, apes, and humans share characteristics that place them in the same kingdom, phylum, class, and order, but they diverge from one another at the level of family.

**1.24**   What are the five kingdoms and the chief distinguishing features of each?

| Kingdom | Distinguishing Characteristics | Examples of Organisms |
|---|---|---|
| 1. Monera | Single-celled, *prokaryotic* organisms: cells lack nuclei and certain other specialized parts | Bacteria |
| 2. Protista | Single-celled, *eukaryotic* organisms: cells contain nuclei and many specialized internal structures | Protozoa |
| 3. Plantae | Multicellular, eukaryotic organisms that manufacture their food | Ferns, trees |
| 4. Fungi | Eukaryotic, plantlike organisms, either single-celled or multicellular, that obtain their food by absorbing it from the environment | Yeasts, molds |
| 5. Animalia | Eukaryotic, multicellular organisms that must capture their food and digest it internally | Fishes, birds, cows |

# Supplementary Problems

**1.25**  Science tends to deal primarily with questions of   (*a*) why.   (*b*) how.   (*c*) ethics.   (*d*) logic.

**1.26**  Induction is involved in   (*a*) testing hypotheses.   (*b*) discovering correlations among facts.   (*c*) developing hypotheses.   (*d*) none of the above.

**1.27**  The scientific method was originated by   (*a*) Darwin.   (*b*) Buffon.   (*c*) Bacon.   (*d*) Lamarck.

**1.28**  A good hypothesis should be   (*a*) falsifiable.   (*b*) consistent with the data.   (*c*) the simplest explanation.   (*d*) all of the above.

**1.29**  A hypothesis that has been confirmed many times is called   (*a*) a theory.   (*b*) a religious law.   (*c*) pseudoscience.   (*d*) none of the above.

**1.30**  Life to a biologist is essentially   (*a*) spiritual.   (*b*) physicochemical.   (*c*) mechanical.   (*d*) none of the above.

**1.31**  The study of animals is called   (*a*) botany.   (*b*) zoology.   (*c*) cytology.   (*d*) evolution.

**1.32**  The fixity (unchangingness) of species is assumed by (*a*) Lamarckians.   (*b*) special creationists.   (*c*) evolutionists.   (*d*) ecologists.

**1.33**  Evolution and natural selection are identical concepts.
(*a*) True   (*b*) False

**1.34**  Lamarck believed in the inheritance of acquired characteristics.
(*a*) True   (*b*) False

**1.35**  Darwin was an acknowledged expert in genetics.
(*a*) True   (*b*) False

**1.36**  Evolution is a process played out upon an unchanging earth.
(*a*) True   (*b*) False

**1.37**  Humans and apes belong to the same species.
(*a*) True   (*b*) False

**1.38**  Bacteria have cells with large nuclei.
(*a*) True   (*b*) False

# Answers

| | | | |
|---|---|---|---|
| **1.25** (*b*) | **1.29** (*a*) | **1.33** (*b*) | **1.37** (*b*) |
| **1.26** (*c*) | **1.30** (*b*) | **1.34** (*a*) | **1.38** (*b*) |
| **1.27** (*c*) | **1.31** (*b*) | **1.35** (*b*) | |
| **1.28** (*d*) | **1.32** (*b*) | **1.36** (*b*) | |

# Chapter 2

## The Chemistry of Life: An Inorganic Perspective

### 2.1 ATOMS, MOLECULES, AND CHEMICAL BONDING

All matter is built up of simple units called *atoms*. Although the word *atom* means something that cannot be cut (*a* = "without," *tom* = "cut"), these elementary particles are actually made up of many smaller parts, which are themselves further divisible. *Elements* are substances that consist of the same kinds of atoms. *Compounds* consist of units called *molecules*, which are intimate associations of atoms (in the case of compounds, different atoms) joined in precise arrangements.

Matter may exist in three different states, depending on conditions of temperature, pressure, and the nature of the substance. The *solid* state possesses a definite volume and shape; the *liquid* state has a definite volume but no definite shape; and the *gaseous* state possesses neither a definite volume nor a definite shape. Molecular or atomic movement is highest among gases and relatively low in solids.

Every atom is made up of a positively charged nucleus and a series of orbiting, negatively charged electrons surrounding the nucleus. A simple atom, such as hydrogen, has only one electron circulating around the nucleus, while a more complex atom may have as many as 106 electrons in the various concentric *shells* around the nucleus. Each shell may contain one or more *orbitals* within which electrons may be located. Every atom of an element has the same number of orbiting electrons, which is always equal to the number of positively charged *protons* in the nucleus. This balanced number of charges is the *atomic number* of the element. The *atomic weight*, or mass, of the element is the sum of the protons and neutrons in its nucleus. However, the atomic weights of atoms of a given element may differ because of different numbers of uncharged *neutrons* within their nuclei. These variants of a given element are called *isotopes*.

**EXAMPLE 1**   Oxygen is an element with an atomic number of 8 and an atomic weight of 16. Its nucleus contains eight protons and eight neutrons. There are eight circulating electrons outside the nucleus. Two of these electrons

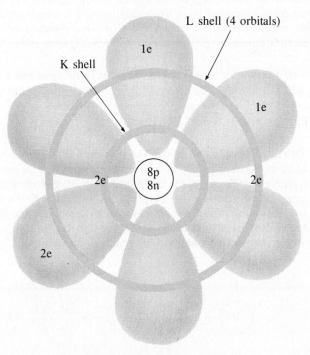

**Fig. 2.1**

are contained in the one spherical orbital of the first (K) shell, or energy level. The second (L) shell, which can accommodate as many as eight electrons, contains the remaining six electrons. They are distributed in orbitals, each of which contains two electrons. In the case of oxygen, one of the four orbitals of the second shell is not occupied by a pair of electrons. (See Fig. 2.1.)

Those electrons that occupy orbitals close to the nucleus have less energy associated with their rapid orbital revolution than the electrons occupying orbitals farther away from the central nucleus. Thus, when an atom absorbs energy, an electron is moved from a lower-energy-level orbital that is close to the nucleus to a higher-energy-level orbital farther away. Since electrons cannot be found between the discrete orbitals of the atom, according to modern theory (see *Schaum's Outline of College Chemistry*), energy exchanges involving the atom can occur only in definite "packets" called *quanta* (singular: quantum), which are equal to the average difference in energy between any two orbitals. When an excited electron jumps back to its original orbital, the energy difference is accounted for by the emission of quanta from the atom in the form of light. Electrons also possess other properties, such as *spin*.

Atoms interact with one another to form chemical communities. The tightly knit atoms making up the communal molecules are held together by chemical bonding. These bonds result from the tendency of atoms to try to fill their outermost shells. Only the noble gases—inert elements like neon and helium—have completely filled outer shells. The other elements will undergo changes that lead to more stable arrangements in which the outer shells are filled with electrons.

One way of achieving this more stable state is for an atom with very few electrons in its outer shell to donate them to an atom with an outer shell that is almost complete. The atom that donates the electrons will then have more protons than electrons and assume a positive charge; it is called a *cation*. The atom receiving the electrons assumes a negative charge and is called an *anion*. These two oppositely charged *ions* are electrostatically attracted to each other and are said to have an *ionic*, or *polar*, bond.

**EXAMPLE 2**   Sodium (Na), a corrosive metal, has an atomic number of 11, so that its third (M) shell has only one electron. (Shell K holds two electrons; shell L can hold eight; this leaves one electron for shell M.) Chlorine (Cl), a poisonous gas, with an atomic number of 17, has seven electrons in its outermost shell ($17 - 2 - 8 = 7$). In the interaction between these two atoms, sodium donates an electron to chlorine. Sodium now has a complete second shell, which has become its outermost shell, while chlorine now has eight electrons in its outermost shell. Na, having given up an electron, has a positive charge of $+1$; Cl, having absorbed an electron, now has a negative charge of $-1$ and will bond electrostatically with sodium to form NaCl, table salt.

A second way in which atoms may join with one another to bring about a filling of their outermost shells is by *sharing* a pair of electrons. The two bonding atoms provide one electron each in creating the shared pair. This pair of electrons forms a *covalent* bond that holds the two atoms together. It is represented by a solid line in the formula of a compound.

**EXAMPLE 3**   Hydrogen (H) contains only one electron in its outer (K) shell but requires two for completing that first shell. Oxygen (O) has six electrons in its outer shell and requires eight for completion. A single hydrogen atom may move within the sphere of influence of the outer shell of an oxygen atom to share its electron with the oxygen. At the same time one of the electrons of the oxygen atom is shared with the hydrogen atom to bring the hydrogen's outer shell up to the required two. If a second hydrogen is used to repeat this process, the oxygen will then have eight electrons and each hydrogen will have two electrons. In this process, two hydrogens have become covalently bonded to one oxygen to produce a molecule of water, $H_2O$ (see Fig. 2.2).

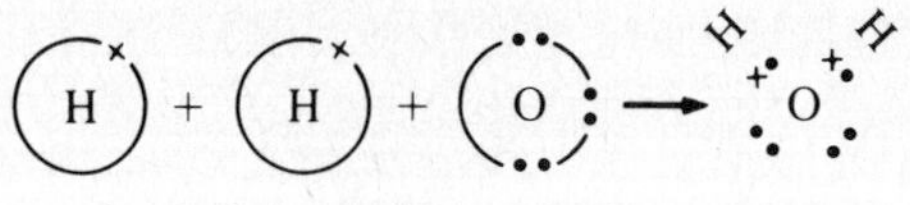

**Fig. 2.2**

In many molecules covalent bonding may occur not just singly (sharing a single pair of electrons), but may involve the formation of double or triple bonds in which two and even three pairs of electrons are shared. These double and triple bonds tend to fix the position of the participating atoms in a rigid manner. This differs from the situation of the single bond in which the atoms are free to rotate or spin on the axis provided by the single bond.

**EXAMPLE 4**   Carbon dioxide ($CO_2$) is a compound in which each of two oxygen atoms forms a double bond with a single carbon (C) atom, which in its unbonded state has four electrons in its outer shell. In this reaction, two electrons from a carbon atom join with two electrons from an oxygen atom to form one double bond, and remaining two electrons in the outer shell of the C atom join two electrons from the outer shell of a second O atom to form a second double bond. In this molecule, the C atom has a full complement of eight electrons in its outer shell, and each of the O atoms has eight electrons in its outer shell, as shown in Fig. 2.3.

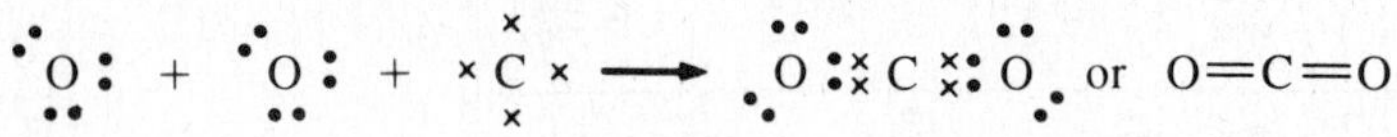

$$\overset{\bullet\bullet}{\underset{\bullet\bullet}{\text{O}}}\!\!: \; + \; \overset{\bullet\bullet}{\underset{\bullet\bullet}{\text{O}}}\!\!: \; + \; \overset{\times}{\underset{\times}{\times\,\text{C}\,\times}} \longrightarrow \; \overset{\bullet\bullet}{\underset{\bullet\bullet}{\text{O}}} \overset{\times}{\underset{\times}{:\!:}} \text{C} \overset{\times}{\underset{\times}{:\!:}} \overset{\bullet\bullet}{\underset{\bullet\bullet}{\text{O}}} \;\; \text{or} \;\; \text{O}\!=\!\text{C}\!=\!\text{O}$$

**Fig. 2.3**

In many covalent bonds, the electron pair is held more closely by one of the atoms than by the other. This imparts a degree of *polarity* to the molecule. Since oxygen nuclei have a particularly strong attraction for electrons, water behaves like a charged molecule, or *dipole*, with a negative oxygen end and a positive hydrogen end. Such molecules are considered to be polar in their activities, and the bond is classified as a *polar covalent* bond. Many properties of water, including its ability to bring about the ionization of other substances, are based on this polarity of the molecule.

Each type of molecule has bonding properties that fall somewhere along a continuous range from the totally polar bonds formed by electron transfer between atoms to the nonpolar situation found in most organic compounds, in which an electron pair is shared equally by the bonded atoms.

Occasionally, a pair of electrons present on a single atom may be shared with a second atom or ion that does not share its electrons. In the formation of an ammonium ion ($NH_4^+$), an ammonia molecule ($NH_3$) may attract a hydrogen ion ($H^+$) to a pair of electrons present on the N atom that are not involved in covalent bonding with the hydrogens already present in the molecule. This type of bond, in which a pair of electrons coming from one of the interacting moieties (parts) provides the "glue," is called a *coordinate covalent* bond. Such a bond is actually no different in its chemical significance from the more common covalent bonds previously described.

The gravitational (attractive) forces between molecules are known as *van der Waals forces*. These attractions do not effect chemical changes but are significant in influencing the physical properties of gases and liquids.

Even more significant in biology is the *hydrogen bond*, in which a proton ($H^+$) serves as the link between two molecules or different portions of the same large molecule. Although H bonds are considerably weaker than covalent bonds and do not lead to new chemical combinations, they play an important role in the three-dimensional structure of large molecules such as proteins and nucleic acids. It is H bonding that accounts for the loose association of the two polynucleotide chains in the double helix structure of DNA. Hydrogen bonding between adjacent water molecules accounts for many of the properties of water that play an important role in the maintenance of life.

The chemical properties of atoms are largely due to the number of electrons in their outer electron shells. All atoms with one electron in their outer shells behave similarly, while those with two electrons in their outer shells share another set of chemical properties. Atoms may be arranged in a table in accordance with their increasing atomic numbers. Each horizontal row starts with an atom containing one electron in its outer shell and ends with an atom containing a full outer shell. Such an arrangement of atoms is demonstrated in Fig. 2.4 and is called the *periodic table of the elements*. The vertical rows of elements have the same number of electrons in their outer shells, so that a periodicity (recurrence) of chemical properties happens as we move through the table from simpler to more complex elements.

Legend:

- 1 — atomic number
- H — symbol
- Hydrogen — name
- 1.0079 — atomic weight

Periodic Table

| 1 | 2 | 3 | 4 | 5 | 6 | 7 | 8 | 9 | 10 | 11 | 12 | 13 | 14 | 15 | 16 | 17 | 18 |
|---|---|---|---|---|---|---|---|---|----|----|----|----|----|----|----|----|----|
| **1**<br>H<br>Hydrogen<br>1.0079 | | | | | | | | | | | | | | | | | **2**<br>He<br>Helium<br>4.00260 |
| **3**<br>Li<br>Lithium<br>6.941 | **4**<br>Be<br>Beryllium<br>9.01218 | | | | | | | | | | | **5**<br>B<br>Boron<br>10.81 | **6**<br>C<br>Carbon<br>12.011 | **7**<br>N<br>Nitrogen<br>14.0067 | **8**<br>O<br>Oxygen<br>15.9994 | **9**<br>F<br>Fluorine<br>18.99840 | **10**<br>Ne<br>Neon<br>20.179 |
| **11**<br>Na<br>Sodium<br>22.98977 | **12**<br>Mg<br>Magnesium<br>24.305 | | | | | | | | | | | **13**<br>Al<br>Aluminum<br>26.98154 | **14**<br>Si<br>Silicon<br>28.086 | **15**<br>P<br>Phosphorus<br>30.97376 | **16**<br>S<br>Sulfur<br>32.06 | **17**<br>Cl<br>Chlorine<br>35.453 | **18**<br>Ar<br>Argon<br>39.948 |
| **19**<br>K<br>Potassium<br>39.098 | **20**<br>Ca<br>Calcium<br>40.08 | **21**<br>Sc<br>Scandium<br>44.9559 | **22**<br>Ti<br>Titanium<br>47.90 | **23**<br>V<br>Vanadium<br>50.9414 | **24**<br>Cr<br>Chromium<br>51.996 | **25**<br>Mn<br>Manganese<br>54.9380 | **26**<br>Fe<br>Iron<br>55.847 | **27**<br>Co<br>Cobalt<br>58.9332 | **28**<br>Ni<br>Nickel<br>58.70 | **29**<br>Cu<br>Copper<br>63.546 | **30**<br>Zn<br>Zinc<br>65.38 | **31**<br>Ga<br>Gallium<br>69.72 | **32**<br>Ge<br>Germanium<br>72.59 | **33**<br>As<br>Arsenic<br>74.9216 | **34**<br>Se<br>Selenium<br>78.96 | **35**<br>Br<br>Bromine<br>79.904 | **36**<br>Kr<br>Krypton<br>83.80 |
| **37**<br>Rb<br>Rubidium<br>85.4678 | **38**<br>Sr<br>Strontium<br>87.62 | **39**<br>Y<br>Yttrium<br>88.9059 | **40**<br>Zr<br>Zirconium<br>91.22 | **41**<br>Nb<br>Niobium<br>92.9064 | **42**<br>Mo<br>Molybdenum<br>95.94 | **43**<br>Tc<br>Technetium<br>(97) | **44**<br>Ru<br>Ruthenium<br>101.07 | **45**<br>Rh<br>Rhodium<br>102.9055 | **46**<br>Pd<br>Palladium<br>106.4 | **47**<br>Ag<br>Silver<br>107.868 | **48**<br>Cd<br>Cadmium<br>112.40 | **49**<br>In<br>Indium<br>114.82 | **50**<br>Sn<br>Tin<br>118.69 | **51**<br>Sb<br>Antimony<br>121.75 | **52**<br>Te<br>Tellurium<br>127.60 | **53**<br>I<br>Iodine<br>126.9045 | **54**<br>Xe<br>Xenon<br>131.30 |
| **55**<br>Cs<br>Cesium<br>132.9054 | **56**<br>Ba<br>Barium<br>137.34 | **57-71**<br>See Lanthanides | **72**<br>Hf<br>Hafnium<br>178.49 | **73**<br>Ta<br>Tantalum<br>180.9479 | **74**<br>W<br>Wolfram<br>183.85 | **75**<br>Re<br>Rhenium<br>186.207 | **76**<br>Os<br>Osmium<br>190.2 | **77**<br>Ir<br>Iridium<br>192.22 | **78**<br>Pt<br>Platinum<br>195.09 | **79**<br>Au<br>Gold<br>196.9665 | **80**<br>Hg<br>Mercury<br>200.59 | **81**<br>Tl<br>Thallium<br>204.37 | **82**<br>Pb<br>Lead<br>207.2 | **83**<br>Bi<br>Bismuth<br>208.9804 | **84**<br>Po<br>Polonium<br>(209) | **85**<br>At<br>Astatine<br>(210) | **86**<br>Rn<br>Radon<br>(222) |
| **87**<br>Fr<br>Francium<br>(223) | **88**<br>Ra<br>Radium<br>226.0254 | **89-103**<br>See Actinides | **104**<br>Rf-Ku<br>(Rutherfordium; Kurchatovium)<br>(261) | **105**<br>Ha<br>Hahnium<br>(262) | **106**<br>(263) | | | | | | | | | | | | |

Lanthanides

| 57 | 58 | 59 | 60 | 61 | 62 | 63 | 64 | 65 | 66 | 67 | 68 | 69 | 70 | 71 |
|----|----|----|----|----|----|----|----|----|----|----|----|----|----|----|
| La<br>Lanthanum<br>138.9055 | Ce<br>Cerium<br>140.12 | Pr<br>Praseodymium<br>140.9077 | Nd<br>Neodymium<br>144.24 | Pm<br>Promethium<br>(145) | Sm<br>Samarium<br>150.4 | Eu<br>Europium<br>151.96 | Gd<br>Gadolinium<br>157.25 | Tb<br>Terbium<br>158.9254 | Dy<br>Dysprosium<br>162.50 | Ho<br>Holmium<br>164.9304 | Er<br>Erbium<br>167.26 | Tm<br>Thulium<br>168.9342 | Yb<br>Ytterbium<br>173.04 | Lu<br>Lutetium<br>174.97 |

Actinides

| 89 | 90 | 91 | 92 | 93 | 94 | 95 | 96 | 97 | 98 | 99 | 100 | 101 | 102 | 103 |
|----|----|----|----|----|----|----|----|----|----|----|-----|-----|-----|-----|
| Ac<br>Actinium<br>(227) | Th<br>Thorium<br>232.0381 | Pa<br>Protactinium<br>231.0359 | U<br>Uranium<br>238.029 | Np<br>Neptunium<br>237.0482 | Pu<br>Plutonium<br>(244) | Am<br>Americium<br>(243) | Cm<br>Curium<br>(247) | Bk<br>Berkelium<br>(247) | Cf<br>Californium<br>(251) | Es<br>Einsteinium<br>(254) | Fm<br>Fermium<br>(257) | Md<br>Mendelevium<br>(258) | No<br>Nobelium<br>(255) | Lr<br>Lawrencium<br>(260) |

Fig. 2.4

Helium, neon, argon, etc., all belong to the *noble gases,* and their particular property of nonreactivity will recur each time we reach the group that has a complete outer shell of electrons. A similar relation holds for the metals lithium, sodium, potassium, etc., all of which have in their outer shell one electron, which tends to be removed in interaction with other atoms. The arrangement of atoms into a table of this type conveys a sense of order among the more than 100 elements and readily demonstrates the relationship of atomic structure to chemical function as we move from simpler to more complex atoms.

## 2.2  CHEMICAL REACTIONS AND THE CONCEPT OF EQUILIBRIUM

Chemical reactions are represented by equations in which the reacting molecules (*reactants*) are shown on the left and the *products* of the reaction are shown on the right. An arrow indicates the direction of the reaction. The participants in the reaction are indicated by *empirical formulas,* which are a shorthand method of indicating the makeup of the molecules participating in the reaction. Each element in the molecule is denoted by a characteristic symbol (e.g., H for hydrogen and O for oxygen), and the number of each of these atoms is given by a subscript to the right of each symbol (e.g., $H_2O$). The number of molecules involved is indicated by a numerical coefficient to the left of each participating molecule (e.g., $2H_2O$).

In some reactions a simple decomposition occurs and is shown as AB → A + B. Other reactions involve a simple combination: A + B → AB. More complex reactions might involve the interaction of two or more molecules to yield products that are quite different from the reactant molecules: A + B → C + D. In all these reactions, the numbers and kinds of *atoms* that appear on the left must be accounted for in the products on the right.

Most reactions do not go to completion; instead, they reach a state of *equilibrium,* in which the interaction of reactants to form products is exactly balanced by the reverse reaction in which products interact to form reactants. The *law of mass action* states that at equilibrium the product of the molar concentrations of the molecules on the right-hand side of the equation divided by the product of the molar concentrations of the reactants will always be a constant. (Molar concentrations are explained below.) If the reaction tends to reach equilibrium with a greater amount of product, then the *equilibrium constant* will be high. If reactants tend to predominate (i.e., the reaction does not proceed very far to the right), then the equilibrium constant will be small. Should any molecules of reactant or product be added to the system, the reaction will be altered to reach a state in which the concentrations once again provide a ratio that is equal to the equilibrium constant. In the equation A + B → C + D, the mass action formulation would be represented as

$$\frac{[C] \cdot [D]}{[A] \cdot [B]} = K_{eq}$$

where [ ] stands for the molar concentration and $K_{eq}$ is the equilibrium constant.

*Concentration* is a measure of the amount of a particular substance in a given volume. Since the tendency of most reactions to occur is based partly on how crowded the reacting molecules are, concentration is a significant factor in the determination of chemical events. A common way to express the concentration of a solution is in moles of solute per liter of solution (molarity). A *mole* (mol), which is the molecular weight of a given molecule expressed as grams, may be thought of as a specific number of atoms or molecules. One mol of any given compound contains $6.02 \times 10^{23}$ molecules. Thus 1 mol of $H_2O$ contains the same number of molecules as 1 mol of $CO_2$, as is true for 2 mol or $\frac{1}{3}$ mol. By similar reasoning, a 1-molar (1 $M$) solution contains twice as many solute molecules as a 0.5 $M$ solution. Since molecules are the units involved in chemical transformations, the molar concentration assures a uniform measurement of interacting units and is more meaningful than absolute weights in assessing chemical interactions.

In some cases, *normality* ($N$)-rather than molarity is preferred as a means of expressing concentration. Since normality is essentially molarity divided by the *valence,* or *chemical power,* of a molecule, it more precisely measures the chemical reactivity of materials in solution. Substances with a combining power of 2 need be present in only half the concentration of those with a valence of 1 to bring about a particular effect.

**EXAMPLE 5**    The base NaOH reacts with the acid $H_2SO_4$ to form water, $H_2O$ and the salt $Na_2SO_4$. A balanced equation for this reaction is

$$2NaOH + H_2SO_4 \rightarrow Na_2SO_4 + H_2O$$

If we were to use one liter of 1 *M* NaOH, we would need one liter of only 0.5 *M* $H_2SO_4$ to provide sufficient acid for the reaction to occur, since the equation shows that only half as many moles of $H_2SO_4$ are required. However, if we were measuring concentration using normality, for one liter of 1 *N* NaOH we would also use one liter of 1 *N* $H_2SO_4$. This is because a 1 *N* solution of $H_2SO_4$ is also a 0.5 *M* solution. Similarly, a 1.0 *N* solution of $H_3PO_4$ is also a 0.33 *M* solution. In the case of ions, a 1.0 *N* solution of $Na^+$ is also a 1.0 *M* solution of $Na^+$, and a 1.0 *N* solution of $Ca^{2+}$ is also a 0.5 *M* solution.

## 2.3    COLLIGATIVE PROPERTIES OF SOLUTIONS

The presence of solutes (dissolved particles) in a solvent tends to lower the *vapor pressure*, or escaping tendency, of the liquid molecules. The freezing point is also lowered, and the boiling point is raised by the solute particles. *Osmotic pressure*, as explained below, is also increased by solute particles. These properties, taken together, are known as the *colligative* properties of a solution. They are influenced only by the number of particles, not by the kinds or chemical reactivity of these particles. If a particular molecule dissociates into several ions, it will influence the colligative properties to the extent of its dissociation; e.g., if a compound dissociates into two ions, a 1 *M* solution of the substance will behave as if it were closer to 2 *M* in terms of its effects on osmosis, freezing-point depression, etc.

If we were to divide a container into two compartments by means of a membrane that was impermeable to solute but allowed solvent to pass through (a *semipermeable membrane*) and were to place different concentrations of a solution on each side of the membrane, the solute molecules would be unable to pass through the membrane but the solvent molecules would move to the region where they are less crowded. Since the more dilute compartment contains more solvent molecules than the more concentrated compartment, water or some similar solvent would move from less concentrated solute concentrations to more concentrated solute concentrations. This phenomenon is known as *osmosis*. The pressure exerted by the tendency of solvent molecules to move across the membrane is called *osmotic pressure*. As the volume increases in one compartment relative to the other, the solution will rise until the gravitational forces associated with the increase in height in the more concentrated compartment equals the osmotic pressure associated with the difference in concentration. If the continued changes in concentration are accounted for, measurement of the rise of a column of liquid in a container may be used to determine the actual osmotic pressure.

## 2.4    THE LAWS OF THERMODYNAMICS

*Thermodynamics* deals with the transformations of energy in all of its forms. Although the word literally means the "movement" or "change of heat," all forms of energy may be degraded to heat, so that those rules that apply to heat transformations may describe energy changes in general.

*Energy* is the capacity to do work. *Work* is traditionally defined as a force operating through a distance. *Force* refers to a push or a pull that alters the motion of a body. In biology energy is used to counter natural physical tendencies, as in the migration of sugar molecules against their concentration gradient.

Energy exists in various forms. *Heat* is the energy associated with the rapid internal movement of molecules of liquids and gases. *Mechanical energy* is the energy found in the motion of bodies; *chemical energy* is the energy contained in the bonds that hold atoms together within molecules; and *radiant* energy is derived from the sun and other sources of wave-propagated energy. All forms of energy may exist either in actualized form, such as the *kinetic* energy of a falling stone, or in *potential* form, such as the potential energy of a stone positioned atop a mountain or of certain organic molecules with high-energy bonds, which will release energy when they are broken.

The three laws of thermodynamics govern all transformations of energy in the natural world. The first law, the *law of conservation of energy*, asserts that energy can be neither created nor destroyed. Physicists now view matter as a special case of energy, so that the reactions associated with atomic fission or fusion may be understood in terms of the first law. In atomic and hydrogen bombs, a small amount of mass is converted to great amounts of energy in accordance with Albert Einstein's equation $E = mc^2$, where the mass lost is multiplied by the velocity of light squared.

The second law of thermodynamics is sometimes stated in terms of the transfer of heat: heat moves from hot bodies to cold bodies. However, this formulation does not provide sufficient insight into the real significance of the second law. A better explanation is that in any transformation, energy tends to become increasingly unavailable for useful work. Since useful work is associated with producing order, we may also express the second law as the tendency in nature for systems to move to states of increasing disorder or randomness. The term for disorder is *entropy*, although this term is also defined as a measure of the unavailability of energy for useful work (a consequence of disorder). The second law may also be viewed in terms of potential energy: in any spontaneous reaction, one in which external energy does not play a role, the potential energy tends to be diminished. All these formulations can be condensed into the somewhat pessimistic conclusion that the universe is running down and that eventually all energy will be uniformly distributed throughout an environment in which no further energy exchanges are possible, because entropy has been maximized.

The third law states that only a perfect crystal, a system of maximum order, at −273 °C (absolute zero) can have no entropy. Since this ideal condition can never actually be met, all natural systems are characterized by some degree of disorder.

All reactions that result in the release of *free energy*, the form of energy associated with the performance of useful work, are classified as *exergonic* reactions. These are reactions that tend to occur spontaneously. In living systems exergonic reactions are usually associated with the breakdown of complex molecules, whose bonds represent a storage of ordered forms of energy, into simpler molecules containing bonds of much lower orders of energy. An analogy that illustrates the nature of such exergonic reactions is a stone rolling down from the top of a hill. The energy that went into placing the stone on the hilltop exists as potential (stored) energy in the stone by virtue of its position. The stone can move downhill without additional energy input and, in doing so, will release its stored energy in mechanical form as it moves to the bottom. The energy of motion is called *kinetic energy*, from a Greek root meaning "movement." Although the stone has a tendency to move down the hill, it may require an initial push to get it over the edge. This represents the *energy of activation* that is required to cause even spontaneous reactions to begin. Not all the stored energy is released as mechanical energy, since a portion of the starting energy will be given off as heat during the movement of the stone as it encounters friction with the hill's surface.

Those reactions that involve a change from a lower energy state to a higher one are called *endergonic* reactions. In this case, free energy must come into the system from outside, much like a stone being rolled uphill by means of the expenditure of energy. In biological systems, endergonic reactions are only possible if they are coupled with exergonic reactions that supply the needed energy. A number of exergonic reactions within living systems liberate the free energy that is stored in the high-energy bonds of molecules like adenosine triphosphate (ATP). This ATP is broken down to provide energy to drive the various endergonic reactions that make up the synthesizing activities of organisms.

## 2.5  THE SPECIAL CASE OF WATER

Water is the single most significant inorganic molecule in all life forms. It promotes complexity because of its tendency to dissolve a broad spectrum of both inorganic and organic molecules. Because of its polar quality, it promotes the dissociation of many molecules into ions, which play a role in regulating such biological properties as muscle contraction, permeability, and nerve impulse transmission.

Water is instrumental in preventing sharp changes in temperatures that would be destructive to the structure of many macromolecules within the cell. It has one of the highest *specific heats* of any natural substance; that is, a great deal of heat can be taken up by water with relatively small shifts in temperature. It also has a high *latent heat of fusion*, meaning that it releases relatively large amounts of heat when it passes from the liquid to solid (ice) phase. Conversely, ice absorbs relatively large amounts of heat when it melts. This quality produces a resistance to temperature shifts around the freezing point. The high *latent heat of vaporization* of water (the heat absorbed during evaporation) serves to rid the body surface of large amounts of heat in conversion of liquid water to water vapor.

**EXAMPLE 6**   Each gram (g) of water absorbs 540 calories (cal) upon vaporization. Calculate the amount of heat lost over 5 square centimeters ($cm^2$) of body surface when 10 g of water is evaporated over that surface.

Since 1 g of water absorbs 540 cal upon vaporization, 10 g of water will take up 5400 cal over the 5-$cm^2$ area, or 1080 cal per $cm^2$. This avenue of heat dissipation is lost if the air is saturated with water so that evaporation cannot occur, which explains the discomfort associated with a hot, muggy day.

The characteristics mentioned above, as well as a high surface tension and water's anomalous property of expanding upon freezing, are largely owing to the tendency of water molecules to cohere tightly to one another through the constant formation of hydrogen bonds between adjacent water molecules.

Finally, water is transparent; thus, it does not interfere with such processes as photosynthesis (at shallow depths) and vision, both of which require light.


## 2.6   MAINTAINING STABLE pH IN LIVING SYSTEMS

Acidity and alkalinity are measured by a standard that is based on the slight ionization of water. Acidity is determined by the concentration of $H^+$, while alkalinity is a function of the concentration of $OH^-$; therefore, the ionization of water $H_2O \rightarrow H^+ + OH^-$ theoretically yields a neutral system. In pure water, dissociation occurs so slightly that at equilibrium 1 mol (18 g) of water yields $10^{-7}$ mol of $H^+$ and $10^{-7}$ mol of $OH^-$. We may treat the un-ionized mass of water as having a concentration of 1 $M$, since its ionization is so small. Thus

$$\frac{[H^+][OH^-]}{[H_2O]} = \frac{[10^{-7}][10^{-7}]}{1} = k = 10^{-14}$$

The meaning of this relationship in practical terms is that the molar concentration of $H^+$ multiplied by the molar concentration of $OH^-$ will always be 1/100,000,000,000,000, or $10^{-14}$, $M$, the equilibrium constant. Thus, as the concentration of $H^+$ increases, the $OH^-$ concentration must decrease. To avoid using such cumbersome fractions or negative exponents, a system has been devised that allows us to express acidity in terms of positive integers. The expression *pH* stands for "power of H" and is defined as the negative logarithm (or 1/logarithm) of the hydrogen-ion concentration. Since pH is a power, or exponential function, each unit of pH represents a 10-fold change in $H^+$ concentration. The lower the pH, the greater the hydrogen-ion concentration (e.g., a pH of 3 represents $10^{-3}$ mol of $H^+$ ions, but a pH of 2 indicates the presence of $10^{-2}$ mol). Neutral solutions have a pH of 7, while the maximum acidity in aqueous solutions is given by a pH of 0. A pH above 7 indicates an alkaline solution, while the maximum alkalinity is given by a pH of 14.

The pH encountered within most organisms and their constituent parts is generally close to neutral. Should the pH of human blood (7.35) change by as much as 0.1 unit, serious damage would result. (Although the digestive fluids of the stomach fall within the strong acid range, the interior of this organ is not actually within the body proper; rather, it represents an "interior external" environment: in essence, during development the body folded around an exterior space, thereby forming an interior tube.) Excess $H^+$ or $OH^-$ ions produced during metabolic reactions are neutralized, or absorbed, by chemical systems called *buffers*. These buffer systems often consist of a weak acid and its salt. Excess $H^+$ ions are captured by the anion of the salt to yield more of the weak acid, which remains relatively undissociated. Excess $OH^-$ will combine with the weak acid and cause it to release its $H^+$ ion into solution. This will prevent a large decrease in hydrogen-ion concentration and consequent rise in pH. Among the buffer systems that maintain relative constancy of pH are the carbonic acid–bicarbonate ion system of the blood and the acetic acid–acetate ion system in some cells. Buffer systems are effective in dealing with moderate pH insults but may be overwhelmed by large increases in acid or base.

# Solved Problems

**2.1**    What is an atom?

An atom is the basic unit of all simple substances (elements). It consists of a positively charged nucleus surrounded by rapidly moving, negatively charged electrons. The number of electrons revolving around the nucleus of an atom in an un-ionized state is equal to the number of positively charged protons within the nucleus.

**2.2**    What is the difference between the atomic number and the atomic weight of the atoms of an element?

The atomic number is equal to the number of protons in the nucleus or the number of orbiting electrons. The *atomic weight* is equal to the number of protons plus the number of neutrons present in the nucleus. The neutron is a nuclear particle with a mass approximately equal to that of the proton but with no electrical charge. The various particles found within the nucleus are known as *nucleons*, but for biologists it is the neutrons and protons that are of principal interest. Physicists believe that many of the nucleons, once thought to be fundamental particles, are themselves composed of much smaller units called *quarks*.

**2.3**    Are all the atoms of an element identical in their structure?

All atoms of the same element share a common atomic number but may differ in their atomic weights. This difference is due to a variation in the number of neutrons within the atomic nucleus. These variants are known as *isotopes*. The standard atomic weights given in chemical tables are derived by averaging the specific isotopes in accordance with their relative frequencies. Many isotopes are unstable because of the changes that additional neutrons produce in the structure of the nucleus. This leads to the emission of radioactive particles and rays. Such radioactive isotopes are of importance in research because they provide a marker for particular atoms.

Since the *chemical* properties of an atom are based on the arrangement of the orbiting electrons, so the various isotopes of an element behave alike in terms of their chemical characteristics.

**2.4**    How are the electrons arranged around the nucleus?

In older theories, the electrons were thought to revolve around the nucleus in definite paths like the planets of the solar system. It is now believed that electrons may vary in their assigned positions, but that they have the greatest probability of being in a specific pathway, or orbital, surrounding the nucleus. In some formulations, the orbitals are shown as clouds (shadings), with the greatest density of these clouds corresponding to the highest probability of an electron's being in that particular region. The position of an electron in the tremendous space around the nucleus of an atom can ultimately be reduced to a mathematical equation of probability.

**2.5**    What would you guess keeps electrons in their orbit around the nucleus?

The stability of electrons traveling in their assigned orbitals is due to the balance of the attractive force between the positively charged nucleus and the negatively charged electron and the centrifugal force (pulling away from the center) of the whirling electrons.

**2.6**     What is the difference between an orbital and a shell?

The shell is an energy level around the nucleus that may contain one or more orbitals. The first shell, designated as the *K shell*, contains one spherical orbital, which may hold up to two electrons. The second shell, farther from the nucleus, contains four orbitals. Since each of these orbitals can hold two electrons, this second, higher-energy shell may hold as many as eight electrons before it is full. This second shell is designated the *L shell*; a third shell, called the *M shell*, may contain from four to nine orbitals. In all, there are as many as seven shells (K through Q) that may be present around the nucleus of successively more complex atoms. The first shell consists of a single spherical orbital. The second shell contains a spherical orbital and three dumbbell-shaped orbitals whose central axes are oriented perpendicularly to one another.

The elegance of atomic structure is based on the stepwise addition of electrons to the concentric shells that surround the nucleus. The simplest atom, hydrogen, contains one electron revolving around its nucleus. Helium contains two electrons in its K shell. Lithium, with an atomic number of 3, has a complete inner K shell and one electron in its L shell. Succeeding atoms increase in complexity by adding electrons to open shells until each of these shells is complete. Generally (but not invariably), the shells closest to the nucleus are completed before electrons are added to outer shells, since atomic stability is associated with the lowest energy level for a particular arrangement of electrons in space.

**2.7**     What is the basis for the interaction of atoms with one another?

All the chemical reactions that occur in nature appear to be due to the necessity of atoms to fill their outer electron shells. Those atoms that already possess a full complement of electrons in their outer shell are chemically unreactive; they constitute a series of relatively inert elements known as the noble gases. Examples are helium, with an atomic number of 2 and a satisfied K shell, and neon, with an atomic number of 10 and a satisfied L shell.

Almost all other atoms interact (react) with one another to produce configurations that result in complete outer shells. Such combinations of atoms are called molecules. Some molecules may be highly complex, consisting of hundreds or even thousands of atoms, while others may have as few as two or three atoms. Just as single kinds of atoms are the units of an element, so combinations (molecules) of different kinds of atoms make up a compound.

**2.8**     Name four types of interactions that occur between atoms or molecules.

Ionic bonds, covalent bonds, hydrogen bonds, and van der Waals forces.

**2.9**     Calcium (Ca) has an atomic number of 20. Given that it readily forms ionic bonds, what charge would you expect calcium to have in its ionic form? What compound would you expect it to form with chlorine (Cl)?

Calcium has two extra electrons in its outer shell $(20 = 2 - 8 - 8 - 2)$. By losing these two electrons it can assume a stable configuration of eight electrons in its outer shell. Therefore, in its ionized form, it has a charge of +2 and is designated $Ca^{2+}$. Since chlorine needs one electron to fill its outer shell, two chlorines each accept one of calcium's electrons and form the ionic compound $CaCl_2$, calcium chloride.

**2.10**    Nitrogen has an atomic number of 7 and forms covalent bonds with itself, yielding $N_2$. Explain the covalent bonding of $N_2$ in terms of electrons.

With a total of seven electrons, nitrogen has five electrons in its second shell and thus needs three more electrons to create a stable outer shell of eight. By forming a triple bond, in which each nitrogen shares covalently three of its electrons with the other nitrogen, both nitrogen atoms achieve stability in their outer shells.

**2.11**    What is the relationship between the chemical reactions that elements undergo and their position in the periodic table?

The periodic table, first developed by Dmitri Mendeleev in 1869, represents an arrangement of all of the elements according to their increasing weights. There are currently about 106 different elements, but in the nineteenth century only about 89 were known. It was found that the chemical properties of the listed elements demonstrated a periodicity, or recurring regularity. If the elements are arranged according to increasing atomic number, a pattern emerges, with horizontal rows of atoms ranging from one electron in the outer shell to a complete outer shell. The first row starts with hydrogen, and helium is the second and last member, since helium is complete with two electrons in its K shell. However, lithium, the element with the next-highest atomic number of 3, again has only one electron in its outer shell. It is followed by six other elements with increasing numbers of electrons in their outer shells. The last of these is neon with an atomic number of 10 and a complete outer shell of eight electrons. The third row then begins with sodium, with an atomic number of 11, and ends with the noble gas argon, with an atomic number of 18.

Each horizontal row of increasing atomic number is known as a *period*. The vertical rows, which are similar in the numbers of outer electrons they contain, constitute a *group*. The noble gases, as the last elements of a series of periods, form one group; all the elements with one electron in their outer shell make up another group. Since the chemical properties of the elements are directly related to the configuration of their outer electrons, all the elements making up a single group will generally have similar chemical properties. This is the basis for the periodicity first observed in the properties of all chemical elements.

**2.12    How are chemical reactions described?**

All chemical reactions involve a reshuffling of bonds. These reactions are usually described in the form of a chemical equation, in which the reactants (molecules undergoing change) are placed on the left and the products to be formed are placed on the right. An arrow denotes the direction of the reaction from reactants to products. A typical reaction might be shown as $A + B \rightarrow C + D$. Each of the molecules (or atoms) participating in the reaction will be written as a formula, a shorthand expression of the kinds and numbers of atoms involved. Thus, if A were water, it would be written as $H_2O$, since H is the symbol for hydrogen and O is the symbol for oxygen; two atoms of hydrogen are covalently bonded to oxygen in a molecule of water.

**2.13    Why must both sides of a chemical equation balance? Balance the equation for the production of water from elemental hydrogen ($H_2$) and oxygen ($O_2$).***

Because the law of conservation of matter tells us that matter can be neither created nor destroyed, all equations must be balanced; that is, the number and kinds of atoms appearing on one side of the equation cannot be destroyed and so must appear in the same number and kind on the other side. In representing the formation of water by the simple addition of hydrogen and oxygen we might select the equation

$$H_2 + O_2 \rightarrow H_2O$$

However, this equation is not balanced, because there are different numbers of atoms on each side of the equation. Balance is achieved by manipulating the coefficients, which indicate how many of each of the molecules are involved in the equation:

$$2H_2 + O_2 \rightarrow 2H_2O$$

Now the equation is balanced.

**2.14    Do all chemical reactions go to completion?**

In actual fact, most chemical reactions do not go to completion. A state of balance, or equilibrium, is reached in which the concentrations of the reactants and the concentrations of the products reach a fixed ratio. This ratio is known as the *equilibrium constant* and is different for each chemical reaction.

---

* Elements such as hydrogen or oxygen tend to occur in nature as molecules of two or more atoms of the elements, rather than as single atoms.

An equation may be viewed as a balance between two reactions—a forward reaction in which the reactants are changed to products and a reverse reaction in which the products interact to form reactants. Most reactions are reversible, and it might be more appropriate to write a chemical equation with the arrows going in both directions:

$$A + B \leftrightharpoons C + D$$

When reactants are first mixed, the forward reaction predominates. As the products are formed, they interact to produce reactants and the reverse reaction will increase.

It should be noted that at equilibrium both forward and reverse reactions continue but there is no *net* change; i.e., the forward reaction is exactly balanced by the reverse reaction. This equilibrium situation is obtained only under specified conditions of temperature, pressure, etc. If these environmental variables are altered, the equilibrium will shift. The formation of a substance that leaves the arena of chemical interaction tends to shift the equilibrium as well. In reactions in which a gas or precipitate is produced, the reaction will be pushed in the forward direction, since the products do not have as much opportunity to interact with one another to produce a reverse reaction. Although some chemists view all reactions as theoretically reversible, there are many reactions in which the forward or reverse reactions are so overwhelming that for all practical purposes they may be regarded as irreversible.

**2.15**   How do molecules or atoms actually interact to bring about chemical changes?

The basis of the "social interactions" of all chemical substances is the tendency of atoms to form bonds that complete their outer electron shell. These bonds may be disrupted and new bonds created just as friendships and marriages may undergo change and realignment. However, most chemical substances will not undergo change unless the participating molecules are in close contact with one another. Solid blocks of substances do not appreciably interact with one another except at their boundaries. Gases and materials that are dissolved in a liquid to form solutions are far more likely to interact with one another. According to the *kinetic molecular hypothesis* of gases, the molecules of a gas are in constant rapid motion and undergo continuous collision. It is these collisions that provide the basis for chemical change. In similar fashion, the dissolved particles (solute) within the liquid (solvent) of a solution are finely dispersed and in rapid random motion and thus have an opportunity for chemical change.

An increase in temperature will speed up the movement and number of collisions of the particles and increase the rate of interaction. So too will the degree of dispersion of the molecules within the medium (fully dissolved molecules interact more often than partially precipitated ones). An increase in the concentration of reacting molecules also tends to speed up the rate of a reaction, since it enhances the possibility of more collisions.

**2.16**   How is concentration measured in a solution?

The concentration of any substance is the amount of that substance in a specific volume of a particular medium. Concentrations of the constituents of blood are often expressed as a percentage denoting the number of milligrams (mg) of a specific substance in 100 milliliters (mL) of blood. Thus, a blood sugar concentration of 95 percent means that there are 95 mg of sugar (usually glucose) in every 100 mL of whole blood.

Percentage by weight is not the best method of expressing concentrations, since the same percentage of a solution containing heavy molecules will have fewer molecules than one containing lighter molecules. This is apparent when we consider that 1000 lb worth of obese people in a room will comprise fewer individuals than the same weight of thin people. Since chemical reaction rates depend on the number of molecules present, it would be preferable to use a standard for concentration that takes only the number of molecules into account.

A *mole* may be defined as the molecular weight of a substance expressed in grams. Thus, a mole of water would consist of 18 g of water, while a mole of ammonia ($NH_3$) would contain 17 g of the gas. Since a mole of any molecule (or atom) contains the same number of molecules (or atoms), molar concentration is more useful in comparing reactants and products in chemical equations. Molar concentration ($M$) is expressed as the number of moles of solute dissolved in one liter of total solution. Equimolar concentrations of any substance will have equal numbers of molecules. The number of molecules present in a 1 $M$ solution of any substance is $6.02 \times 10^{23}$, also known as *Avogadro's number*. This is also the number of molecules present in 22.4 liters of any gas at standard temperature and pressure.

Some molecules consist of atoms or ionic groups with a capacity to unite with more than one simple atom such as hydrogen. Thus, oxygen can form two covalent bonds with two different atoms of hydrogen. Similarly, the sulfate ion ($SO_4^{2-}$) can ionically bind to two sodium ions. This combining capacity of atoms or ions is known as *valence*. Obviously, an atom with a valence of 3 will be as effective in chemical combination as three atoms with a valence of 1. To account for the difference in combining power, concentrations are sometimes expressed in terms of normality ($N$). This unit is the number of gram equivalent weights per liter of solution. A *gram equivalent weight* is the molar weight divided by valence. Similar normalities of various solutions will always be equivalent to one another when the volumes involved are the same.

**2.17**   *Diffusion* is the tendency of molecules to disperse throughout a medium or container in which they are found. How does diffusion differ from osmosis; how is it similar?

Diffusion involves movement of *solute* particles in the absence of a semipermeable membrane. Osmosis is a special case of diffusion involving movement of *solvent* molecules through a semipermeable membrane. The two processes are similar in that movement of the molecules in each is driven by their collisions and rebounds with their own kind and proceeds toward areas in which collisions are less likely, namely, areas with fewer molecules of their kind (from crowded to less crowded regions).

**2.18**   Why does putting a lettuce leaf in water make the leaf crisper?

When living cells are placed in a medium, they may be in osmotic equilibrium with their surroundings, in which case there will be no net flow of water into or out of the cell. Such a medium is designated as *isotonic*, or *isosmotic*. If the concentration of the solutes of the medium is greater than that of the cell, the surroundings are *hypertonic*, and water will be drawn from the cell by the more concentrated medium, with its higher osmotic pressure. If the cell is placed in an environment that is more dilute than the cellular interior, it will draw water from this *hypotonic* environment and tend to swell. The crisping of lettuce by conscientious salad preparers is achieved by placing the leaves of lettuce in plain water, causing the cells to absorb water and swell against the restraining cell wall, thus producing a general firmness. Another osmotic phenomenon is the tendency of magnesium salts to draw water into the interior of the intestine and thereby act as a laxative.

**2.19**   Describe the laws that govern exchanges of energy.

The laws dealing with energy transformations are the three laws of thermodynamics. The first law (conservation of energy) states that energy can be neither created nor destroyed, so that the energy input in any transformation must equal the energy output.

The second law states that energy as it changes tends to become degraded to scattered states in which the capacity for useful work diminishes. *Entropy* is a measure of the disordered, random property of energy, and the second law may be phrased in terms of the natural tendency for entropy to increase in a transformation. Thus, while the total energy input is always equal to the total energy recovered, the ability of this energy to be utilized for useful work continuously decreases. In living systems, which must maintain a high degree of complex order, the enemy that is continuously resisted is entropy, or the drift to disorder.

The third law states that a perfect crystal at a temperature of absolute zero possesses zero entropy; i.e., it is in a state of maximum order. This law is not as useful for the biologist as the first two laws, but it does emphasize the prevalence of disorder in almost all natural states, which clearly do not involve ideal crystalline states or the unattainable temperature of absolute zero in which no molecular movement may occur.

**2.20**   Why doesn't the apparent discrepancy between energy input and output in nuclear reactions contradict the first law of thermodynamics?

The release of tremendous amounts of energy in nuclear transformations such as fission or fusion (as occurs in atomic and hydrogen bombs) is accounted for by the disappearance of mass during these reactions and the conversion of this mass to energy in accordance with Einstein's equation $E = mc^2$. Matter (mass) is now regarded as a special case of energy, and the mass lost during nuclear reactions is multiplied by $c^2$, which is the speed of light squared, to yield the awesome energy releases associated with nuclear devices.

**2.21**    What is meant by an exergonic reaction?

An *exergonic reaction* is one in which energy is released during the course of the reaction. The potential energy of the initial state is greater than that of the final state, so this reaction will tend to occur spontaneously, much as a stone atop a hill will tend to roll down. Although an exergonic reaction will tend to occur, it may require an activation process for its initiation, just as a stone must be pushed over the edge of a hill before it can begin its descent. The role of enzymes in initiating reactions or in altering reaction velocities is discussed in Chap. 3.

A mathematical analysis may be helpful in fully understanding the concept of an exergonic reaction. The total energy of a system is designated as $H$ (heat). In any reaction there will be a change in the total energy relative to the starting system. Since the Greek letter delta ($\Delta$) refers to change, the symbol $\Delta H$ represents this change in total energy (also known as change in *enthalpy*). The change in total energy consists of two components. One is the change in free energy of the system, represented by $\Delta G$. Free energy is that component that can perform useful work or be stored for later performance of such work.

The second component of the total energy is the change in entropy $\Delta S$. If entropy increases, then the total amount of energy made available also increases because the system is moving "downhill." Since entropy change is related to temperature, the entropy factor is denoted as $T\Delta S$. Now we have an equation for the total heat (or energy) change of any transformation:

$$(1)\quad \Delta H = \Delta G + T\Delta S$$

If $\Delta H$ is negative, heat will be given off to the surroundings and the reaction is an exothermic one. However, not all exothermic reactions are also exergonic (capable of doing work). In order for a reaction to be considered exergonic, it must liberate free energy ($\Delta G$ must be negative). From Eq. 1, it is clear that $\Delta H$ can be negative (exothermic) even though $\Delta G$ is positive (*end*ergonic) if the change in entropy ($\Delta S$) is negative and sufficiently large. Perhaps a better way to express Eq. 1 is in the form

$$(2)\quad \Delta G = \Delta H - T\Delta S$$

Here we see clearly that $\Delta G$ may also be negative (liberate free energy) even though $\Delta H$ is positive (absorb heat) provided that the increase in entropy is high enough. If ether is applied to the skin, it will evaporate, although the evaporation is accompanied by an absorption of heat from the surroundings ($\Delta H$ is positive). This exergonic reaction is a downhill, or spontaneous, phenomenon even though heat (energy) is absorbed. The increase in entropy involved in the formation of a gas is so great that the value of $G$ is negative.

**2.22**    What is meant by an endergonic reaction, and how does it differ from an endothermic reaction?

*Endergonic* reactions are essentially uphill reactions and are characterized by positive $G$ values. In endergonic reactions, free energy is taken up in the reaction process; in chemical reactions, this free energy may be stored in high-energy bonds in the products. Since this free energy cannot be created, it must come from an accompanying *ex*ergonic reaction in which free energy is liberated to drive the endergonic process. The various endergonic, or building-up processes within an organism are always associated with an exergonic process in which energy-rich molecules are degraded. In an automobile, mechanical movement is achieved by the degradative conversion of energy-rich fuel to energy-poor by-products such as water and carbon dioxide.

Although most endergonic processes are also endothermic, in that heat will be absorbed by the system, this is not necessarily the case. Once again, the change in entropy must be taken into account. In biology we are usually interested in whether a reaction is exergonic, occurring spontaneously, or whether it is endergonic, requiring the infusion of free energy. The transfer of heat is generally of secondary significance.

**2.23**    What characteristic of the water molecule endows water with so many qualities essential to life?

Those properties of water that promote life functions are largely due to the arrangement of the bonds between hydrogen and oxygen within the molecule and the consequent distribution of electrons. Although the hydrogens and oxygen in water form covalent bonds, the shared electron pairs lie more closely within oxygen's sphere of influence and thus form a dipole. The hydrogens of any one $H_2O$ molecule are the positive ends of the dipole, while the oxygen end is a double-negative pole. The two hydrogen ends of one $H_2O$ molecule are attracted to the oxygen ends of two other water molecules, while the double-negative charge of the oxygen end attracts hydrogen ends from two more water molecules. This hydrogen bonding to four other water molecules produces the properties of water that tend to stabilize aqueous systems. The hydrogen bonds are continuously formed and broken, a process that allows water to flow while simultaneously maintaining a strong cohesion that keeps it a liquid through a broad range of temperatures and pressures.

**2.24**  In what specific ways does the dipole nature of water promote the maintenance of life?

Living material is extremely complex. Any medium that supports such complexity would have to accommodate a broad variety of substances. Since water is a universal solvent, taking up more different kinds of solute than any other known liquid, it is the ideal medium for supporting complexity. Water is also one of the most stable substances in existence. This assures a long-term continuation of water-based substances. Water's tendency to remain a liquid also assures that drying out or freezing will not readily occur. This feature is enhanced by the influence of dissolved solutes, which raise the already high boiling point and lower the freezing point of liquid water. Water also has a tendency to adhere to the sides of a containing vessel. In the case of thin tubes (capillaries) the water will actually rise to considerable heights as the adherent molecules haul other molecules of water up with them because of hydrogen bonding. This property plays a significant role in bringing water through minute spaces in the soil to the roots of plants. The great cohesiveness existing between adjacent molecules of water also accounts for the high surface tension of water, enabling some insects actually to walk on the compacted surface molecules. Surface tension when high may be reduced by a variety of surface-active substances, called *surfactants*. This lowering of surface tension may facilitate some necessary movement within organisms. Perhaps most peculiar in the behavior of water is its tendency to expand upon freezing. Like all other substances, water tends to shrink as its temperature drops. Thus, its density (weight per unit volume) increases with a lowering of temperature. But at 4 °C water begins to expand as the temperature decreases further. At freezing there is a further expansion, so that ice is even less dense than liquid water at similar temperatures. There are several practical results of this anomalous expansion at low temperatures. When ice forms it tends to float at the surface of a pond or stream, so that these bodies of water will freeze from the top down and form an insulating cap of ice at the surface, which allows aquatic organisms to survive and maintain their activities below it during freezing weather. Also, the surface and bottom waters of lakes and ponds will undergo an exchange (vertical convection) twice each year that brings nutrient material to the surface and carries oxygen to the lower layers. This is directly due to the increased density of water associated with a plunge in temperature followed by the expansion below 4 °C that causes extremely cold water to move back toward the surface as it nears 0 °C and begins to freeze. Finally, water offers organisms both internal and external stability against temperature fluctuations (see Prob. 2.25).

**2.25**  How does water provide internal and external temperature stability for organisms?

Water plays a most significant role in the maintenance of temperature both within the organism and in its supporting external environment. Since extremes of temperature threaten the structural components of cells and may also alter the tempo of chemical reactivity, the role of water as a temperature buffer within and without living organisms is vital to life. Water has one of the highest specific heats of any common substance, a property referring to the amount of heat absorbed in comparison with the rise in temperature accompanying this heat absorption. One gram of water absorbs one calorie to bring its temperature up by one degree, whereas a substance like aluminum will show a similar rise of one degree with only a fraction of a calorie. In this aspect, water acts like a heat sink; it absorbs a great deal of heat with only a modest rise in temperature. Because of this buffering, land areas that are near large bodies of water tend to have more moderate temperatures than those in the interior of continents.

Water also has a high latent heat of fusion, a phenomenon involving the liberation of heat when liquid water forms solid ice. Thus, the freezing of water produces heat that counteracts a further drop in temperature. A mixture of ice and water constitutes a temperature-stable system—a drop in temperature will produce freezing and heat will be released; a rise in temperature will cause the ice to melt and approximately

80 cal of heat will be absorbed. In the environment, these transformations resist sharp changes in temperature and permit organisms to adjust more readily to temperature fluctuations with the changing seasons.

Water also has the highest latent heat of vaporization of any common natural substance. This property, closely associated with the strong attractive forces between water molecules, refers to the amount of heat energy required to convert 1 g of liquid water to 1 g of water vapor at the boiling temperature (100 °C). Water, however, evaporates or vaporizes at lower temperatures as well, and whenever this occurs, a great deal of heat is taken up to change liquid molecules into the more rapidly moving vapor molecules. This explains why sweating and the ensuing evaporation tend to cool the surface of the body and prevent excessive buildup of heat on a hot day. On humid days evaporative cooling is inhibited.

**2.26**   Why is there a lower limit of 0 and an upper limit of 14 in the range of pH values?

This range of 0 to 14 is associated with aqueous systems. A pH of 0 indicates a $[10^0]$, or 1 $M$, concentration of $H^+$, which is the maximum encountered with even the strongest acids dissolved in water. Although stronger concentrations of acid can theoretically be obtained, they will not dissociate beyond the 1 $M$ level of $H^+$. A similar situation exists with regard to strong bases at high pH levels.

**2.27**   If there are $6.02 \times 10^{15}$ molecules of $OH^-$ per liter of aqueous solution, what is the solution's pH?

Since pH is based on molar concentrations, it is first necessary to determine the moles per liter of $OH^-$ ions:

$$6.02 \times 10^{15} \text{ molecules} \times \frac{1 \text{ mol}}{6.02 \times 10^{23} \text{ molecules}} = 10^{-8} \, M$$

The solution therefore contains a $10^{-8} \, M$ concentration of $OH^-$ ions. However, pH is based on the concentration of $H^+$ ions. This concentration can be determined from the equation for the equilibrium constant of water.

$$[H^+][OH^-] = 10^{-14} \, M$$

Therefore
$$[H^+][10^{-8}] = 10^{-14} \, M$$

$$[H^+] = 10^{-6} \, M$$

and
$$pH = -\log [H^+] = -\log 10^{-6} = -(-6) = 6$$

**2.28**   When carbon dioxide ($CO_2$) is released into the extracellular fluid by the cells as a by-product of metabolism, much of it combines with water to form carbonic acid:

$$CO_2 + H_2O \rightleftharpoons H_2CO_3$$

Given the narrow range of pH in which cells can function properly, why does this introduction of an acid not harm the organism?

The extracellular fluid in higher animals is buffered by, among other things, a carbonic acid–bicarbonate ion system. The salts of the bicarbonate ion ($HCO_3^-$), such as sodium, potassium, magnesium, and calcium bicarbonate, buffer the fluids against the introduction of $H^+$ ions caused by the dissociation of carbonic acid and thus prevent an appreciable lowering of pH.

# Supplementary Problems

**2.29**   The chemical properties of an atom are most closely associated with its   (a) atomic number.   (b) atomic weight.   (c) number of neutrons in the nucleus.   (d) all of the above.   (e) none of the above.

**2.30**   Atoms with the same atomic number but different atomic weights are called _____.

**2.31**   The second shell of electrons contains (at maximum)   (*a*) a total of two electrons.   (*b*) a total of eight electrons.   (*c*) two orbitals.   (*d*) four orbitals.   (*e*) both (*b*) and (*d*).

**2.32**   The noble gases readily unite with other elements.
(*a*) True   (*b*) False

**2.33**   Polar bonds arise from a sharing of a pair of electrons between two atoms.
(*a*) True   (*b*) False

**2.34**   Hydrogen bonds are relatively weak but play a significant role in the three-dimensional structure of proteins and nucleic acids.
(*a*) True   (*b*) False

**2.35**   The atoms constituting a single group in the periodic table share similar chemical properties.
(*a*) True   (*b*) False

**2.36**   A high value for the equilibrium constant indicates that at equilibrium the reaction lies farther to the right, since the products of the reaction (C, D) exceed the reactants (A, B) in concentration.

$$\frac{[C]\cdot[D]}{[A]\cdot[B]} = K_{eq} = {>}1$$

(*a*) True   (*b*) False

**2.37**   A 3 $M$ concentration of any substance will be one-third as concentrated as a 1 $M$ concentration of that substance.
(*a*) True   (*b*) False

**2.38**   The colligative properties of a solution are due to the number of solute particles in the solution.
(*a*) True   (*b*) False

**2.39**   Lettuce many be crisped by placing it into a hypertonic solution.
(*a*) True   (*b*) False

**2.40**   A 1 $N$ solution of $H_2SO_4$ will contain approximately 0.5 mol of the compound per liter of solution.
(*a*) True   (*b*) False

**2.41**   What useful property of water permits light reactions like those of photosynthesis to occur in the ocean?

**2.42**   If an amoeba is isotonic with a solution that is hypertonic for a crab, into which organism will water show a net flow when these organisms are immersed in the solution?   (*a*) amoeba.   (*b*) crab.   (*c*) neither.   (*d*) both.

**2.43**   Why does the addition of solutes to water act as an antifreeze?

**2.44**   If 3 g of $H_2O$ evaporate from a surface, the number of calories absorbed will be _____.

**2.45**   What is the pH of a 0.001 $M$ acid solution?

**2.46**   Do fish obtain their oxygen from the water molecules of the medium?

# Answers

| | | | | |
|---|---|---|---|---|
| **2.29** | (a) | **2.35** | (a) | **2.41** Transparency |
| **2.30** | Isotopes | **2.36** | (a) | **2.42** (c) |
| **2.31** | (e) | **2.37** | (b) | **2.43** Additional solute lowers the freezing point. |
| **2.32** | (b) | **2.38** | (a) | **2.44** 1620 calories |
| **2.33** | (b) | **2.39** | (b) | **2.45** pH $= -\log 0.001 = -\log 10^{-3} = -(-3) = 3$ |
| **2.34** | (a) | **2.40** | (a) | **2.46** No, their oxygen comes from oxygen in the air that is dissolved in the water. |

# Chapter 3

## The Chemistry of Life: The Organic Level

### 3.1 INTRODUCTION

*Organic compounds* are the relatively complex compounds of carbon. Since carbon atoms readily bond to each other, the backbone of most organic compounds consists of carbon chains of varying lengths and shapes to which hydrogen, oxygen, and nitrogen atoms are usually attached. Each carbon atom has a valence of 4, which significantly promotes complexity in the compounds that can be formed. The ability of carbon to form double and even triple bonds with its neighbors further enhances the possibility for variation in the molecular structure of organic compounds.

**EXAMPLE 1**  Among the organic compounds found in nature are the *hydrocarbons*, the molecular associations of carbon and hydrogen, which are nonsoluble in water and are widely distributed. *Aldehydes* are organic molecules with a double-bonded oxygen attached to a terminal carbon atom; this carbon-oxygen combination is referred to as a *carbonyl group*. *Ketones* contain a double-bonded oxygen attached to an internal carbon atom. An *organic alcohol* contains one or more hydroxyl (OH) groups, and an *organic acid* contains *a carboxyl group* (a hydroxyl and a double-bonded oxygen attached to a terminal carbon atom). The many classes of organic compounds were once thought to arise only from living organisms, but with the synthesis of urea in 1828 (Wohler), it was apparent that organic compounds could be synthesized from simpler inorganic compounds.

Among the organic compounds most closely associated with basic life processes are carbohydrates, proteins, lipids, and nucleic acids (polynucleotides). This last class of compounds will be discussed in Chap. 7, The Nature of the Gene, because the polynucleotides are centrally involved in the processing of information within the cell.

### 3.2 CARBOHYDRATES

*Carbohydrates* are hydrates of carbon with a general *empirical formula* of $C_x(H_2O)_y$. Carbohydrates include sugars. The basic sugar unit is a *monosaccharide*, or simple sugar. It may contain from three to seven or more carbon atoms, but the most common monosaccharides contain six carbon atoms and are known as *hexoses*.

**EXAMPLE 2**  A typical hexose monosaccharide such as *glucose* (also called *dextrose*) consists of a carbon chain to which are attached hydroxyl groups. (Figure 3.1 shows the *structural formula* for glucose and for another hexose, fructose.) These —OH groups confer both sweetness and water solubility upon the molecule. An =O is attached

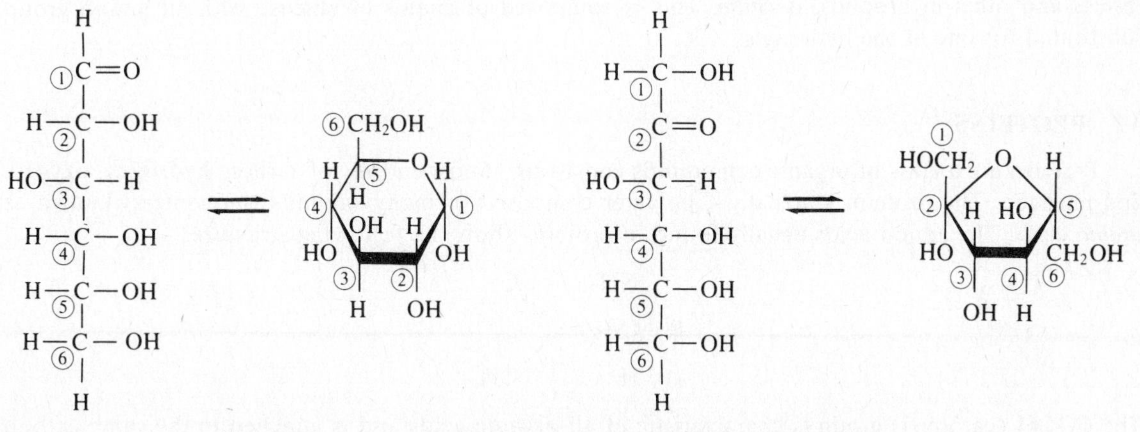

Glucose                    Fructose

**Fig. 3.1**

29

to a terminal carbon in glucose, making it an *aldo sugar*. If an internal C=O group is present (as in fructose), the monosaccharide is designated a *keto sugar*.

Monosaccharides may fuse through a process known as *condensation*, or *dehydration synthesis*. In this process two monosaccharides are joined to yield a *disaccharide*, and a molecule of water is liberated (an —OH from one monosaccharide and a —H from the second are removed to create the C—O—C bond between the two *monomers*, or basic units). Common table sugar is a disaccharide—formed by the condensation of glucose and fructose. Condensation may occur again to yield trisaccharides and eventually polysaccharides.

Glycogen is the major polysaccharide associated with higher animal species. The principal polysaccharides of plants are starch and cellulose. All these polysaccharides are composed of glucose units as the basic monomer.

*Glycogen* is a highly branched chain of glucose units that serves as a calorie storage molecule in animals, principally in liver and muscle. The straight-chain portion is associated with 1 → 4 linkages of the glucose units [that is the C-1 (first carbon) atom of one glucose unit is joined to the C-4 of a second glucose], while the branched portions are produced by 1 → 6 linkages. The enzyme *glycogen synthase* promotes formation of the straight-chain portion of glycogen, and *amylo-(1,4 → 1,6)-transglycosylase* catalyzes branch formation. Glycogen degradation is effected by two enzymes, one that cleaves the 1 → 4 linkages, *glycogen phosphorylase*, and one that cleaves the 1 → 6 linkages, *α (1 → 6)-glucosidase*. Breaking the bonds is accomplished by a reversal of the condensation process: water is added back to the molecule. Thereby the —OH and —H are restored and the bond is broken. The process is called *hydrolysis*.

Enzymes promoting the synthesis of glycogen from glucose are increased by insulin, a hormone released into the bloodstream when blood glucose levels begin to rise. Glycogen may be broken down to its constituent glucose molecules by enzymes such as phosphorylase, which are activated by the hormones epinephrine and glucagon.

In plants, *starch* is the primary storage form of glucose. It occurs in two forms: α-amylose, which consists of long, unbranched chains, and amylopectin, a branched form with 1 → 6 linkages forming the branches. The primary structural component in plants is *cellulose*, a water-insoluble polysaccharide that forms long, unbranched chains of 1 → 4 linkages. These chains are cemented together to form the cell walls of plants. Their parallel structure and lack of branching give them strength and resistance to hydrolysis. Because of the variation in cellulose's β(1 → 4) linkages, animal enzymes normally associated with polysaccharide digestion are ineffective with cellulose. Ruminants and other animals that digest cellulose are able to do so because of symbiotic bacteria in their digestive tracts that have the enzyme *cellulase*, which can degrade cellulose.

A structural polymer similar to cellulose, but commonly found in fungi and in the exoskeletons of insects and other arthropods, is *chitin*. This is composed of chains of glucose with an amino group substituted for one of the hydroxyls.

## 3.3  PROTEINS

Proteins are a class of organic compounds consisting almost entirely of carbon, hydrogen, oxygen, and nitrogen. The protein is actually a polymer composed of many subunits (monomers) known as *amino acids*. The amino acids usually found in proteins show the following structure:

$$\text{R}-\underset{\underset{\text{H}}{|}}{\overset{\overset{\text{NH}_2}{|}}{\text{C}}}-\text{C}\overset{\displaystyle O}{\underset{\displaystyle OH}{}}$$

The COOH (carboxyl) group is characteristic of all organic acids and is attached to the same carbon as the NH$_2$ group. This carbon is designated the *α-carbon* atom; the entire amino acid is known as an *α-amino acid*. The *R* is a general designation for a variety of side groups that differentiate the 20 different amino acids found in nature. Such properties of a protein as its water solubility or charge are due to the kinds of R groups found in its constituent amino acids.

In a manner similar to the way monosaccharides join to form higher-order polysaccharides, amino acids join by expelling a molecule of water. An —OH is removed from the carboxyl group of one amino acid, and a —H is removed from the amino group of a second. The resultant bond between the C and N atoms of the carboxyl and amino groups is called a *peptide bond*, and the compound formed is a *dipeptide*. A dipeptide may unite with another amino acid to form a second peptide bond, and this will yield *a tripeptide*. If many amino acids are joined in this condensation process, the result is a *polypeptide*; such chains of amino acids may range from less than 100 amino acids to as many as 1000.

## PRIMARY, SECONDARY, TERTIARY, AND QUATERNARY STRUCTURE

The linear order of amino acids in a protein establishes its *primary structure*. This primary structure is actually encoded in the genetic blueprint that is preserved and passed on from parent to child in the DNA of the chromosomes.

The interactions between the bonded amino acids of the primary structure may lead to folding, kinking, or even pleating of the protein chain. Hydrogen bonding is largely responsible for these changes in the configuration of the protein chain, which constitute the *secondary structure* of the protein molecule. Among the shapes assumed in secondary structure is the *α helix*—a configuration similar to a winding staircase or stretched spiral. Another kind of secondary structure is the *pleated sheet* arrangement, in which side-by-side polypeptide chains are cross-linked by hydrogen bonds to form a strong but flexible molecule that tends to resist stretching. A third type is the *triple helix* structure of collagen.

Superimposed on the secondary structure may be striking alterations, consisting of superfolding or a complex twisting that yields highly intricate spheres or globules, in the three-dimensional shape of the molecule. This constitutes the *tertiary structure* of a protein. Such characteristic folding is particularly associated with proteins like myoglobin and many of the enzymes—proteins that function as catalytic and carrier molecules. Many of these proteins are influenced in their final tertiary configuration by disulfide bridges and charge interactions as well as by hydrogen bonding. The three-dimensional configuration of a protein is also called its *conformation*.

Finally, some proteins are actually composed of two or more separable polypeptide chains. The aggregation of multiple polypeptides to form a single functioning protein is called the *quaternary structure* of a protein. Many of the enzymes that function in metabolism consist of as many as four to six polypeptide subunits. Changes in the kinds or arrangements of these subunits lead to alternative forms of the enzyme called *isozymes*.

**EXAMPLE 3**  The diversity of amino acids and their interactions have led to many different types of proteins. *Fibrous proteins* (hair, silk, tendons) consist of long chains, frequently comprising repeating patterns of particular amino acids, a feature of primary structure that is reflected in the *α*-helical and *β*-pleated sheet configurations of the secondary structure. They are most often involved in structural roles. *Globular proteins* lack the regularity in primary and secondary structure seen in fibrous proteins, but exhibit complex folding patterns that produce a globular tertiary structure. Although they do in some instances serve structural functions, as in the case of microtubules, they more often occur as enzymes, hormones, and other active molecules.

Environmental insults, such as from heat or an appreciable change in pH, can lead to alterations in the secondary, tertiary, and quaternary structure of a protein. This is known as *denaturation*. Denatured proteins generally lose their enzymatic activity and may demonstrate dramatic changes in physical properties. This demonstrates the importance of conformation in the properties of a protein.

Many proteins are intimately attached to nonprotein organic or inorganic groups to form *conjugated proteins*. Among such nonprotein (prosthetic) groups commonly encountered are carbohydrates (glycoproteins), lipids (lipoproteins), and such specialized compounds as the heme portion of hemoglobin. These prosthetic groups may profoundly change the properties of proteins to which they are bound.

## PROTEINS AS ENZYMES—ACTIVE SITE AND CONFORMATION

Proteins are important to living organisms both as basic structural units and as enzymes. Their structural role will be considered in subsequent chapters. As *enzymes*, proteins serve as catalysts that regulate the rates of the many reactions occurring in the cell and thus control the flow of molecular traffic necessary for cell viability. Enzymes are generally complex globular proteins with a special region in the molecule known as the *active site*. The substance that the enzyme acts upon, the *substrate*, fits into a cleft, or groove, of the active site and attaches to areas of complementary charge found in the active site. As a result of this attachment, the substrate is stretched or otherwise distorted and thus more readily undergoes appropriate chemical change. Essentially the union of substrate and enzyme lowers the resistance of the substrate to alteration and promotes the reaction in which the substrate is changed to its products.

Formerly the active site was regarded as a rigid region of adjacent amino acids within the protein molecule, into which the substrate would fit like a key in a lock. Later investigation by Daniel Koshland and his research group at Berkeley revealed that the three-dimensional structure of the active site is rather flexible and its final conformation occurs as the substrate attaches to the enzyme, in much the same way that a hand determines the final form of a glove into which it is inserted. The active site may include nonadjacent regions of the protein molecule's primary structure, since the ultimate folding of the molecule may bring once-distant regions of the protein into close apposition to produce the final enzyme-substrate complex.

## ESSENTIAL AMINO ACIDS

Proteins are made up of 20 naturally occurring amino acids. The kinds and number of amino acids vary with each protein. In many cases, an organism can convert one amino acid into another, so that the food ingested by that organism need not consist of proteins containing every one of the 20 amino acids. However, eight or nine of the amino acids, such as tryptophan and phenylalanine, are not synthesized from other amino acids, particularly in animal species. These are called *essential amino acids* and must be supplied from the food taken in. Many of the essential amino acids are richly supplied in meat and dairy products but tend to be less balanced in vegetables.

A broad variety of proteins in the diet will ensure adequate provision of all the amino acids necessary for the manufacture of proteins. A dietary deficiency of the essential amino acids will result in the production of faulty proteins or may even prevent the building of some proteins entirely. This means that key enzymes and structural proteins may be nonfunctional or absent.

The body normally degrades (by deamination and oxidation) a certain amount of its own protein into the constituent amino acids. When this loss is not countered by a compensatory intake of protein, more nitrogen leaves the body than comes in; this results in a *negative nitrogen balance*. Gradually this leads to a wasting of muscle and other vital organs and, ultimately, to death.

## 3.4  STRUCTURE AND FUNCTION OF LIPIDS

*Lipids* are a class of organic compounds that tend to be insoluble in water or other polar solvents but soluble in organic solvents such as toluene or ether. They consist largely of carbon, hydrogen, and oxygen, but they may contain other elements as well.

Triglycerides and other lipids have much more energy associated with their bonding structure than do the carbohydrates or proteins. One gram of most carbohydrates yields approximately 4.3 cal upon oxidation, 1 g of protein yields 4.6 cal, while the oxidation of 1 g of triglyceride produces more than 9 cal. Fats as energy storage media also take up much less room and involve less weight than carbohydrates do. This is because carbohydrates incorporate water during their storage, while fats do not require water in their final storage form or in the intermediate conversions that produce storage molecules.

Besides serving as media of energy storage, certain kinds of lipids cushion and protect the internal organs of the body, while others, in the form of a layer of fat just below the skin in many mammals, provide insulation against possible low environmental temperatures.

Lipids are more difficult to categorize than the carbohydrates or proteins, since there is such diversity in the lipid group. Among the major classes of lipids functioning within living organisms are the neutral fats (triglycerides), the phospholipids, and the steroids. Waxes are found as protective layers on the surfaces of many plants and animals.

## TRIGLYCERIDES AND PHOSPHOLIPIDS

The *neutral fats*, or *triglycerides*, are the most common and familiar of the lipids. They are composed of three fatty acids joined to each of the three hydroxyl groups of the triple alcohol glycerol (see Fig. 3.2). Since the union of an acid and an alcohol yields an ester, triglycerides are also known as *triesters*.

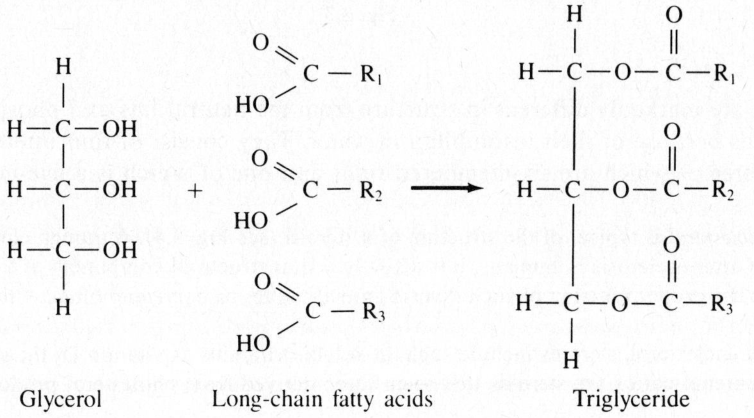

Glycerol          Long-chain fatty acids          Triglyceride

**Fig. 3.2**

Fatty acids can be classified according to their level of saturation. *Saturation* refers to the amount of hydrogen in the long carbon chains of the fatty acids found in neutral fats. If the carbon chain of each fatty acid is holding a maximum number of H atoms, it is said to be saturated; beef and pork contain saturated fats. However, if there are double or triple bonds between any of the carbon atoms so that there is consequent reduction in the amount of H atoms held by those carbon atoms, the fat is considered to be *unsaturated*. Unsaturated fats tend to have kinked chains, rather than the straight chains of saturated fats, because of the multiple bonding. Should there be many such double and triple bonds in the fatty acid chains, the fat is classified as *polyunsaturated*; fish and vegetables are rich in polyunsaturated fats. If a triglyceride is solid under ordinary conditions, it is called a *fat*. If it is a liquid under such conditions, it is called an *oil*. Both fats and oils are extremely rich sources of energy.

*Phospholipids* are similar in chemical makeup to the triglycerides. The first two hydroxyl groups of glycerol are joined in ester linkage to two fatty acids, but the third position is occupied by a phosphate group. Most phospholipids also contain another charged group attached to the phosphate portion. A typical phospholipid is shown in Fig. 3.3. Note the charges occurring in the phosphate portion. It is these charges that give phospholipids their unique properties—one end is polar and soluble in water, while the bulk of the molecule is nonpolar and insoluble in water. Phospholipids play an important role in the cell membrane in maintaining the polar-nonpolar layering structure. Phospholipids are also useful in the transport of lipid material within such aqueous media as blood.

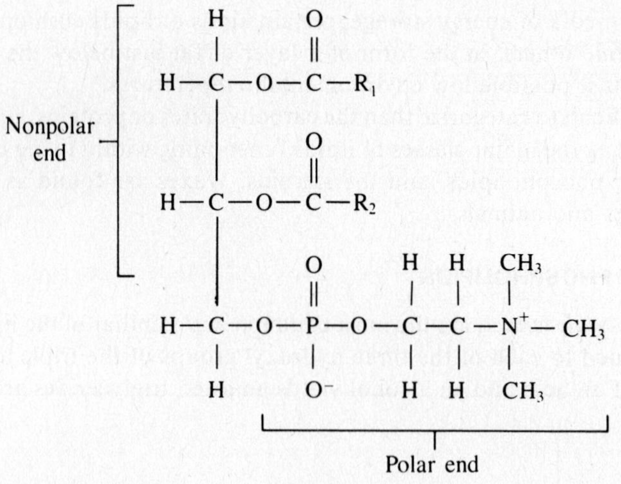

**Fig. 3.3**

## STEROIDS

The *steroids* are markedly different in structure from the neutral fats and phospholipids. They are classified as lipids because of their insolubility in water. They consist of four interconnecting rings of carbon atoms, three of which are six-membered rings and one of which is a five-membered ring.

**EXAMPLE 4** *Cholesterol* is typical of the structure of a steroid (see Fig. 3.4). Although cholesterol is associated with the advent of arteriosclerosis in humans, it is actually a vital structural component of the cell membrane and plays a key role in the proper function of such diverse animal tissues as nerve and blood. Cholesterol is not found in plants.

In addition to cholesterol, steroids include such fat-soluble vitamins as vitamin D; the sex hormones and the hormones of the adrenal cortex are steroids that seem to be derived from cholesterol produced within the body.

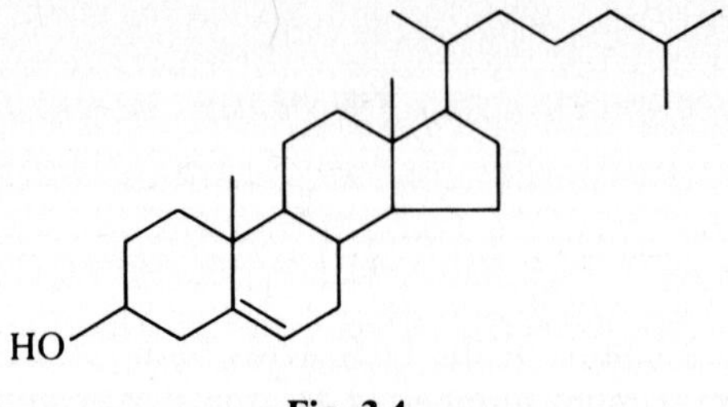

**Fig. 3.4**

## WAXES

A *wax* is a lipid because of its nonpolar solubility characteristics as well as its extremely hydrophobic (water-hating) properties. Waxes are composed of a single, highly complex alcohol joined to a long-chain fatty acid in a typical ester linkage. Waxes are important structural lipids often found as protective coatings on the surfaces of leaves, stems, hair, skin, etc. They provide effective barriers against water loss and in some situations make up the rigid architecture of complex structures such as the honeycomb of the beehive. They serve a commercial use as well, in furniture polish, automobile coating compounds, and floor finishes.

## 3.5 THE CHEMICAL BASIS OF LIVING SYSTEMS

All living organisms are composed of complex systems of organic and inorganic compounds. The boundary between a highly complex nonliving system and the simpler emergent life forms is somewhat arbitrary.

**EXAMPLE 5**   Viruses may be classified as either "living" or "nonliving," depending on one's point of view. The complexity of living systems is one necessary condition, but the ability of complex systems to grow, to reproduce themselves, to maintain internal order, and to process information is even more crucial to the status of "living entity." (For an in-depth discussion of this issue, see Probs. 1.11 to 1.13.)

# Solved Problems

**3.1**   What is the difference between the empirical and structural formulas for a monosaccharide?

An empirical formula merely sums up the number and kinds of atoms present in a molecule but does not show the arrangement of these atoms. The empirical formula $C_6H_{12}O_6$ refers to many different kinds of monosaccharides, including glucose, fructose, mannose, and galactose. The structural formula provides an insight into the number, kinds, and *arrangements* of the atoms making up the molecule. Thus, glucose may be differentiated from its isomer fructose by the structural formula, given either as a straight chain or in the form of a ring (see Fig. 3.1). Monosaccharides exist in both forms.

**3.2**   What are the major differences among the common polysaccharides glycogen, starch, and cellulose?

All three of these polysaccharides are composed of long, often branching chains of glucose molecules. Both glycogen and starch are principally used as storage forms of energy and are readily broken down by enzymes, which liberate the glucose monomers for further metabolic degradation. Cellulose is significant as a major structural macromolecule found in the cell walls of most plants and is not easily degraded to its constituent monosaccharides. Those organisms that can subsist on wood or grass can do so only because of microorganisms within their digestive tracts that are capable of digesting cellulose. If humans could develop such a symbiotic relationship with those cellulose-digesting microorganisms, the world food crisis could be eased considerably as we could all go out and graze.

Cellulose differs from both starch and glycogen in that it forms long unbranched chains that confer both strength and rigidity to the polymer. Further, the chemical makeup of cellulose and the nature of its bonds tend to produce long strands that are linked at particular points to yield a strong, fibrillar structure, much like the cables that are used on bridges to provide interlocking tensile strength. Starches contain both straight-chain and branched units, while glycogen has considerable branching. The branching within both starch and glycogen confers some slight solubility upon these molecules and also leads to their greater vulnerability to enzymic degradation.

**3.3**   What is the composition of chitin?

Chitin is a major constituent of the exoskeleton of insects and other arthropods and is also found among the fungi. It is a tough, water-resistant polymer consisting of long chains of a glucose derivative to which a nitrogen-containing group has been added. Although not, strictly speaking, a polysaccharide, it may be viewed as a modified polysaccharide. The modification consists of the substitution of

$$\begin{array}{c} -NH \\ | \\ O{=}C \\ | \\ CH_3 \end{array}$$

for the hydroxyl (—OH) group on the second carbon atom of each glucose.

**3.4**   Describe the interplay of hormones and enzymes in controlling glycogen levels.

The discovery and investigation of glycogen by Claude Bernard, an eighteenth-century French physiologist, led to a recognition of the role that antagonistic processes play in maintaining a constant internal environment in living organisms. This concept, later called *homeostasis*, was key to the realization that function, in both health and disease, involves a balance between mechanisms which tend to increase a particular constituent in body fluids and those which tend to decrease that constituent. Glycogen levels are controlled by the interaction of hormones and enzymes. Glycogen production is enhanced by high levels of glucose 6-phosphate, a precursor of glycogen, which stimulates glycogen synthase while inhibiting glycogen phosphorylase. *Insulin* stimulates the activation of glycogen synthase and, thus, the production of glycogen. Opposing insulin is *glucagon*, which along with *epinephrine* promotes the degradation of glycogen into glucose. Thus, the maintenance of an equilibrium for blood sugar is achieved by the careful elaboration of hormones: some promote the storage of glucose as glycogen when monosaccharide levels are high; others induce the breakdown of glycogen when blood glucose levels fall.

**3.5**    The genetic information in DNA comprises a code for the primary structure of a protein. What determines the important arrangements of secondary, tertiary, and quaternary structure for the protein?

Once the primary structure, which constitutes the linear array of amino acids making up the protein, is determined, the higher orders of structure are assumed automatically. These changes, involving alterations in the three-dimensional configuration of the protein, are produced by charge interactions within the molecule or by the coming together of hydrophobic or hydrophilic regions. In addition, the formation of H bonds between adjacent or even initially distant regions of the protein chain contributes to the folding, kinking, pleating, etc., involved in higher levels of structure. Particularly important in the association of individual polypeptides making up the quaternary structure is the formation of S—S bonds from the sulfhydryl groups (—SH) found on single polypeptide molecules. These S—S bonds are found, for example, as links between the two polypeptide strands of the insulin molecule.

**3.6**    How are proteins similar and different?

All proteins share certain common properties. They are assemblages of amino acids joined by peptide bonds to produce long chains known as polypeptides. They all undergo modifications in the shape of these polypeptide chains, which produce the secondary structure, and further alterations of configuration, which produce the superfolding or complex bending of tertiary structure. A number of polypeptide chains may be joined to produce the quaternary structure. However, beyond these commonly shared properties, proteins may demonstrate great differences, particularly in their degree of complexity.

*Fibrous proteins*, insoluble proteins particularly significant as structural entities, generally exist as long chains with regular sequences of particular amino acids. This regularity in amino acid composition imparts a regularity in the configuration of the fibrous protein. Many of the fibrous proteins show an $\alpha$-helical secondary structure ($\alpha$-keratins such as hair), while others show the pleated sheet arrangement ($\beta$-keratins such as silk). Tertiary structure is generally simpler and more regular than in other types of proteins. *Collagen* is a fibrous protein which exhibits a third type of structure, made up of three polypeptide chains that are tightly wound around each other to form a complex helix. These chains, whose internal "glue" is largely composed of H bonds, are extremely strong and contribute to the toughness of tendons and ligaments.

*Globular proteins* demonstrate far greater complexity than that found in the fibrous proteins, yet they tend to be relatively soluble. They generally lack the regularity in primary structure associated with the fibrous proteins and are highly irregular in their secondary structure. Their tertiary structure is particularly striking, involving complex folding patterns that yield a globular conformation. The quaternary structure involves complex interdigitation of already highly folded polypeptide chains.

Some globular proteins serve a structural role, as in the case of $\alpha$- and $\beta$-tubulin, which aggregate to form the microtubules of the cell; however, most are involved in more dynamic physiological processes. For example, the proteins of the blood, enzymes, and the protein hormones are all globular proteins. Many of these proteins are conjugated; i.e., they consist of a nonprotein molecule attached to the protein moiety. Hemoglobin is a conjugated protein in which four prosthetic heme groups are associated with the four independent, but intertwined, globular polypeptide chains.

**3.7**    Given the fact that many enzymes and blood proteins are conjugated proteins, what role might the prosthetic groups of such proteins play in the life of the cell?

The prosthetic group of a conjugated protein is usually an organic molecule loosely or tightly bound to the protein, but it may be a much simpler molecule or even a single atom or ion. The presence of the prosthetic group may profoundly alter the properties of the protein to which it is attached. In the case of metabolic enzymes, the prosthetic group may actually provide an attachment point for some of the constituents involved in the reaction. For example, many of the enzymes involved in dehydrogenation reactions contain a prosthetic group to which the hydrogens attach during the course of the reaction. These prosthetic groups are also called *coenzymes* because of the helper role they assume in the overall catalytic reaction.

In the terminology originally developed for enzyme reactions, the protein portion is called the *apoenzyme*, while the prosthetic group and apoenzyme together are termed the *holoenzyme*. The apoenzyme, as a complex protein, is sensitive to heat and does not readily diffuse across a membrane; the prosthetic group is generally resistant to heat and does readily diffuse.

**3.8**   What is the likely result of a diet that lacks one or more of the essential amino acids?

Essential amino acids, by definition, cannot be produced by the body. Thus, if they are not taken in through diet, they will be unavailable for incorporation during protein synthesis. This in most cases would either stop production of any protein requiring a missing amino acid or yield proteins of diminished or altered function—in both cases with dire consequences for the organism. Other amino acids might even be excreted because of the absence of a particular amino acid. The lack of a single essential amino acid in the diet may produce protein malnutrition.

**3.9**   What is the result of a diet lacking adequate protein?

A lack of suitable amounts of protein or the incompleteness of the protein ingested could lead to a state of negative nitrogen balance, a serious condition in which more nitrogen (a measure of protein) leaves the body than is taken in. Eventually the wasting of muscle and other vital tissues associated with protein depletion would lead to death. Such ravaging diseases as kwashiorkor (lack of protein) and marasmus (lack of protein and other nutrients) are the obvious signs of dietary protein deficiency in famine-bound regions. Unfortunately, restoring adequate protein to the diet is more difficult and expensive than providing calories from carbohydrates or even lipids.

Healthy adults are usually in nitrogen balance. Nitrogen lost through protein breakdown is replaced by an intake of nitrogen-containing protein. During growth and in convalescence from disease, a person may show positive nitrogen balance, more protein being taken in than lost.

**3.10**   Why are the steroids classified as lipids although their structures are so different from those of neutral fats (oils) and phospholipids?

The inclusion of the steroids in the lipid group is based entirely on their solubility. They share with other lipids a tendency to dissolve in such fat solvents as chloroform and toluene and to remain undissolved in water. It is also true that the metabolic pathways for the degradation of fats may interact with those for the conversion of cholesterol. Diseases that manifest themselves in the inability to handle neutral fats may sometimes be accompanied by symptoms of poor cholesterol metabolism as well.

Since steroids are relatively soluble in lipid, steroid hormones may accumulate in fat-rich tissues such as adipose tissue. This may pose a health problem for those who eat meat that comes from animals treated with steroids. Such treatment is a rather common practice among farmers seeking to increase the muscle mass of their livestock.

**3.11**   Saturated triglycerides tend to form fats, whereas polyunsaturated triglycerides tend to be oils. What is a possible explanation for this fact, and what are the implications in terms of health?

Polyunsaturated chains, because they are kinked, cannot lie next to each other as well as the straight-chain saturated fats can. This means they cannot form hydrophobic bonds as readily as saturated fats can, and therefore they tend to be less cohesive and heat-stable. This explains why they occur more often as liquid oils than as solid fats. Because they tend to remain liquid, they may be less likely to solidify in arteries. There is some evidence that diets high in saturated fat may be linked to atherosclerosis and cardiovascular diseases—a fact that has led some nutritionists to urge that fish and vegetable sources of protein be substituted for the red meat that makes up a significant component of the diet in affluent societies.

**3.12**  Solid vegetable shortenings are produced by treating plant oils with a stream of gaseous hydrogen in the presence of metal catalysts such as platinum. How does this process work?

Under these conditions the double bonds of the unsaturated oils open, hydrogen is covalently bound to the carbons, and a solid saturated (hydrogenated) fat is produced.

**3.13**  Plants have much lower levels of fats than animals do. Why is this?

Plants are generally not placed at a disadvantage in using starch as a principal energy storage macromolecule, because of their sedentary lifestyle. In situations where efficient (and therefore lightweight) storage of calories is advantageous in plants, triglycerides may be found. Seeds, which must store a maximum of calories in a limited space, are often rich in fats and oils. Oil-rich seeds, such as those from the cotton plant, are valuable commercial sources of oil.

Animals, because of their increased locomotion relative to plants, benefit from lighter-weight sources of energy and, hence, have evolved mechanisms to produce and store higher levels of fats than plants do. An illustration of the importance of fat storage in advanced vertebrate animal forms is their gradual development of a special organ, the *adipose tissue mass*. Among fish, fat is stored in an irregular fashion within muscle. Amphibians possess fat bodies, and reptiles have a rudimentary adipose organ. In birds and mammals, however, a metabolically competent adipose organ exists that is not merely a storage depot for lipid but is actively involved in the synthesis of lipid from carbohydrate. In mammals, the fat tissue is exquisitely sensitive to a variety of hormones and may even be used diagnostically to test for the activity of such hormones. Insulin, the major hormonal influence in the conversion of carbohydrate to fat, is assayed by measuring its effect on the uptake of glucose by adipose tissue.

**3.14**  Soaps are formed by the treatment of triglycerides or fatty acids with strong bases (usually NaOH or KOH). They are basically the salts of long-chain fatty acids, e.g.,

$$R-C{\overset{\displaystyle O}{\underset{\displaystyle O^-}{\Big\langle}}} \quad + \ Na^+$$

Given the disparate nature of the two ends of a soap molecule, what is the mechanism that makes soap an effective detergent?

The basis of soap's action is the ready association of the nonpolar portion (the long carbon chain) with oils, grease, and other lipids (the main constituents of dirt), which are then washed away with the entire molecule of soap by water, which associates with soap's charged polar end. This "dual allegiance" of soap to the world of oil and the world of water explains its function in the removal of dirt and contaminants.

**3.15**  Could we expect a mixture of complex chemicals, in the same proportions as found in living material, to possess the characteristics of life?

No, we could not expect an aggregate of inorganic and organic materials to take on such properties even if they were identical with the materials found in living cells. Life *is* considered to be a natural phenomenon and *is* susceptible to understanding in terms of its physical and chemical components. However, the materials in living organisms are not just put together haphazardly like the ingredients in a stew. Instead they are organized into complex structures with a variety of interactions that are possible only because of the *highly ordered positioning* of the materials. Localized concentrations of materials are maintained by membrane-bound compartments that permit specialization within the cell. A hierarchy of structural and functional organization that developed slowly over millions and even billions of years exists within the cell, and within multicellular organisms as well. Scientists can simulate some of the simpler levels of complexity in living matter, but we are still a long way from understanding all the complex interlocking parts that provide life with its pulse and its integrity.

# Supplementary Problems

**3.16** Name the following organic compounds:

(a) $R-\overset{H}{\underset{}{C}}=O$   (b) $R-\overset{}{\underset{O}{C}}-R$   (c) $R-C\overset{O}{\underset{OH}{}}$   (d) $R-OH$

**3.17** A pentose will contain how many carbon atoms?

**3.18** The union of three monosaccharides to form a trisaccharide yields how many molecules of water?

**3.19** In what way do glucose and dextrose differ?

**3.20** Which two hexoses are found in common table sugar, a disaccharide?

**3.21** Which polysaccharide is associated with animal organisms?

**3.22** The great strength and resistance to hydrolysis of cellulose is due to which of its structural features?

**3.23** How many amino acids exist in nature?

**3.24** Which two atoms are joined in a peptide bond?

**3.25** The linear order of amino acids in a protein, which constitutes its primary structure, is maintained by what type of bonding?

**3.26** The union of several polypeptide chains to form a complete protein is regarded as what level of protein structure?

**3.27** A healthy infant will be in what state of nitrogen balance?

**3.28** Lecithin is an example of a(n)_____?

**3.29** Hormones produced in the ovary belong to which class of lipid?

**3.30** The protective cuticle of a leaf is particularly rich in what substance?

**3.31** Which two lipids should be reduced in the diets of people suffering from cardiovascular disease?

**3.32** Which derived polysaccharide is found in the cell walls of fungi and the exoskeletons of insects?

**3.33** Name two globular proteins that play a primarily structural role.

**3.34** Which portion of an enzyme is resistant to boiling?

# Answers

**3.16** (a) Aldehyde (b) Ketone (c) Organic acid (d) Alcohol
**3.17** Five
**3.18** Two
**3.19** None, they are the same.
**3.20** Glucose and fructose

**3.21** Glycogen

**3.22** Parallel arrangement of fibers and lack of branching

**3.23** 20

**3.24** Carbon and nitrogen

**3.25** Covalent

**3.26** Quaternary

**3.27** Positive

**3.28** Phospholipid

**3.29** Steroids

**3.30** Wax

**3.31** Saturated fats and cholesterol

**3.32** Chitin

**3.33** $\alpha$- and $\beta$-tubulin

**3.34** The prosthetic group

# Chapter 4

# The Cellular Organization of Life

## 4.1 THE CELL DOCTRINE

All living things are made up of *cells*. The cell is the unit of life. In the kingdoms Monera and Protista the entire organism consists of a single cell. In most fungi and in the animal and plant kingdoms, the organism is a highly complex arrangement of up to trillions of cells. The human brain alone contains billions of cells. So vital are cells and their activities to an understanding of life that the cell doctrine has become a central organizing principle in the field of biology. Today, the concept of the cell is taken for granted, but this has not always been true.

Although the *cell doctrine* is generally credited to Matthias Schleiden (1838) and Theodor Schwann (1839), it is actually a result of the efforts of many biologists. Highlights of its development are as follows:

- van Leeuwenhoek refines the making of lenses and microscopes (middle seventeenth to early eighteenth centuries).
- Hooke publishes a paper on the cellular nature of cork (1665).
- Lamarck states that all living organisms must possess cellular tissue (1809).
- Dutrochet asserts that all living matter is made up of tiny globular cells, which increase in both size and number (1824).
- Brown describes the nucleus (1831).
- Schleiden publishes studies of plant cells (1838).
- Schwann publishes studies of animal cells (1839).
- Virchow concludes that all cells come from preexisting cells (1858).

By the nineteenth century the cellular organization of all living material was apparent, and the resulting cell doctrine maintained that:

1. All living things are made up of cells and the products formed by cells.
2. Cells are the units of structure and function.

In 1858, Virchow added a third statement:

3. All cells arise from preexisting cells.

By the close of the nineteenth century, cells were also recognized as the basis for understanding disease; i.e., when people were sick, it was because their cells were sick. Until the middle of the twentieth century, *pathology* (the study of disease) largely involved a cellular (cytological) approach.

## 4.2 CELLULAR ORGANIZATION

In the nineteenth century, the cell was described merely as having a limiting outer membrane, an interior nucleus, and a large mass of cytoplasm surrounding the nucleus. Little, beyond its actual existence, was known of it. However, ever sharper methods of probing the cell gradually brought its internal structure to light. Early microscopes used thinly sliced specimens through which light could be shone to illuminate cellular features. Improved staining methods enabled researchers to heighten the visibility of cellular structures selectively. These methods killed the cells being studied; however, later techniques, using the way that certain characteristics of light (such as polarization) are affected by the density and regularity of cellular structures, achieved even greater contrast between those structures, without killing the specimens. Vital stains were also employed to study living cells. By this point, many subcellular features had been studied, including the chromosomes and nucleolus of the nucleus. With the development in the 1930s of the electron microscope, which employs electrons rather than light, the elaborate membrane architecture of the cytoplasm was more completely resolved.

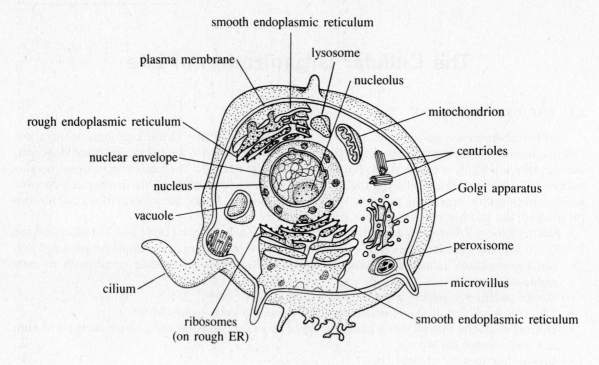

(a) Animal Cell

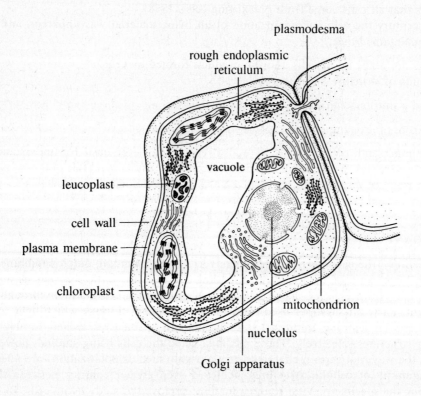

(b) Plant Cell

Fig. 4.1

With the elucidation of the numerous subcellular structures, it became clear that the functions of the cell are accomplished by specialized structures comparable to the organs of the body; consequently, these structures were called *organelles*. These organelles provide regional separateness, like the rooms of a house, and permit specialization.

The arrangement of the organelles is compatible with a division of labor within the cell. Generally, organelles develop as organized systems of membranes within the cell and may be seen as compartments in various shapes. The form assumed by an organelle is related to its functions. This intimate association of structure (form) and function is universal in biology and can be observed at every level—cell, multicellular organism, and even the ecosystem.

**EXAMPLE 1**  Nerve cells are a particularly apt illustration of the form-function relationship. These cells maintain a communication network throughout the body by the transmission of electrical impulses to muscles and glands and the similar movement of electrical signals into the "switchboard" of brain and spinal cord. The carrying of electrical signals can be best accomplished by long thin processes that extend throughout the body. The nerve cells consist of such long extensions of their cytoplasm, resembling the network of wires in a telephone or electrical system.

## 4.3  CELLULAR ORGANELLES

Many cellular organelles appear to be derived from membranes, which are thin sheets of living material within a surrounding amorphous medium. Some organelles, however, do not have a membrane structure. These include ribosomes, microtubules and microfilaments, flagella, cilia, and centrioles. The prokaryotes have only a limited repertoire of organelles in their cytoplasm, and these are generally nonmembranous, such as the ribosomes. They lack cilia, centrioles, microfilaments, and microtubules. Eukaryotes are rich in the numbers and kinds of organelles that are present, and they include both membranous and nonmembranous types (see Fig. 4.1). In both prokaryotic and eukaryotic cells a cell membrane (plasma membrane) is always present.

The *cell membrane* is the outer layer of the living cell. It controls the passage of materials into and out of the cell. An older view of the cell membrane, the *unit membrane hypothesis*, describes the membrane as an inner and outer dense protein layer surrounding a thicker but less dense phospholipid layer. This sandwich structure was indicated by electron microscope studies of many membranes. Channels were also seen to run through the membrane to the exterior.

More recently S. J. Singer and G. L. Nicholson have introduced the *fluid mosaic model* (see Fig. 4.2). Like the earlier model, it proposes a double layer of phospholipids, with their polar ends facing

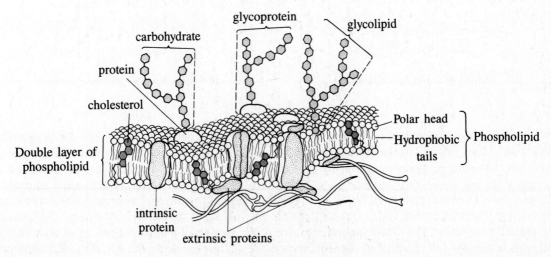

**Fig. 4.2**

the inner and outer surfaces and the hydrophobic, nonpolar ends apposed at the center of the bilayer. However, the fluid mosaic model better explains the dynamic nature of the membrane proteins. According to this model, these proteins may reside on the exterior or interior face of the lipid bilayer (*extrinsic proteins*) or may be located in the phospholipid matrix (*intrinsic proteins*); some may be embedded in the bilayer but project through to the exterior, the interior, or both surfaces of the membrane. The primary and tertiary structures of the proteins are compatible with their locations on or in the membrane. Intrinsic proteins tend to have predominantly hydrophobic amino acids, and they assume conformations that segregate any hydrophilic amino acids from the hydrophobic bilayer; extrinsic proteins, conversely, tend to have hydrophilic residues, which can bond with the polar end of the phospholipids and interact with the surrounding aqueous solution. According to this model, some lateral circulation of phospholipid and protein occurs.

In many cells the outer membrane is surrounded by a rigid wall or a tough *pellicle*. These extraneous structures are nonliving additions to the membrane surface that do not substantially affect the permeability of the cell. In plants, fungi, and bacteria, this outer coat is called a *cell wall* and is composed, respectively, of cellulose, chitin, or a variety of complex carbohydrate and amino acid combinations. The cell wall provides support and may even keep the cell from bursting in hyposmotic environments. Most animal cells contain an outer layer of short carbohydrate chains covalently bonded to the membrane and called a *glycocalyx*. This covering contains *receptors* that bind with external substances controlling internal cell activity. The glycocalyx also contains the antigenic glycoproteins that give cells their immunological identities. Other external structures serve to anchor cells tightly together or may provide intercellular spaces for transport between cells. The shells that surround and protect various vertebrate and invertebrate eggs are another form of barrier outside the cell membrane.

The *endoplasmic reticulum* (ER) is a series of membranous channels that traverse the cytoplasm of most eukaryotic cells (see Fig. 4.3). It forms a continuous network extending from the cell membrane to the nuclear membrane. In some regions of the cell it may appear as a series of flattened disks or sacs. In many parts of the cell the endoplasmic reticulum is associated with small dense granules lying along the outer border of its membrane. These structures are known as *ribosomes*. They impart a rough appearance to the endoplasmic reticulum, so that the ER is called the *rough endoplasmic reticulum* (RER) in these regions, which are usually associated with active protein synthesis. The *smooth endoplasmic reticulum* (SER) does not contain ribosomes and is associated with cellular regions which are involved in the synthesis and transport of lipids or the detoxification of a variety of poisons.

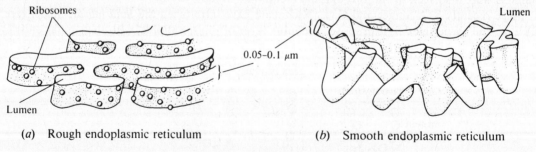

(*a*)  Rough endoplasmic reticulum          (*b*)  Smooth endoplasmic reticulum

**Fig. 4.3**

Similar in their membrane structure to the ER are the organelles known as the *Golgi apparatus* (Fig. 4.4*a*). They exist as stacks of flattened sacs, or vesicles, that are continuous with the channels of the SER. Their major function is the storage, modification, and packing of materials produced for secretory export, since these organelles are particularly prominent in secretory cells such as those of the pancreas. The outer portion of the Golgi apparatus releases its secretory material within membrane-enclosed globules (secretory vesicles) that migrate to the surface of the cell. It may also provide material for the cell membrane. The Golgi apparatus may actually be part of a dynamic system of membranous channels within the cell in which all elements such as the nuclear envelope, the ER, the Golgi apparatus, and the cell membrane are connected to each other without sharp boundaries.

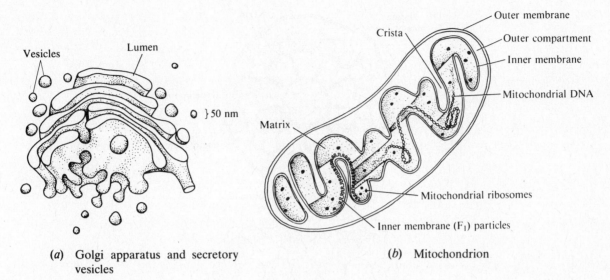

(*a*)  Golgi apparatus and secretory
       vesicles

(*b*)  Mitochondrion

**Fig. 4.4**

*Mitochondria* are rounded or cigar-shaped organelles (see Fig. 4.4*b*) that are particularly prominent in cells with high metabolic activity (see Chap. 5). Their name derives from their threadlike appearance (Greek *mitos*, "thread") under the light microscope. Mitochondria have a double wall: an outer smooth membrane which forms the outer boundary and an inner membrane which is extensively folded. The folds, or *cristae*, project into the interior of the organelle and have a variety of enzymes embedded in them. These enzymes are involved in the systematic degradation of organic molecules to yield energy for the cell. Like the chloroplasts of plants, the mitochondria contain their own DNA and ribosomes; they replicate independently of the rest of the cell and appear to control the synthesis of their membranes.

*Lysosomes* are similar in shape to mitochondria but are smaller and consist of a single boundary membrane. They contain powerful enzymes that would digest the cellular contents if they were not contained within the impermeable lysosomal membrane. Rupture of this membrane releases these enzymes. The lysosome plays a role in intracellular digestion and may also be important in the destruction of certain structures during the process of development. In the metamorphosis of a frog, lysosomal enzymes help destroy those structures of the tadpole that are no longer useful in later developmental stages. The raw materials arising from degradation of such regions as the tail are then used in the formation of more mature parts. Lysosomes are also involved in such autoimmune diseases as rheumatoid arthritis.

*Peroxisomes* are similar to lysosomes except that the enzymes contained in these organelles are oxidative in function. Peroxisomes are involved in the oxidative deamination of amino acids, a reaction vital to the conversion of proteins to other kinds of compounds.

*Flagella* and *cilia* (see Fig. 4.5*a*) are hairlike projections anchored at one end and capable of various movements at the other. They are encased in a membrane continuous with the cell membrane. They play an important role in cell motility because of their coordinated beating motion.

**EXAMPLE 2**  A human sperm cell has an active beating flagellum at its posterior end. The energy for this flagellar activity is derived from a mitochondria-rich midpiece, to which the flagellum is attached. The payload of the sperm cell, the anterior nucleus, is propelled by the flagellum along the female reproductive tract, where it may eventually fertilize an egg in the oviduct. Failure of the sperm to achieve sufficient motility may lead to inability to fertilize the egg.

Flagella and cilia may be treated as a single kind of organelle: if the structures are few in number and relatively long, they are called flagella; if short and numerous, they are considered cilia.

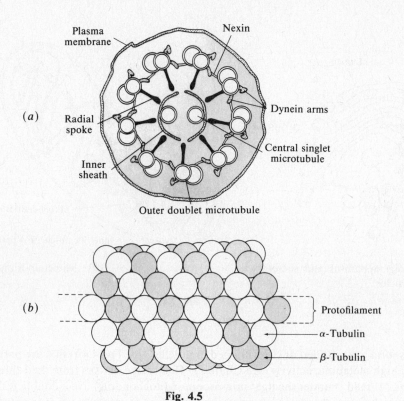

Fig. 4.5

Both flagella and cilia derive their motility from structures called *microtubules*. As their name implies, these are elongated, hollow, cylindrical structures; they are assembled from two protein subunits called $\alpha$- and $\beta$-*tubulin*, which combine to form a unit building block for the microtubule. These unit pairs are stacked end on end to form a long string called a *protofilament*. Thirteen of the protofilaments are aligned parallel to each other in a circle like the posts of a circular fence and thus form the microtubule (see Fig. 4.5*b*).

The flagella and cilia of eukaryotes contain an outer circle of nine pairs of microtubules that surround a central core of two microtubules. A *basal body*, to which each flagellum or cilium is attached, lies just under the surface of the cell. It is similar in its underlying structure to the cilium or flagellum that extends from it except that it contains a peripheral arrangement of nine *triplet* microtubules with *no* central dyad of microtubules (see Fig. 4.6).

Fig. 4.6

*Centrioles* are present as a pair of cylindrical rods in many eukaryotic cells. They lie just above the nuclear envelope (membrane), and since their longitudinal axes are perpendicular to one another, they form a cross. The microtubular structure of the centriole is the same as that of the basal body and may have arisen from primitive basal bodies during cellular evolution. Centrioles probably play a role in the formation of the spindle apparatus, which is an essential feature of both mitosis and meiosis (see Chap. 8).

Functions of cellular support and movement are supplied by the cell's *cytoskeleton*, which comprises three different types of protein—microtubules, microfilaments, and intermediate filaments—together with a number of auxiliary proteins. Microtubules, in addition to being constituents of cilia and basal bodies and the spindle apparatus, form a structural network around the periphery of many protozoans. *Microfilaments* are formed from the protein *actin*. The actin subunit is polymerized into a long strand, and two of these strands unite in a double helix to form the microfilament. Microfilaments radiate throughout the cytoplasm and thus maintain cell shape. In combination with the protein *myosin*, they form sliding filaments associated with muscle contraction and changes in cell shape. The *intermediate filaments*, so called because their diameter is intermediate to the other two structural filaments, are longer-lived than the other two and are found, for example, in dense meshworks that give tensile strength to epithelial tissue sheets.

*Vacuoles* are discrete, clear regions within the cell that contain water and dissolved materials. The vacuole may act as a reservoir for fluids and salts that might otherwise interfere with metabolic processes occurring in the cytoplasm. The membrane surrounding the vacuole is called a *tonoplast*. Many protozoans have a contractile vacuole, which periodically contracts and forces fluid and salts out of the cell. The structure serves to prevent an accumulation of fluids in organisms that live in fresh water. Vacuoles containing digestive enzymes may also be formed around ingested food particles in a variety of cells. In the cells of many plants, a large central vacuole is a prominent feature; this vacuole may swell, press against the rigid cell wall, and give the cell a high degree of rigidity, or *turgor*.

In almost all plant cells a variety of tiny membrane-enclosed sacs are found that contain pigments or provide storage space for starch. These organelles are called *plastids*. Chloroplasts are included in this group. The plastids move passively along with the cytoplasm that streams within healthy plant cells.

The *nucleus* is a round or oval body lying near the center of the cell. It is surrounded by a double membrane, the nuclear membrane or envelope. These membranes coalesce in certain portions of the nuclear envelope, and in these regions, pores (openings) may be formed that provide a route for materials to leave the nucleus directly. Since the outer membrane of the nuclear envelope is continuous with the endoplasmic reticulum, the pores may actually permit passage from the interior of the nucleus to the channels of the endoplasmic reticulum.

Within the nucleus, one or more *nucleoli* may be seen. These are dense bodies containing the subunits for the ribosomes, the cytoplasmic organelles involved in the synthesis of protein. The nucleolus is involved in the assembly and synthesis of ribosomes. It is usually attached to a specific chromosome in the nucleus. Each *chromosome* exists as a tiny individual rod or string throughout the life of the cell, but in the resting (nondividing) cell the chromosomes look like a single network of thin threads. The gene material of the cell is found in the chromosomes.

## 4.4  PLANT AND ANIMAL CELLS: TISSUE ORGANIZATION

Eukaryotic cells occur in all animals and plants, but there are a number of significant differences between the cells of organisms in these two kingdoms (see Fig. 4.1*a* and *b*). Plant cells almost always contain an extracellular cell wall, which is made up of cellulose. Animal cells do not generally possess a cell wall. Cell walls are also found in fungi and bacteria, but they are not composed of cellulose in these organisms. Plastids are a feature of most plant cells but are not found in the cells of animals. Vacuoles are quite prominent in plant cells, but are far less significant in or absent from animal cells. While animal cells invariably demonstrate a pair of centrioles lying just outside the nucleus, centrioles are not usually found in plants. As will be discussed in Chap. 8, plants differ markedly from animals in specific details of the process of cell division (mitosis), although the general features of this reproductive function are similar in the two groups.

In both plants and animals, groups of similar cells are organized into loose sheets or bundles called a *tissue*. Tissues carry out a specific activity. A variety of different tissues, in turn, are arranged in discrete structures of definite shape known as *organs*. Organs carry out a specific function within the organism; e.g., the kidney is an organ that removes wastes from the blood and excretes them as urine.

A number of organs may be associated as an *organ system*, a complex that carries out some overall function. Thus, stomach, mouth, intestine, etc., form the digestive system of animal organisms. The sequence of cell → tissue → organ → organ system → organism represents a hierarchical structure in which higher levels integrate the processes derived from lower levels.

## PLANT TISSUES

One method of classification for advanced plants divides all plant tissues into *meristematic* tissue and *permanent* (differentiated) tissue. The meristematic tissue, which is found at the growing ends of roots and stems as well as in peripheral areas of stems, tends to have cells that are undifferentiated, small, compact, and packed with metabolically active cytoplasm.

The permanent tissue is subdivided, in one scheme, into *lining tissues*, *fundamental tissues*, and *vascular tissues*. The lining tissue is generally *epidermis*, a thin-walled layer of cells with prominent vacuoles. On some epidermal layers a thick cuticle of wax is secreted that protects the plant from water loss. In older perennial plants the surface of roots and stems consists of *periderm*, a tissue composed of several layers of corky cells that are resistant to water.

Fundamental tissues are found in the interior of the plant. The most widely distributed is *parenchyma*, thin-walled cells active in photosynthesis and highly diffuse in arrangement. Cells with irregular thickening make up the *collenchyma*. Thick-walled cells which impart structural support for the plant make up the *sclerenchyma*, a fibrous layer which can also exist as the hard part of many seed shells and pits.

The vascular tissues of the plants are primarily *xylem* and *phloem* (see Fig. 4.7). Xylem serves as a continuous passageway for the transport of water and dissolved solutes, primarily in an upward direction. Two types of cells associated with xylem are *vessels* and *tracheids*. These cells are arranged in bundles, and upon maturity they lose their living material. The cell walls then serve as tubular containers for the transport of fluids and also provide structural support for the plant. Wood consists

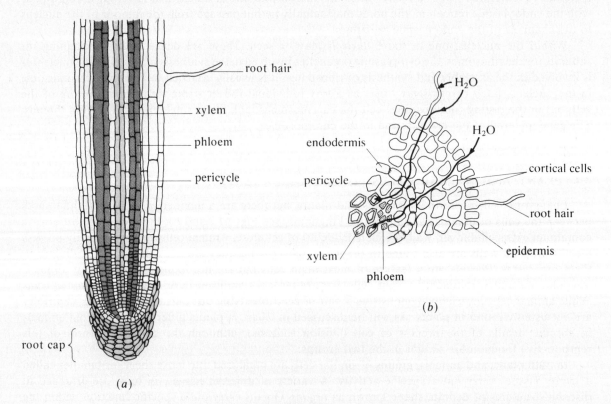

**Fig. 4.7**

largely of xylem vessels and tracheids, many of which have become plugged so that they no longer function in fluid transport.

*Phloem* functions in the transport of carbohydrates, amino acids, oligopeptides (i.e., peptides having fewer than 10 amino acids; *oligo* means "few"), and some lipids. Unlike xylem, it remains alive while performing its transport function, although some of its cell types lose their nuclei.

These three major tissue types are restricted mainly to the higher land plants, and their hierarchical ordering into organs and systems is not apparent.

## ANIMAL TISSUES

Animal tissues are of four major types (see Fig. 4.8):

1.  Epithelia—a covering tissue
2.  Connective—a connecting and supporting tissue
3.  Nerve—a conducting tissue
4.  Muscle—a contracting tissue

*Epithelial tissue* consists of densely packed cells that line the surfaces of the body. It often functions as a barrier, regulating absorption of materials or offering protection from dehydration, cold, microbial invasion, etc. Skin, for example, is made up largely of epithelial tissue; the digestive tract and the other cavities of the body, as well as the ducts and blood vessels, are also lined with epithelial tissue. The three cell types constituting epithelial tissue (the thin, flat *squamous cells*; the *cuboidal cells*; and the *columnar cells*) usually have one end anchored in a fibrous *basement membrane*, while the other end varies with function.

*Connective tissue* occurs in diverse forms but is characterized by the extracellular matrix in which its cells lie. For example, bone consists primarily of an extracellular matrix, with the relatively few bone cells residing in *lacunae* (hollow spaces) in the solid matrix. Other forms of connective tissue are *blood, cartilage* (the tough, flexible support material of the ear, for example), and the various types of support fibers that impart strength and, sometimes, elasticity to the body and frequently connect tissues (for example, tendons, which attach muscle to bone).

*Nervous tissue* is made up of *neurons* (nerve cells), some of which may be over a meter long. Nerve impulses pass *from* the cell body of the neuron along its *axon* and *toward* the cell body of the neuron along one of its *dendrites. Sensory neurons* are often highly specialized to respond to specific stimuli (touch, sound, smell, etc.). *Motor neurons* are responsible for activating muscular response and are usually coordinated with sensory neurons through *association neurons.* Nervous tissue is found throughout the body, most notably in the brain and spinal cord.

*Muscle tissue* is a contractile tissue comprising three distinct types: *skeletal* or *striated* muscle produces voluntary movement; *smooth muscle* effects most involuntary movement, for example, the peristalsis of the digestive tract; and *cardiac muscle* forms the muscle of the heart.

The organization of tissues into organs and organ systems is readily observable in almost all animal groups.

## 4.5  CELL SIZE AND ITS CONSTRAINTS

With a few exceptions, cells tend to be very small, rarely exceeding microscopic dimensions. This is particularly true for cells with high metabolic rates. The needs of the cell for oxygen and food are met by the passage of these substances from the exterior across the surface membrane into the interior of the cell. However, as with all spheres, as the radius of the cell increases, the contents (volume) of the cell increase proportionally with the *cube* of the radius but the surface area of the cell increases only by the *square* of the radius. Therefore, the ability of the surface to pass materials across its membrane increases by the square of the radius, but the needs of the cell grow by the cube of the radius. A point is reached in cell size at which the surface is too small to support the metabolic needs of a more rapidly expanding interior.

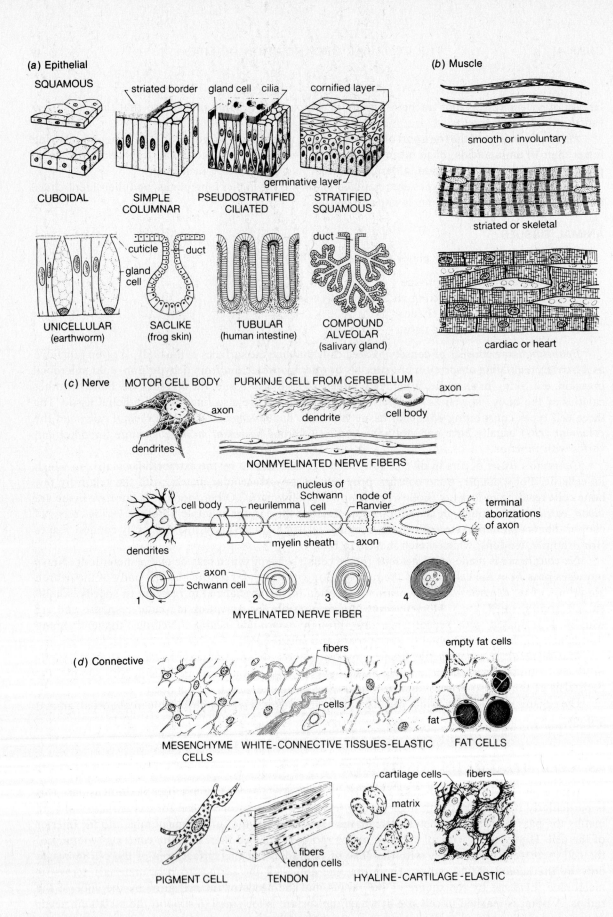

(a) Epithelial

SQUAMOUS

CUBOIDAL

striated border     gland cell    cilia

cornified layer

germinative layer

SIMPLE
COLUMNAR

PSEUDOSTRATIFIED
CILIATED

STRATIFIED
SQUAMOUS

cuticle     duct

gland
cell

UNICELLULAR
(earthworm)

SACLIKE
(frog skin)

TUBULAR
(human intestine)

duct

COMPOUND
ALVEOLAR
(salivary gland)

(b) Muscle

smooth or involuntary

striated or skeletal

cardiac or heart

(c) Nerve     MOTOR CELL BODY     PURKINJE CELL FROM CEREBELLUM

axon

axon

dendrites

dendrite     cell body

NONMYELINATED NERVE FIBERS

cell body     neurilemma

nucleus of
Schwann
cell

node of
Ranvier

terminal
aborizations
of axon

dendrites

myelin sheath     axon

axon
Schwann cell

1          2          3          4

MYELINATED NERVE FIBER

(d) Connective

fibers

empty fat cells

cells

fat

MESENCHYME
CELLS

WHITE-CONNECTIVE TISSUES-ELASTIC

FAT CELLS

cartilage cells     fibers

matrix

fibers
tendon cells

PIGMENT CELL     TENDON     HYALINE-CARTILAGE-ELASTIC

**Fig. 4.8**

Another factor involved in cell size is the necessity for the nucleus to maintain control over cellular activity. As the cell increases in size, its outlying parts are farther from the central nucleus. This makes the integration of cell function by the nucleus extremely difficult. Even the passage of materials from one region of the cytoplasm to another takes longer in large cells and further slows the integration of response.

## 4.6  MOVEMENT INTO AND OUT OF THE CELL

The property of membranes that permits movement across their surface is called *permeability*. The internal environment of the cell is carefully maintained by the selective permeability of the cell membrane. Many materials pass across the membrane in accordance with their concentration gradients. At one time the membrane was thought to play only a passive role in the movement of solutes and water into and out of the cell through diffusion and osmosis. We now know that a variety of mechanisms exist within the membrane to initiate or speed up the transport process. If the movement of molecules across the membrane is in accordance with the concentration gradient and energy is not used, then transport is *passive*. If the flow is opposed to the concentration gradient, energy must be expended and the term *active transport* is used.

A variety of nonlipid materials, such as $Na^+$ and $K^+$, probably pass across the membrane through special *channels* or pores. These channels may be transient or relatively permanent and are thought to facilitate the passage of particular molecules or ions on the basis of their diameter, charge, or ability to form weak bonds between the migrant species and some constituent of the channel. Channels that tend to be a permanent part of the membrane may exist in an "open" or "shut" condition depending on the state of a protein gate; these gates provide a means for altering permeability with a change in environmental conditions.

A variety of components within the membrane function as *carriers*. In *facilitated transport* systems, the carriers combine with small molecules or ions that are plentiful at one surface to form a complex; this complex then moves along its concentration gradient to the other surface, where it releases the transported molecule. In *active transport* systems the carrier system may be an enzyme (permease) that undergoes conformational change when it combines with its passenger molecule and resumes its original shape when the passenger is discharged. Unlike facilitated transport carriers, however, the complex can move *against* solute concentration gradients. The enzyme ATPase has been suggested as an enzymic carrier in the movement of $Na^+$ and $K^+$ across the membrane.

Carriers that are specific for the transport of ions are known as *ionophores*.

**EXAMPLE 3**  The antibiotic valinomycin provides an instructive example of ionophore action. This organic molecule combines with $K^+$, as well as $Na^+$, to form a small circle with an interior polar region and an exterior hydrophobic ring. This arrangement provides for an interior hold on the ion while the entire complex passes easily across the hydrophobic region of the membrane.

Substances that are of *macromolecular* size generally do not penetrate the membrane. However, large particles may get into the cell in a bulk transport phenomenon known as *endocytosis*. Such particles may attach to specialized receptors on the membrane. The particle-membrane complex lengthens and invaginates and then pinches off to form vesicles within the cytoplasm. The surface engulfment of large structures such as bacteria by protozoans or white blood cells is a type of endocytosis known as *phagocytosis* (see Fig. 4.9a). *Pinocytosis* is a form of endocytosis in which the particles are relatively small; tiny indentations of the surface permit an interior migration of the particles and their surrounding fluid, which are then collected in tiny vacuoles or vesicles at the blind end of the indentation (see Fig. 4.9b). Many of the vesicles arising in pinocytosis show a fuzzy coating at an early stage of their formation. This coating, made up of a highly ordered array of protein, may be involved in the interaction of the vesicles with other components of the cytoarchitecture of the cell.

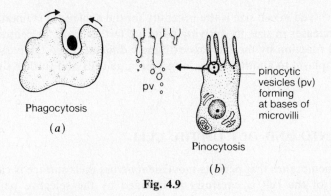

Phagocytosis

(a)

pinocytic
vesicles (pv)
forming
at bases of
microvilli

pv

Pinocytosis

(b)

**Fig. 4.9**

In *exocytosis*, materials enclosed in a membranous vesicle are brought to the surface membrane. Fusion occurs with the surface membrane, and the vesicular surface bursts to release the enclosed material to the exterior. Materials of large size which are no longer needed by the cell or even substances which are produced by the cell to be used in other parts of the organism may regularly be extruded by exocytosis.

# Solved Problems

**4.1**    Describe the evolution of the cell doctrine.

Robert Hooke was the first scientist to describe cellular structure. He studied thin sections of cork and noted its boxlike structure in a paper published in 1665. The honeycomb arrangement of these box units reminded him of the tiny rooms of a monastery, which are called *cellulae* in Latin.

Hooke could not have seen what were actually the nonliving cell walls of his cork preparation were it not for the *microscope*, an instrument that uses a system of magnifying lenses to reveal the features of objects too small to be seen by the unaided eye. As early as 300 B.C. the Greeks had used curved glass containers filled with water to magnify nearby objects, but it was not until the seventeenth century that Anton van Leeuwenhoek refined the grinding process to produce lenses that could be used effectively in simple microscopes.

In 1809, Lamarck recognized that all living things show cellular structure. In 1824 Dutrochet stated unequivocally that all living tissues are made up of tiny globular cells. Further, he realized that growth involved both an increase in the size of existing cells and an increase in the number of cells. In 1831, Robert Brown described the nucleus, which is a feature of almost all eukaryotic cells. In 1838, Schleiden published his studies of the cellular structure of plants, and Schwann released his parallel findings on the cellular makeup of animal tissue the following year. Because of the clarity of their description and the vigor with which they campaigned for acceptance of their notions, Schleiden and Schwann are generally credited with formulating the cell doctrine and placing the cell in the center of investigations into the nature of life. When Rudolph Virchow asserted in 1858 that all cells come from preexisting cells, the cell assumed the role of a continuous living chain in time by which life was to be understood.

**4.2**    Trace the evolution of the microscope in the study of cell structure.

The structural components of the cell are best seen with a microscope. Early microscopes utilized light that passed through extremely thin sections of cellular material. These sections were first fixed with substances such as formaldehyde that stabilized the components of the cell. Appropriate stains were then applied that could render visible specific organelles. The cells were, however, killed by the preparation process. With the development of the *interference* and *phase-contrast* microscopes, living material could be studied without the intervention of fixing and staining. These microscopes make use of the fact that light traveling through materials of different densities can be altered by special devices that tend to emphasize the contrast between adjacent structures. *Polarizing* microscopes utilize a beam of polarized light (wave movement in a single plane rather than in all directions) to distinguish those areas of a cell in which there is a regular alignment of constituent parts.

Perhaps most fruitful in the study of cellular organization is the *electron microscope*, a device employing a beam of electrons rather than light. Magnifying unresolved regions does not enhance clarity. Higher magnification is effective only with an enhancement of *resolution*. The use of electrons permits a resolution (separation of neighboring particles) that is more than 1000 times better than that of the light microscope. Thus, two points that are only 0.2 nm apart can be seen distinctly with the electron microscope.

The electron microscope requires a vacuum chamber, since electron beams are dissipated by air. A magnified image from the electron beam is visualized on a fluorescent screen and may also be permanently recorded on a photographic plate. The preparation of the specimen for electron microscopy is more tedious and requires even greater care than that for light microscopy.

A new generation of electron microscope, the scanning electron microscope (SEM), scans the surface of opaque preparations and produces an image of the scatter pattern that emerges. Some investigators use highly energized electron beams to penetrate thicker sections of material, which may be more representative of actual cellular structure than very thin sections.

**4.3**    With *centrifugation*, a heterogeneous solution is exposed to centrifugal force by being spun, and its constituent parts are separated according to their relative densities. How and why would this technique have been used in cell studies?

The first step involves the disruption and dispersal of cell preparations in a fluid to produce a *homogenate*. The homogenate contains all the material of the intact cell but in a scattered arrangement. Subsequent sequential centrifugation of the homogenate produces *fractions* containing specific organellar material that can be studied and tested. At comparatively low centrifugal force, a sediment will accumulate containing mostly nuclei, unbroken cells, and other large, dense fragments. Higher speeds of the centrifuge will next remove mitochondria and lysomes. Extremely high speeds will then produce a pellet rich in a complex of ribosomes and endoplasmic reticulum (microsomes). The remaining fluid of the homogenate will contain a variety of soluble molecules and some miscellaneous low-density materials. The various fractions obtained can be analyzed biochemically, stained for enzyme activity, or observed in the light or electron microscope. Radioactive tracer molecules may be particularly useful in clarifying the role of many of these fractions.

**4.4**    Describe the unit membrane hypothesis.

The unit membrane hypothesis of J. D. Robertson was based on electron micrographs taken of red blood cells and the outer myelin sheath that usually surrounds the nerve fibers of all vertebrates. This sheath is derived from the winding growth of Schwann cells around the axon. These micrographs suggested that the membrane possessed a trilaminar (three-layer) structure, in which two electron-dense layers surrounded a broader, less dense layer, much as two slices of bread enclose an inner layer of cheese in a sandwich. The dense inner and outer "bread" layers were postulated to be protein, while the lighter interior layer was described as a double layer of lipid. This lipid layer was believed to be largely phospholipid with some cholesterol.

Robertson's model of the cell membrane tended to support an earlier hypothesis of membrane structure developed by J. F. Danielli of London in 1940. Danielli reasoned that the permeability characteristics of red blood cells and other types of cells were consistent with a membrane containing a double layer of lipid and surrounded by an inner and an outer protein coat. Danielli further suggested that the double layer of lipid was oriented so that the outermost, polar ends of each phospholipid layer faced the inner or outer protein surfaces while the innermost, nonpolar ends of each phospholipid layer lay next to one another. The membrane was also traversed by nonlipid channels scattered randomly throughout its structure. In a number of ways the electron micrograph interpretations of Robertson were consistent with the theoretical model developed by Danielli; this led to the popularization of the unit membrane hypothesis as the basis for all membrane structure within the cell.

**4.5**    Describe the fluid mosaic model.

By the late 1960s evidence was accumulating that the unit membrane hypothesis was inadequate to account for the dynamic nature of membrane proteins, although it was clearly in harmony with the known distribution of membrane lipids. In 1972, a new hypothesis of membrane structure was developed by Singer and Nicholson. Known as the fluid mosaic model, it views the membrane as a double layer of fluid phospholipid into which proteins are inserted in various ways (a mosaic) rather than maintained as a continuous layer. The proteins that are associated with the exterior or interior face of the lipid mosaic are called *extrinsic* proteins. In different membranes they are highly variable and may even be absent. The proteins found within the lipid bilayer are known as *intrinsic* proteins. They may be confined entirely to the lipid bilayer, or they may project through either the interior or exterior surface. In some cases a large intrinsic protein may extend through both surfaces. The proteins found within the lipid matrix tend to be rich in hydrophobic amino acids, which permit a maximum amount of interaction between these proteins and the surrounding medium; the extrinsic proteins, on the other hand, tend to be rich in hydrophilic groups, which promote interaction with the surrounding water and the ions it may contain. In many cases, single proteins associated with the membrane twist and fold so that their hydrophobic portions remain embedded in the lipid matrix, while their charged, or hydrophilic, regions tend to project past the surface into the surrounding aqueous medium.

Separate proteins in the membrane may interact with one another to form a complex unit such as a channel or pore. The linking of proteins within the membrane may also provide some measure of stability for the protein arrangement, a necessary condition to assure the functional continuity of the membrane. The fluid phospholipid layer is relatively free to move in a direction parallel to the plane of the membrane itself, since the lipid molecules are generally held together by weak forces rather than by covalent bonding. However, both cholesterol and intrinsic proteins may inhibit such movement within the membrane and impose a modest degree of rigidity.

Although recent studies with the electron microscope tend to confirm the predictions of the fluid mosaic model, the unit membrane hypothesis has been of great value in guiding membrane studies and should not be viewed as a false step in our gradually increasing understanding of membrane structure. The fluid mosaic model itself may well be supplanted by a more useful concept in the future.

**4.6**    Are plasma membranes the outer boundaries of all cells?

No, in many cells the plasma membrane is encased within an extraneous coat of nonliving material that usually confers some rigidity on the cell it surrounds. Although not affecting permeability, this rigid coat affords considerable protection to the underlying cell, especially in dilute solutions in which the cell might take in water and tend to swell. The coat is known as a *cell wall* in plant cells, fungi, and bacteria. This wall is composed of cellulose (see Chap. 3) in all plant cells. Cellulose is the major component of wood and certain other commercially useful plant products. In fungi, the walls are composed of *chitin*, a complex carbohydrate rich in amino-containing sugars. Bacterial cell walls consist of complex carbohydrates and linked short peptides, but there is considerable variation in the chemical makeup of these mixed polysaccharide and amino acid chains in different bacterial strains.

In most animal cells there may be a layer of carbohydrate lying outside the plasma membrane, but this is not a separable coat. It is called a *glycocalyx* and exists in intimate association with the membrane. The carbohydrates themselves are generally short chains but are covalently bound to the lipids or proteins of the membrane to produce a thin, furlike cover for the cell. The glycocalyx contains receptors for a variety of substances with which the cell may interact. The blood types of humans are based on the antigenic properties of the glycocalyx of the red blood cell. The sites on many cells that signal "self" or "foreigner" to the immune system of host organisms occur within the glycocalyx as well. Thus, in dealing with the possibility of whether a transplanted tissue or organ will be rejected, a major focus is on the glycoproteins formed within the glycocalyx. Collectively, the chief set of such glycoproteins is known as the *major histocompatibility antigens*, and these antigens are coded for by a group of genes known as the *major histocompatibility complex* (MHC).

In many cells, special structures are formed that anchor the cells firmly together. These structures, particularly associated with epithelial tissue, include *tight junctions*, in which there is virtually no intercellular space, and *desmosomes*, in which a highly layered, narrow space can be discerned. Such structures may also play a role in the transport of materials from one cell to another. Animal cells involved in absorption, such as the cells that line the intestine, often have filaments extending out from the plasma membrane.

These filaments, which are rich in carbohydrates, are known as *microvilli*. They increase the absorptive surface of the cell and may also contain enzymes that function in digestion.

Another type of extracellular coat is formed by the layers of polysaccharide that surround the eggs of many vertebrate and invertebrate species. These coats, usually added to the egg proper as it passes along the reproductive tract before hatching, must be penetrated by the sperm at the time of fertilization. The enzymes associated with the *acrosome* found in the head of most sperm cells aid in carrying out this task. Among many protista a highly elastic *pellicle* overlies the plasma membrane, but the function of this structure is not completely clear. In *Euglena* it consists of flexible protein strips.

**4.7**    A continuous system of membranous channels is believed to connect the nucleus with the cell membrane. Describe the organelles prominent in this system.

Keith Porter first described a membranous network extending throughout the cytoplasm in 1945. He called this organelle the *endoplasmic reticulum* (ER). These membranes, now recognized as a universal feature of all eukaryotic cells, can be seen in electron micrographs as tiny canals, flattened sacs, or parallel tubules within the cytoplasm. The channels of the ER are fluid-filled spaces segregated from the *cytosol*, or fluid lying outside the ER.

Attached to the outer surface of some sections of the ER are tiny granules called *ribosomes*. These ribosomes are concerned with the synthesis of proteins. The ER that is partnered with ribosomes assumes a rough appearance and is called the rough *endoplasmic reticulum* (RER). The ER that is free of ribosomes has a smooth appearance and is called the *smooth endoplasmic reticulum* (SER).

Closely associated with the SER are the membrane-lined, parallel sacs of the Golgi apparatus. The Golgi apparatus is similar in structure to the SER, but it tends to be limited to particular regions of the cell. Although first described by Camillo Golgi in 1898, this organelle was long shrouded in controversy about whether it was real or artifact. With the application of electron microscopy the reality of this organelle of localized, flattened stacks of sacs was clearly established. It was also shown that the Golgi is closely associated with the secretory process in many cells.

**4.8**    Describe the interrelationship of function among the organelles of this membranous network.

The functional relationship of these interconnecting organelles has been fairly clearly established. Risosomes found free in the cytosol are apparently involved in the synthesis of proteins that are destined to function within the cytosol, especially enzymes that are not membrane-bound. On the other hand, proteins produced either for export outside the cell or for incorporation into membrane organelles of the cell are found within the RER. Signals probably exist that direct the newly synthesized protein from the ribosomes (attached to the outside of the RER) through tiny channels in the RER wall into the fluid matrix of the RER sacs. The protein is then moved through the RER into the smooth ER and thence into the Golgi apparatus, which abuts the SER. Some modification of the transported protein may occur within the RER and SER, but it is in the Golgi apparatus that the major modification of the protein occurs. For example, protein may be conjugated with other molecules, or molecules may be readied for export by packaging them in membrane-enclosed secretory vesicles. These secretory vesicles may coalesce with the cell surface through exocytosis, which extrudes the secretory product and also adds new membrane to the plasma membrane of the cell.

The constituents of all these organelles are in a continuous state of movement and exchange. The outer layer of the double membrane surrounding the nucleus is continuous with the membrane lattice of the endoplasmic reticulum, so that a continuous pathway from the interior of the nucleus through the cytoplasm and on to the exterior of the cell is present. These membrane channels not only provide a unique transport system within the cell but also extend the surface area along which key enzymes may influence the chemical makeup of molecules passing through the network.

**4.9**    Compare lysosomes and peroxisomes.

*Lysosomes* have been well known since they were described by Christian de Duve in the early 1950s. They are small, usually oval organelles that contain powerful digestive enzymes in an acidic environment. The membrane surrounding the organelle is single and functions to maintain a high degree of internal acidity. The potentially destructive enzymes of the lysosome are probably synthesized by ribosomes, which then pass the enzymes through the endoplasmic reticulum to the Golgi apparatus. The lysosomes appear to bud off the Golgi apparatus in some cells. Lysosomes play a role in cellular digestion. They may fuse with other bodies within the cell to effect the digestion of a variety of materials. In phagocytosing white cells, such as the mammalian neutrophil, lysosomes bring about the breakdown of bacteria and other foreign material engulfed by the cell. Lysosomes are also involved in the destruction of cells that have been injured or are no longer viable. During the development of the frog, the long tail, which is a feature of the tadpole, is resorbed through the action of lysosomes and its molecules used for newly emergent structures. In the thyroid gland, the lysosomes act on thyroglobulin to produce the actual active hormone thyroxin. As might be expected, the degradative enzymes of lysosomes work best at low (acid) pH.

*Peroxisomes*, originally called *microbodies*, are small organelles bounded by a single membrane and usually contain a fine granular matrix. In plants they play a significant role in photorespiration (see Chap. 6). Those peroxisomes that participate in the metabolic cycle involving the formation of glyoxylate have been termed *glyoxysomes*. All of the peroxisomes thus far studied contain enzymes for oxidizing materials, in contrast to the hydrolytic (digestive) enzymes of the lysosomes. Generally, hydrogen peroxide ($H_2O_2$) is produced during these oxidations, but the enzyme *catalase* quickly decomposes the $H_2O_2$ to prevent a harmful buildup of peroxides. The precise steps by which peroxisomes are assembled are not completely clear, but they appear to be closely associated with the endoplasmic reticulum.

**4.10**    The bundle of microtubules of the cilium (or flagellum) is called an *axoneme*. Compare the axoneme with the pattern of microtubules found in basal bodies and centrioles.

In cross section, the cilium is bounded by a membrane (extended from the cell membrane) and within its cytoplasmic matrix are nine pairs of fused microtubules arranged along the periphery and two separate microtubules lying in the center. This "9 + 2" arrangement appears to be a universal feature of cilia and flagella. The ciliary beat appears to arise from within the cilium itself and probably involves a sliding of ciliary tubules.

The basal body (granule) into which the cilium is anchored presents a slightly different cross-sectional structure. Here nine *triplet* tubules are peripherally distributed, and there are no central microtubules. In each triplet of microtubules the central microtubule is complete, but the inner lining of the adjacent microtubules is shared with the central microtubule. The cilium (or flagellum) that arises from the basal body is continuous with the elements of that structure. However, the doublets in the cilium are formed as extensions of two of the three microtubules of the basal body triplets. The basal body seems to control the development of the microtubular structure of the cilium. However, the two central single tubules of the axoneme may arise from the tip of the cilium, since no central counterpart of microtubules exists within the basal granule.

The centrioles, which lie just above the nuclear membrane in most animal cells, have the same microtubular structure (9 + 0) as the basal body. Evidence exists that these two structures are interchangeable. In some sperm cells, structures that begin as centrioles migrate to the periphery and assume a basal body role. The centrioles generally occur in pairs oriented at right angles to one another. They are usually embedded in a region of the cytoplasm known as the *centrosome*, a zone rich in radiating microtubules.

**4.11**    The actin monomer can add monomers (grow) at one end faster than at the other end. At the slower end, monomers also tend to dissociate (break off) more readily. How might this be useful in cell movement?

Given proper physiological conditions, a state can be reached in which new monomers are being added at one end at the same rate as monomers at the other end are being lost. This assembly-disassembly phenomenon is called *treadmilling* and can produce lateral movement in the direction of the growing end, rather like the progress of a tank, tread. Among the factors that are probably involved in the regulation of the assembly-disassembly process are levels of $Ca^{2+}$, availability of cyclic AMP or other cyclic nucleotides, phosphorylation of proteins, and concentration of the actin monomer.

**4.12**    The endosymbiotic hypothesis, advanced most cogently by Lynn Margulis of Boston University, states that certain organelles of the eukaryotic cell arose as prokaryotic invaders of eukaryotic precursors. What organelles would you guess are included in this hypothesis?

Mitochondria, chloroplasts, and even basal bodies demonstrate characteristics that are entirely consistent with this view. Both mitochondria and chloroplasts, for example, contain their own DNA, as well as their own ribosomes. They replicate independently of the rest of the cell and appear to control the synthesis of their membranes. Further, the characteristics of the DNA and ribosomes of these organelles are very similar to those of present-day prokaryotes. The existence of many one-celled organisms that contain smaller photosynthetic prokaryotes within their cytoplasm also supports the theory. However, some mitochondrial enzymes are coded for by the genetic information of nuclear chromosomes, so that the independence of the mitochondrial organelle is limited. Nevertheless, there seems to be a consensus favoring the endosymbiotic hypothesis for the origin of certain organelles.

**4.13**   Why is the nucleus centrally positioned in most eukaryotic cells?

The nucleus is the chief organelle involved in cellular reproduction. During the life of the cell, the nucleus also directs the metabolic activity of the cell and helps shape the cell into its final form. Instructions for the synthesis of key proteins continuously pass from the nucleus to the cytoplasm, particularly through the channels of the endoplasmic reticulum. Even the basic material of the protein-synthesizing ribosomes are synthesized within the nucleolus of the nucleus, stored, and then passed to the ribosomes.

These key functions of the nucleus suggest that it should receive a maximum degree of protection. A position deep within the interior of the cell provides such protection. Those cells that developed this arrangement presumably were better adapted and tended to survive.

**4.14**   Why are plant tissues so much harder to classify than animal tissues?

Tissues are groups of similar cells that carry out a particular activity. Animal tissues each show a marked distinctiveness; however, boundaries between tissue types in plants tend to blur, and one type may change to another in the course of development. Also, structural features, so distinct in animal cells, are very indefinite among plant cells.

**4.15**   How does connective tissue differ from the three other tissue types in animal organisms?

Muscle, nerve, and epithelial tissue are all characterized by the properties of the cells they contain. The various types of muscle tissue show contractility because of the contractile properties of skeletal, smooth, and cardiac muscle tissue cells. The lining properties afforded by the various types of epithelial tissue are likewise derived from the closely apposed and densely packed layer of epithelial cells. Nerve cells, with their long axonal processes and often branching dendrites, provide the basis for the conduction properties of the tissue.

In connective tissue an entirely different situation obtains. The properties of the tissue are a function of the extracellular material produced by the cells rather than of the cells per se. The variation in connective tissue types far exceeds that found in each of the other tissue classes.

If we examine these variant connective tissue types, the significance of the extracellular component becomes clear. *Vascular tissue* is derived from and contains cells, but the bulk of it consists of extracellular fluid such as plasma and lymph. In *yellow elastic tissue* the elastic fibers that are responsible for tissue elasticity lie outside the cells. Most striking are the properties of *cartilage* and *bone*. In the latter case, a network of bone cells is moored within a matrix of hard, calcareous material that is laid down by the bone cells. Even in fat cells, a large oil globule, carried within a thin rim of cytoplasm, could be viewed as nonliving material lying outside the living portion of the cell.

**4.16**   How does the organ level of organization differ in plants and animals?

The organ level of organization is much less definite in plants than it is in animals. At most, we might distinguish roots, stems, leaves, and reproductive structures. Clear-cut functions, the distinguishing features of organs, can be assigned to each of these structures. *Roots* are involved in anchoring the plant and procuring water and minerals. The *shoot* or *stem* supports the entire plant, while the *leaves* are primarily organs for food procurement. *Flowers* or other reproductive structures are involved in producing progeny for the next generation.

In animals, organ development is far more complex and defined. Organs are parts of organ systems where total functions are carried out. Thus, excretion is handled by the kidneys, ureters, bladder, and urethra. The nervous system consists of the brain, spinal cord, and peripheral nerve network. Associated with the complexity and discreteness of the organ systems of animals is a far greater range of functions and activities than is found in plants.

**4.17**   What challenges face a cell that undergoes a great increase in size?

The cell must exchange materials with the environment across the surface membrane. An increase in size will result in a relatively greater increase in volume and mass than in surface area, so that the cell will lose effective exchange capacity. This will impose restrictions on the amount of food and oxygen that can move across the membrane to service the metabolic needs of the increased living mass in the interior. Waste materials, produced in greater abundance, will be excreted with greater difficulty.

The distribution of materials by diffusion will also take longer as the cell grows larger. Many portions of the cell may well suffer deprivation of vital fuel materials or other essential molecules as a result of enlargement.

Nuclear control of the activities of cytoplasmic structures depends on chemical communication between the nucleus and cytoplasm. In an enlarged cell the efficiency of such communication will be inhibited by the increase in distance involved.

**4.18**   Lipids, small molecules, and uncharged particles pass into and out of the cell with relative ease. What characteristics of the cell membrane can be inferred from these observations?

Permeability exists in all membranes and is a function of the structure of the membrane. The tendency of lipid-soluble materials to move readily across most membranes is thought to be owing to the large amount of lipid found in the membrane. Similarly, the greater ease with which smaller molecules enter the cell compared with larger molecules of the same general type suggests that there are tiny openings or channels in the membrane that confer the properties of a sieve on the outer cell boundary. Membranes usually have a positive charge on their outer surface, and this may explain why uncharged particles can enter the cell more readily than charged particles, which tend to interact with the charged field of the membrane.

**4.19**   List six ways in which substances are transported into a cell.

(1) Osmosis, (2) diffusion through pores, (3) facilitated transport, (4) active transport, (5) endocytosis, (6) pinocytosis.

# Supplementary Problems

**4.20**   According to the cell doctrine, all cells arise from   (*a*) inorganic material.   (*b*) organic material.   (*c*) preexisting cells.   (*d*) petri dish cultures.

**4.21**   Cells were named by   (*a*) Dutrochet.   (*b*) Schleiden.   (*c*) Schwann.   (*d*) Hooke.

**4.22**   A true nucleus is absent in cells of kingdom _____.

**4.23**   The rough endoplasmic reticulum (RER) differs from the smooth endoplasmic reticulum (SER) in its association with _____.

**4.24**   A double membrane encloses which two organelles?

**4.25**   A reservoir and possible regulation center for fluid and electrolytes is provided by the _____.

**4.26**   A(n) _____ is a plastid containing chlorophyll.

**4.27**   Cellular digestion is associated with which organelle?

**4.28**   Oxidations may occur in both mitochondria and peroxisomes.
(a) True   (b) False

**4.29**   Meristematic tissues are unspecialized.
(a) True   (b) False

**4.30**   The sliding filament hypothesis explains the conduction of nerve impulses.
(a) True   (b) False

**4.31**   The surface of a sphere is proportional to $r^2$, while the volume of a sphere is proportional to $r^3$.
(a) True   (b) False

**4.32**   Semipermeable membranes are involved in osmotic phenomena.
(a) True   (b) False

**4.33**   Membranes usually have a negative charge on their outer surface.
(a) True   (b) False

**4.34**   Robertson is best known for the fluid mosaic model of membrane structure.
(a) True   (b) False

# Answers

| | | |
|---|---|---|
| **4.20**  (c) | **4.25**  Vacuole or contractile vacuole | **4.30**  (b) |
| **4.21**  (d) | **4.26**  Chloroplast | **4.31**  (a) |
| **4.22**  Monera | **4.27**  Lysosome | **4.32**  (a) |
| **4.23**  Ribosomes | **4.28**  (a) | **4.33**  (b) |
| **4.24**  Nucleus and mitochondria | **4.29**  (a) | **4.34**  (b) |

# Chapter 5

# Energy Transformations

## 5.1 ENERGY AND LIFE

The highly complex organization of living systems requires a constant infusion of ordered energy for development and maintenance. The source of this energy is the sun—a giant hydrogen bomb. Green plants, the initial trappers of the sun's energy, convert solar radiant energy into chemical energy that is stored in the bonds of organic substances such as glucose. This energy moves through the living world as green plants are eaten by herbivorous animals, which in turn are eaten by carnivorous predators, in a progression of dining relationships known as a *food chain* or *web*. Each organism in the chain transforms the energy contained in its meal and creates a store of chemical energy that undergoes continuous transformations at the cellular and subcellular levels. Because transformations increase the disorder (entropy) of the energy, some of the energy's capacity to perform useful work is lost from the organism's system in each conversion and must be replaced by new supplies of ordered energy. Thus each link in the food chain passes on less energy than it received from the preceding link.

*Metabolism* encompasses those processes by which organisms both extract energy from the chemical bonds of foods and synthesize important compounds. The general nature of metabolism was suggested by Antoine Lavoisier's recognition in the eighteenth century of a basic similarity between combustion in nonliving materials and respiration in animals. The chemistry of metabolism, however, can be understood only against a backdrop of thermodynamics.

## 5.2 THERMODYNAMICS

As we saw in Chap. 2, the first law of thermodynamics dictates that the total energy of the products of a reaction will be equal to the total energy of the entering substances (reactants). However, the second law indicates that the *potential* (useful) energy of the products will be *less* than the potential energy of the reactants, i.e., energy tends to be degraded. In nature, this usually means that the randomness (entropy) of any system will tend to increase. Therefore, the reactions that are spontaneous may be viewed as downhill events in which entropy (disorder) increases and energy is released. These reactions are classified as *exergonic*. In contrast, *endergonic* reactions are uphill reactions in which the randomness of the system decreases. Endergonic reactions can occur only with the input of free energy from an outside source, thus increasing the disorder within that source. Thus the many endergonic reactions occurring in living organisms, which create highly ordered systems, must "pay" for this order by promoting exergonic reactions as well, such as the *catabolism* (breakdown) of glucose molecules.

The relationship between free energy and entropy is expressed by

$$\Delta G = \Delta H - T \Delta S$$

where $\Delta$ stands for "change (in)," $G$ is free energy, $H$ is heat, $T$ is the absolute temperature, and $S$ is the entropy (randomness, disorder).

Confusion sometimes occurs between the terms *exergonic* and *exothermic*. *Exergonic* refers to the release of free energy ($\Delta G$ is negative), while *exothermic* refers to the release of heat ($\Delta H$ is negative). Exothermic reactions are generally, but not always, exergonic. A few *endothermic* (heat-absorbing) reactions, such as the evaporation of water, are also exergonic, in spite of a positive $\Delta H$ value. This is because the change in structure from the liquid form to the more amorphous gas produces a sufficiently great increase in entropy to offset the heat absorption, and thus results in a negative value for $\Delta G$.

**EXAMPLE 1** Why is the free-energy change of the combustion of glucose greater than the change in the heat content of the system?

The total free energy available for useful work is determined by the equation

$$\Delta G = \Delta H - T\,\Delta S$$

The change in heat content $\Delta H$ has been determined by calorimetry to be $-673$ Cal (kcal) per mole of glucose. Under standard conditions the $T\,\Delta S$ term is 13 Cal. Then

$$\Delta G = \Delta H - T\,\Delta S$$
$$= -673\,\text{Cal} - 13\,\text{Cal}$$
$$= -686\,\text{Cal}$$

The change in free energy is determined by both the change in heat *and* the change in entropy. Note that the change in entropy is *subtracted* from the change in heat. In this reaction, therefore, the change in entropy augments the release of heat.

## 5.3   CELL METABOLISM

Life essentially involves order and organization in a universe that appears to be governed by a tendency to disorder. The reactions that characterize metabolism maintain order by supplying energy and raw materials for the struggle against disorder.

Cellular combustion of fuels such as carbohydrates, fats, and even proteins releases the potential energy stored in these molecules by disrupting their chemical bonds. The energy released is stored in *ATP molecules*, which are in turn broken down by the cell and the released energy used for such activities as movement, reproduction, and synthesis. Since the release of a great amount of energy at one time would harm the cell, the breakdown of fuel molecules is controlled and occurs in a series of coordinated steps known as a *metabolic pathway*.

### ANAEROBIC PATHWAYS

The first stage in the breakdown of a typical cellular fuel such as glucose involves a metabolic pathway called *glycolysis* (also called the *Embden-Meyerhof pathway*, in honor of its discoverers). Interestingly enough, glycolysis does not involve the intervention of molecular oxygen. It is therefore an *anaerobic* process and probably served the needs of cells long before the earth's atmosphere contained molecular oxygen.

Glycolysis and *fermentation* refer to essentially the same process, although their end products differ. The former is usually applied to animal cells, while the latter is used for the reaction in bacteria and yeast. Although glycolysis involves the breakdown of glucose or related carbohydrates to pyruvic acid, it should be borne in mind that other foodstuffs may "plug in" to the process at a variety of key places. The details of glycolysis are shown in Fig. 5.1.

Four major events which characterize glycolysis are:

1. Preliminary phosphorylation—uses up two ATP molecules
2. Splitting of the molecule
3. Oxidation and formation of a high-energy phosphate bond—produces NADH and ATP
4. Molecular rearrangement to form a high-energy phosphate bond—produces another ATP

1. *Preliminary phosphorylation*:   The initial step of glycolysis involves a breakdown of ATP to the less energetic ADP molecule during the formation of glucose 6-phosphate. This may be viewed as an investment of ATP by the cell to yield a greater return of ATP at a later stage. Another investment of ATP occurs in the formation of fructose 1,6-diphosphate.
2. *Splitting of the molecule*:   The doubly phosphorylated hexose (i.e., fructose 1,6-diphosphate) is cleaved and two trioses are formed: phosphoglyceraldehyde (PGAL) and dihydroxyacetone phosphate (DHAP). Note that DHAP may be converted to PGAL, so that the products of the ensuing reactions are doubled.

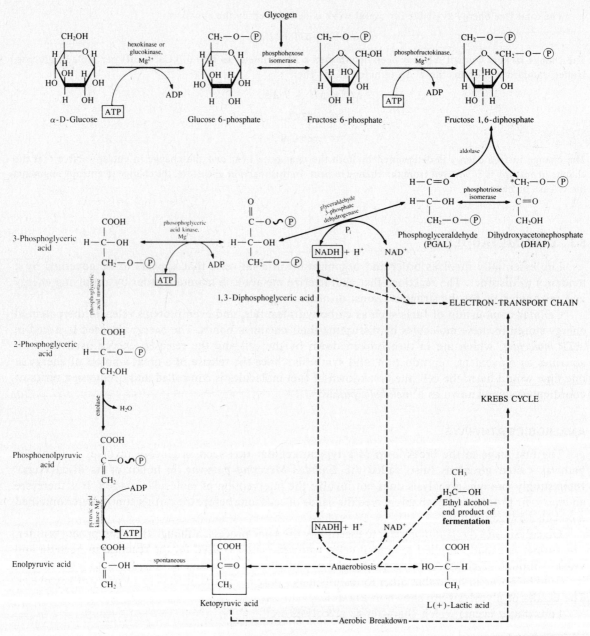

**Fig. 5.1** Glycolysis

3. *Oxidation and formation of a high-energy phosphate bond*:  The PGAL then undergoes oxidation (accomplished without molecular oxygen) to 1,3-diphosphoglyceric acid* and in the process accomplishes two important changes: (1) a pair of electrons and hydrogen ions are passed to the coenzyme NAD$^+$ to form the more highly energetic NADH and (2) inorganic phosphate (P$_i$) is picked up from the cytoplasm to form a second phosphate bond that will produce an ATP for the cell in the next reaction.

4. *Molecular rearrangement to form a high-energy phosphate bond*:  From the internal molecular rearrangements that follow, another molecule of ATP is generated. The end product is pyruvic acid.

---

*In many texts, the term *glycerate* is used rather than *glyceric acid*, since under cellular conditions, the free acid is more likely to exist as the salt.

Under anaerobic conditions, pyruvic acid generally reacts with hydrogen to form either ethyl alcohol (most plants and bacteria) or lactic acid (animals and some bacteria). (This reduction of pyruvic acid is essential so that the $NAD^+$ necessary in the glycolytic conversion of PGAL to 1,3-diphosphoglyceric acid is regenerated. Without it, glycolysis would stop.) However, glycolysis proper ends with the formation of pyruvic acid.

Because each starting glucose yields two trioses at the aldolase step of glycolysis, proper bookkeeping dictates a net product of two pyruvic acids and therefore a return of four ATPs to the cell. Since the cell used up two ATPs in the process, the actual energetic yield of glycolysis is two ATPs. In some situations the DHAP formed upon cleavage of the hexose is used to synthesize fat rather than to provide PGAL for further ATP production, so that the pathway does not always yield the same net output of ATP.

Although many of the enzymatic steps of this pathway are reversible, some key reactions, such as the formation of ketopyruvic acid from enolpyruvic acid, are not. Therefore, other routes are utilized within the cell to shunt (bypass) the irreversible steps when it is necessary to move *toward* the formation of glucose from metabolites such as lactic acid or the various amino acids. These alternative routes collectively constitute the gluconeogenic pathway (see Prob. 5.32).

## AEROBIC PATHWAYS—THE KREBS CYCLE

With the formation of molecular oxygen in the atmosphere billions of years ago, it became possible to degrade fuels still further to tap the enormous energy stored within their bonds. In these pathways, oxygen serves as the final electron acceptor in cellular oxidations.

The *Krebs cycle* is the major aerobic pathway for oxidative degradation of the products of glycolysis. It is also known as the *tricarboxylic acid (TCA) cycle* because of the involvement in the pathway of organic acids with three carboxyl groups. The Krebs cycle components are found within the mitochondria.

The Krebs cycle may be summarized as follows (see Fig. 5.2):

1. Formation of a 6-carbon molecule by combining a 4-carbon and a 2-carbon molecule
2. Oxidation of the 6-carbon molecule to form a 5-carbon molecule
3. Oxidation of the 5-carbon molecule to form a 4-carbon molecule
4. Molecular rearrangement to form the starting 4-carbon molecule

Before entering the Krebs cycle, pyruvic acid (the end product of glycolysis) is degraded to acetaldehyde (a 2-carbon molecule), with the loss of $CO_2$. The acetaldehyde is then oxidized to acetic acid and attached to coenzyme A (CoA), with $NAD^+$ being reduced to NADH in the process. The acetyl-CoA then enters the Krebs cycle.

1. *Formation of a 6-carbon molecule*:   Acetyl-CoA is highly reactive, and the 2-carbon acetyl group combines with the 4-carbon molecule, oxaloacetic acid, to form citric acid, a 6-carbon compound.
2. *Oxidation of the 6-carbon molecule*:   Citric acid is oxidized, with loss of $CO_2$, to a 5-carbon substance, $\alpha$-ketoglutaric acid.
3. *Oxidation of the 5-carbon molecule*:   $\alpha$-Ketoglutaric acid is oxidized, with loss of $CO_2$, to a 4-carbon molecule, succinic acid.
4. *Molecular rearrangement and oxidation*:   In the subsequent reactions, oxaloacetic acid is regenerated, and the cycle begins anew with another acetyl-CoA produced from pyruvic acid.

For each pyruvic acid entering the cycle, three molecules of $CO_2$ are generated. The reduced coenzymes produced within the cycle, as well as the ATP eventually generated from the GTP (guanosine triphosphate) formed in the transformation of succinyl-CoA, result in a net production of 36 molecules of ATP for each glucose molecule entering glycolysis and proceeding through the Krebs cycle and the electron-transport chain (see next section). Many texts give 38 molecules of ATP as the total yield; they do not include in their count the loss of ATP which may occur when the NADH produced during glycolysis is brought across the mitochondrial membrane (see Prob. 5.30).

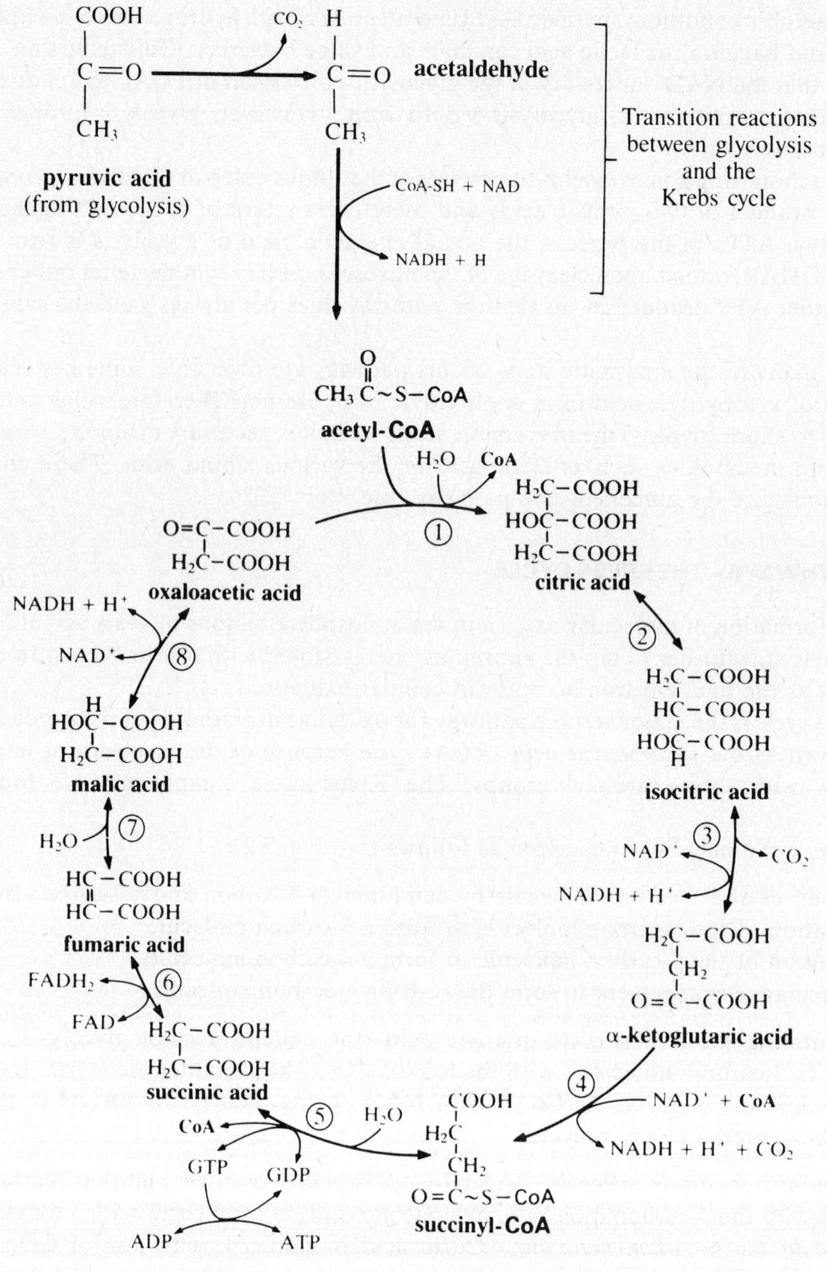

**Fig. 5.2** Krebs cycle

**EXAMPLE 2** As can be seen from Figs. 5.1 and 5.2, $P_i$ is involved in some reactions. Of what significance is $P_i$ in these metabolic pathways?

$P_i$ denotes inorganic phosphate in ionized form in the cytoplasm. Its incorporation in these reactions constitutes *substrate-level phosphorylation*. This process involves the direct production of ATP from inorganic phosphate in association with several reactions in either glycolysis or the Krebs cycle. The formation of ATP during the conversion of phosphoenolpyruvic acid to pyruvic acid is an example.

## AEROBIC PATHWAYS—ELECTRON-TRANSPORT CHAIN AND OXIDATIVE PHOSPHORYLATION

The electron-transport chain (ETC) represents a series of respiratory pigments of the mitochondrion that function as a "bucket brigade" for the passage of electrons from reduced coenzymes (NADH,

FADH$_2$) to oxygen (see Fig. 5.3). Since electrons associated with the reduced coenzymes are at a comparatively high energy level, whereas their union with oxygen constitutes a markedly lower energy level, the passage of electrons down the ETC results in the release of energy. The energy released is used to form ATP: two ATP molecules are produced when FADH$_2$ is the electron donor, and three ATP molecules are produced when the more highly energetic NADH contributes its electrons.

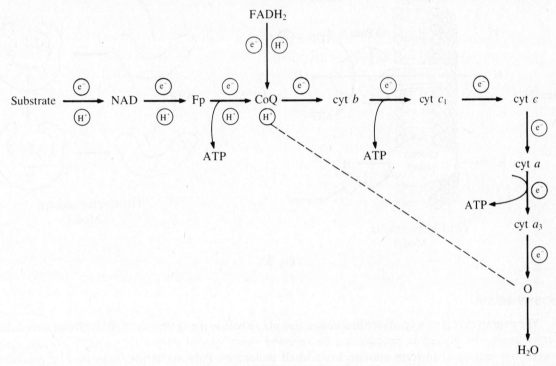

**Fig. 5.3** Electron transport chain

Electron transport may be envisioned as a functional spoke radiating from each oxidative step of the Krebs cycle. The stepwise passage of the electrons and their accompanying hydrogen ions (protons) permits efficient formation of ATP from the energy released. The pigments (cytochromes) of the ETC are embedded in the cristae of the mitochondria; this ensures ready access of the reduced coenzymes produced in the Krebs cycle to the ETC.

The redox (oxidation-reduction) reactions occurring during electron transport couple oxidation with phosphorylation and are known as *oxidative phosphorylation*. The localization of oxidative phosphorylation in the mitochondrion has earned that organelle the title of "powerhouse of the cell." The efficiency of that powerhouse is considerably enhanced by the precision of the repetitive arrangement of the pigments of the ETC within the inner membrane of the mitochondrion. Each electron-transport unit is considered to act as an independent respiratory assembly. Most of the enzymes of the Krebs cycle, on the other hand, are found loosely dissolved within the matrix of the mitochondrion. The fixed spatial arrangement of the ETC may also account for the creation of a "proton potential" across the inner membrane in the chemiosmotic theory of oxidative phosphorylation.

There are currently three major theories about how the energy of the transported electrons is transferred to form ATP:

1. *Chemical model*:   The phosphorylation of ADP to ATP is accomplished enzymatically (much like glycolytic phosphorylation) and is coupled to the oxidation of the electron carriers.
2. *Chemiosmotic model*:   The electron flow along the ETC pumps protons across the mitochondrial membrane, against a proton gradient. As the protons "fall" back to their original position, the energy released is used by a membrane-bound protein to phosphorylate ADP to ATP (see Fig. 5.4*a*).

3. *Conformational model*:   The energy released by the flow of electrons changes the conformation of the large molecules residing in the mitochondrion. The return of these molecules to their original shape releases sufficient energy for phosphorylation to occur (see Fig. 5.4b).

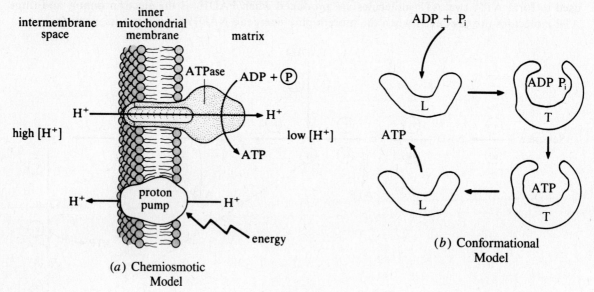

(a) Chemiosmotic Model

(b) Conformational Model

**Fig. 5.4**

## BIOSYNTHESIS

The degradative pathways described above also play a role in the synthesis of vital cellular molecules. For example, the glycolytic process can be reversed (with special enzymatic steps to overcome the irreversible reactions) to form glucose from small molecules. Polysaccharides may also be produced from such metabolites as glucose 6-phosphate through the action of such enzymes as glycogen synthetase.

The synthesis of neutral fats requires $\alpha$-glycerophosphate, which is supplied by glycolysis or by phosphorylation of glycerol. The fatty acids of such lipids may be enzymatically chopped into 2-carbon fragments (acetyl-CoA) and enter the Krebs cycle for further degradation. However, acetyl-CoA molecules may also be utilized to synthesize long-chain fatty acids by a special cluster of enzymes and ancillary molecules lying outside the mitochondria. This synthesis requires, among other factors, appreciable amounts of NADPH. This reduced coenzyme is supplied by several oxidative reactions, chief among them being a pathway called the *pentose shunt*, which involves the direct oxidation of glucose 6-phosphate in association with $NADP^+$. Reduced $NADP^+$ is required for many reductive syntheses, e.g., the formation of steroids, production of milk in the mammary gland, and photosynthesis. As can be seen, a vital link exists between catabolism and anabolism within the cell.

Amino acids and proteins may also participate in the metabolic mill. The amino groups can be stripped from their basic carbon skeletons (deamination) and attached to different ones. Such metabolites as pyruvic acid or ketoglutaric acid can be formed from the amino acids alanine or glutamic acid.

Generally, the major fuel molecules are interconvertible. The direction of molecular traffic, then, is not limited by the starting molecular types, but by the actions of hormones on enzymes, the ratio of ATP to ADP, the availability of coenzymes, the ratio of their reduced to oxidized forms, etc. See Fig. 5.5.

## 5.4  BIOENERGETICS

The energy economy of the whole organism in its interaction with the environment constitutes the subject matter of bioenergetics. Bioenergetics is fundamental in understanding how organisms stay alive and maintain a dynamic "steady state," an internal balance known as *homeostasis* (Chap. 15).

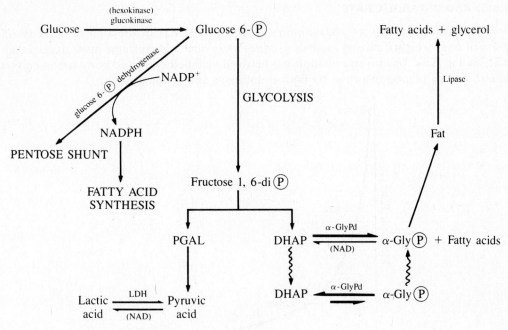

**Fig. 5.5**

However, bioenergetics is also involved in a host of more practical concerns, such as evaluating the cost-effectiveness of livestock by balancing the expense of feed against the return in work or profit from the sale of the animals. In clinical terms, obesity may be regarded largely as a bioenergetic defect—too many calories taken in as opposed to too few expended in work.

The energy economy of the whole organism is mediated by the cellular metabolic pathways described earlier. Generally, only about 40 percent of the energy of metabolized fuel can be captured in the high-energy bonds of ATP or similar energy currencies. The remainder is liberated as heat. Under standard dietary conditions, the relationship between oxygen uptake and heat release is rather constant.

## ENDOTHERMY AND ECTOTHERMY

Most organisms are *ectotherms* (poikilotherms)—that is, they demonstrate fluctuating body temperatures depending on the temperature of their surroundings. Although they may move to a more congenial environment, they cannot alter this dependency. Since temperature has a profound effect on metabolic activity [the rate doubles with each 10 °C increase in ambient (surrounding) temperature], ectotherms are severely restricted by the circumstances of their habitat in their attempts to find food, seek mates, and move about: low temperatures cause them to react sluggishly, while high temperatures produce an increase in activity that may not always be advantageous.

In birds and mammals, internal body temperature is independent of the ambient temperature. These organisms are called *endotherms* (homeotherms). Endothermy, which appears imperfectly even in some reptiles, involves the utilization of metabolic heat as well as the insulating action of epidermal modifications (hair, feathers). Such organisms maintain a uniformity of metabolic activity and are free to explore a wide variety of environments to satisfy their needs. The more uniform temperatures of the seas and even freshwater environments suggest that endothermy was not an adaptive imperative until dry land was invaded by the early reptilian ancestors of birds and mammals. The major source of heat for endotherms is the large muscle mass covering the entire body. Contractions of the musculature may account for more than 80 percent of the total heat produced by these organisms. Ectotherms may also take advantage of muscle contractions for sudden surges of heat production preceding expected activity.

## BODY SIZE AND METABOLIC RATE

Heat production and oxygen consumption depend on the size of the animal. On a *specific basis* (per gram of body weight), smaller animals produce more heat and consume more oxygen than larger animals (see Fig. 5.6). This inverse relationship between metabolic rate and body size is universal and is demonstrated by plants, as well as by both endotherms and ectotherms.

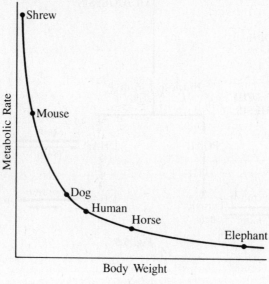

**Fig. 5.6**

Early investigators hypothesized that an organism's metabolic rate was proportional to its surface area, across which heat is lost to the environment. Because relative surface area tends to decrease as an organism increases in size (refer back to cell size and its constraints in Chap. 4), smaller animals have to compensate for their relatively greater heat loss with a higher metabolic rate. Later investigators showed that other factors were also involved, including a greater proportion of low-metabolizing supporting tissue in larger organisms.

**EXAMPLE 3** The summated respiration of individual tissues removed from an organism may not account completely for the total respiration in the intact animal, because such tissues, when isolated, may not reflect their in vivo metabolic rates. However, it is of interest that such tissues from larger animals demonstrate lower activities of respiration than the corresponding tissues of small animals. It has also been demonstrated that a variety of specific respiratory enzymes are lower in activity in tissues from larger animals.

The shrew is probably the smallest mammal possible, given the great metabolic intensity associated with its tiny size. Because it must eat its own weight in food each day merely to stay alive, the shrew is constantly foraging and feeding. Other small mammals modify their activities through a process called *hibernation*, a state of metabolic torpor close to death, which offers some respite from the grim necessity of constantly supplying fuel to the raging metabolic fires associated with small body size.

# Solved Problems

**5.1**    What is the relationship between the laws of thermodynamics and the nature of living systems?

As described in Chap. 3, energy can be understood as the capacity to perform useful work. Living systems are highly ordered, and their great degree of organization is subject to decay into randomness, in accord with the universal tendency toward disorder (entropy) in the universe. Only through the capture and utilization of new highly ordered forms of energy, whose energy can be used to build highly ordered living systems, can this drift toward randomness be resisted. The kinetic energy of heat that leaves the organism must be replaced by the potential energy of ingested ordered molecules such as carbohydrates, proteins, and lipids to sustain life. Living systems must work (expend energy) to stay alive.

**5.2**    If energy can be neither created nor destroyed, why do we require continuous supplies of fresh energy to stay alive? Why don't we simply recycle the energy we have?

The laws of thermodynamics apply essentially to closed systems, i.e., systems that are sealed and not in contact with the outside. Living organisms, such as human beings, are open systems that exchange matter and energy with an outside environment and hence constantly lose their supplies of both. Actually, even if we were isolated from the environment, we would fall prey to the second law of thermodynamics, which stipulates that in any energy transformation there is a tendency toward greater disorder of the energy produced. With their lowered potential, such disordered forms of energy as heat given off by living organisms are no longer capable of producing useful work under ordinary circumstances.

**5.3**    If there is a universal tendency toward disorder and randomness, how can one explain growth, development, and increasing complexity in living organisms?

The tendency toward randomness occurs in terms of a total picture. Within living organisms an uphill increase in order in one part is coupled with a downhill decrease in order in another part. The randomness of the *entire* system is increased, but an isolated portion of that coupled system may demonstrate increased order. In thermodynamic terms, endergonic reactions (in which new patterns of order are produced) are coupled with exergonic reactions (those that yield decreased order) to permit specific kinds of syntheses.

**5.4**    What is the role of ATP in energy transformations within living organisms?

This universally occurring molecule of the world of life represents an energy currency, a medium of energy exchange, that simplifies the complexity of metabolic cycles within the cell. Thus, the many different mechanisms by which energy is released from fuel foods all lead to the creation of ATP from precursors like ADP. Similarly, diverse energy-consuming processes, such as muscle contraction, and active transport, all utilize ATP as an energy source and degrade it back to ADP. Occasionally, related nucleotides (see Chap. 7) such as cytosine triphosphate (CTP) and guanosine triphosphate (GTP) may substitute for ATP. The energy associated with these nucleotides is stored in the last two phosphate bonds of the molecule. These high-energy bonds are usually designated with a tilde ($\sim$), e.g., adenosine-P$\sim$P$\sim$P.

**5.5**    What is glycolysis and how does it contribute to the welfare of the cell?

*Glycolysis* literally means the "splitting, or degradation, of glucose." It involves a series of approximately nine related reactions in which glucose is gradually broken down to two molecules of the simpler compound pyruvic acid. The pyruvic acid may be converted anaerobically to one of several reduced forms, e.g., lactic acid or ethyl alcohol, or it may enter the Krebs cycle.

Each step in glycolysis is catalyzed by a specific enzyme. The presence of these enzymes in close proximity to one another permits glycolysis to occur as an integrated metabolic pattern called a *pathway*. All the enzymes of glycolysis are soluble proteins that have been isolated and crystallized, so that scientists know a great deal about their structure and function. Glycolysis occurs within the cytoplasm of the cell rather than within a specific organelle. The reactions of glycolysis produce a net gain of two molecules of ATP for the activities of the cell, as well as two molecules of reduced $NAD^+$, which can ultimately generate additional amounts of ATP. Thus, glycolysis is a pathway that taps some of the energy locked within the starting glucose molecule.

**5.6**    Since glycolysis has evolved as a mechanism for *producing* energy (generating ATP) from the partial degradation of glucose, why are two molecules of ATP used up early in the process?

A molecule of ATP is spent in the initial generation of glucose 6-phosphate from glucose, and shortly thereafter still another molecule of ATP is expended in producing fructose 1,6-diphosphate from fructose 6-phosphate. Since ATP is formed through the transfer of phosphate, this phosphorylation is essential. Thus, the expenditure of ATP in these early steps of glycolysis can be compared to an investment that leads to a later payoff. The triose (3-carbon sugar) that arises from the splitting of fructose 1,6-diphosphate will be oxidized and then eventually converted to pyruvic acid. These later steps result in the production of two ATP molecules for each triose—part of the return on the initial investment. Since two trioses are generated from each glucose molecule, a total of four ATP molecules will be generated in the process. However, if we subtract the two ATPs invested initially, we arrive at a net yield of two ATPs for the entire process.

**5.7**    How is glucose phosphorylated?

The terminal phosphate group of ATP is transferred to the carbon in the sixth position of glucose. This is an exergonic reaction, and some of the energy released is conserved in the chemical bond that links the phosphate to the glucose.

**5.8**    When glucose is phosphorylated, what happens to ATP?

The ATP donating its terminal phosphate group becomes ADP, which is less energetic. The glucose molecule, however, assumes a higher energy state.

**5.9**    What regulates phosphorylation of glucose?

The enzymes *hexokinase* or *glucokinase* control the reaction. These enzymes are similar in function but differ in their specificity for the substrate. Note that each reaction of glycolysis is regulated by specific enzymes.

**5.10**    Why is glucose phosphorylated?

The transformed glucose will eventually be used to convert ADP to ATP; this will require the donation of phosphate to ADP. The phosphorylation of glucose energizes the molecule and makes it more reactive. Also, the cell membrane is less permeable to the phosphorylated molecule than it is to the free sugar; thus, the phosphorylated molecule is prevented from escaping the cell before it has been properly processed.

**5.11**    After phosphorylation, what happens to the glucose?

The structure of the glucose molecule is altered from a six-sided ring to the five-sided ring of fructose. Note that the number of carbon atoms is the same in both, since fructose is an isomer of glucose. An *isomer* is an alternative arrangement of the same number and kinds of atoms of a given molecule.

**5.12** Why is fructose 6-phosphate phosphorylated in the next reaction?

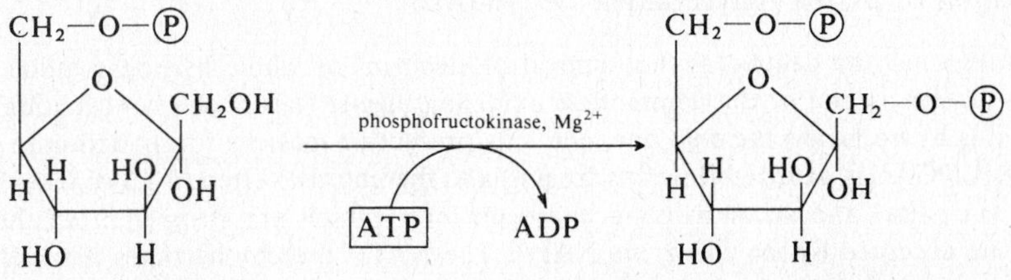

Fructose 6-phosphate Fructose 1,6-diphosphate

Through phosphorylation, fructose gains a second phosphate. This enables it to supply two phosphorylated trioses upon cleavage. Each triose thereby becomes capable later of yielding ATP.

**5.13** What regulates the phosphorylation of fructose 6-phosphate?

*Phosphofructokinase* catalyzes the phosphorylation of fructose 6-phosphate. It interacts *allosterically* with ATP; that is, it binds ATP at a site other than its active site, incurring a change in shape and a consequent change in its ability to catalyze the phosphorylation. In this case, the ATP inhibits the reaction by binding to the enzyme—the change in shape induced in phosphofructokinase renders it unable to bind fructose 6-phosphate and thereby effect the phosphorylation.

**5.14** Is the allosteric interaction between ATP and phosphofructokinase important?

Yes, because this interaction controls ATP production in glycolysis in accordance with the ATP needs of the cell. If cellular ATP levels are sufficient to bind to the enzyme (and thus prevent phosphofructokinase from interacting with fructose 6-phosphate), glycolysis stops. As the cell depletes its store of ATP, the ATP bound to the phosphofructokinase is released. The enzyme then reverts to its active configuration and glycolysis proceeds.

**5.15** Of what significance is the aldolase step?

The aldolase splits the doubly phosphorylated fructose into two 3-carbon sugars (trioses).

Fructose 1,6-diphosphate

Dihydroxyacetone phosphate (DHAP)

Phosphoglyceraldehyde (PGAL)

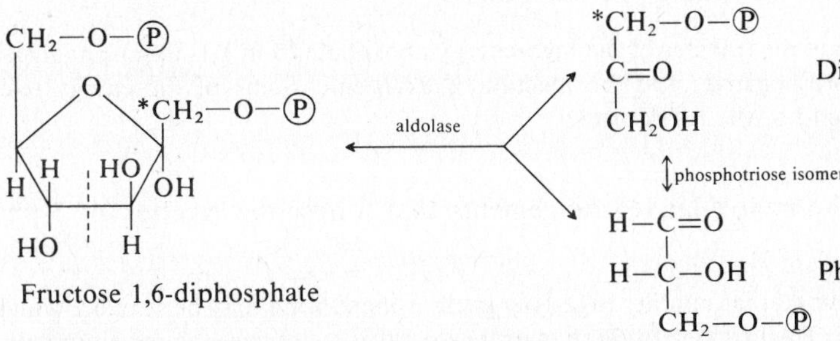

Through subsequent reactions, each PGAL yields two molecules of ATP. Because DHAP may eventually be converted to PGAL, all subsequent reaction yields are doubled.

**5.16** Why is the next step, the oxidation of PGAL, so important?

This oxidation-reduction reaction is the first to net energy for the cell, in the form of a high-energy phosphate bond and a molecule of NADH formed from the oxidation.

Phosphoglyceraldehyde (PGAL) 1,3-Diphosphoglyceric acid

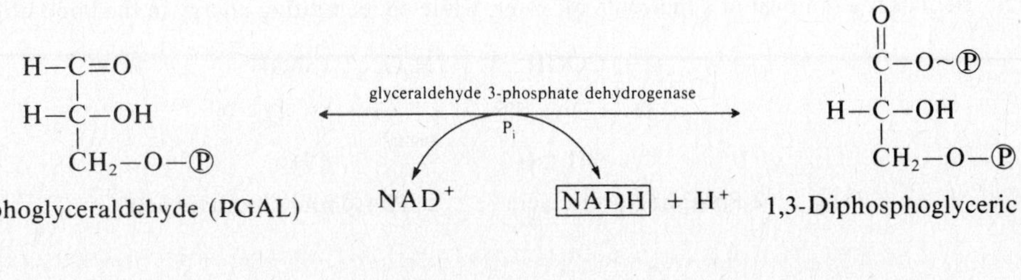

**5.17** How can glycolysis proceed under anaerobic conditions when a key step in the process involves the oxidation of phosphoglyceraldehyde (PGAL)?

*Oxidation* may be defined as the removal of electrons or whole hydrogen atoms from a compound. Although oxygen is one of the commonest oxidizing agents (hence the contribution of its name to the process), it is by no means the only one, nor is its presence necessary for oxidation to occur. In a key step in glycolysis, PGAL is oxidized (loses hydrogen) and thereby converted to 1,3-diphosphoglyceric acid. The electrons (in pairs) and accompanying hydrogen ions (which are stripped from the aldehyde to yield the acid) are accepted by the coenzyme $NAD^+$. The $NAD^+$ thereby becomes reduced to NADH and the aldehyde is oxidized. Since the free energy liberated in this highly exergonic reaction is so great, a phosphate group is picked up from the surrounding medium and attached as a high-energy phosphate (~P) to the end of glyceric acid. This high-energy phosphate will be used to form ATP in the next reaction.

**5.18** What is the NADH used for?

NADH may be used for further production of ATP through the electron-transport chain, with each NADH yielding three ATP molecules in this aerobic pathway. In the absence of oxygen, NADH may be used to form either ethanol (bacteria) or lactic acid (animals). The $NAD^+$ consequently formed is then recycled for use in glycolysis.

**5.19** What happens to the 1,3-diphosphoglyceric acid?

It loses its high-energy phosphate to ADP and forms ATP.

1,3-Diphosphoglyceric acid 3-Phosphoglyceric acid

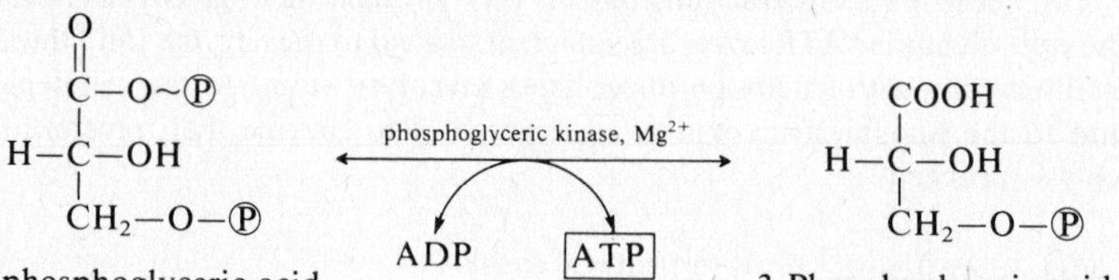

**5.20** Is the above an exergonic or an endergonic reaction?

Much energy is released in the transfer of the high-energy phosphate from 1,3-diphosphoglyceric acid to ADP. The $\Delta G$ is therefore negative, and the reaction is exergonic. Some of the energy released is conserved in the chemical bond of the ATP formed.

**5.21** What is the purpose of the molecular rearrangements that follow the creation of 3-phosphoglyceric acid?

The rearrangements occur so that another high-energy phosphate bond may be formed, which again will yield an ATP for the cell. The first step in these rearrangements is to transfer the remaining phosphate to the 2 position:

3-Phosphoglyceric acid 2-Phosphoglyceric acid

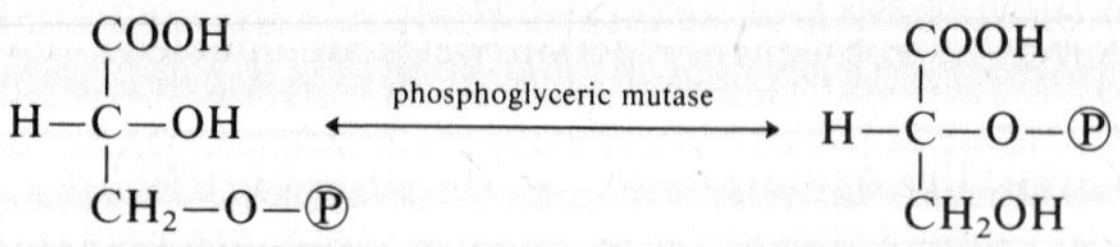

This permits the removal of a molecule of water, while concentrating energy in the bond of the phosphate:

2-Phosphoglyceric acid Phosphoenolpyruvic acid

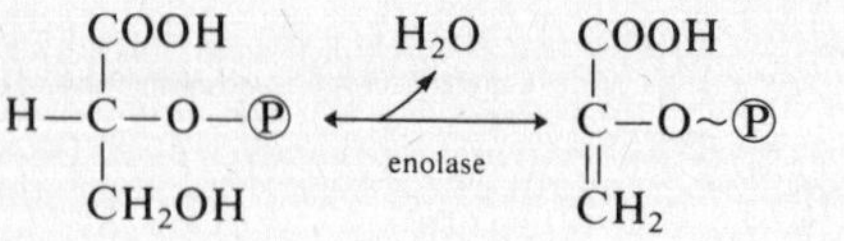

The high-energy phosphate is then transferred to ADP and forms another molecule of ATP.

$$
\begin{array}{ccc}
\text{COOH} & \text{COOH} & \text{COOH} \\
| & | & | \\
\text{C}-\text{O}\sim\text{\textcircled{P}} \xrightarrow[\text{Mg}^{2+}]{\text{pyruvic kinase}} & \text{C}-\text{OH} \xrightarrow{\text{spontaneous}} & \text{C}=\text{O} \\
\| & \| & | \\
\text{CH}_2 \quad \text{ADP} \quad \boxed{\text{ATP}} & \text{CH}_2 & \text{CH}_3
\end{array}
$$

| Phosphoenolpyruvic | Enolpyruvic | Ketopyruvic |
| acid | acid | acid |

**5.22** Summarize the reactions of glycolysis in a diagram.

Glucose → Phosphorylation of glucose ↔ Structural change to fructose ⌐

⤷ ATP   ADP

⌐→ Cleavage into two interconvertible trioses ↔ Phosphorylation of fructose ←

⤷ ADP   ATP

⌐→ Oxidation and phosphorylation ↔ Dephosphorylation ↔ Molecular rearrangement ⌐

⤷ NAD⁺   NADH        ADP   ATP

Pyruvic acid ← Dephosphorylation ←

⤷ ATP   ADP

**5.23** What are the likely mechanisms of control of the rate of glycolysis in the cell?

A number of factors play a role in determining both the rate of glycolysis and the direction of many of the reversible reactions in the pathway. As was seen in Chap. 2, increasing concentrations of particular metabolites will exert a "push" in promoting the reactions for which they serve as substrates, while diminishing concentrations of other metabolites will exert a "pull" on the enzyme reactions of which they are the products. For example, in the reaction A → B, an increasing amount of A will push the forward reaction, while a decreasing amount of B will pull the forward reaction. Diminishing amounts of A or increasing amounts of B will exert a braking effect.

Since one step in glycolysis, the oxidation of PGAL to 1,3-diphosphoglyceric acid, requires $NAD^+$, the availability of this coenzyme in the oxidized state will also exert a regulatory action. Similarly, the amount of $P_i$ (inorganic phosphate) within the cytosol may become a limiting factor, although this is rare. Perhaps the most significant control mechanism for glycolysis and a variety of other metabolic pathways is the existence of allosteric enzymes within the system. These enzymes contain within their three-dimensional structure (conformation) both an active site for the attachment of the substrate and an additional, allosteric region where modifying substances may attach. These modifying substances change the overall conformation of the enzyme and thereby alter its catalytic activity. As seen in Prob. 5.13, one such allosteric enzyme in glycolysis is phosphofructokinase (PFK), the enzyme that catalyzes the formation of fructose 1,6-diphosphate from fructose 6-phosphate. ATP tends to bind to the allosteric site of the enzyme and inhibit its activity. Thus, in a cell that contains sufficient amounts of ATP, the inhibition of PFK will tend to shut down glycolysis until there is once again a need to produce more energy currency.

**5.24** Under anaerobic conditions, degradation of the end product of glycolysis (pyruvic acid) produces lactic acid or ethyl alcohol. Why is this step beyond glycolysis necessary?

Glycolysis always produces pyruvic acid as an end product. In the one oxidative step in the process in which PGAL is converted to 1,3-diphosphoglyceric acid, an $NAD^+$ is reduced to NADH. If glycolysis is to continue, this $NAD^+$ must continually be regenerated through the oxidation of NADH. Under anaerobic conditions, this is accomplished in animals by "dumping" the electrons and hydrogen ions of NADH onto pyruvate to form lactic acid, and thus $NAD^+$ is regenerated (see Fig. 5.1). The enzyme lactic dehydrogenase plays a key role in the reaction in many organisms. In bacteria, a carbon atom is first removed from the pyruvate, and the resulting 2-carbon compound is reduced to ethyl alcohol, with a consequent regeneration of $NAD^+$. The carbon is actually removed as a molecule of $CO_2$. (This gas acts as the leavening agent that causes bread to rise when the dough contains fermenting yeast.) In some insects, one of the products arising from the splitting of fructose 1,6-diphosphate, namely, dihydroxyacetone phosphate (DHAP), receives the electrons and hydrogen ions from NADH to form an important reduced product known as $\alpha$-glycerol phosphate, a substance that may be used in the synthesis of neutral fats. Thus a variety of metabolic mechanisms have evolved to reoxidize NADH, with the consequent formation of characteristic reduced products.

**5.25**    What is the significance of the Krebs (tricarboxylic acid) cycle in the economy of the cell?

The Krebs cycle is a metabolic mill or merry-go-round in which the pyruvic acid arising from glycolysis is aerobically degraded, or oxidized, to carbon dioxide and water with the generation of great amounts of ATP. Like glycolysis, the Krebs cycle is a metabolic pathway in which a variety of enzymatic steps are integrated into a cohesive pattern. The enzymes of the Krebs cycle are found within the mitochondria; many of these enzymes are loose in the matrix, while others are attached to the inner membrane of this organelle. Glycolysis, which can proceed in the absence of oxygen, may be viewed as the anaerobic, initial phase of carbohydrate degradation; the Krebs cycle can be considered the terminal, aerobic phase of this process.

**5.26**    Does pyruvic acid enter the Krebs cycle directly?

No, pyruvic acid is first acted upon by a multienzyme complex within the mitochondria, where it is oxidatively decarboxylated (loses its carboxyl group as $CO_2$) to acetic acid; the acetic acid is attached to a molecule of coenzyme A and, as a highly reactive acetyl-CoA, enters the Krebs cycle proper. In the formation of acetyl-CoA, a molecule of NADH is produced. Acetyl-CoA then unites with oxaloacetic acid $(C_4)$ to form citric acid $(C_6)$. In elaborating the specific steps of the cycle in his early experiments, Sir Hans Krebs was aware of the initial formation of citrate, and this led to the use of *citric acid cycle* as the common descriptive name of the pathway.

**5.27**    Why is the Krebs pathway known as a cycle?

Since the citric acid produced in the initial condensation reaction of acetyl-CoA with oxaloacetic acid runs through the "mill" and eventually yields another molecule of oxaloacetic acid, the sequence may be seen as returning to its starting point in typical cyclical fashion. A new acetyl-CoA will then join the oxaloacetic acid to initiate another turn of the merry-go-round. An apt characterization of the Krebs cycle would be a metabolic system in which the products of glycolysis are first modified and then ground down to low-energy products with the help of ancillary pathways closely connected to the main cycle.

**5.28**    What is the relationship between the Krebs cycle and the electron-transport chain (ETC)?

Within the Krebs cycle, oxidations of involved metabolites occur at specific steps. Each of these steps is catalyzed by an enzyme that works with a helper called a *coenzyme*, usually $NAD^+$ or FAD. It is actually the coenzyme which acts as an acceptor for the electrons and hydrogen ions which are stripped from specific substrates, the metabolites of the Krebs cycle, and it is the coenzyme that is the connecting link between the Krebs cycle and the ETC. The action of the Krebs cycle is indeed one of completely oxidizing the pyruvic acids that are continuously being fed into it. However, while the individual carbons of the entering molecules are oxidized to lower energy levels, the electrons and hydrogens attached to the reduced coenzymes (NADH and $FADH_2$) are still in a highly energized state. This energy is tapped through the intervention of the ETC, a series of pigments that acts like a bucket brigade in carrying electrons and hydrogen ions to lower energy levels. The complete participation of this chain, starting with NADH, results in the creation of three ATP molecules. If this chain is entered farther down, as is the case with $FADH_2$, only two molecules of ATP are produced from the energy released in the downhill path. Since ATP is produced through the repeated removal of electrons from the pigments of the electron-transport chain, the total process is called *oxidative phosphorylation*. The components of the ETC are embedded in the cristae of the inner mitochondrial membrane, closely apposed to the enzymes of the Krebs cycle. They are arranged for maximum efficiency of electron transport.

**5.29**  What is the actual mechanism by which oxidations with the ETC are coupled with the synthesis of ATP?

This problem has been the subject of considerable investigation and extensive speculation. Older notions stressed the participation of specific phosphorylating enzymes along the ETC, enzymes that were thought to be concentrated at the ends of little tufts embedded in the inner membrane. The more generally accepted idea at present involves a pumping of hydrogen ions across the inner mitochondrial membrane to the outside compartment during the migration of electrons along the ETC (see Fig. 5.4). This imbalance of hydrogen ions, or protons, creates an energized state. The ensuing movement of protons to an equilibrium state releases the stored energy, which is utilized by membrane enzymes to create ATP. This view has been termed the *chemiosmotic model*. Still another theory focuses on the possibility of changes in the three-dimensional structure of membrane proteins during the migration of electrons along the chain. The return of these proteins to their original shape, much like the resumption of the spherical shape of a squeezed rubber ball, releases enough energy for ATP production.

**5.30**  What is the total yield of ATP from the complete oxidation (respiration) of a molecule of glucose?

Although some difference of opinion exists about the exact total of ATP produced, the consensus of opinion among biochemists is that 36 ATPs are produced. Glycolysis alone results in four ATPs, but two are used up in the early phosphorylating steps to yield a net of two. The NAD reduced in the formation of 1,3-diphosphoglyceric acid will yield three ATPs, or a total of six ATPs, since one glucose molecule will produce two molecules of 1,3-diphosphoglyceric acid. However, in bringing these NADH molecules into the mitochondria, where they will join the ETC, one molecule of ATP may be used up for each of the two NADH molecules transported, so the net yield from glycolysis under aerobic conditions is six ATPs $[(4 - 2) + (6 - 2)]$. The oxidative decarboxylation of pyruvic acid produces an NADH within the mitochondrion; hence another six ATPs (two pyruvic acids are generated for each glucose molecule) are produced. Finally, the generation of 12 ATPs for each turn of the Krebs cycle means that a starting molecule of glucose will produce 24 ATPs, for a grand total of 36 $(6 + 6 + 24)$. Thus, the availability of aerobic mechanisms for the degradation of carbohydrates like glucose enhances the energy yield by 18-fold. However, organisms like the tetanus bacterium continue to live today although they are confined to the relatively inefficient modes of an anaerobic metabolism. Within human organisms great bursts of energy may be supported by glycolysis alone when the muscle demands for energy exceed the capacity of the bloodstream to bring sufficient oxygen for aerobic pathways. Under these conditions, lactic acid will accumulate as an "oxygen debt," to be repaid when normal conditions resume, at which time the lactic acid can be reconverted to pyruvic acid, which will enter the usual aerobic pathways.

**5.31**  Are carbohydrates such as glucose the only source of energy for the cell?

No, the cell can utilize fats and proteins for fuel as well as sugars. Proteins to be used as a fuel are degraded to their constituent amino acids. These amino acids are then stripped of their amino group in a process called *deamination*. The deamination may involve an oxidation or a transfer to a keto acid. The resultant products may then enter the pathways utilized for carbohydrates. Thus, alanine may be oxidatively deaminated to pyruvic acid and become a metabolite for the Krebs cycle:

$$
\begin{array}{ccc}
\underset{\text{Alanine}}{\overset{\overset{\displaystyle CH_3}{|}}{\underset{\overset{\displaystyle |}{\underset{\overset{\displaystyle C}{\nearrow}}{\underset{\displaystyle OH}{}}}}{HC-NH_2}}} + H_2O & \overset{(O)}{\longleftrightarrow} & \underset{\text{Pyruvic acid}}{\overset{\overset{\displaystyle CH_3}{|}}{\underset{\overset{\displaystyle |}{\underset{\overset{\displaystyle C}{\nearrow}}{\underset{\displaystyle OH}{}}}}{C=O}}} + NH_4OH\,(NH_3\cdot H_2O)
\end{array}
$$

The ammonia may be excreted as such or may combine with $CO_2$ to form urea.

In similar fashion, the 5-carbon glutamic acid may be converted to $\alpha$-ketoglutaric acid, an intermediate of the Krebs cycle. Polysaccharides can be hydrolyzed to simple sugars. In the case of fats, most of the caloric potential resides in the fatty acids attached to the glycerol backbone of the molecule. These fatty acids may be split from the alcohol groups of glycerol. They then undergo a process called $\beta$ oxidation, in which 2-carbon fragments are continually chipped from the carboxyl end of the fatty acid chain and attached to coenzyme A. This compound is the acetyl-CoA, which serves as fodder for the Krebs cycle. Since many more acetyl-CoA molecules are generated in the oxidative breakdown of fats than arise from glycolysis, fats are a richer source of energy than carbohydrates.

**5.32**  Can glycolysis or the Krebs cycle be reversed to produce glucose from metabolites arising in these processes?

Theoretically, this could occur and, in general, does take place. However, many of the steps in both glycolysis and the Krebs cycle are almost irreversible, so that "end runs" around these blocking steps are necessary in the building up (*anabolism*) of complex substances such as glucose. An example of such an anabolic modification is the conversion of glucose 6-phosphate to free glucose. The initial step of glycolysis, you will recall, involves the phosphorylation of glucose to glucose 6-phosphate at the expense of an ATP and with the catalytic assistance of hexokinase or glucokinase. This step cannot easily be reversed, and an enzyme called glucose 6-phosphatase is required to split off the phosphate group and generate free glucose. This enzyme and others used in succeeding steps in the reversal process form part of a pathway known as *gluconeogenesis* ("forming glucose anew"). The enzymes of gluconeogenesis are stimulated by cortisone and other hormones of the adrenal cortex; this accounts, in part, for the elevation of blood sugar level under conditions of anger or fear, since these emotional states heighten activity of the adrenal gland. Gluconeogenesis generally takes place in the liver and kidneys.

**5.33**  How are proteins and fats synthesized within the cell?

The amino acids for the synthesis of proteins may be produced by adding amino groups to metabolites generated within the glycolytic pathway, the Krebs cycle, or still other pathways. The amino acids are then joined by enzymes promoting the formation of peptide bonds to yield protein.

Fats are formed in a two-step process. The significant first step, known as *lipogenesis*, involves the formation of a long-chain fatty acid. This occurs outside the mitochondria and involves the participation of a multienzyme complex. NADPH, a reduced coenzyme that is active in a variety of syntheses, plays a significant role in the generation of the highly reduced fatty acid. In the second step, the fatty acids are joined to an activated glycerol molecule to form a triglyceride in a process known as *esterification*. In both processes, the products of carbohydrate metabolism play an important role in promoting the synthesis. This has led some biochemists to the conclusion that fats are forged in a carbohydrate flame. In diabetes an insufficient supply of insulin, the major regulator of carbohydrate metabolism, occurs. Advanced diabetics cannot convert their carbohydrate to fat, nor can they efficiently degrade and oxidize fat. Death in diabetics often results from the accumulation of incompletely oxidized metabolites of fat, called *ketone bodies*, since these substances drastically increase the acidity of blood and body fluids. Although most of the various pathways in metabolism are interconnected, the linkage between the transformations of carbohydrate and fat is especially crucial.

**5.34**  What are the advantages of endothermy in birds and mammals?

Since metabolic activity, as well as the general movement, of an organism is a function of temperature, ectotherms are limited in their lifestyle by the temperature of the environment. In extreme cold they become remarkably sluggish as life processes slow; as the temperature rises, they become increasingly active. By contrast, the endothermic birds and mammals maintain an internal temperature that is relatively constant owing to their evolution of mechanisms for both the retention and dissipation of heat. It should be recalled that cellular respiration (involving glycolysis, the Krebs cycle, and the electron-transport chain) is about 40 percent efficient. The energy that is not effectively stored as ATP (60 percent) is given off as heat, which, if effectively hoarded, contributes to the maintenance of the constant body temperature of endotherms. Consequently, birds and mammals enjoy the luxury of a lifestyle that is not subject to extreme fluctuations in tempo. Since the body temperature of these endotherms is maintained at relatively high levels (37–40 °C), their activity levels are uniformly high and they can have more intensive levels of movement, foraging, exploration of the environment, etc. They are also more flexible in choosing their habitats, since they can maintain their internal temperature in a variety of settings. This provides a competitive edge in their struggle with other kinds of organisms.

**5.35** Why are smaller mammals much more active than larger mammals?

Among all organisms there is a tendency for the specific metabolic rate to vary inversely with body size. *Specific metabolic rate* is the metabolic activity per unit mass of living organism. This is particularly evident among the small mammals, in which heart rate, specific oxygen consumption, breathing rate, etc., are extremely high. In fact, there is a lower limit of size among mammals, since too small an organism would not be able to provide enough food and oxygen to sustain its intense metabolic activity. A shrew (~3.5–5 cm or 1.4–2 in) actually eats its own weight of food in little more than 24 hours just to stay alive.

The inverse relationship between size and metabolic rate is related to the tendency of any structure to increase in volume or weight much more rapidly than in surface area as it increases in overall size. In mathematics we learn, for example, that the volume of a sphere is proportional to the *cube* of its radius, while its surface is proportional to the *square* of its radius. Thus, smaller bodies have a relatively greater surface area than larger ones. Since heat is lost at the surface and most of the other significant exchanges between an organism and its environment occur at the surface, it is readily apparent that smaller organisms must work harder to maintain constant temperature and other equilibrium conditions than larger ones. Many years ago, Rubner advanced the surface law to explain these differences associated with body size. Now we realize that the situation is even more complex, since specific metabolic rate is not merely a function of surface-volume ratios. The fact that larger organisms have a relatively greater amount of less active supporting tissue also plays a role in the decreasing metabolic intensity of these groups.

**5.36** Are the differences in metabolic intensity among mammals of different sizes expressed at the cellular level or do they involve central controls operating at the organismic level?

There is evidence that the cells in the active tissues of smaller mammals do have higher levels of both some enzymes of the Krebs cycle and pigments of the electron-transport chain. Researchers have noted an increased population of mitochondria in the liver cells of these small mammals, as well as an increased oxygen consumption by slices of liver. However, the total oxygen consumption of the intact organism is not merely the summated oxygen consumption of extirpated individual structures, so the significance of these tissue studies has not been clearly established. The role of central factors such as hormones and nervous stimulation must be significant, since tissues with a characteristic metabolic apparatus are capable of wide variation in response under the influence of specific hormones. Cellular metabolic features probably impose limits on the range of metabolic responses, while the central control mechanisms impose specific activity levels.

# Supplementary Problems

**5.37** What is the source of the energy that flows through the living world? (*a*) photosynthesis. (*b*) chemical bonds. (*c*) green plants. (*d*) the sun.

**5.38** What eighteenth-century scientist is credited with overthrowing the phlogiston theory and establishing the actual nature of chemical oxidations or combustions and their similarity to respiration in animals? (*a*) Joseph Priestley. (*b*) Antoine Lavoisier. (*c*) Jean Baptiste van Helmont. (*d*) Gregor Mendel.

**5.39** An exergonic reaction is (*a*) a spontaneous reaction in which energy is given off. (*b*) an uphill reaction requiring input of energy. (*c*) an oxidation reaction. (*d*) an anaerobic reaction.

**5.40**   Exergonic reactions are usually   (a) endothermic.   (b) exothermic.   (c) neither of the above.   (d) both of the above.

**5.41**   In the combustion of glucose, what is the change in the heat content of the system?

**5.42**   In the same combustion of glucose, what is the change in the free energy of the system?

**5.43**   Why is the free-energy change in the combustion of glucose greater than the change in heat content?

(a)   The heat loss is subtracted from the total change in free energy.

(b)   Most of the reactions of glucose combustion are reversible and therefore spontaneous, exergonic reactions.

(c)   The increase in entropy augments the heat loss of the system.

(d)   Free energy is added to the system by the enzymes involved in the breakdown of glucose.

**5.44**   How do green plants store the radiant energy of the sun that they capture?

**5.45**   What is the balanced equation for the complete cellular degradation of glucose?

**5.46**   Glucose enters glycolysis by the action of hexokinase (glucokinase) to yield what compound?

**5.47**   In the one oxidative step of glycolysis, electrons and hydrogen ions are accepted by what substance?

**5.48**   Fructose 1,6-diphosphate is split in glycolysis to yield what two trioses?

**5.49**   Glycogen may enter glycolysis without ATPs being used to form a glucose phosphate ester. What would be the net ATP yield from glycolysis if we start with glycogen as the substrate?

**5.50**   What is the advantage of glycolysis, since it taps only a small fraction of the energy available in the glucose molecule?

(a)   It may be used when oxygen is unavailable.

(b)   It is cyclical, so that less substrate is required.

(c)   It requires no investment of ATP.

(d)   It is composed only of spontaneous reactions.

**5.51**   The pentose shunt, or direct oxidative pathway for the degradation of glucose, provides a metabolic alternative. What is the advantage of its prominence in tissues where synthesis of lipid is high?

(a)   It can occur in the absence of oxygen.

(b)   It provides the NADPH necessary for this synthesis.

(c)   It is faster.

(d)   It recycles $NAD^+$ for glycolysis.

**5.52**   The iron-containing cytochrome pigments are involved in which metabolic pathway?   (a) Glycolysis, (b) Krebs cycle,   (c) gluconeogenesis,   (d) electron-transport chain.

**5.53**   Which component of the electron-transport chain is lipid soluble? (a) Coenzyme Q, (b) flavoprotein,   (c) cytochrome a,   (d) $FADH_2$.

**5.54**   A cell can carry out its constant and intensive metabolic activities with relatively small amounts of such vital components as ATP, $NAD^+$, coenzyme A, etc., because it can

(a)   bypass reactions that require these compounds.

(b)   rapidly recycle these compounds.

(c)  incur an oxygen debt until these compounds become available.

(d)  utilize substitute molecules in place of these compounds.

**5.55**  The primary advantage of hibernation to small homeotherms is that it

(a)  eliminates the need for seasonal migration.

(b)  enhances survival through regeneration of cells.

(c)  reduces the risk of predation.

(d)  reduces the need for food.

**5.56**  $Q_{10}$ is the ratio of biological activity at a particular temperature to the same activity at a temperature $10\,°C$ lower. If the metabolic rate of a typical mammal is denoted as $X$ at $0\,°C$, what will be the metabolic rate at $20\,°C$ if $Q_{10}$ is 2?

# Answers

**5.37**  (d)

**5.38**  (b)

**5.39**  (a)

**5.40**  (b)

**5.41**  $\Delta H = -673$ Cal/mol

**5.42**  $\Delta G = -686$ Cal/mol

**5.43**  (c)

**5.44**  Chemical bonds

**5.45**  $C_6H_{12}O_6 + 6O_2 \rightarrow 6CO_2 + 6H_2O + energy$

**5.46**  Glucose 6-phosphate

**5.47**  $NAD^+$

**5.48**  Phosphoglyceraldehyde and dihydroxyacetone phosphate

**5.49**  three ATPs

**5.50**  (a)

**5.51**  (b)

**5.52**  (d)

**5.53**  (a)

**5.54**  (b)

**5.55**  (d)

**5.56**  A $Q_{10}$ of 2 means that a doubling of metabolic activity occurs with each $10\,°C$ increase in ambient temperature. So, at $10\,°C$, the metabolic rate would be $2X$; at $20\,°C$, it would be $2(2X)$ or $4X$

# Chapter 6

# Photosynthesis—The Basic Energy-Capturing Reaction of the Living World

## 6.1 AN OVERALL VIEW

*Photosynthesis* is the process by which high-energy, complex food molecules are produced from simpler components by green plants and other autotrophic organisms in the presence of light energy. In the process of photosynthesis, photons (unit packets) of light are captured by specific pigment molecules. Electrons within these pigment molecules are excited by the absorbed photons, and these excited electrons eventually liberate their energy to the cell as they fall back to the unexcited ground state. Many cells use this energy to reduce carbon dioxide to a carbohydrate.

Photosynthesis is the major endergonic reaction of the living world—an uphill process in which low-energy molecules such as carbon dioxide and water interact to form high-energy carbohydrates and eventually lipids and proteins. The photosynthetic reaction is essentially a reversal of the exergonic process of cellular respiration.

**EXAMPLE 1** In respiration, energy is released when molecules such as glucose ($C_6H_{12}O_6$) are oxidized to $CO_2$ and $H_2O$. The released energy is stored as ATP. Most of the ATP formed during respiration is derived from reactions that occur in the mitochondria. In photosynthesis, energy from the sun is absorbed by pigment systems within the chloroplast, and this energy is utilized first to produce ATP and then to form a sugar molecule. In the process, oxygen gas is liberated.

The wavelengths of light used for photosynthesis are only a small fraction of the total spectrum of electromagnetic radiation. In the higher plants, violet, blue, and red appear to be most effective in promoting photosynthesis. These colors of the visible spectrum have wavelengths ranging from approximately 380 to 750 nanometers (nm), respectively. The ability of light to dislodge electrons is related to the wavelength rather than to the brightness (intensity) of the light beam. Only a small percentage of light reaching the plant is actually absorbed. Most of the light either passes through the plant (is *transmitted*) or is *reflected* from the plant's surface.

The process of photosynthesis consists of a series of complicated metabolic pathways that may be summarized as follows: a light-dependent reaction sequence produces both NADPH and ATP, which are then used to reduce $CO_2$ to a carbohydrate in the "dark" reaction. The light reaction requires light

**Table 6-1.  Evolution of Photosynthetic Theory**

| Event | Date |
|---|---|
| Joseph Priestley demonstrates that green plants give off oxygen | 1771 |
| Jan Ingenhousz shows that photosynthetic evolution of $O_2$ requires sunlight and occurs only in green parts of plants | 1779 |
| Nicholas Theodore de Saussure shows that water is necessary in sugar production by plants | 1804 |
| James Clerk Maxwell develops the wave model of light, leading to recognition that light is source of energy in photosynthesis | 1864 |
| F. F. Blackman demonstrates that photosynthesis has both a light-dependent, temperature-independent step and a light-independent, temperature-dependent step | 1905 |
| C. B. van Niel proposes that $O_2$ evolved in photosynthesis is derived from $H_2O$ not $CO_2$ | 1930s |
| Melvin Calvin and his colleagues use carbon 14 to trace the conversion of $CO_2$ to carbohydrate | 1940s |

for only one or two steps in its sequence. Table 6-1 presents the milestones in unlocking the pathways of photosynthesis.

## 6.2  THE LIGHT REACTION

In the light reaction of photosynthesis (see Fig. 6.1), a photon of appropriate wavelength is absorbed by various pigment molecules (chlorophyll *a* and *b*, carotenoids) of *photosystem I* and its energy is transferred to a particular chlorophyll *a* molecule in the *reactive site*. An electron of this chlorophyll *a* molecule is boosted to a high-energy state, combines with a receptor molecule, moves down a free energy gradient, and eventually returns to its starting point. In the process, a small amount of ATP is generated; as in mitochondria, the generation of ATP is believed to occur chemiosmotically through the creation of a $H^+$ gradient. Because ATP is formed in association with the absorption of light, the reaction is termed *photophosphorylation*. The chlorophyll electron that is energized eventually completes a circuit, so this particular pathway is known as *cyclic photophosphorylation*.

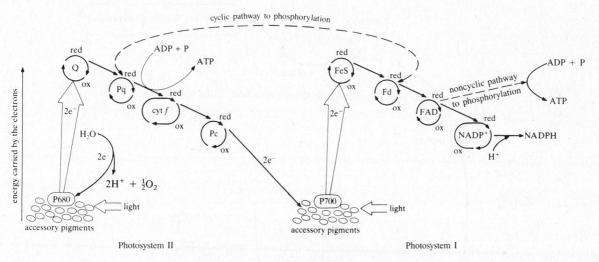

**Fig. 6.1** A summary of the light reaction

A second option, probably developed later in the evolution of photosynthesis, is available to the excited electrons produced by the absorption of light in photosystem I. These electrons may combine with a receptor molecule, but then, instead of returning to their starting point, they move along a chain that terminates in the coenzyme $NADP^+$ and convert it to NADPH. In their passage to $NADP^+$, the electrons, as with the first option, move down an energy gradient and are involved in the generation of ATP. Both ATP and NADPH will be used in the dark reaction to reduce $CO_2$ to a carbohydrate.

The electrons removed from chlorophyll and accepted by $NADP^+$ are eventually returned through the intervention of a second light event involving a different pigment system from the one used in the first. In this second light event, chlorophyll molecules of *photosystem II* absorb light and transfer excited electrons to an acceptor that initiates their passage along an energy gradient to photosystem I. This restores the electron "hole" of photosystem I, but creates a hole in photosystem II. This hole is filled from a reaction in which water is split by the energy of absorbed photons to form electrons, $H^+$, and oxygen. The electrons are taken in by the chlorophyll molecules of photosystem II to restore them to their original condition. The $H^+$ will move along with the electrons captured by $NADP^+$ to effect the reduction of $CO_2$ in the dark reaction, with $O_2$ being liberated in the process. The overall pathway in which electrons move from water to photosystem II, to photosystem I, to $NADP^+$ is called *noncyclic photophosphorylation*.

There are many variants of the photosynthetic theme. In all probability the earliest autotrophs did not have a synthetic dark reaction. They merely absorbed light energy, using pigment molecules whose electrons were raised to a high-energy state, and then passed this energy indirectly to other pathways in the cell. Also, certain bacteria use $H_2S$ instead of $H_2O$ as a source of replacement of electrons and $H^+$; they liberate $S_2$ instead of $O_2$.

## 6.3  THE DARK REACTION (CALVIN-BENSON CYCLE)

The dark reaction is the pathway by which $CO_2$ is reduced to a sugar. Its components are found in the stroma of the chloroplast. This reaction does not actually require darkness; it is merely light-independent. Since $CO_2$ is an energy-poor compound, its conversion to an energy-rich carbohydrate involves a sizable jump up the energy ladder. This is accomplished through a series of complex steps involving small bits of energy (see Fig. 6.2).

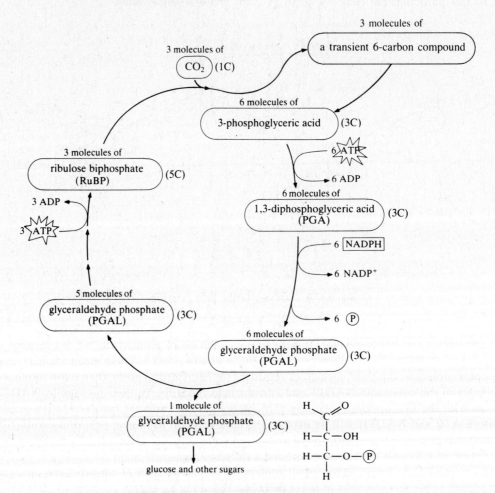

**Fig. 6.2** The dark reaction of photosynthesis

The initial reaction involves the union of $CO_2$ with a 5-carbon compound called *ribulose biphosphate* (*RuBP*). An elusive 6-carbon compound is probably formed that breaks down to two molecules of the 3-carbon compound phosphoglyceric acid (PGA), which we encountered in the discussion of glycolysis in Chap. 5. Each of the molecules of PGA is then reduced to the highly energetic phosphoglyceraldehyde (PGAL), an actual sugar, which is the first stable end product of photosynthesis.

**EXAMPLE 2**   In glycolysis, a critical step is the oxidation of PGAL to diphosphoglyceric acid, with $NAD^+$ acting as the electron acceptor. In photosynthesis, essentially a reversal of the degradation of carbohydrate represented by glycolysis, PGA is reduced to PGAL, with NADPH serving as the electron donor. In many reductive syntheses, NADPH is the coenzyme involved, while in the degradation reactions for energy liberation, $NAD^+$ is usually the active coenzyme. The ratio between PGAL and PGA may be a crucial measure of the balance between synthesis and breakdown in the cell.

For every six molecules of PGAL produced, five will be used to form new RuBP so that $CO_2$ can be continually fixed and indirectly converted to PGAL. Although the remaining molecule of PGAL can be converted to glucose through a reversal of the usual glycolytic pathway, it is not stored as such in the cell. Instead, a disaccharide such as sucrose may be formed, or, more usually, starch will accumulate at the site of the photosynthetic activity. The plant cell may also convert the PGAL to the lipids and proteins it may require.

## 6.4   PHOTORESPIRATION

The enzyme that incorporates $CO_2$ into RuBP in the first steps of the dark reaction is called *ribulose biphosphate carboxylase*. Under conditions of low $CO_2$ and high $O_2$ concentrations, this enzyme binds with $O_2$ (instead of $CO_2$) and thereby catalyzes the oxidation of RuBP first to glycolic acid and then to $CO_2$ in the peroxisomes. This oxidative breakdown of a vital intermediate of the dark reaction is called *photorespiration*. Although, through a complex series of reactions, much of the $CO_2$ may be salvaged by the cell rather than lost, photorespiration appears to be very wasteful. It does not serve any known useful purpose to the plant and may reduce the food yield of many crops. Photorespiration occurs during very hot, sunny, dry days when the stomata are closed to prevent water loss and $O_2$ tends to accumulate within the leaf. It does reduce those oxygen levels but with no benefit to the plant. Photorespiration is inhibited by the $C_4$ pathway.

## 6.5   THE $C_4$ PATHWAY

The *$C_4$ pathway* is an alternative to the Calvin-Benson cycle of the dark reaction and was first clearly delineated by M. D. Hatch and C. R. Slack. Known as the *Hatch-Slack pathway* of $C_4$ photosynthesis (Fig. 6.3), it is based on the formation of a 4-carbon compound acid, oxaloacetic acid, by a union of $CO_2$ with the 3-carbon phosphoenolpyruvic acid (PEP). Just as RuBP carboxylase catalyzes the carboxylation ($CO_2$ fixation) of RuBP, so a PEP carboxylase is involved in the addition of a $CO_2$ to PEP.

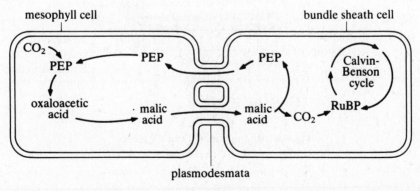

Fig. 6.3

The oxaloacetic acid arising from the union of PEP and $CO_2$ has several possible fates, but it is usually reduced to malic acid by NADPH. Such a reduction parallels the conversion of PGA to PGAL in the Calvin-Benson cycle. Malic acid moves into the cells that surround the vascular elements of a variety of plants—the *bundle-sheath cells*. These cells are particularly rich in starch but have a paucity of grana.

**EXAMPLE 3**   The chief organs of photosynthesis in vascular plants are the leaves (see Fig. 6.4). Many leaves of the familiar land plants consist of a thin stalk called a *petiole* to which a broad, thin blade is attached. The upper and lower surfaces of the blade consist of epidermis. The epidermal cells are coated with a layer of wax, called a cuticle, which is generally quite heavy on the upper surface. Between the upper and lower epidermal layers a great number of parenchymal cells can be seen. These thin-walled cells, forming the *mesophyll*, are particularly rich in chloroplasts and hence contribute greatly to the total photosynthetic activity of the leaf.

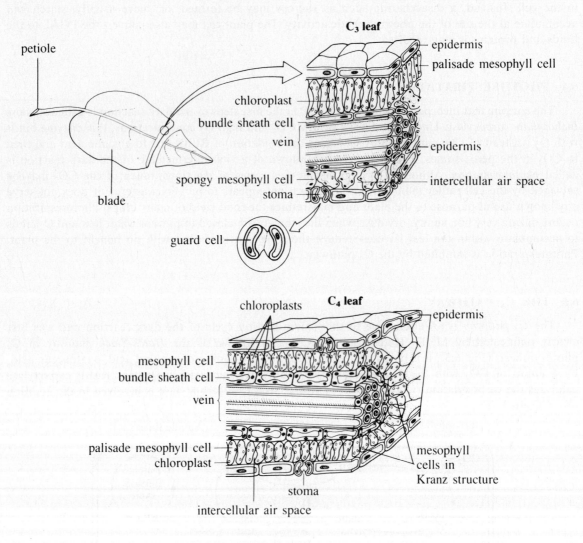

**Fig. 6.4** Anatomy of $C_3$ and $C_4$ leaves

Generally the mesophyll is arranged in two distinct layers. An upper columnar layer consisting of a regular arrangement of cylindrical cells is aptly called the *palisade mesophyll*. Below the columns of the palisade mesophyll are the irregularly scattered *spongy mesophyll*. Ramifying throughout the blade of the leaf is a system of veins (tubules) making up the vascular system of the leaf. Each vein contains both xylem and phloem and is often surrounded by a dense covering of bundle-sheath cells.

In the bundle-sheath cells the malic acid, strangely enough, breaks down into $CO_2$ and PEP. The PEP passes back to regular mesophyll cells in the leaf, but the $CO_2$ is picked up by RuBP in the bundle-sheath cells and enters the Calvin-Benson cycle.

Since the Calvin-Benson cycle appears to be the only significant producer of carbohydrate, investigators wondered why a cycle exists in which a 4-carbon intermediate is formed in the fixation of $CO_2$ before the standard Calvin-Benson cycle. The answer seems to lie in the fact that plants in hot, dry environments, where the Hatch-Slack pathway is particularly active, tend to keep their *stomata* (openings in the epidermis of leaves and other plant parts, see Fig. 6.4) partially closed most of the time to prevent water loss. This closure reduces the influx of $CO_2$ and might act as a brake on photosynthetic activity. The $C_4$ pathway, a mechanism for binding $CO_2$ and transporting it more deeply into the leaf, provides a "sink" for holding onto $CO_2$ and assures a ready supply of $CO_2$ for the Calvin-Benson cycle. It should also be noted that PEP carboxylase more avidly binds $CO_2$ than does the RuBP carboxylase, especially when $CO_2$ levels are low and $O_2$ levels are high. This assures that free $CO_2$ levels in the mesophyll cells will always be low, and hence $CO_2$ in the environment will tend to move into the leaf when the stomata are open. The Hatch-Slack pathway thus appears to be a mechanism for survival in plants that have evolved in hot, dry climates.

## 6.6  STRUCTURE AND FUNCTION IN THE CHLOROPLAST

The outer and inner membranes of the choloroplast (Fig. 6.5) are congruent in shape and lie close to each other. The unstructured portion of the fluid-filled space enclosed by the inner membrane is called the *stroma*. Distributed throughout the stroma is the *thylakoid membrane system*. This is a network of channels that occasionally forms *grana*, stacks of flattened sacks or disks. The components of the light reaction are generally tightly bound to this thylakoid membrane (see Fig. 6.6) and are accessible to molecules found in the stroma, where the reactions of the Calvin-Benson cycle occur.

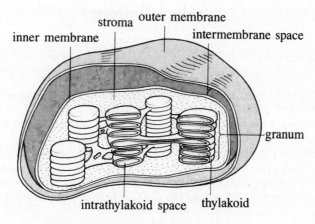

**Fig. 6.5** Structure of a chloroplast

A chemiosmotic system is believed responsible for the generation of ATP during the course of the light reaction. Protons are pumped into the inner thylakoid compartment, and a proton gradient is thereby created between the inner thylakoid space and the external stroma. A difference of up to 4 pH units may exist between these two compartments. It is currently felt that $H^+$ ions tend to migrate out of the thylakoid space through an ATP synthetase system attached to the thylakoid membrane. The energy of the electrochemical gradient is used to generate ATP at a specific site of the complex called the *CF$_1$ region*.

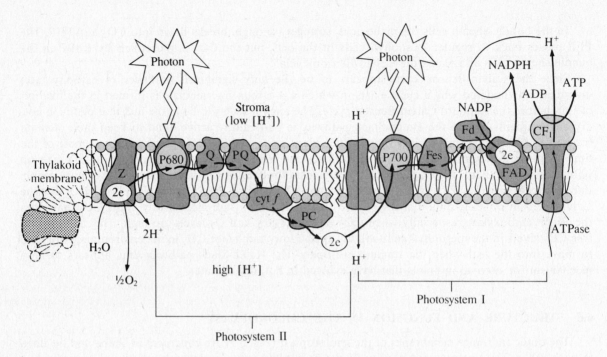

**Fig. 6.6** Structure of the thylakoid apparatus

# Solved Problems

**6.1** Criticize the following statement: Plants carry on photosynthesis; animals carry on respiration instead.

This is an incorrect view. Both plants and animals carry on respiration as a major source of energy for vital functions. Only plants, however, are also capable of carrying out photosynthesis. The components of photosynthesis are sequestered within the chloroplast, whereas the major steps of respiration occur within the mitochondrion. Thus, both processes may be occurring simultaneously in any plant cell, since the compartmentalization of these functions precludes their interfering with each other. Many of the sequences in the pathways of respiration are found as part of the photosynthetic process; this suggests a close evolutionary association of the two processes. The balance between respiration and photosynthesis will determine whether there is a net uptake of $O_2$ or $CO_2$ from a particular plant region.

**6.2** What is the history of our understanding of the photosynthetic process?

In 1771 an English pastor, Joseph Priestley, who is credited with the discovery of oxygen, demonstrated that over several days a green plant could restore the air that had been "injured" by a burning candle. Without recognizing the fact, Priestley had demonstrated that green plants produce $O_2$ during photosynthesis and that this evolution of $O_2$ balances the uptake of $O_2$ for respiration within the living world.

In 1779 Jan Ingenhousz reported his investigations which showed that sunlight is necessary for the evolution of oxygen in photosynthesis and that this process occurs only in the green parts of a plant. In 1804 the Swiss scientist Nicholas Theodore de Saussure showed that water is necessary for green plants to produce sugar.

Later in the nineteenth century James Clerk Maxwell developed the wave model of light, which led to the recognition that the energy for photosynthesis derives from absorbed visible light.

In the first decade of this century, F. F. Blackman demonstrated that photosynthesis is a twofold process. The first set of reactions is light-dependent but temperature-independent, while the second sequence (dark reactions) does not require light but shows a strong dependence on temperature.

In the 1930s C. B. van Niel conducted experiments suggesting that the oxygen evolved during photosynthesis comes from water, not carbon dioxide as was previously believed (see Prob. 6.3).

In the 1940s a special radioactive isotope of carbon, $^{14}C$, was used to follow the fate of $CO_2$ labeled with this isotopic carbon. In the late 1940s Melvin Calvin and his group utilized $^{14}C$ to elucidate the mechanisms involved in the conversion of $CO_2$ to a carbohydrate—a pathway now called the *Calvin-Benson cycle.*

Present research is focused on the nature of the carriers involved in noncyclic photophosphorylation, the relationship between the proton gradient within the chloroplast compartments and the formation of ATP, and the precise ultrastructure of the thylakoid membrane apparatus.

**6.3**  Some bacteria split $H_2S$ instead of $H_2O$ during photosynthesis and evolve sulfur. It was in the study of these bacteria that van Niel first realized that the oxygen evolved in more highly evolved plants during photosynthesis is derived from water. What reasoning, would you say, led him to this conclusion?

Before van Niel proposed his theory in the 1930s, plant physiologists thought that oxygen was split from the carbon of $CO_2$ and liberated, while the carbon joined with water to form a carbohydrate. On the basis of his studies with bacteria, van Niel questioned this interpretation and proposed the following equation:

$$CO_2 + 2H_2S \xrightarrow{\text{light}} CH_2O + H_2O + S_2$$

Clearly, the $S_2$ evolved comes from the $H_2S$ molecule. Since in higher plants $H_2O$ is substituted for $H_2S$, it seemed, by parallel reasoning, that the $O_2$ evolved came from the $H_2O$ molecule, not from the $CO_2$. Likewise, for bacteria, since there is no oxygen in $H_2S$, the oxygen incorporated into the carbohydrate must come from $CO_2$; in all likelihood, therefore, it probably comes from $CO_2$ in higher plants as well. Van Niel theorized that a variety of compounds similar to $H_2O$ and $H_2S$ could provide hydrogens for the reduction of $CO_2$. A general equation for photosynthesis thus becomes:

$$CO_2 + 2H_2A \xrightarrow{\text{light}} (CH_2O) + H_2O + A_2$$

In all higher plants, water is the source of reducing equivalents (electrons and $H^+$) for $CO_2$, and oxygen is liberated.

The broad spread of photosynthetic organisms more than 2 billion years ago led to the oxygen revolution—the creation of large amounts of free oxygen in the atmosphere. Although oxygen can be toxic to many organisms, its presence sparked the development of the aerobic pathways in opportunistic survivors, which provided so much more energy from the degradation of organic fuel molecules.

**6.4**  Describe the basic characteristics of a photosystem and explain how photosystem I differs from photosystem II.

Both photosystems consist of arrays of pigment closely associated with the thylakoid membrane. They each contain up to 400 macromolecules consisting of chlorophyll *a*, chlorophyll *b*, and carotenoid pigments. In some algae, still other types of chlorophyll may be found. The actual reaction center of either photosystem is a chlorophyll *a* molecule that can be excited by the absorption of an appropriate photon of light. The many pigment molecules surrounding the reactive site act as antennae to absorb light and pass the energy of the photon to the chlorophyll *a* of the reaction center. An electron is then raised to a higher energy level and bound to an acceptor molecule. This leaves the chlorophyll *a* of the active center an electron short. What transpires then will be dealt with in the next solved problem.

In photosystem I the chlorophyll *a* molecule has an absorption peak at 700 nm and is called pigment 700 (P700). In photosystem II the central chlorophyll *a* absorbs light maximally at 680 nm and so is known as P680.

Photosystem II, which plays a role in noncyclic photophosphorylation, probably evolved after photosystem I had become established. Both cyclic and noncyclic photophosphorylation involve photosystem I.

**6.5**    How does cyclic photophosphorylation differ from noncyclic photophosphorylation?

In cyclic photophosphorylation, involving photosystem I, an electron from P700 is raised to a high-energy state and is attached to a specialized receptor enzyme (FeS), as shown in Fig. 6.1. It then passes down an energy gradient which includes a series of redox carriers which eventually bring it back to its starting point. Although each step is exergonic and free energy is released, only one of the steps [plastoquinone (PQ) to cytochrome $f$] is used by the cell for the production of ATP. Since a cycle is completed in this process, it is appropriately labeled *cyclic photophosphorylation*. It probably represents the more ancient technique for converting captured light energy into ATP.

In noncyclic photophosphorylation an electron is also boosted from P700 to a high-energy state and trapped by FeS. But following the transfer to ferredoxin (Fd), the electron, with much of its energy still intact, is transferred to $NADP^+$ and produces reduced NADP. The electron taken from P700 is eventually restored by the second light event, in which a molecule of P680 is energized and its excited electrons are passed first to acceptor Q and then along an energy gradient to P700. But what of the electron gap now in P680? The absorption of light by P680 is associated with the splitting of $H_2O$ by an enzyme complex called Z into $\frac{1}{2}O_2$, $2\ e^-$, and $2\ H^+$. The released electrons restore the condition of P680, the $H^+$ ions travel with reduced NADP to effect the reduction of $CO_2$ in the dark reaction, and the $O_2$ is liberated. Because the entire route of noncyclic photophosphorylation can be diagrammed as a $Z$, this pathway has also been called the *Z scheme* in some texts.

**6.6**    The *action spectrum* is a graph that shows the efficiency of photosynthetic activity at the various wavelengths of light. An *absorption spectrum* is a graph that shows the degree to which various wavelengths of light are absorbed by a given material. Why does the action spectrum for photosynthesis (Fig. 6.7) differ from the absorption spectrum for chlorophyll $a$ if chlorophyll $a$ must be excited to initiate the photosynthetic process?

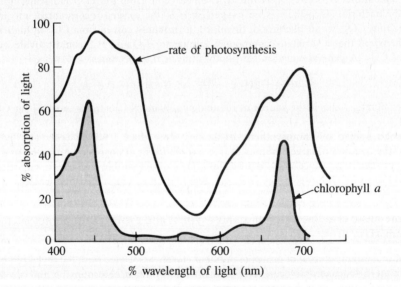

**Fig. 6.7** Action and absorption spectra

If chlorophyll $a$ were the only pigment involved in the capture of photons, the action spectrum should coincide with the actual absorption of light (*absorption spectrum*) by chlorophyll $a$. However, a variety of accessory pigments are present in each photosystem, and these extend the potential for carrying out photosynthesis even when chlorophyll $a$ has not absorbed light directly. Other chlorophylls, as well as the carotenoid pigments, may absorb light in accordance with their absorption spectra and pass the energy acquired along the pigments of the photosystem until it reaches chlorophyll $a$ in the reaction center. This is why even light in the yellow and green range can be used for photosynthesis, although it is much less effective than either red or blue light.

**6.7**   It has been estimated that of the total light that falls on a cornfield only 0.1 percent can be utilized for photosynthesis. Why is this?

Most of the light may not be usable because of geometric considerations related to the plant surfaces available. Of the light that does actually reach the plant, three alternatives exist: (1) The light may pass through the cells, especially in the case of algae for which cells may exist singly or in thin layers. This is called *transmitted light*; in solution chlorophyll *a* is green in transmitted light. This suggests that green light is not of a wavelength used in photosynthesis, since it is not absorbed. (2) Light may strike the surface and bounce back to yield *reflected light*. Such reflected light from a plant surface containing chloroplasts or from a chlorophyll *a* solution is also green. (3) Light may be absorbed, with the consequent boosting of electrons to higher energy levels.

The light absorbed by photosystems may be utilized for useful work by the plant if all other conditions are attuned to photosynthetic activity. If more light is captured by the plant's photosystems than can be handled by its photosynthetic pathways, some of the light energy will be lost as heat and the rest reemitted to the exterior at a longer wavelength (less energy). This is called *fluorescence*. Fluorescence in chlorophyll solutions generally results in the emission of red light.

**6.8**   In what ways do the reactions of the Calvin-Benson cycle resemble glycolysis?

Both of these pathways occur in the nonmembranous space of an organelle. In the case of glycolysis, the enzymes are soluble and are associated with the cell sap rather than assembled along a membranous structure. In the Calvin-Benson cycle the enzymes are also soluble and are found in the unstructured stroma of the chloroplast. In both processes an equilibrium between PGA and PGAL is a key to relative rates. Similar steps occur in each process but with a marked difference in terms of the direction of the reaction.

**6.9**   By the early 1940s a radioactive isotope of carbon ($^{14}C$) was made available to the research community by Samuel Rubin and Martin Kamen. How do you suppose Calvin and Benson were able to work out the steps of the dark reaction, using this research tool?

Calvin and his collaborators used the isotope to label the C of the $CO_2$ they were studying. Stopping the photosynthetic reaction at various times, they were able to trace the intermediates formed in the dark reaction cycle by following which metabolites contained the $^{14}C$. If they stopped the reaction in a matter of seconds after introducing the $^{14}CO_2$, only the early compounds in the cycle were labeled with $^{14}C$. If they waited for an appreciable interval, a greater variety of substances were found to be labeled. PGA was found to be an early intermediate, but the first stable product of the reaction, accumulating over time, was PGAL. Although many of the intermediate steps have been identified, the nature of the 6-carbon compound arising from the union of $CO_2$ and RuBP is still uncertain.

**6.10**   At the beginning of the present century, plant anatomists noted that a seed plant native to hot, dry climates, rather than exhibiting a palisade mesophyll, has a ring of mesophyll cells surrounding the bundle-sheath cells. The arrangement has the appearance of a crown—hence the German word *Kranz* for the structure (see Fig. 6.4). This ring of Kranz cells exhibits a high level of $C_4$ activity. What might the function of the Kranz structure be?

Under conditions of high temperature and the threat of water loss, the stomata of desert plants remain closed for long periods of time. As $CO_2$ is used up in photosynthesis, its levels will fall within the leaf and $O_2$ levels will rise. ($O_2$ is a waste product of photosynthesis.) This situation would encourage the wasteful diverting of RuBP into photorespiration, literally an unfixing of $CO_2$. But the mesophyll cells that wreath the bundle sheath are characterized by $C_4$ photosynthesis. They are able to fix $CO_2$ and form a $C_4$ intermediate. The $C_4$ intermediate formed in the mesophyll cells then passes into the bundle-sheath cells and releases its $CO_2$ to the Calvin-Benson cycle operating there. The Kranz anatomy maximizes contact between the mesophyll and bundle-sheath cells and therefore maximizes the pumping of $CO_2$ into the bundle-sheath cells so that photosynthesis may continue under the adverse conditions of a hot, dry environment. It is essentially a structural mechanism for sequestering the $CO_2$-capturing step from the carbohydrate-generating process to maintain photosynthetic efficiency.

Corn, which grows in a temperate climate but can trace its ancestry to desert plants, possesses the Kranz anatomy; so too, do sugar cane and that perennial lawn pest, crabgrass.

**6.11**    Compare the ultrastructure and function of the chloroplast with those of the mitochondrion.

In many respects the chemical events associated with the fine structure of the mitochondria are paralleled within the chloroplast. Both organelles have a double membrane; a membrane-linked electron-transport chain, in which are embedded such pigments as the cytochromes; and an elaborate chemiosmotic apparatus for building up an electrochemical gradient of $H^+$ ions between two compartments within the organelle. The subsequent movement of $H^+$ ions through an enzyme complex releases energy and results in the formation of ATP. In the chloroplast, however, the inner membrane is not folded and indented but tends to remain close to the outer membrane. A third membrane system instead provides a convoluted set of membranous vesicles—the thylakoid membrane system. It is across this membrane that the $H^+$ gradient is developed. The energy of this electrochemical gradient is used to generate ATP at a specific site of the complex called the *CF$_1$ region*. This is analogous to the $F_1$ region of the inner mitochondrial membrane, which is similarly involved in the generation of ATP from the flow of protons. It is felt that the chemiosmotic apparatuses in these two organelles have a common evolutionary origin.

# Supplementary Problems

**6.12**    What are organisms called that manufacture their food from simple chemical substances?

**6.13**    The "oxygen revolution" involved    (*a*) the formation of metal oxides from organic acids.  (*b*) the formation of free oxygen in the atmosphere as a result of photosynthesis.  (*c*) Lavoisier's death in the French revolution.  (*d*) the origin of the Krebs cycle.  (*e*) none of the above.

**6.14**    Oxygen in photosynthesis is evolved from the splitting of $CO_2$.
(*a*) True   (*b*) False

**6.15**    Light travels in discrete energy packets called *photons*.
(*a*) True   (*b*) False

**6.16**    The green color of chlorophyll indicates that green is the most efficient wavelength of light in terms of photosynthetic activity.
(*a*) Trúe   (*b*) False

**6.17**    Carbon 14 was used by van Niel to show that $H_2O$ was the source of the oxygen evolved during photosynthesis.
(*a*) True   (*b*) False

**6.18**    Chlorophyll *a* is the reaction center for photosystems I and II.
(*a*) True   (*b*) False

**6.19**    Only light that is absorbed can be used for photosynthesis.
(*a*) True   (*b*) False

**6.20**    Which is longer   (*a*) the wavelength of blue light or  (*b*) the wavelength of red light?

**6.21** Which is greater  (*a*) the energy of blue light or  (*b*) the energy of red light?

**6.22** Which is greater  (*a*) the molecular weight of chlorophyll *a* or  (*b*) the molecular weight of RuBP?

**6.23** Which has greater NADPH production  (*a*) cyclic photophosphorylation or  (*b*) noncyclic photophosphorylation?

**6.24** Which is greater  (*a*) the energy in the bonds of PGA or  (*b*) the energy in the bonds of PGAL?

**6.25** When chlorophyll is extracted from the leaf, it may still absorb light but will shortly reemit it at a lower energy level. This phenomenon, which also occurs to a limited extent in the intact leaf, is called (*a*) phosphorescence.  (*b*) fluorescence.  (*c*) red shift.  (*d*) photorespiration.  (*e*) none of these.

**6.26** P700 is  (*a*) found in photosystem I.  (*b*) a form of chlorophyll *a*.  (*c*) the reaction center for photosystem I.  (*d*) all of the above.  (*e*) none of these.

**6.27** Pigments such as ferredoxin and plastocyanin are unique to photophosphorylation, but pigments encountered in both photophosphorylation and the electron-transport chain of respiration are the (*a*) cytochromes.  (b) flavoproteins.  (*c*) coenzyme Q.  (*d*) both *a* and *b*.  (*e*) both *b* and *c*.

**6.28** Cyclic photophosphorylation is probably an older, more primitive pattern of photosynthesis. (*a*) True  (*b*) False

**6.29** Carbon fixation (union of $CO_2$ with an organic compound) occurs during the light reaction. (*a*) True  (*b*) False

**6.30** An electron hole created during noncyclic photophosphorylation is filled by an electron coming from photosystem II. (*a*) True  (*b*) False

**6.31** The ultimate source of electrons and $H^+$ for the reduction of $NADP^+$ in an initial phase of photosynthesis is the splitting of water. (*a*) True  (*b*) False

**6.32** All the oxygen liberated during photosynthesis derives from the water that is split during the light reaction. (*a*) True  (*b*) False

**6.33** The grana consist of stacks of thylakoid membranes. (*a*) True  (*b*) False

**6.34** A hydrogen-ion gradient between the thylakoid space and the stroma of the chloroplast provides the energy for ATP formation in photophosphorylation. (*a*) True  (*b*) False

**6.35** In the Calvin-Benson cycle, five of the six PGAL molecules formed by three turns of the cycle are used for (*a*) the formation of sugar.  (*b*) the formation of more RuBP.  (*c*) both *a* and *b*.  (*d*) none of these.

**6.36** Photorespiration involves the oxidation of  (*a*) PGA.  (*b*) RuBP.  (*c*) chlorophyll *a*.  (*d*) both *a* and *b*. (*e*) none of these.

# Answers

| | | | | | | | |
|---|---|---|---|---|---|---|---|
| **6.12** | Autotrophs | **6.19** | (a) | **6.25** | (b) | **6.31** | (a) |
| **6.13** | (b) | **6.20** | (b) | **6.26** | (d) | **6.32** | (a) |
| **6.14** | (b) | **6.21** | (a) | **6.27** | (d) | **6.33** | (a) |
| **6.15** | (a) | **6.22** | (a) | **6.28** | (a) | **6.34** | (a) |
| **6.16** | (b) | **6.23** | (b) | **6.29** | (b) | **6.35** | (b) |
| **6.17** | (b) | **6.24** | (b) | **6.30** | (a) | **6.36** | (b) |
| **6.18** | (a) | | | | | | |

# Chapter 7

# The Nature of the Gene

## 7.1 THE CONCEPT OF INFORMATION PROCESSING IN THE CELL

The complex activities of single cells, and of multicellular organisms as well, are controlled and guided by a set of blueprints (genes) locked in the chromosomes. Essentially the blueprints contain information for the synthesis of specific proteins. Since proteins make up both the basic cytoarchitecture and the enzymes that direct metabolism, the nature of the proteins made will determine the structural and functional characteristics of the cell or organism.

Modern *molecular biology* focuses on (1) how the information is arranged (encoded) in the chromosome, (2) how the information is processed, (3) how the blueprint replicates itself whenever the cell divides so that each cell may have a copy of the blueprint, and (4) how the information can be modified to provide new message material. Regulation and control of each of these information transactions constitute yet another aspect of molecular biology, a discipline that provides an understanding of life through an understanding of the roles played by macromolecules.

We now realize that the properties of a single macromolecule, DNA, can explain the entire agenda of information encoding, processing, replication, and modification (mutability). DNA is a blueprint that directs the destiny of the cell during its lifetime. It also is, in a certain sense, the material of immortality, since it is DNA that passes from one generation to the next to maintain genetic continuity between parent and progeny. In some viruses RNA, a close relative of DNA, takes on the central informational role.

Like a mold for a bronze casting, DNA creates a messenger molecule of complementary structure rather than perform the job of creating proteins directly. The necessity of using a messenger lies in the importance of DNA, the primary gene material, and the dangers that lurk in the cytoplasm where protein synthesis occurs. Although DNA is a relatively stable, long-enduring macromolecule, it can be degraded by enzymes called *DNAses* or altered by changes in its immediate environment. Just as a general does not enter the thick of battle during a war, so too, DNA (in eukaryotes) is sequestered behind the lines of the nuclear membrane, from which protected site it can produce one, none, or many copies of the messenger molecule that will effect the production of a specific protein.

The production of a messenger molecule from a DNA template is called *transcription*. The messenger molecule is a particular species of RNA called *messenger RNA* (*mRNA*). The mRNA will later join with the ribosomes of the cytoplasm and some other accessory molecules to synthesize a protein. This step is called *translation*. Thus, the specialized information or code for the assembling of amino acids into a protein is first transcribed from DNA onto an RNA "tape" and then translated into a protein. The DNA itself undergoes replication whenever a cell divides so that the information of one generation may be passed along to each member of the next generation.

In the process of encoding information, transmitting that information along until a protein is assembled, and replicating the blueprint, a high degree of fidelity is required. If the message becomes garbled at any of these steps, a lack of coherency in producing proteins will occur. However, some change in message is possible within the conservative nature of the information apparatus. Physical alteration in the substructure of DNA leads to changes in the coded sequence called *mutations*. These mutations, relatively rare events, provide new genetic sequences to be tested in the course of evolutionary modifications. Sex shuffles the deck; mutations add new cards to the old.

## 7.2 THE SEARCH FOR THE CHEMICAL BASES OF INHERITANCE

### IMPLICATING THE CHROMOSOMES AND DNA

At the same time that Gregor Mendel (see Chap. 9) was performing his genetics experiments on the garden pea (1865-1870), a young Swiss biochemist named Friedrich Miescher was isolating a hitherto unknown substance from the nuclei of pus cells and, later, fish eggs. He called this acidic material *nuclein*. Later it was named *nucleic acid* because of its properties and location. The sticky material that Miescher painstakingly extracted from the smelly, pus-soaked bandages he collected from hospital patients was actually DNA, the very material which comprised the genes of Mendel's crosses and which provides the basis for modern molecular biology.

In 1882, Walther Flemming stained the chromosomes of cells undergoing mitosis (see Chap. 8) and described the complex process of cell division by which a constancy of chromosome number is maintained. Later, Theodor Boveri described the meiotic process that results in a precise reduction of chromosome number from two sets (2n) to one set (n) in the formation of gametes (sperm/eggs).

Despite the seemingly obvious significance of the chromosomes in the life of the cell and the mechanisms of heredity, most biologists tended to dismiss their importance until 1918, when a series of experiments in sex linkage (see Chap. 9) conducted by C. B. Bridges showed conclusively that chromosomes are the physical basis of inheritance; i.e., they carry the *genes*, the units of heredity. Quite often in the sciences there is a lag of many years between a discovery and its recognition and acceptance by the general research community.

At the start of World War I, the German chemist Robert Feulgen developed a method for staining DNA and later showed that DNA is found virtually exclusively in the chromosomes. In the 1920s the chemical properties of DNA were worked out by P. A. Levene. He showed that the basic unit of this macromolecule is a nucleotide (see Fig. 7.1), a molecule consisting of a phosphate group, a 5–carbon sugar called a deoxyribose, and a single nitrogenous base (see Fig. 7.2) selected from a pool of two purines (adenine or guanine) and two pyrimidines (thymine or cytosine).

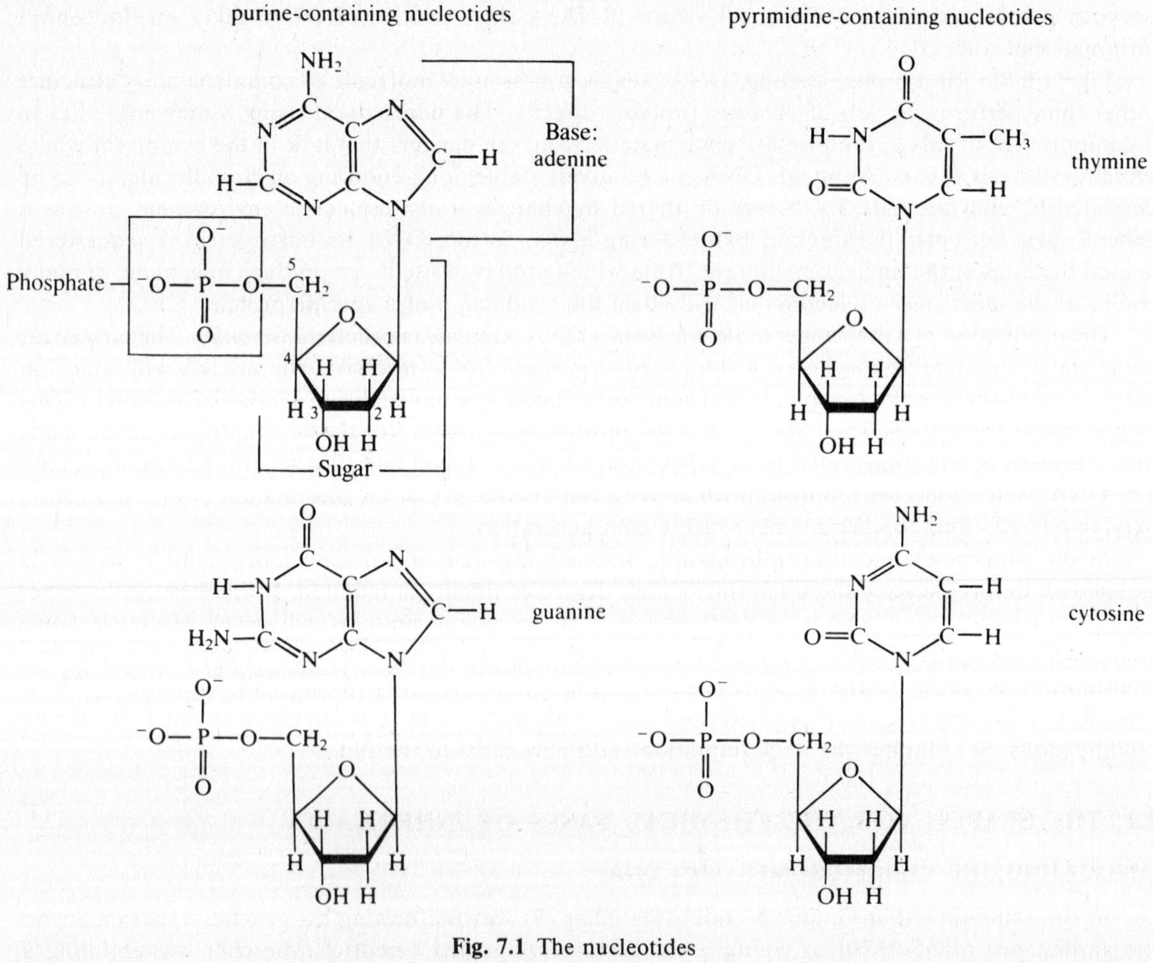

**Fig. 7.1** The nucleotides

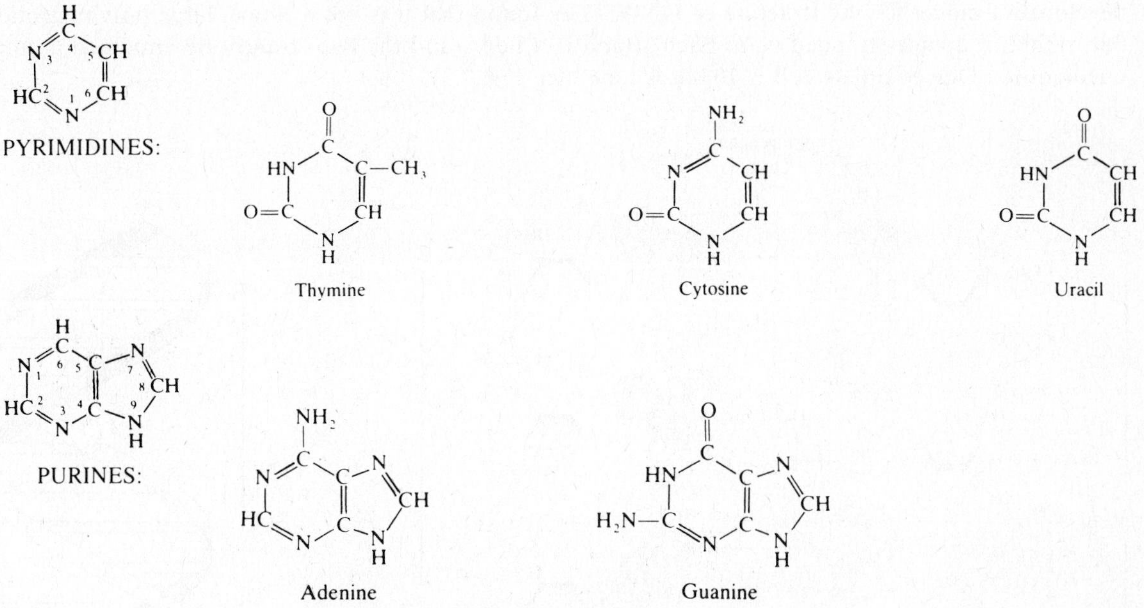

**Fig. 7.2** N-bases

Levene went on to propose that DNA was a *tetranucleotide*, that the macromolecule consisted of four nucleotides joined together. Each nucleotide of the tetranucleotide was thought to contain usually a different nitrogenous base, so that the four bases would then be present in approximately equal amounts in all DNA. Such a simple structure for the molecule led scientists to mistakenly conclude that DNA could not be the primary material for heredity and the control of cellular metabolism. A molecule possessing the properties of a genetic language would have to be capable of carrying a great many variations within its fundamental structure in order to account for the many bits of information required to regulate cell structure and function.

A greater complexity of the DNA was suggested in the 1940s by the elegant series of experiments carried out by Erwin Chargaff and his associates. Chargaff showed that the four common bases of DNA are not present in equal parts but instead demonstrate considerable variation. The repertoire of variation is consistent with the possibility that DNA is the blueprint material of the cell. Chargaff also showed that within the pattern of variability for the bases, one constant rule obtains: the amount of adenine (A) always equals the amount of thymine (T) and the amount of guanine (G) always equals the amount of cytosine (C). It follows from this that the total amount of purines (A + G) is equal to the total amount of pyrimidines (T + C). These regularities were particularly revealing when added to the finding that the amount of DNA in all somatic (nonsex) cells of an organism are always the same but that the gametes (cells of reproduction) have half the total found in somatic cells.

Evidence continued to accumulate implicating DNA as the fundamental hereditary material: pneumococci (bacteria that cause pneumonia) of one strain could be permanently altered (the process of *transformation*) by the DNA of a second strain; viruses were shown to multiply by capturing the protein-synthesizing machinery of a bacterial cell and subverting it to the needs of the viral DNA; and the sexual recombination of bacteria clearly involved the manipulation of DNA. What was needed was the establishment of the structure of DNA in order to explain in terms of molecular architecture how DNA could encode information, process this information, replicate itself, and even alter its own structure on occasion to produce new message units.

## DNA'S SECRET IS UNRAVELED: THE DOUBLE HELIX

Using the x-ray diffraction photographs of DNA by Rosalind Franklin and the findings of many other investigators, as well as their own brilliant interpretations of the available data, Watson and Crick developed a model for the structure of DNA. They found that it is not a single, large polynucleotide but, rather, a double-stranded twin. Each strand is a helix, and the two strands are entwined around one another. One complete coil is 10 bases long (see Fig. 7.3).

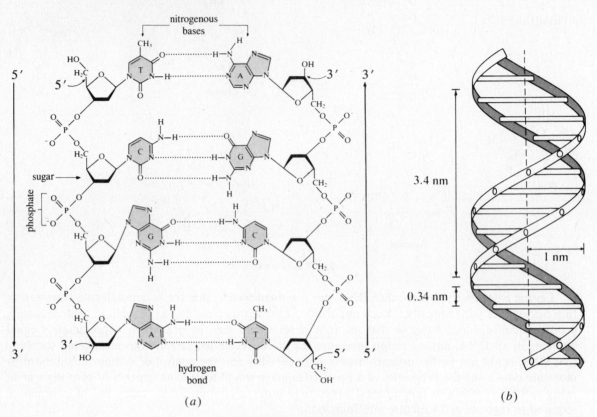

Fig. 7.3 Structure of DNA

**EXAMPLE 1** The easiest way to visualize the structure of DNA is to regard it as a twisted ladder. Each side of the ladder is made up of alternating sugars and phosphate groups. Jutting out of each sugar are the bases. A base from one strand is hydrogen bonded with a base from the adjoining strand. These pairs of bases, one from each strand, make up the rungs of the ladder and thus fasten one strand to another. DNA may exist in at least four geometric formations. The form described by Watson and Crick is *B-DNA* (Fig. 7.4*b*). A partially unwound form is known as *A-DNA*, while a tightly coiled, extremely compact configuration is known as *C-DNA*. *Z-DNA* is a form in which the helix is wound opposite in direction to that of B-DNA and the backbone of the molecule demonstrates a zigzag shape (Fig. 7.4*a*).

The hydrogen bonding of the bases (along with hydrophobic interactions between them) holds the double helix together. Although each H bond is relatively weak, in the aggregate, with so many H bonds, the attachment of the two strands is quite strong. Because of the respective sizes and chemical makeup of the bases, adenine is always cross-linked with thymine (by two hydrogen bonds) and cytosine is always linked with guanine (by three hydrogen bonds). Adenine and thymine are therefore said to be *complementary bases*, as are guanine and cytosine. This complementary relationship explains Chargaff's findings that in DNA, A = T and G = C.

Each sugar forming the backbone of the strand is joined by a phosphate molecule at carbon atom 5 (C-5) to the sugar above; a phosphate at C-3 joins it to the sugar below. One end of the strand, having a sugar with no sugar above it and thus a free or phosphorylated carbon at the C-5 position, is called the *5' end*; the other end, at which the terminal sugar has no sugar below it and thus has a free or phosphorylated carbon at C-3, is called the *3' end* (see Fig. 7.3). The prime (') is used in

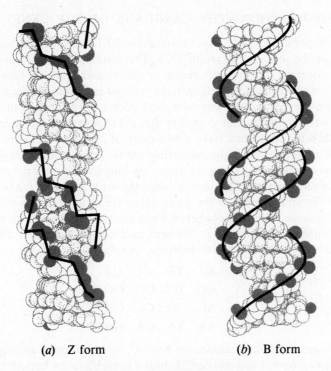

(a)  Z form                              (b)  B form

**Fig. 7.4** Alternate forms of DNA

numbering the carbons in the sugar to distinguish them from the carbons in the attached base. Each strand runs opposite in polarity to its complementary strand. The 5' end of one strand is at the same end as the 3' end of the associated strand. They are thus said to be *antiparallel*. This suggests that the strands may actually have been synthesized in opposite directions.

It should be noted that since a relatively large purine base will always be associated with a relatively small pyrimidine, the long DNA molecule will not bulge or constrict but will tend to maintain a uniform width throughout the double helix. Since the average cell contains about 2 m of DNA, there is a great necessity to compact the molecule within the nucleus. The fundamental packing unit is the *nucleosome* (Fig. 7.5), which contains both DNA and various types of a small basic protein called a *histone*. The nucleosome is a beadlike structure that considerably diminishes the extended length of the DNA.

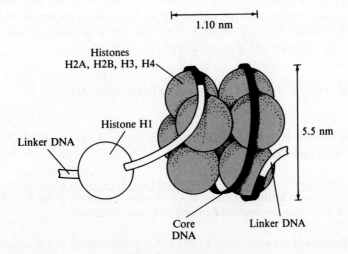

**Fig. 7.5** The nucleosome

## 7.3  ENCODING INFORMATION—THE LANGUAGE OF THE GENE

Watson and Crick published their hypothesized model for the structure of DNA in a one and one-half page paper in the journal *Nature* in 1953. This landmark event began the era of molecular biology. Shortly thereafter an active search began to understand precisely how the structure of DNA could account for information encoding, information processing (transcription and translation), replication of the information-carrying macromolecules, and occasional alteration of the message itself.

George Gamow was among the first to realize that if DNA carried the code for assembling amino acids into proteins, then the code must have a minimum of 20 different "words" (*codons*) to account for the arrangements of the 20 naturally occurring amino acids. Since phosphates and sugars are arranged in an unvarying alternating sequence along the long axis of each polynucleotide strand, these moieties of the nucleotide cannot represent a language with a broad range of structural options. Obviously, then, it is the arrangement of the bases along the strands that is the encoding mechanism. Could this language, or code, with its alphabet of bases consist of codons with one letter? No, because four different bases could provide only four different codons. Similarly, four different bases taken two at a time would yield only 16 different combinations, as follows:

$$AT \quad TT \quad CT \quad GT$$
$$AG \quad TG \quad CG \quad GG$$
$$AC \quad TC \quad CC \quad GC$$
$$AA \quad TA \quad CA \quad GA$$

If we take three bases as the unit codon, we have $4^3$, or 64, different arrangements possible, more than enough words to control the assembly of 20 amino acids. Gamow hypothesized the triplet codon, and later Crick demonstrated its operative validity. (To determine the 64 triplet codons add each of the four bases in turn to the 16 doublets shown above.)

The genetic code, consisting of a linear sequence of three bases, appears to be universal for all living forms, from bacteria to humans. It is also degenerate (redundant) in that a number of different codons may represent the same amino acid. For example, from Table 7.1, we can see that leucine has six codons: UUA, UUG, CUU, CUC, CUA, and CUG. Since there are many more codons than are necessary for the insertion of 20 amino acids into a protein chain, some of the codons may be used as punctuation, i.e., starting or stopping signals, during the synthesis of a protein.

Meaningful information about the synthesis of a protein can be contained in only one of the strands of the double helix. Any codon on the opposite strand in that region is the complement of its respective codon on the *sense strand*. For example, reading both strands from their respective 5′ ends, a triplet sequence of CTG on the sense strand would create the sequence CAG on the opposite strand. Even though the complementary sequences may well code for amino acids, and not gibberish, they will probably not be the same amino acids as coded for by the codons on the sense strand, and will thus place the wrong amino acids into the protein. However, for a different protein, the complementary strand may become the sense strand in another portion of the DNA.

## 7.4  PROCESSING THE INFORMATION—PROTEIN SYNTHESIS

### TRANSCRIPTION

The information contained in the sequence of bases in the sense strand of DNA is impressed upon a deputy molecule, *messenger RNA*.

**EXAMPLE 2**  *RNA* stands for *ribonucleic acid*. RNA resembles DNA, but is single-stranded and has ribose sugar rather than deoxyribose sugar (a hydroxyl instead of a hydrogen is attached at the 2′ carbon of the ribose sugar). It also differs from DNA in using the pyrimidine *uracil* in place of thymine.

Transfer of the information is accomplished by the DNA strand acting as a template for the arrangement of RNA bases with a sequence complementary to that of the DNA. The sequence begins with bases

**Table 7.1.  Codon–Amino Acid Assignments of the Genetic Code**

| First Position (5′) | Second Position | | | | Third Position (3′) |
|---|---|---|---|---|---|
| | U | C | A | G | |
| U | Phe | Ser | Tyr | Cys | U |
| | Phe | Ser | Tyr | Cys | C |
| | Leu | Ser | (CT) | (CT) | A |
| | Leu | Ser | (CT) | Trp | G |
| C | Leu | Pro | His | Arg | U |
| | Leu | Pro | His | Arg | C |
| | Leu | Pro | Gln | Arg | A |
| | Leu | Pro | Gln | Arg | G |
| A | Ile | Thr | Asn | Ser | U |
| | Ile | Thr | Asn | Ser | C |
| | Ile | Thr | Lys | Arg | A |
| | Met(CI) | Thr | Lys | Arg | G |
| G | Val | Ala | Asp | Glv | U |
| | Val | Ala | Asp | Gly | C |
| | Val | Ala | Glu | Gly | A |
| | Val(CI) | Ala | Glu | Gly | G |

The codons shown are for mRNA, so uracil (U) replaces thymine. C = cytosine; A = adenine; G = guanine. The amino acids encoded for are: Ala = alanine; Arg = arginine; Asn = asparagine; Asp = aspartate; Cys = cysteine; Gln = glutamine; Glu = glutamate; Gly = glycine; His = histidine; Ile = isoleucine; Leu = leucine; Lys = lysine; Met = methionine; Phe = phenylalanine; Pro = proline; Ser = serine; Thr = threonine; Trp = tryptophan; Tyr = tyrosine; Val = valine. CI = chain initiation; CT = chain termination.

that will later permit the mRNA to attach to a ribosome. The balance of the relatively long mRNA molecule codes for the amino acid sequence of the protein that will be constructed during the translation process and ends with a codon that will signal termination of the synthesis of the protein.

*Unwinding* proteins begin transcription by breaking the H bonds between complementary bases, so that the DNA opens, in sporadic fashion, along its length. The enzyme *RNA polymerase* attaches to the *promoter* region of DNA. Synthesis begins at this 3′ position of DNA with the creation of the 5′ end of RNA (an antiparallel arrangement).

Each base of the exposed DNA template attracts a complementary base from a pool of free bases in the nuclear sap. These free bases exist as the nucleotide triphosphates ATP, GTP, UTP, and CTP (adenosine, guanosine, uridine, and cytidine triphosphate) and so contain a great deal of energy, which facilitates their assembly (polymerization) into a polynucleotide. The nitrogenous base of each nucleotide hydrogen bonds (according to the base pairing rule) to its complementary base on the DNA sense strand. It is the RNA polymerase that catalyzes the attachment of the nucleotides to each other to form a strand. Thus, complementary RNA codons represented by a sequence of free bases are first built upon a DNA template, and then these bases are joined into a *single*-stranded RNA molecule. A *termination* signal marks the 3′ end of the finished mRNA molecule, which is released and soon migrates out of the nucleus. Note that the direction of RNA synthesis is 5′→3′.

Depending on the number of protein molecules to be manufactured, many copies of mRNA will be formed. Some mRNAs will be long-lived, while others will have only a transitory existence and produce only one or a few proteins.

The transcription process in eukaryotes is complicated by the existence of long sequences of bases in DNA (*introns*) that do not contain meaningful information for protein synthesis. During transcription, both the introns and the DNA sections that *are* subsequently translated (*exons*) are transcribed into the mRNA. In the nucleus, this slurry of primary transcripts is referred to as *heterogeneous nuclear RNA*, or *hnRNA*. Before the mRNA leaves the nucleus, the sections representing the introns are excised and the remaining sections are annealed; this yields functional mRNA, which migrates out of the nucleus for the next stage of protein synthesis, translation.

## TRANSLATION

The next step in the processing of the information is the actual production of protein, the *translation* of the code. Translation requires the integrated functioning of ribosome, mRNA, and a small species of RNA called *transfer RNA* (*tRNA*), which is responsible for carrying amino acids to the mRNA for incorporation into the protein chain. One end of each tRNA binds to a specific amino acid; the other end contains an *anticodon* (Fig. 7.6), a base triplet sequence that is complementary to one of the mRNA codons for the amino acid at the other end. Thus, there are at least as many tRNAs as there are codons for amino acids.

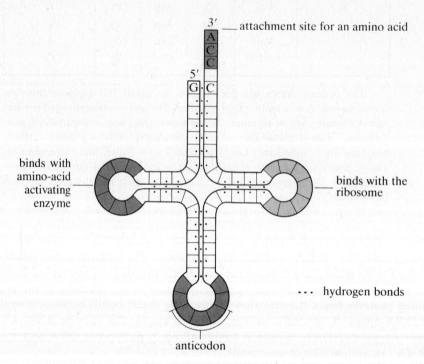

**Fig. 7.6** Transfer RNA

The process of translation (Fig. 7.7) begins with the migration of mRNA to the cytoplasm, where the 5′ end binds to the smaller (30 S) of the two subunits of a ribosome. (*S* stands for *Svedberg unit*, a specific measure of sedimentation in a gravitational field.) A tRNA with the appropriate anticodon is attracted to the starting position of the mRNA; at the same time the tRNA attaches to the first of two binding sites on the larger (50 S) ribosomal subunit. When the second binding site is occupied by a second tRNA–amino acid complex, the two respective amino acids are joined with a peptide bond produced by an enzyme (peptidyltransferase) located in the 50 S subunit. The first tRNA molecule is then freed as the ribosome shifts over to the next codon. This brings the second tRNA–amino acid

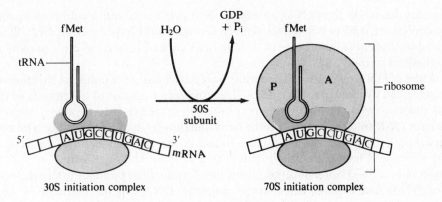

Initiation of polypeptide synthesis.

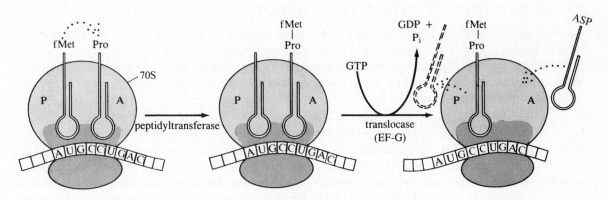

Elongation of polypeptide synthesis.

**Fig. 7.7** Translation

complex into the first (P) site on the larger subunit and permits yet another tRNA to bring a third amino acid into position. Each time the ribosome shifts to a new codon on the mRNA, the second (A) position is made available to, and filled by, the tRNA–amino acid complex appropriate to the codon being read. The new amino acid is joined to the polypeptide chain, and the process is repeated. When the ribosome reaches a termination signal, it releases the polypeptide extending from its 50 S subunit, which may then move through the channels of the endoplasmic reticulum.

Several ribosomes may be simultaneously translating a single mRNA, so that several copies of a protein are in various stages of production along a single strand of mRNA. Such a configuration of a single mRNA with a number of ribosomes is known as a *polysome*.

The fundamental dogma of molecular biology is summed up as DNA → RNA → protein. It assumes the primacy of DNA as initiating the entire process and being free of influence from either RNA or protein. This one-way flow of information was accepted until the discovery of *reverse transcriptase*, an enzyme associated with certain RNA viruses (*retroviruses*). This enzyme catalyzes the formation of DNA on an RNA template and thus reverses the usual flow of information. Although a special case, the situation in the retroviruses challenged the universality of the fundamental dogma of Watson and Crick.

## 7.5  REPRODUCTION OF INFORMATION—DNA REPLICATION

Watson and Crick recognized that each strand of DNA could serve as a template for a complementary strand. If the double helix could unwind and separate, the individual strands could attract their complementary bases (as in mRNA synthesis); each original strand would thus again be associated with its complement, and two identical double helices would have been created. This is considered *semiconservative replication* because each double helix created consists of one "parent" strand and one newly synthesized strand.

This model of DNA replication is generally accepted today. Unwinding of the double helix appears to be accomplished by a number of proteins. *Topoisomerases* cut one of the strands so that it can begin to unwind and also serve to relieve pressures in the coil caused by its unwinding. *Single-stranded DNA binding protein* (*SSB protein*) destabilizes the helix and thus facilitates unwinding. The actual unwinding is accomplished by *helicases*, which can be found at the replication fork (the fork created by the unwinding strands). Chain growth is initiated by a *primase* and is extended by a *DNA polymerase*. Growth occurs in a $5' \rightarrow 3'$ direction (i.e., from the $3'$ end of the template). *Exonucleases* can remove sections of DNA, and abutting segments of unjoined DNA can be annealed by *ligases*. Although variations exist, this general pattern seems to hold for both prokaryotes and eukaryotes.

## 7.6 MODIFICATION OF INFORMATION—MUTATION

The message in DNA is read in terms of the linear array of bases in one direction along the sense strand. Each word or codon consists of a triplet of bases. A change in the sequence of triplet codons is termed a *mutation*. One kind of mutation is caused by the addition of an extra base. This insertion of an extra base drops the reading frame back one letter (base) from the point of insertion and thus changes all the codons subsequent to it.

$$\boxed{abc}\ \boxed{def}\ \boxed{ghi}\ \boxed{jkl} \xrightarrow{\text{addition}} \boxed{xab}\ \boxed{cde}\ \boxed{fgh}\ \boxed{ijk}\ l$$

If the insertion occurs at the beginning of the gene message, it will probably halt production of the protein. This is the kind of mutation that may underlie a number of genetic diseases in humans in which an enzyme is missing (Tay-Sachs disease, galactosemia). Should the insertion occur toward the end of the gene "tape," then only one or several terminal amino acids would be affected and a functional, but slightly altered, protein would be produced.

A similar shift in the reading frame would occur if a base were deleted from the gene. In this case the reading frame would be advanced one letter, and all the subsequent codons would be changed.

$$\boxed{abc}\ \boxed{def}\ \boxed{ghi}\ \boxed{jkl} \xrightarrow{\text{deletion}} \boxed{bcd}\ \boxed{efg}\ \boxed{hij}\ kl$$

Here, too, the strong likelihood would be that no functional protein would be produced.

Less likely to eradicate the synthesis of a protein would be a mutation called *substitution*, in which one base is substituted for another. In such an event one amino acid is substituted for another, because only a single codon is changed. If the new amino acid is similar in its properties to the original one, no damage would be expected to result. In humans, the blood disease sickle-cell anemia is caused by the substitution of valine for the usual glutamine. Although only one base substitution and one amino acid change occur, the properties of valine (hydrophobic) are sufficiently different from those of the polar amino acid glutamine that a hemoglobin is produced that cannot adequately carry out its oxygen-transport function.

Recently, the phenomenon of transposition (jumping genes) has become better understood as a source of genetic variation. In this situation relatively long stretches of DNA jump from one chromosome to another in some still unexplained manner.

**EXAMPLE 3**  In 1983, Barbara McClintock was awarded the Nobel Prize for Physiology or Medicine for having demonstrated the phenomenon of transposition. This involves the wholesale movement of genetic material from one area of a chromosome to a completely different area within the family of chromosomes. This notion that genes could jump from site to site was actually ridiculed when first introduced by McClintock in the late 1940s to explain unusual color patterns in corn. At present we recognize that the *transposon* is probably one or several genes that not only can jump from one location to another but also may bring about larger changes in other parts of the chromosome, such as gene duplication. Transposons are generally transferred as circular strips of DNA and are capable of producing the enzyme that enables them to intrude within the new chromosomal site. In a later chapter we will discuss alterations in the genetic message ascribed to changes in the chromosome.

The agents that cause mutations are known as *mutagens*. Among the most potent of mutagens are various chemicals and ionizing radiation, such as x-rays and cosmic rays; alpha, beta, and gamma rays, which are ionizing radiations produced by a variety of radioactive elements; and ultraviolet light, which is a high-energy radiation, although not energetic enough to strip electrons from an atom.

## 7.7  GENETIC ENGINEERING

*Genetic engineering* refers to a broad group of procedures by which the machinery of genetic information is intentionally altered. The biologist becomes an engineer and reconstructs the DNA molecule or the whole *genome* (entire set of genes) within the nucleus for purposes of ameliorating specific genetic diseases or gaining a better understanding of the genetic apparatus.

*DNA recombinant* procedures (*gene splicing*) are the best-known examples of genetic engineering. DNA from a foreign organism, usually an entirely different species, is introduced to and integrated with an existing genome. A new, hybrid genome is obtained with characteristics of the donor organism reflected in the recipient.

**EXAMPLE 4**  The introduction of the gene for synthesizing human insulin into bacteria illustrates this procedure. These bacteria are, in a certain sense, part human because they are capable of producing a human protein. Since bacteria divide every 20 min or so, many billions of bacteria with the human gene can be cultured and the insulin they produce harvested. Individual genes for human growth hormone and interferon (an animal protein that inhibits viruses) have also been introduced into bacteria.

For recombinant DNA studies the major tools are restriction enzymes (endonucleases), plasmids, and viruses. *Restriction enzymes* were isolated at the same time that the retroviruses were discovered. They act like a pair of scissors to permit the DNA to be cut at precise regions. A *plasmid* is a small circular piece of DNA lying outside the chromosome in bacteria and some yeast. It may contain a single gene, a few genes, or many genes. In many bacteria the plasmid contains genes that confer resistance to many kinds of antibiotics.

In one recombinant technique, foreign DNA is incubated with plasmids that have been opened by the restriction enzymes. These enzymes create in the plasmids "sticky ends" that readily permit a reassembling of DNA strands. The foreign DNA is incorporated into the plasmids, which then close and are taken up by recipient bacteria. When the bacteria divide, the plasmid replicates and passes on to each daughter cell.

Another technique for bringing foreign DNA into a bacterium is to use viral particles as a vector. Viral DNA is incubated with foreign DNA fragments and incorporates this foreign DNA into its genome. The virus then invades a bacterial cell and integrates its total genome, foreign and viral DNA, into the bacterial chromosome. This technique is essentially *transduction*, a phenomenon described in 1952 and used to demonstrate the blueprint role of DNA within the cell.

Genetic engineering involving the whole genome has been carried out in the melding (union) of nuclei from different species. Usually the chromosomes of one species remain functional, while those of the other species tend to break down. However, some chromosomes of the second species may continue to function within the combined nucleus. These experiments are most useful for delineating the specific functions of the genes on the few chromosomes of the second species that continue to function.

*Cloning* is a technique in which many copies of a single gene, chromosome, or whole individual may be produced. *Clone* comes from a Greek root meaning "twig." Nonreproductive tissues are used in cloning entire individuals; thus sexual recombination is not involved. Cloning in carrots was carried out by Fred Steward in the late 1950s. He used fully differentiated cells from the vascular tissue of the plant, which are ordinarily not capable of producing a new organism. By manipulating the medium in which these cells were grown, Steward's group "fooled" these mature cells into returning to the embryonic condition in which they could produce all the components of a new carrot. In the case of vertebrates, the cloning of frogs has been achieved, but only by placing nuclei of mature cells into eggs from which the nucleus had been removed. The clone develops into a frog with all the characteristics of the organism from which the transplanted nucleus was taken. The only purpose of the egg cytoplasm seems to be to provide a hospitable environment for growth and development. Mammals have not been cloned, although unverified claims have been made for the cloning of mice. In mammals, not only must a clone be produced by placing a nucleus from a mature cell into an egg "casing," but that future embryo must be placed within a uterus to assure successful development.

# Solved Problems

**7.1**    What is a gene?

A *gene* is a unit of information that directs the activity of the cell or organism during its lifetime. It passes its message along to the progeny when the cell or organism divides or reproduces, so that the gene is also a unit of inheritance. In the classic experiments of George Beadle and Edward Tatum (1940) the gene was shown to carry the information required to make a single enzyme or, as the hypothesis was later modified, a single protein. Actually a single gene carries the information for the synthesis of a single polypeptide chain, whereas many proteins consist of a number of polypeptide chains (quaternary structure).

At one time, the genetic unit for creating a functional protein (the cistron) was believed to be different from the length of DNA required to produce a mutation (muton) or to effect a recombination of genetic alignment along the chromosome (recon). This separation of various gene functions, however, has not been seen in the recent literature. We may view the gene as a length of DNA that codes for a specific product. Usually this product is a polypeptide chain, but it may also be a type of RNA, such as tRNA or ribosomal RNA (rRNA). A given gene may exist in one of several alternative forms, or *alleles* (see Chap. 9).

**7.2**    What was the evidence that led to the acceptance of DNA as the primary hereditary material?

The eukaryotic chromosome, long recognized as the bearer of life's blueprint, was known to be rich in both DNA and protein. Either of these molecules could have been the primary blueprint material. But in 1928 Fred Griffith showed that dead, virulent pneumococci (containing a capsule), when injected into mice, could transform live, nonvirulent pneumococci (without a capsule) into live, virulent bacteria with a capsule. Since the capacity for virulence, associated with the capsule, was governed by hereditary factors, Griffith had demonstrated the possibility of effecting hereditary changes in live bacteria by presenting them with some substance found in dead bacteria. Later, Dawson repeated these experiments in vitro. In 1943 Oswald Avery and his collaborators carefully extracted the transforming principle (substance) from the bacteria and found it to be DNA. Thus it was DNA that could literally reach out from the grave and produce a permanent hereditary change in a living organism. A variety of traits were soon found to be susceptible to transformation in bacteria; this lent strong support to the concept that DNA was the significant information molecule.

In 1952, Alfred Hershey and Martha Chase showed with radioactive labeling that when protein and DNA of an infectious virus (bacteriophage) were administered to host bacteria, it was only the DNA of the virus that got into the cell and brought about the production of new viral material. Since the protein of the virus did not even enter the bacterial cell, it was concluded that DNA was the genetic material of the bacteriophage.

Recombination of genetic traits occurs in bacteria, an organism once thought to be free of sexual exchanges. Careful study of the recombination process showed that it involved the creation of a protoplasmic bridge (*pilus*) between two or more bacteria along which the opened chromosomal ring of one bacterium could be transferred into a recipient bacterium. This *conjugation* process, clearly seen in electron micrographs, would, during replication, result in the creation of new chromosomes containing genetic material from two sources. Since the chromosomes of bacteria are naked lengths of DNA, the sexual recombination of bacteria further strengthened the concept of DNA as the primary hereditary material.

Another line of evidence pointing to DNA as the primary hereditary material comes from *transduction*. This phenomenon, similar in some ways to transformation, involves the exchange of genetic information between two bacterial species with a virus acting as a vector to effect the exchange. Viral DNA enters one bacterium, where it subverts the bacterial information processing system to produce new viral material. In the process some of the DNA of the bacterial cell is integrated with the viral genome. When the virus leaves the first cell and infects another cell, it brings the DNA from the first cell along with it to mix with the DNA of the second cell. At the same time, hereditary characteristics of the first cell appear in the second one. Since it is only DNA that passes into the bacterium from the virus, this transfer of genetic information from one bacterial cell to the next implicates DNA as the primary blueprint material. Transduction represents another technique for genetic engineering. A virus, treated with foreign gene material, provides a probe for bringing genes into a recipient cell.

**7.3**  What are introns, and how do they affect the processing of genetic information?

Introns, a feature of eukaryotic DNA, are intervening stretches of DNA lying between the exons. The exons represent message material that actually gets translated into protein, whereas the introns must eventually be excised from the final mRNA product. Initially, both introns and exons are coded into an mRNA transcript. The existence of introns and exons was demonstrated when it was discovered that only about one-third of the primary transcript of DNA was used to make protein. In a highly complicated process the intron regions are removed from the rough primary transcript on the DNA template to produce a secondary, mature mRNA that codes only for the exons. It is now clear that initiator and terminator signals must mark the beginning and end of each intron within a gene.

Despite the fact that the intron is not directly involved in shaping the final protein product, it appears to be essential to the synthesis of a functional mRNA. The process of excising the intron regions involves the participation of RNA as an enzyme. This catalytic RNA is part of a protein-RNA particle known as the *small nuclear ribonucleoprotein particle*.

**7.4**  If a DNA codon, reading from the 5' end, is C-A-T, what will the base sequence be, reading from the 5' end, of the corresponding anticodon? (Hint: tRNA alignment with mRNA is antiparallel.)

Since the 3' end of mRNA aligns with the 5' end of DNA in transcription, the complementary base sequence of the mRNA, reading from the mRNA's 3' end, will be G-U-A. Note that uracil is substituted for thymine in RNA. Since the 5' end of tRNA aligns with the 3' end of mRNA, the tRNA base sequence, from the tRNA's 5' end, will be C-A-U (see Fig. 7.8). Except for the substitution of uracil, anticodons are the same as the original DNA triplets they are translating.

**7.5**  The 50 S ribosomal subunit has two sites that receive tRNA. Describe how they relate to each other.

The first site, designated the *peptidyl site*, or *P* site, receives the initial tRNA–amino acid complex. A second tRNA–amino acid complex moves into the second site, designated the *aminoacyl site*, or *A site*. An enzyme, *peptidyl transferase*, catalyzes the formation of a peptide bond between the two amino acids, and the tRNA in the P site is released from the resulting dipeptide and leaves. The ribosome then moves to the next codon on the mRNA and thereby brings the second tRNA and its attached dipeptide from the A site into the P site. The A site is now free to receive a third tRNA–amino acid complex, and the elongation process can be repeated.

**7.6**  Would you expect the polypeptide chains of a polysome to be longer at the 5' or the 3' end of the mRNA?

Translation begins at the 5' end of the mRNA molecule. As the ribosome moves along the mRNA toward the 3' end, it adds amino acids to the protein chain. Therefore, the farther the ribosome has progressed toward the 3' end, the more amino acids it has added to the protein and the longer its protein will be.

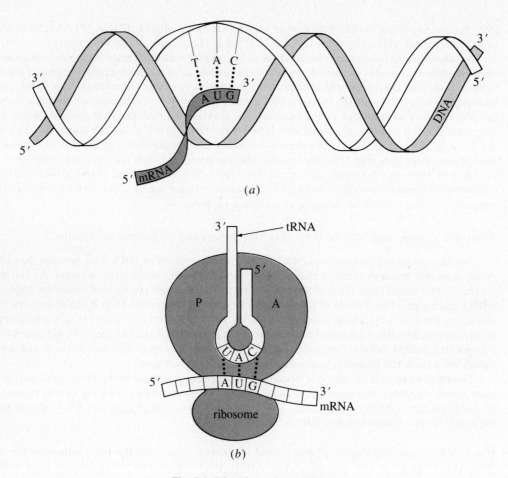

Fig. 7.8 Direction of translation

**7.7**    In eukaryotic cells does the type of protein being produced determine where it is produced?

Yes, there are spatially separate processes involved in the synthesis of different classes of protein. Those proteins that are destined to remain in the cytoplasmic compartment of the cell are translated in the cell sap through the association of mRNA and free ribosomes. However those proteins that will be incorporated into ER or secreted from the cell are translated on the RER. In the latter case, the mRNA involved begins the translation process by associating with the two ribosome subunits necessary for translation. Translation then pauses until the mRNA and its associated ribosome bind to the RER. At this point translation resumes, and the polypeptide chain emanating from the ribosome is extruded into the channel of the endoplasmic reticulum, where it passes through the smooth ER and the Golgi apparatus and on out of the cell. If designed for the ER itself, the protein will be incorporated directly into the RER channels.

**7.8**    Discuss the different kinds of RNA present in the cell.

There are more than three types of RNA present, and each kind is coded for by a specific gene in the chromosome. All three types are usually single-stranded, but in some cases the strand may fold upon itself to produce a local region of double-strandedness.

One type of RNA is found in the ribosome (rRNA). It is produced on the chromosome, stored in the nucleolus, and eventually becomes part of the ribosome. A second type, usually extremely long, is the mRNA that represents the product of transcription. These transcripts vary in their usable life in accordance with the number of copies of a given protein that are to be made. Transfer RNA (tRNA) represents a third type of RNA; it is also coded for by the DNA of the nucleus. There must be at least 20 different species of tRNA, corresponding to each of the 20 different amino acids. Actually there are as many as 40 specific tRNAs. Each transfer RNA is a single strand but folded back on itself to produce a cloverleaf. Four loops are produced, and the free tip of the second loop contains a triplet known as the anticodon, which binds to a specific codon on the mRNA. At the 3' end of the molecule a terminal CCA codon is bound to the activated amino acid that is carried by the tRNA.

A number of specialized types of RNA are also found in the nucleus, such as those components that play a role in the removal of introns. Some of the RNA produced from DNA in organelles (mitochondria, chloroplasts) differs slightly in coding sequences from the RNA produced by nuclear DNA.

**7.9**    Starting with activated nucleotides, a segment of DNA, and cell-free extracts containing requisite enzymes, Arthur Kornberg was able to effect the in vitro synthesis of new DNA, identical with the original segment; this supports the belief that DNA acts as a template in its own replication. However, these experiments did not elucidate whether replication was semiconservative or *conservative* (i.e., kept the original DNA intact, not hybridized with the new, but somehow effected the synthesis of an identical double helix). The semiconservative nature of DNA replication was later demonstrated by Matthew Meselson and Franklin Stahl, using two isotopes of nitrogen. How would you guess this was done?

Meselson and Stahl used two isotopes of nitrogen, $^{14}$N and $^{15}$N, to specifically delineate the source of each chain of DNA during periods of replication. When chains containing bases that were derived from $^{15}$N (heavy chains) were incubated in a medium in which only $^{14}$N was available for nucleotide synthesis, a single replication cycle produced DNA of intermediate chain weight. All the DNA, on rigorous analysis, was found to contain one heavy chain (derived from a parent $^{15}$N strand) and one light chain (newly built from $^{14}$N). This clearly showed that DNA replication occurs in semiconservative fashion.

**7.10**    Construct a flowchart showing the generalized events in DNA replication and the enzymes and other proteins involved in bringing these about.

Double helix $\xrightarrow{\text{topoisomerase}}$ Nicking of a strand $\xrightarrow{\text{helicases}}$ Unwinding $\xrightarrow[\text{topoisomerase}]{\text{SSB protein}}$ Destabilization and

relief of pressure $\xrightarrow{\text{primase}}$ Initiation $\xrightarrow{\text{DNA polymerase}}$ Elongation $\xrightarrow{\text{DNA ligase}}$ Closing of nicks

**7.11**    Because DNA replication occurs only in the 5'→3' direction of the *new* DNA strand and because the two strands of the double helix are antiparallel, uninterrupted replication would seem possible only on one of the template strands (the 3'→5' template strand). How do you suppose replication occurs on the 5'→3' template strand?

Fork movement (opening of the helix) for the second template strand is in the 5'→3' direction, and therefore the polymerases, which can elongate a growing strand in the 5'→3' direction only and so must always move in a 3'→5' direction on a template, must replicate DNA in a direction opposite to fork movement. Replication cannot be continuous on this strand (the *lagging* strand), since the polymerases must wait for new stretches on the template to be exposed by the unwinding helix and then work backward from the fork. The result is a series of unconnected segments (*Okazaki fragments*) that eventually are joined to form an uninterrupted strand.

**7.12**    Replication is far slower in eukaryotes (40–50 base pairs per second) than in prokaryotes (500 base pairs per second), yet the eukaryotic genome is much more complex. What mechanism would you guess has evolved in eukaryotes to compensate for their slow rate of replication?

Replication in prokaryotes tends to proceed from a single origin, or *initiation* point. Eukaryotes begin replication at *multiple* initiation sites; thus, though the rate of replication at each *replicon* (replication segment) is slower, the replicons are much smaller and so replication takes less time overall.

**7.13**   How might chemical mutagens and ionizing radiation lead to mutations?

The chemical mutagens may mimic the structure of the naturally occurring bases and become incorporated into the DNA strand where they block transcription. In some cases a chemical mutagen may act on the bases to convert one into another and thereby create a substitution. The mutagenic capability of a substance may be related to its carcinogenicity, since the induction of cancerous properties into the cell may involve an alteration in the regular information processes of the cell.

The basis for the action of ionizing radiation is, in all probability, the breaking of DNA strands. Secondarily this may lead to additions, deletions, or base substitutions. In the case of ultraviolet radiation, the mechanism for mutation seems to be the creation of bonds between two adjacent thymine bases along the same strand. These thymine bases can no longer bond to their complementary bases (adenine) on the adjacent strand; this effectively blocks both transcription and replication at the site of dimer formation (T-T).

**7.14**   Although *cloning* is most often associated with complete organisms, at the macromolecular level specific genes can be cloned. Suggest one possible way, using reverse transcriptase, to produce multiple copies of a specific gene.

Reverse transcriptase catalyzes the formation of DNA on an RNA template. In a cell actively producing the protein coded for by the desired gene, the corresponding mRNA can be isolated. This mRNA can be used as a template with reverse transcriptase to fashion a DNA molecule containing the desired gene. This molecule can then be inserted into a plasmid and used to make multiple copies of the gene within a bacterial culture that has absorbed the plasmid as the cells divide.

**7.15**   The development of genetic engineering has raised various concerns, moral and scientific. What might some of these be?

In many bacteria, the plasmids used in recombinant studies contain the gene conferring resistance to antibiotics. It was feared, during early attempts to effect DNA transplantation, that bacteria would be created that would be resistant to all known antibiotics. The possibility of an uncontrolled plague that would sweep through the human population was only one of many concerns that arose from consideration of the consequences of tinkering with the genetic apparatus.

Cloning also presents scientific and ethical questions. Since the cloning technique does not involve the usual sexual process for recombining genetic material, it short-circuits a mechanism for providing variability, which in turn deprives the organism of adaptive potential and thus leaves it more vulnerable to selection pressures. Questions also have arisen about the legal status of a possible human clone, since it would be a genetic replica of one individual but would be born to a woman who has provided a foster womb without making a genetic contribution to the clone. Questions exist about who the parents of the clone would be, what family relationships the cloned person would have, etc. These considerations involve the legal and religious communities as well as biologists. In some countries, special panels have been set up to deal with the ethical and biological issues involved in genetic engineering.

# Supplementary Problems

**7.16**   In 1964, Marshall Nirenberg broke the genetic code by showing that UUU is the mRNA codon for the amino acid   (*a*) glycine.  (*b*) tryptophan.  (*c*) valine.  (*d*) threonine.  (*e*) none of these.

**7.17**   The virulence of pneumococcus is associated with its   (*a*) chromosome.  (*b*) ribosome.  (*c*) capsule.  (*d*) lysosomes.  (*e*) none of these.

**7.18**   DNA was first discovered by   (*a*) Feulgen.  (*b*) Griffith.  (*c*) Mendel.  (*d*) Miescher.  (*e*) none of these.

**7.19**    The eukaryotic chromosome contains   (*a*) DNA.  (*b*) protein.  (*c*) RNA.  (*d*) nucleosome spools.  (*e*) all of these.

**7.20**    There are three hydrogen bonds between complementary adenine and thymine bases.
(*a*) True  (*b*) False

**7.21**    One complete coil of the DNA helix is 10 bases long.
(*a*) True  (*b*) False

**7.22**    A significant piece of data influencing Watson and Crick was the x-ray diffraction photographs of B-DNA taken by Rosalind Franklin.
(*a*) True  (*b*) False

**7.23**    The two helical strands of DNA are held together by covalent bonds.
(*a*) True  (*b*) False

**7.24**    DNA polymerase is involved in the formation of mRNA.
(*a*) True  (*b*) False

**7.25**    Meselson and Stahl showed clearly that DNA replication was semiconservative.
(*a*) True  (*b*) False

**7.26**    The promoter region plays a role in transcription.
(*a*) True  (*b*) False

**7.27**    The primary transcript is much longer than mature mRNA because of the presence in the former of introns.
(*a*) True  (*b*) False

**7.28**    Translation of the genetic code takes place on the ribosome.
(*a*) True  (*b*) False

**7.29**    Introns are found in the circular chromosomes of prokaryotes.
(*a*) True  (*b*) False

**7.30**    Transfer RNA is double-stranded throughout its length.
(*a*) True  (*b*) False

**7.31**    Which is longer  (*a*) tRNA or  (*b*) mRNA?

**7.32**    Which is more stable  (*a*) DNA or  (*b*) mRNA?

**7.33**    Which are larger  (*a*) the purines or  (*b*) the pyrimidines?

**7.34**    Which is more complex  (*a*) the human chromosome or  (*b*) the bacterial chromosome?

**7.35**    Which weighs more  (*a*) ribose or  (*b*) deoxyribose?

# Answers

| | | | | | | | | |
|---|---|---|---|---|---|---|---|
| **7.16** | (*e*) | **7.21** | (*a*) | **7.26** | (*a*) | **7.31** | (*b*) |
| **7.17** | (*c*) | **7.22** | (*a*) | **7.27** | (*a*) | **7.32** | (*a*) |
| **7.18** | (*d*) | **7.23** | (*b*) | **7.28** | (*a*) | **7.33** | (*a*) |
| **7.19** | (*e*) | **7.24** | (*b*) | **7.29** | (*b*) | **7.34** | (*a*) |
| **7.20** | (*b*) | **7.25** | (*a*) | **7.30** | (*b*) | **7.35** | (*a*) |

# Chapter 8

# Cell Reproduction

## 8.1 CELL CYCLES AND LIFE CYCLES

### CELL CYCLES

Most eukaryotic cells reproduce at a regular rate to produce new daughter cells that contain the distributed materials of the original cell. The distribution of nuclear materials, particularly the chromosomes, is known as *mitosis*. The apportionment of cytoplasm is called *cytokinesis*. These processes (see Fig. 8.1) are part of a larger sequence of events known as the *cell cycle*—a rhythmic recurrence of growth within the cell followed by reproduction of the cell. In many cells the entire cycle, from $G_1$ to cytokinesis, is completed within a few hours; in other cells, it takes many days, while in still other cells it is permanently arrested in one phase.

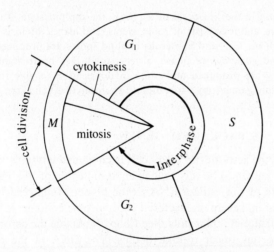

**Fig. 8.1** Cell Cycle

The significant vegetative (nonreproductive) functions of the cells occur during $G_1$. These include growth, increase in the number of organelles, and production of materials for both export and intracellular use. Those cells that do not continue to divide are usually arrested in $G_1$ of their cycle.

During the *S period*, the DNA of the nucleus doubles (in preparation for the later division of the nucleus and its chromosomes). Proteins associated with the chromosomes may also be formed at this time, while metabolic activity of the cell is considerably reduced.

Following this synthesis stage a second gap, $G_2$, occurs during which materials for the specialized structures required for chromosome movement and cell replication are organized.

The cell begins the active process of division as $G_2$ gives way to mitosis proper. The chromosomes become visible and undergo an ordered movement, and then a set of replicates (chromatids) migrate to each pole of the cell. The cytoplasm divides, and two new cells are formed from the contents of the original cell. The cell cycle is now ready to begin again with a new $G_1$.

### LIFE CYCLES

In those rare single-celled organisms that reproduce only asexually, the cell cycle represents the entire sequence of events occurring during the life of the cell. In organisms, whether single-celled or multicellular, that reproduce sexually the cell cycle is incorporated into a more complex sequence of events. In the sexual process two significant events invariably occur. One of these is the union of special sex cells (*gametes*), which incorporate genetic material from two different sources. The fusion product is the *zygote*. In some organisms the gametes that fuse are alike, but in most organisms the gametes are dimorphic, specialized structures known as *sperm* and *egg*. They each have only half the number of chromosomes found in the nonreproductive (*somatic*) cells of the organism. Whether gametes are alike (*isogametes*) or not (*heterogametes*), their union, called *fertilization*, results in a doubling of the chromosome number. Each gamete generally contains one set of chromosomes, so fertilization will produce a zygote (usually a fertilized egg) with two sets of chromosomes. This latter condition is called the *diploid* state (designated 2n), while possession of only a single set of chromosomes is known as the *haploid* state (1n). A process of chromosome reduction called *meiosis* intervenes between one sexual event and another. Meiosis, occurring in diploid cells, produces up to four haploid cells. A series of successive sexual unions could not occur without an intervening meiotic stage, because the chromosome number would double with each fertilization.

In multicellular organisms there is usually a characteristic organismic form associated with haploid and diploid phases. This alternation between haploid and diploid phases brought about by meiosis and fertilization makes up the *life cycle* of all organisms.

**EXAMPLE 1**   The prominent phase in the life cycle of a fern is the diploid state. The fern plant made up of diploid cells is called the *sporophyte*, and it consists of roots, stems, and leaves. Meiosis occurs within a group of specialized sacs on the underside of the leaf, and numerous haploid spores are produced. These spores leave the sac and divide to form the haploid *gametophyte* phase, a very small, thin heart-shaped structure. Within this gametophyte plant, sperm and egg are produced and unite in fertilization to give rise eventually to another sporophyte. This swing from a haploid gametophyte to a diploid sporophyte is called the *alternation of generations*.

## 8.2   THE CHROMOSOMES AS PACKAGED GENES

Information for carrying out the activities of every cell is contained within a vital blueprint molecule. Since the 1940s, DNA has been known to be this fundamental molecule of life. (Some strains of viruses, which are only on the borderline of life, utilize RNA as the blueprint material.) In bacteria, the single circular strand of naked DNA that makes up the bacterial chromosome contains almost all the DNA of the cell. Some DNA is found as small circlets (plasmids; See Chap. 7) outside the chromosome. In eukaryotes, the chromosomes exist as numerous linear rodlike bodies. The DNA of the eukaryote chromosome exists in intimate association with small basic proteins called *histones*. A set of eight histone molecules (see Chap. 7, Fig. 7.5) forms a unit called a *nucleosome*, around which several hundred base pair lengths of DNA are wound. The entire chromosome is made up of repeating nucleosome units. Further coiling of the DNA brings the nucleosomes closer together until the chromosome appears to be a dense band 30 nm wide called chromatin fibers. During mitosis, the bands themselves are coiled, producing the characteristic thickness seen in chromosomes during this phase of the cell cycle.

In both prokaryotes and eukaryotes the genes exist as specific stretches of DNA along the chromosome. In prokaryotes, the structural genes lie in a continuous array along the length of the chromosome, separated only by regulatory regions and the starting and stopping signals associated with turning transcription on or off (see Chap. 7). In eukaryotes, the DNA is packaged into coiled nucleosome units. This chain of spooled structures is not neatly separable into specific gene units, but the gene material does exist as a specific region within the chromosome. It is estimated that a typical gene comprises about four nucleosome units along the chromosome. It has also been shown that less than one tenth of the DNA within the eukaryotic chromosome is actually translated into protein. The prokaryotic chromosome exists as a single unit (is haploid), and so each gene may be expressed without interaction with a corresponding gene on a homologous chromosome. The eukaryotic chromosomes exist as homologous pairs (are diploid) during much of the life cycle of the organism, and so their gene expression is subject to modulation from gene-gene interaction.

Most of the genes of prokaryotic forms are contained within the single circular chromosome, with a few more possibly on a plasmid. No introns are present. In eukaryotes the genes are scattered among the complement of chromosomes making up the characteristic set for the species. Every species has a particular number and characteristic morphology (*karyotype*) for its chromosome sets, and particular genes are generally found in the same locus of a specific chromosome. Alteration in the number of chromosomes or in their structure is usually associated with genetic abnormalities.

**EXAMPLE 2**   The normal karyotype for humans consists of 23 pairs of chromosomes. The chromosome pairs are numbered according to relative size. The twenty-third pair constitutes the sex chromosomes, which consist of a pair of X chromosomes in the female or an X and Y chromosome in the male. In some people a third chromosome 21 is found in the cells; this situation, known as *trisomy*, is associated with *Down's syndrome*. Such people are retarded and may show developmental abnormalities; they have a characteristic rounded face and epicanthic eyefolds, and often develop leukemia in later life.

## 8.3   MITOSIS

Mitosis is the process during which the chromosomes are distributed evenly to two new cells that arise from the parent cell undergoing division. During the S phase of interphase before mitosis proper, each chromosome will have replicated. The two chromosomal strands (*chromatids*) are identical in their genetic material and are joined at a constricted region called the *centromere*. Within the centromere is one or more rings of protein known as *kinetochores*. The kinetochores will play a significant role in the attachment of the spindle fibers to the chromosomes and in the subsequent migration of the chromosomes.

Mitosis has four main stages—prophase, metaphase, anaphase, and telophase (see Fig. 8.2). In *prophase*, the relatively long first stage of division, the nuclear membrane breaks down and the spindle forms.

**EXAMPLE 3**   In most animal cells and some fungi a *centrosome* region exists just outside the nuclear membrane. Within the centrosome are two pairs of centrioles. *Centrioles* occur in pairs, each being a cylindrical rodlet that lies perpendicular to its partner. The centriole pairs migrate to opposite poles of the cell, where they appear to be associated with the *spindle apparatus* of the cell. The centrioles replicate during interphase. When centrioles are present, an array of microtubular rays known as *asters* may form around them at each pole in a starlike cluster. Centrioles and asters are rarely present in higher plants. The microtubular threads making up the spindle apparatus produce a barrel-shaped structure in plants, but in animals each end of the spindle is pointed. This difference may be associated with the influence of the centrioles.

The chromosomes condense and begin to move toward the *equatorial* (middle) *plane* of the cell. *Metaphase* is characterized by the precise lineup of the chromosomes along the equatorial plane. At the start of *anaphase* the centromeres of each chromosome split so that each chromatid now exists as a separate chromosome. Guided by the mooring spindle fiber, one chromatid of each pair is moved to one pole, while the other chromatid is moved to the opposite pole.

**EXAMPLE 4**   The precise migratory movements of the chromosomes during mitosis (and meiosis) are largely owing to the spindle apparatus. This rigid, semisolid network of microtubules consists of at least three types of hollow fibers. One type of fiber extends continuously from either pole, almost reaches the opposite pole, and overlaps with fibers from the opposite pole. A second type is attached to kinetochore rings within the centromere and may even arise from these kinetochore structures. A third type occurs within the space that separates migrating chromatids as they move to opposite poles.

The basic subunits of the spindle fibers are already present in the cell before mitosis. Thus, assembly of the spindle apparatus, as well as its disassembly, may occur extremely rapidly.

Despite the fact that a doublet chromosome early in mitosis has a single centromere region, the two chromatids each have their own kinetochore from which spindle microtubules may extend. The kinetochore of one chromatid is associated by microtubules to one pole, while the kinetochore of the sister chromatid is associated with the other. This accounts for the ability of each chromatid to move in a direction opposite to that of its sister chromatid. Analysis of the centromere indicates that it has a unique base composition.

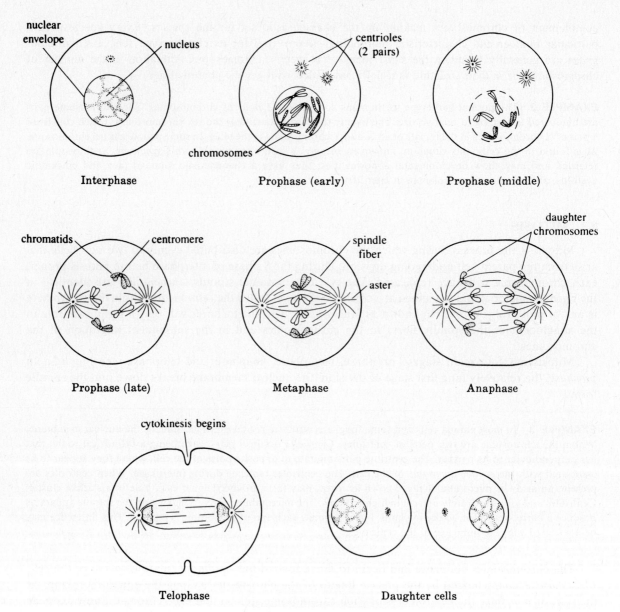

**Fig. 8.2** Mitosis (Animal)

Once the chromosomes reach opposite poles, the last phase of mitosis, *telophase*, begins. The chromosomes gradually lose their stainability as they begin to lose their supercoiling. The nucleolus, which tended to disappear during prophase, begins to re-form at specific nucleolar organizing regions of certain chromosomes. The spindle apparatus breaks down to its constituent macromolecules, and a new nuclear membrane begins to form around each of the two clumps of chromosomes aggregated at their respective poles. Telophase may be regarded as a prophase run backwards.

With the completion of the nuclear division events, the cytoplasm usually begins its division—a process known as *cytokinesis*. Although accomplished differently in animals and plants, the results are the same: the creation of two separate cells.

In animal cells a cleavage furrow begins as a puckering along the surface of the cell in the region of the equatorial plane. It extends and deepens until the original cell is completely cut in two, with each part containing a complement of chromosomes as a result of the antecedent mitosis. In some primitive plants, an inward growth of both plasma membrane and cell wall material from the surface occurs until a complete partition is formed. However, in all higher plants, cytokinesis begins near the middle of the cell with the formation of an internal *cell plate* along the line of the equatorial plane, the center of the old spindle. A ringed structure made up largely of protein, the *phragmoblast*, is observed first and then the creation of the flattened cell plate. This plate gradually extends to the surface of the cell and partitions it into two new cells. New cell wall material is brought to the partition in membranous sacs derived from the Golgi apparatus to complete the process of cell division. In prokaryotes (bacteria and blue-green algae) cell division is less complex. A single circular strand of DNA fastened to the plasma membrane at one end constitutes the complete karyotype of the cell. At the time of division, the chromosome begins to replicate. By the time replication is complete, the resulting chromosomes are both attached to the plasma membrane. The entire cell undergoes elongation even before chromosome replication so that cytokinesis occurs immediately following the attachment of the replicate chromosome to the membrane. Newly synthesized membrane and even cell wall material extends from the surface along the midline of the cell and completely partitions it into two new daughter cells, which separate into independent entities.

## 8.4  MEIOSIS

A union of gametes in sexual reproduction always yields a doubling of the chromosome number. To maintain homeostasis in terms of chromosome number, the uniting gametes may be haploid rather than diploid owing to a unique pair of cell divisions that segregate homologous chromosomes into separate cells. The process by which this is accomplished is called *meiosis,* from a Greek word meaning "to diminish." Meiosis probably evolved as a modification of mitosis and incorporates many of its features.

In animals, the two divisions of meiosis produce the haploid gametes, which eventually unite to form a diploid zygote. In many algal cells, meiosis occurs immediately after fertilization to produce haploid cells that constitute the dominant life form of these organisms. In many plant cells meiosis occurs some time after fertilization, with alternating haploid and diploid phases occurring in the plants. More primitive plants spend a greater portion of their life cycle in the haploid (gametophyte) stage, whereas more advanced plants are characterized by a dominant diploid (sporophyte) stage. These variations in meiotic timing within the living world should not obscure the significance in all organisms of an intervening reduction process to compensate for the increase in chromosome number from sexual union.

### THE FIRST MEIOTIC DIVISION

Meiosis begins in similar fashion to mitosis (see Fig. 8.3): each chromosome replicates in the S phase of interphase, and prophase begins after $G_2$ with an increasing coiling and condensation of each of the doublet chromosomes. As in mitosis, the nuclear membrane begins to break down, centrioles move to opposite poles of the cell, and the chromosomes begin to migrate toward the equatorial plane. Spindle fibers start to aggregate from microtubules, and nucleoli disappear. Important differences, however, soon become apparent. The prophase of meiosis I is a very much longer and more extensive process than the prophase of mitosis and is actually divided into substages. The most dramatic difference occurs early in the prophase when the homologous chromosomes mysteriously start to come together in pairs (*synapsis*). Homologues touch at one or several points, then the chromatids appear to zip together to form an intimate four-stranded structure known as a *tetrad.* When the tetrad begins to loosen late in prophase, the individual chromosomes from each tetrad start to separate. At this point there may still be a few physical links between chromatids of one homologous chromosome and those of another. These clinging structures that seem to defy the tendency of the homologues to separate are called *chiasmata* (sing., *chiasma*). Each of the chiasmata formed along the various homologues represents a point at which a section from one chromatid has physically broken off and been exchanged with the corresponding chromatid section on the homologous chromosome. Such an exchange of chromosome parts between the chromatids of two homologous chromosomes is known as *crossing over*

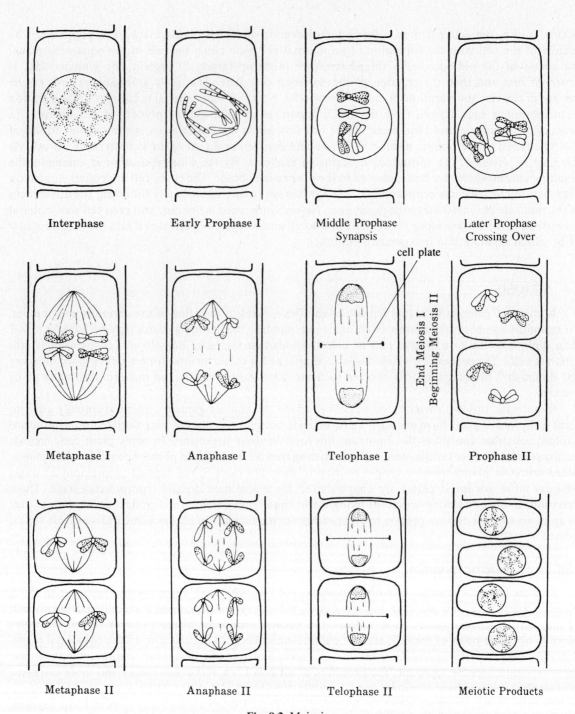

**Fig. 8.3** Meiosis

and results in the formation of hybrid chromosomes with mixed genetic material (see Sect. 8.6 and Fig. 8.5).

The metaphase of meiosis I consists of a lining up of pairs of homologous chromosomes, now largely separated, at the equatorial plane. These structures continue to be identified as tetrads, since the homologues are still closely apposed to one another. However, instead of a single line of centromeres at the equator, which is characteristic of mitotic metaphase, there will be a double line of centromeres. The total number of tetrads at the equator will be equal to the haploid (1n) number.

During the anaphase of meiosis I no splitting of centromeres occurs. Instead, whole chromosomes separate, with one homologue moving to one pole and the other to the opposite pole. This results in single sets of chromosomes (with two chromatids) aggregating at each of the poles and effectively reduces the diploid (2n) condition to the haploid (1n) condition. This first division of meiosis is consequently called the *reduction division*.

In the ensuing telophase it is chromosomes with two chromatids that slowly lose their density, a new nuclear membrane forms around each haploid set of doublet chromosomes, and the usual events of telophase ensue. A short stage called *interkinesis* occurs between telophase I and prophase II. However, no synthesis of genetic material occurs, and in some cases the chromosomes do not completely lose their condensed configuration before moving into the second meiotic division.

## THE SECOND MEIOTIC DIVISION

In the second meiotic division, called the *equational division*, a haploid set of replicate chromosomes in each new cell migrates to the equatorial plane and lines up in a single line of centromeres. The centromeres now split, and the former chromatids of each chromosome migrate to opposite poles. Each of the two cell products of meiosis I will produce two new cells, a total of four haploid cells during the full meiotic process. In some cases, only one functional cell arises from the meiotic process, since in many species each of the two meiotic divisions produces one functional cell and one very tiny polar body, which quickly degenerates. The first polar body may even undergo the second meiotic division before disintegrating. The production of gametes (*gametogenesis*) in females (*oogenesis*) is similar to gamete production in males (*spermatogenesis*) in terms of the behavior of chromosomes. However, in the apportionment of cytoplasm to the resultant cells and their modification, differences often arise between the sexes (see Fig. 8.4).

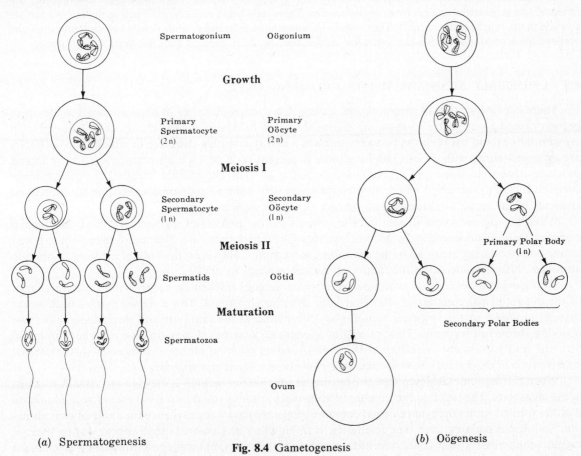

Spermatogonium  Oögonium

**Growth**

Primary Spermatocyte (2n)   Primary Oöcyte (2n)

**Meiosis I**

Secondary Spermatocyte (1n)   Secondary Oöcyte (1n)

Primary Polar Body (1n)

**Meiosis II**

Spermatids   Oötid

Secondary Polar Bodies

**Maturation**

Spermatozoa

Ovum

(*a*) Spermatogenesis

**Fig. 8.4** Gametogenesis

(*b*) Oögenesis

**EXAMPLE 5**  Eggs, or *ova*, are produced from diploid cells called *oogonia*, which are found in *follicles* within the *ovary*. These oogonia may enlarge and undergo modification preparatory to meiotic division, at which time they are designated *primary oocytes*. Following the first meiotic division, a large *secondary oocyte* is produced (containing most of the original cytoplasm) along with a tiny *polar body*. Both the secondary oocyte and polar body have a haploid complement of chromosomes, but the oocyte contains most of the food material of the cell. During the second meiotic division, a large cell and a second polar body are produced. This large cell develops into a mature egg, and the polar body disintegrates. Should the first polar body undergo meiosis II, its polar body products would quickly break down as well. The result of meiosis in female animals is the production of a single, large haploid egg.

In males, diploid *spermatogonia* proliferate mitotically within the testes in special structures called, appropriately enough, *seminiferous tubules*, i.e., "seed-bearing sacs." At maturity these spermatogonia undergo modification to become *primary spermatocytes*, the cells which undergo the first meiotic division to form haploid *secondary spermatocytes*. The secondary spermatocytes each undergo meiosis II to form a total of four haploid cells with single-stranded chromosomes, since the second meiotic division is equational and merely separates the chromatids of haploid sets of chromosomes.

All four cells are viable and generally have equal amounts of cytoplasm, unlike the polar-body phenomenon in oogenesis. The cells produced at the end of meiosis are called *spermatids*. They must undergo considerable cytoplasmic modification before they can be released as functional *sperm* (*spermatozoa*). In this process, using humans as an example, almost all the cytoplasm is modified to form a midpiece motor apparatus and a long tail. The midpiece is rich in mitochondria, which supply the energy necessary for this substructure to whip the tail, the basis for sperm motility. The head of the sperm, which is analogous to the explosive payload of a rocket, is essentially the naked sperm nucleus. Near the head of the sperm is a specialized structure, derived from sperm Golgi vesicles, called the *acrosome*. It contains a variety of hydrolytic enzymes that enable the sperm to penetrate the protective jelly that surrounds the ovum. A short neck region separates the head from the midpiece. The tail is formed as an extension of the midpiece.

In humans, the female fetus begins creating cells (oogonia) that, upon her birth and development, will ultimately be sequestered as eggs (ova) in her ovaries. These cells begin meiosis while still in the embryo but are arrested in the prophase of meiosis I. They remain in "suspended animation" until shortly before fertilization and undergo the second meiotic division only *after* fertilization. In males, meiosis does not begin until maturity.

## 8.5  A POSSIBLE MECHANISM FOR CROSSING OVER

There is evidence that in chromosomes undergoing synapsis a series of axial elements made up of protein extend along each chromosome to provide a thin backbone for the chromatids. Later, protein crossbridges extend between the two axes to form a highly complex structure in which loops of DNA are arranged along with RNA. The longitudinal protein rods of each chromosome and the lateral processes that weld them into a complex connecting all four chromatids is called the *synaptinemal complex*. It is in its formation and continued influence within the paired homologues that the individual elements of the tetrad are brought into perfect alignment with one another.

When the chromosomes disjoin later in meiosis (in the process of *disjunction*), the synaptinemal complex begins to disassemble. Some have attributed the tendency of twin chromatids to remain attached to one another during meiosis I to the synaptinemal complex. However, the bulk of cytological opinion attributes this phenomenon to the failure of the centromeres to divide.

Strong evidence exists that in some species proteinaceous thickening occurs within the crossbands of the synaptinemal complex at sites that later develop chiasmata. These *recombination nodules* are thought to play a part in snipping homologous chromatids at the same site and interchanging the two resulting chromatid segments. This splicing of a length of maternal chromatid on a paternal chromatid "stump" and subsequent annealing of the corresponding paternal chromatid segment on a maternal stump produce the hybrid chromosomes that contribute to genetic variability.

When chromosomes begin to separate late in the prophase of meiosis I, they tend to remain attached at the chiasmata. The fact that the number of chiasmata is about equal to the number of recombination nodules formed within the synaptinemal complex is regarded as strong evidence for a role of recombination nodules in crossing over. The regularity in the number of chiasmata that appear during meiosis suggests that crossing over is not a fortuitous or accidental event, but rather an established mechanism for increasing genetic variability.

## 8.6 SEXUAL REPRODUCTION AND GENETIC VARIABILITY

One explanation for the ubiquity of sexual reproduction is the variability it provides for the forces of evolution to act on. If all the organisms of a single species were exactly alike, they might all be destroyed should the environment become inhospitable. However, if individual members of that same species were more variable in their characteristics, then changes in the environment which destroyed some of the variants would leave others unharmed. Variation, in providing flexibility within a species, enhances the potential for survival of the species in the face of environmental challenge.

We have already seen that genetic mutation is a source of variation; however, it is not necessarily dependent on sexual reproduction for its expression. Sexual reproduction, by bringing together two entirely different genomes, greatly increases genetic variability. Meiosis introduces two sources of genetic recombination that add to this variability.

The first involves the recombination of whole chromosomes. Remember that in the diploid zygote, for each pair of homologous chromosomes one was provided by the male (paternal chromosome) and one by the female (maternal chromosome); half the chromosomes are thus paternal and half maternal. However, the organism that develops from the zygote will produce haploid gametes. Because during meiosis I paternal and maternal homologues line up opposite each other randomly on either side of the equatorial plane, each resulting daughter cell will contain a haploid mixture of maternal and paternal chromosomes (and maternal and paternal genes and traits). Some of these combinations may never before have occurred in the haploid state. The more pairs of chromosomes an organism possesses, the greater potential variability there is in the gametes. The significance of this mixing of chromosomes will be studied in detail in Chap. 9 on the mechanism of inheritance.

**EXAMPLE 6**   In a cell with only one pair of homologous chromosomes, there will be two general classes of gametes produced at the end of meiosis—cells with the maternal chromosome and cells with the paternal chromosome. In the large numbers of gametes produced by gametogenesis, approximately half will contain the maternal and half the paternal homologue. In those cells in which the diploid number is 4, two tetrads will be formed, $M_1P_1$ and $M_2P_2$. Therefore, four different classes of gametes may be produced in the course of meiosis: $M_1M_2$, $M_1P_2$, $P_1M_2$, and $P_1P_2$. The formula giving the number of classes of gametes from a cell undergoing meiosis is $2^n$, where $n$ represents the haploid number (number of tetrads). For humans, with an $n$ of 23, there will be $2^{23}$ (over 8 million) different classes of gametes produced, based only on the recombination of maternal and paternal chromosomes in the gametes. A sperm or egg may contain all maternal or all paternal chromosomes, or some combination of the two.

The second source of variation through chromosomal combination is crossing over. Again mixing occurs between paternal and maternal chromosomes, but within one set of homologues. As with the independent assortment of whole chromosomes, crossing over increases the types of gametes that can be produced and, thus, variability.

**EXAMPLE 7**   Suppose a chromosome has an allele ($B$) for blue blossoms and at another gene locus an allele ($C$) for curly leaves (see Fig. 8.5). Further assume that the homologous chromosome has at these respective gene loci alleles for red blossoms ($R$) and fringed leaves ($F$). Without crossing over, only two types of gametes could

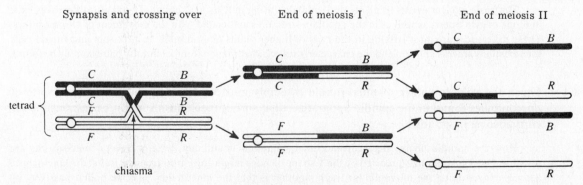

**Fig. 8.5** Crossing over

be produced with respect to these alleles: *BC* and *RF*. However, because of crossing over, two other gamete types are possible: *BF* and *RC*.

When the thousands of gene loci are considered, the potential for variation through crossing over is enormous. Heightening the effect of this variation is the fact that gene loci on a chromosome often interact and thereby produce effects greater than the simple addition of the activities of the individual loci involved.

# Solved Problems

**8.1**     For single-celled, asexually reproducing organisms the cell cycle and the life cycle are one and the same. Explain.

Asexual, single-celled organisms do not undergo fertilization; consequently, they require no homeostatic process beyond mitosis to maintain a constant number of chromosomes. Sexually reproducing organisms would double their complement of chromosomes with each fertilization if they could not first halve that number through meiosis. This alternation of generations between a diploid stage and a haploid gametic stage, which is in addition to the cell cycle, is unnecessary in asexual organisms. Thus, the life cycle includes only the events of the cell cycle.

**8.2**     How does cellular reproduction differ in prokaryotes and eukaryotes?

Cell division in prokaryotes is less complex than in eukaryotes, in part because of the presence of only a single chromosome. The complex network of ancillary structures that aid in separating the homologous chromosomes of eukaryotes is unnecessary. As the single, circular strand of DNA replicates, the resulting two circlet strands merely attach to the plasma membrane. The cell then undergoes cytokinesis, with the membrane and wall growing in, and the two daughter cells separate.

In eukaryotes, the often numerous threadlike chromosomes are each replicated and then go through the more complicated process of mitosis. In this process, considerable modification of the chromosomes occurs, and the spindle apparatus and other structures effecting chromosome migration arise.

The special form of cell reproduction that results in a reduction of chromosome number (meiosis) is not found in prokaryotes, since they exist only in the haploid condition. Sex does occur in prokaryotes, at least in bacteria, but it does not involve the formation and union of gametes. Instead, bacteria conjugate and in that process, which is not clearly associated with reproduction, exchange gene material (see Chap. 7).

**8.3**     How does mitosis differ in plant and animal cells?

The basic nuclear events in plants and animals during mitosis are similar. The major differences exist in terms of cytokinesis as well as in the ancillary structures associated with the movement of chromosomes. These differences are most striking in the cells of higher plants and animals. In protistan and fungal cells the variations are less uniform and the characteristics of either plant or animal mitotic patterns may be found.

**8.4**     Given the volatile nature of microtubule assembly and disassembly and the various sorts of microtubules found in the spindle apparatus, what sort of mechanisms might be responsible for chromosome migration?

How the spindle fibers effect chromosomal migration is still not certain. Several mechanisms are possible. One explanation, supported by the V shape of most migrating chromosomes, is that the kinetochore actively moves along the microtubular track; another is that the kinetochore may be pulled passively by its shortening fiber as the fiber loses subunits at the poles. Yet another notion is that elongation of the original spindle forces the poles apart and carries the chromosomes along with the extending spindle apparatus. Finally it is considered possible the extension of the spindle fibers that form between the migrating chromatids could also produce migration. All or some of these mechanisms may be operative in the migration process in a variety of cells. A vital part of several of these mechanisms involves the continuous dynamic state of assembly and disassembly of the microtubular units of the spindle fibers.

**8.5**    During mitosis, sister chromatids separate and migrate to opposite poles. Why does this not occur during meiosis I?

During mitosis, the two respective kinetochores of a chromosome are attached through microtubules to different (opposite) poles. Thus they move in opposite directions during migration. In meiosis I, the two chromatids of a chromosome are attached at the centromere to only one pole (with the homologous chromosome attached to the other pole); therefore, at anaphase they move in the same direction.

**8.6**    How does the prophase of meiosis I differ from the prophase of mitosis?

The prophase of the first meiotic division is of much longer duration than the prophase of mitosis. It is subdivided into the following functional stages:

1.  *Leptotene.*   Chromosomes have begun to coil and condense and are visible as long, slender threads. Replication has, of course, occurred during the S period of interphase, but individual chromatids cannot yet be seen. The nuclear membrane begins to break down, and centrioles, when present, move toward the poles. Nucleoli disappear as discrete structures.
2.  *Zygotene.*   Homologous chromosomes begin the process of synapsis by making contact at several points along their length. Spindle fibers begin to appear.
3.  *Pachytene.*   Synapsis is complete, with each gene locus of one homologue closely apposed to the corresponding gene locus of the other homologue. Since two chromosomes are joined, the resultant configuration is called a *bivalent* (two parts).
4.  *Diplotene.*   The two chromatids of each of the paired chromosomes are now distinctly visible, and the entire structure is known as a *tetrad* because of its four-stranded appearance. It is during this period that a great deal of RNA synthesis may occur along the length of the chromosomes. Often associated with such RNA synthesis is the formation of loops of DNA to which RNA and histone are usually adhered. These loops may cause the chromosomes to assume a "lampbrush" appearance when examined microscopically.
5.  *Diakinesis.*   The chromosomes have undergone maximum compacting and are highly coiled. At this stage the synapsed chromosomes seem to be pulling apart and are held together only at specialized places called *chiasmata*. The thickened chromosomes of diakinesis migrate toward the equatorial plane; the long prophase of meiosis I becomes metaphase when the chromosome pairs are lined up along the equatorial plane.

The major distinctions of the prophase of meiosis I are its great length, synapsing (joining) of homologous chromosomes, exchange of complementary regions between homologous chromosomes (crossing over), and attachment of spindle fibers along one side only of whole chromosomes so that each chromosome moves to the pole opposite that to which its homologue migrates. Nevertheless, many of the features of the first meiotic prophase are similar to mitotic prophase, especially the behavior of the nuclear membrane, nucleoli, and centrioles and the mechanism of formation of the spindle.

**8.7**    Are there differences between meiosis II and an ordinary mitosis?

Despite the frequent comparison of meiosis II to an ordinary mitotic division, certain differences should be recognized. Immediately preceding a mitotic division, the DNA, as well as the chromosome strands, replicates. In meiosis II there is no replication of DNA or chromosomal material following the completion of meiosis I. Meiosis II prophase may begin, at least in some organisms, right after the telophase of meiosis I with no breakdown and ensuing reconstitution of visible chromosomes, whereas in mitosis the cell starts with a diffuse chromatin distribution. Meiosis II always involves a haploid complement of chromosomes produced during the first meiotic division, whereas mitosis may occur within haploid or diploid cells. Also, each chromosome present during meiosis II may contain a mixture of maternal and paternal chromosome material because of crossing over. None of these differences change the basic similarity in mechanism of chromosome movement during the two processes or the parallel techniques for cytokinesis.

**8.8**    What are some possible mechanisms that could cause homologous chromosomes to recognize one another and synapse during meiotic prophase?

The mechanisms that permit synapsis are unclear. A molecular basis for synapsis could involve complementary base pairing, the concurrence of repetitive sequences of bases along the homologues, or the action of protein molecules along the chromosomes (as with the synaptinemal complex).

**8.9**    Which chromatids may actually participate in crossing over and chiasma formation?

Either of the chromatids of one homologue may exchange corresponding parts with either of the two chromatids of the other homologue. In the usual diagram of a tetrad,

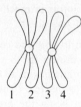

1  2  3  4

which is represented as a two-dimensional construct, it would appear that only strands 2 and 3 could exchange parts. Actually, strand 1 can exchange with either strand 3 or strand 4, while strand 2 can likewise exchange with either strand 3 or 4. Any exchange, if it occurs at all, between sister chromatids would be of no genetic consequence because these strands are genetically identical.

**8.10**    Chromosomal proteins are made in the cytoplasm and must move back into the nucleus. What mechanisms, would you guess, are used to get these proteins into the nucleus?

Some of these proteins, such as the histones, are small enough to pass across the pores of the nuclear membrane. Other proteins, generally larger than the histones, may have more difficulty getting into the nucleus and perhaps require a form of endocytosis involving the nuclear envelope to achieve it.

**8.11**    Abnormalities in chromosome number give rise to diseases of karyotype. How might these aberrations occur?

Under usual conditions of meiotic division each tetrad separates into its constituent homologous chromosomes. One homologue migrates to one pole and the other homologue to the opposite pole during anaphase of the first meiotic division. If this separation does not occur, all the tetrads may move to one pole, while the opposite pole may receive no chromosomes at all. This would eventually produce diploid cells as gametes (in those organisms in which meiosis is involved in gamete formation). Should such a diploid gamete unite with a more typical haploid gamete, a zygote would be produced with three sets of chromosomes, a *triploid* individual. In plants, the formation of triploid and even higher orders of *polyploidy* represents a mechanism for producing new species of the organism in the course of evolution. This alteration of ploidy is less common in animals.

More commonly, a single tetrad will fail to separate into its constituent chromosomes. This will eventually result in gametes that have a double dose of one chromosome and others that have no representative for that particular chromosome. In Down's syndrome, a sperm or egg with two chromosomes 21 unites with a normal haploid sperm or egg to produce a zygote with three such chromosomes. The failure of tetrads to disjoin is called *nondisjunction*, and disorders arising from the phenomenon are known as diseases of nondisjunction. They include Down's, Klinefelter's (XXY male genotype), Turner's (XO, i.e., having only one X chromosome, female genotype) syndromes. Fragmentation, deletion, and internal inversion of chromosomes or chromosome parts may also produce diseases of karyotype.

**8.12**  We have seen that an important contribution to genetic variation is provided by the various combinations in the gamete of maternal and paternal chromosomes, either in their entirety or as segments (through crossing over). Since these chromosomes existed completely mixed together in the diploid somatic cells, how, do you suppose, recombination during meiosis can produce greater variation than this?

The answer to this problem lies in the way alleles of a given gene interact with each other and with the alleles of other genes to produce the characteristics we see in the organism. For example, on a maternal chromosome a particular allele, say, for curly leaves, may for biochemical reasons *always* be expressed, no matter what the paternal allele would express if it were not dominated by the curly allele. If recombination did not occur, all other genes on the maternal chromosomes (for example, one with an allele coding for blue flower color) would *always* be linked to plants with curly leaves. However, through recombination, the maternal chromosome containing the gene for blue flower color may be included in the gamete with the paternal chromosome containing the allele, say, for smooth leaves (instead of curly). If this gamete unites with another one containing a chromosome having a "smooth" allele and thus allows smooth leaves to be produced, the blue color will have escaped its bondage to curly leaves and a new type of plant will have been produced, even though no new alleles have been introduced into the population. Given the thousands of characteristics coded for in the genome, recombination permits millions of juxtapositions that would not occur without it, and thus millions of variations in physical characteristics with which to face selection pressures. Such recombination will be considered in detail in Chap. 9.

# Supplementary Problems

**8.13**  The DNA doubles and chromosomes replicate during which phase of the cell cycle?  (*a*) $G_1$.  (*b*) metaphase.  (*c*) S period of interphase.  (*d*) $G_2$.  (*e*) cytokinesis.

**8.14**  Sperm and egg are  (*a*) isogametes.  (*b*) isomorphs.  (*c*) heterogametes.  (*d*) diploid cells.  (*e*) all of these.

**8.15**  In meiosis, the chromosomes replicate  (*a*) during interkinesis.  (*b*) during the S phase of interphase.  (*c*) only once during the entire process.  (*d*) both *a* and *b*.  (*e*) both *b* and *c*.

**8.16**  In the fern, the sporophyte is  (*a*) the dominant form.  (*b*) diploid.  (*c*) the source of haploid spores.  (*d*) each of these.  (*e*) none of these.

**8.17**  In many organisms, particularly evident within the plant kingdom, an alternation of haploid and diploid forms (generations) occurs. This two-stage life style is known as  (*a*) the reductive system.  (*b*) the life cycle.  (*c*) synthesis.  (*d*) synapsis.  (*e*) neotony.

**8.18**  The basic structural unit of a chromosome is  (*a*) the centromere.  (*b*) the nucleosome.  (*c*) the telomere.  (*d*) all of these.  (*e*) none of these.

**8.19**  Some human males have three sex chromosomes (XXY) and suffer from a genetic disease known as Klinefelter's syndrome. The symptoms include a failure to develop sexually and an impairment of intelligence. This is an example of a disease of  (*a*) karyotype.  (*b*) point mutation.  (*c*) homeostasis.  (*d*) bacterial origin.  (*e*) old age.

**8.20**   In humans, the number of tetrads formed during mitosis is   (*a*) 23.   (*b*) 46.   (*c*) 0.   (*d*) 4.   (*e*) none of these.

**8.21**   The centromere, or primary constriction of the chromosome, contains rings of protein that are intimately associated with a spindle fiber. These rings are called   (*a*) secondary constrictions.   (*b*) centrioles.   (*c*) asters.   (*d*) kinetochores.   (*e*) somites.

**8.22**   The two sets of chromosomes present in the cells of diploid organisms are derived from   (*a*) doubling of a haploid cell.   (*b*) the contribution of one haploid set from each parent.   (*c*) a reduction process within a tetraploid cell.   (*d*) all of these.   (*e*) none of these.

**8.23**   In both mitosis and meiosis the replication of DNA and the chromosomes occurs at the same stage.
(*a*) True   (*b*) False

**8.24**   Crossing over is accompanied by chromosome breakage.
(*a*) True   (*b*) False

**8.25**   Polar bodies lack chromosomes.
(*a*) True   (*b*) False

**8.26**   An organism with a diploid number of 10 would contain 20 tetrads during the prophase of meiosis I.
(*a*) True   (*b*) False

**8.27**   The organelles that provide the power for the motorlike midpiece of the sperm are the mitochondria.
(*a*) True   (*b*) False

# Answers

| | | | | | | | |
|---|---|---|---|---|---|---|---|
| **8.13** | (*c*) | **8.17** | (*b*) | **8.21** | (*d*) | **8.25** | (*b*) |
| **8.14** | (*c*) | **8.18** | (*b*) | **8.22** | (*b*) | **8.26** | (*b*) |
| **8.15** | (*e*) | **8.19** | (*a*) | **8.23** | (*a*) | **8.27** | (*a*) |
| **8.16** | (*d*) | **8.20** | (*c*) | **8.24** | (*a*) | | |

# Chapter 9

# The Mechanism of Inheritance

The fact that children resemble their parents is an example of *inheritance* or *heredity*. Thus, cats have kittens, dogs have pups, and people have babies. Within these groups the specific hereditary traits of the parents are found to varying degrees in the children. The mechanism of heredity is signified by the modern term *genetics*.

## 9.1 PRE-MENDELIAN CONCEPTS

Early folklore reflects the ideas that primitive people held about heredity. Among some isolated tribes there was no clear connection made between human or animal sexual intercourse and pregnancy. In Egypt and Babylon, however, selective plant and animal breeding was developed more than 2500 years ago.

At the time that sperm (seventeenth century) and egg (nineteenth century) were discovered, a lively discussion developed over the notion that a tiny human being was actually preformed within these gametes. Thus, the hereditary characteristics were thought to be present before fertilization had occurred. In more recent times the feeling that blood bore the hereditary determinants was popular. Aristocrats were spoken of as bearing "blue blood," and deleterious traits were thought to be the result of "bad" blood. During World War II the blood of different ethnic groups was segregated because of the popular notion that race and ethnicity were rooted in the blood. This idea is still part of the vocabulary of popular sociology.

Both Aristotle and Darwin believed, at least in part, in *pangenesis*, the concept that representatives, called *gemmules*, from every part of the body gathered in the gametes and were involved in directing the construction of the new individual. This "parliament of genetic parts" was weakened as a theory by the evidence that even humans who lacked an arm or a leg from birth could produce perfectly normal offspring.

## 9.2 MENDEL'S LAWS

Gregor Mendel was born in Austria in 1822 and later became a monk in a monastery located in what is now Brno, Czechoslovakia (then a part of the Austro-Hungarian Empire). He was given a small garden in which to conduct experiments on garden peas. His efforts to solve the riddles of the genetics of his day succeeded beyond his wildest scientific dreams, but, even so, he had no idea that he had laid a permanent foundation for what is called *classical genetics*. Only after his death did his observations and theories become the recognized fundament of modern genetics.

In Mendel's time, heredity was believed to be the additive outcome of maternal and paternal influence, a blending of lineages much like the mixing of paints. What Mendel showed was that heredity involved the interaction of discrete separable factors—a *particulate* theory of inheritance rather than a blending process. He was a painstaking scientist whose success was due to many factors, but especially to his quantification of the data and to his maintenance of careful records.

Pea plants are extremely useful for genetic studies because they are cheap and easy to procure, produce children in a single growing season, and cover their sexual parts with modified petals so that philandering will not intrude upon a careful experiment. Specific, known pollen from the male can be sprinkled upon the female pistil and eventually will send a sperm nucleus to unite with the egg. Such a genetic cross can then be analyzed. Mendel chose seven distinct traits to study, e.g., stem length, seed surface, and seed color. The significant experiments that produced his two major laws were completed in less than five years, but he continued to refine his experimental probes until age, obesity, and administrative responsibilities caught up with him.

## MENDEL'S FIRST LAW

Mendel was an expert plant breeder, and he was able to produce varieties of plants that always self bred true, generation after generation. He selected strains that were true breeders but of opposite types and made crosses between them for each of the seven different characteristics he had selected.

**EXAMPLE 1**  Mendel bred a strain of peas that was always tall. He crossed them genetically with a strain that bred true for shortness. In the first filial ($f_1$) generation [progeny of the original ($P_1$) parental cross] he found that all the offspring were tall, exactly like the tall parent. This certainly was inconsistent with the conventional wisdom that would have predicted that the progeny would all be intermediate in size. But Mendel did not stop there. He crossed all the $f_1$ progeny with themselves. This is relatively easy in garden peas, since the plants contain both male and female parts in the flower and generally are self-fertilizing. In the $f_2$ generation that arose from the $f_1$ cross, Mendel found that there were both tall and short individuals. The short trait, which was not expressed in the $f_1$ generation, reappeared as if by magic in the $f_2$ generation. Of the 1064 young pea plants in the $f_2$ generation, 787 were tall and 277 were short—close to a 3:1 ratio—with none of the offspring intermediate in size.

Mendel studied his results and realized that a trait like pea-plant height did not run a spectrum of values but existed in two clearly distinguished classes, tall or short. He hypothesized that the determinants of a trait also existed as discrete separable factors, a factor for tallness and a factor for shortness. (We would now call these factors *genes*.) Tall plants had the tall factor, and short plants the short factor.

Confronted with the fact that the tall plants of the $f_1$ generation were able to produce both tall and short $f_2$ plants, Mendel realized that individuals did not have a single factor for any inheritable trait, but a pair of factors. This accords with the fact that all sexually reproducing beings, including pea plants and people, have a pair of parents each of whom contributes a factor for a trait. One form of the factor (tallness, in this case) tends to be *dominant* over the other (shortness), which is said to be *recessive*; so when both are present, only the dominant one is expressed in the way the offspring looks (its *phenotype*). This accounts for the fact that in the $f_1$ generation all the offspring were tall; however, lurking in the genetic underpinnings is the factor for shortness.

By keeping track of the alleles in the parents' gametes, it is possible to explain Mendel's results.

**EXAMPLE 2**  The gene for height in Mendel's peas exists in two allelic (alternate) forms. The allele for tall stature will be designated $T$ and that for short stature $t$. The genotype (kinds of alleles present) for the homozygous (same alleles) tall parent would be $TT$, while that for the homozygous short parent, $tt$.

$$P_1 \text{ cross:} \quad TT \times tt$$

The gametes of the tall plants would all have the $T$ allele; the gametes of the short plant, all $t$ alleles, since each of the parents would have only one kind of allele to contribute to the haploid gamete.

In the $f_1$, the $T$ and $t$ gametes unite to produce individuals with a $Tt$ (heterozygous) genotype. Each of these progeny produces gametes, some of which contain the $T$ allele and some the $t$ allele. These gametic classes are produced in equal numbers. The various combinations of the $f_2$ can be shown by placing classes of sperm along one axis and classes of egg along its perpendicular. All possible crosses are then shown as the intersections of gametic columns and rows:

|       | $T$  | $t$  |
|-------|------|------|
| $T$   | $TT$ | $Tt$ |
| $t$   | $Tt$ | $tt$ |

The square formed is called a *Punnett square* after the English geneticist who used that device to illustrate genetic crosses.

The $f_2$ shows a 1 $TT$ : 2 $Tt$ : 1 $tt$ *genotypic* ratio, but a 3:1 *phenotypic* ratio of dominant to recessive pea plants, since the presence of the $T$ allele confers the dominant trait upon the individual.

Mendel's first law is called the *law of segregation*. Simply stated it affirms the existence of a pair of particulate factors (genes) which control each trait and which must segregate (separate) at gamete formation and then come together *randomly* at fertilization. Further, one of these factors tends to be expressed at the expense of the other when both are present.

Given that Mendel was not a cytologist and knew nothing of mitosis and meiosis, his first law is a remarkable reflection of the behavior of chromosomes. However, it was not until 35 years after he presented his results that this remarkable connection was laid bare.

As discussed in Chap. 8, the gene is actually a stretch of DNA representing a blueprint unit. The gene for height, or any other trait, may exist in two or more alternative forms known as *alleles*, e.g., tallness and shortness. If a pair of alleles in an individual are the same, the individual is said to be *homozygous* for the trait in question. An individual with a pair of contrasting (different) factors is *heterozygous*, or hybrid. The alleles that are present in the genome make up an individual's genotype; they are a significant but not the sole determiner of the individual's appearance, or phenotype. The genotype interacts with the environment to produce a final phenotype. Thus, if the genotype of a pea plant were for tallness but such a plant were deprived of sun and nutrients, it would not become tall.

**EXAMPLE 3**   Siamese cats are dark only at their extremities, such as their face and paws. An enzyme for dark pigment production is produced by all cells in a Siamese cat, but this enzyme is destroyed by the relatively high temperature of the cat's body. Only the extremities are cool enough to permit this enzyme to function and produce dark patterns.

## MENDEL'S SECOND LAW

Mendel determined that the law of segregation applied to all seven of the contrasting traits he studied. He then began a study to determine whether two traits examined simultaneously were inherited independently or were influenced by one another's hereditary patterns. If we cross a pure-breeding (homozygous) tall plant that is also homozygous for yellow seed color (dominant) with a plant that is homozygous recessive for both traits, i.e., short height and green seed color, will these traits sort out independently or will offspring that are dominant for one trait also be dominant for the second trait? In other words, does the inheritance of one trait influence, or is it otherwise linked to, the inheritance of a second (or third, fourth, etc.) trait?

When Mendel performed crosses similar to his first group of experiments, but involving two traits at a time, he found that the assortment of alleles was completely "democratic." A gamete that received the dominant allele for height during the segregation process could also receive either the dominant or recessive allele for seed color. All possible combinations of traits could appear in the $f_2$ generation.

**EXAMPLE 4**   The $f_1$ progeny are heterozygous for height (one "tall" allele $T$ and one "short" allele $t$) and for seed color (a $Y$ allele for yellow and a $y$ allele for green). If independent assortment occurs, each heterozygote can produce four classes of gametes—$TY$, $Ty$, $tY$, and $ty$. A Punnett square can be used to show the possible combinations of these gametes.

|    | TY | Ty | tY | ty |
|----|------|------|------|------|
| TY | TTYY | TTYy | TtYY | TtYy |
| Ty | TTYy | TTyy | TtYy | Ttyy |
| tY | TtYY | TtYy | ttYY | ttYy |
| ty | TtYy | Ttyy | ttYy | ttyy |

As can be seen from the Punnett square, there are four possible phenotypes that can result from the independent assortment of the genes for height and seed color—tall plants with yellow seeds, tall plants with green seeds, short plants with yellow seeds, and short plants with green seeds. The Punnett square shows that these are expected to occur in the ratio 9:3:3:1.

Mendel's second law, called the *law of independent assortment*, or the *law of unit characters*, expresses the concept that traits are inherited *independently* (recall Prob. 8.12 in the preceding chapter). Furthermore, the ratios of the different phenotypes are easily calculated from the laws of probability for each class.

**EXAMPLE 5**   In statistics, chance events are readily assessed in terms of their probability, since each event has the same probability of appearing as any other event (there are no favorites for the gods of chance). The probability of any event is always one over the total number of events possible. A coin can land either heads or tails. The chance of producing a head then is 1/2 (1 out of 2). These same laws of probability affirm that the chance of two independent events occurring at the same time is the chance of the first event's occurring multiplied by the chance of the second event's occurring, i.e., the probability of two or more events happening concurrently is the product of their respective probabilities. Thus the chance of obtaining three heads if we toss three coins in the air is 1/8. [Since the chance for heads for each coin is 1/2, we merely multiply (1/2) (1/2) (1/2) for the three coins.]

Mendel used the same logic to predict the frequency of classes obtained when *dihybrid* (heterozygous for two different traits) $f_1$ crosses were made in the studies of the inheritance of two or more traits. Contiguous independent events could always be predicted by multiplying the individual probabilities for each single event.

**EXAMPLE 6**   In a monohybrid cross, three-fourths of the progeny demonstrate the dominant character and one-fourth the recessive trait, the "3/4, 1/4 rule." Since unlinked traits are inherited independently, the chances of two traits' appearing together will be the products of the individual probabilities. Thus, if 3/4 of the offspring are dominant in appearance in monohybrid crosses for each one of the two traits, then (3/4) (3/4) = 9/16 of the offspring will be dominant for both traits in a dihybrid cross. In the same way (3/4) (1/4) = 3/16 of the offspring will be dominant for the first trait and recessive for the second. Those recessive for both traits will be 1/16 of the total [(1/4) (1/4)].

This same approach permits us to predict ratios obtained for particular combinations of traits in trihybrid crosses. A mere 27/64 of progeny will show the three dominant traits in such a cross. Only 1/64 will be recessive for all three traits. If one selects 3/4 to represent the frequency of any one dominant phenotype and 1/4 to represent the frequency of its recessive phenotype, then any problem involving hybrid crosses is readily solvable.

In later years Mendel turned to a common plant called hawkweed to determine whether the laws of heredity worked out for garden peas would apply for other test organisms. Alas, for Mendel, his crosses did not produce similar results, and he died a bitterly disappointed man, although he had risen to abbot of his monastery. Hawkweed is capable of reproducing *parthenogenetically*—an egg develops into a new individual without the genetic contribution of the sperm arising from the pollen. Mendel's crosses with hawkweed were not really crosses at all but more a "double cross" by his obstreperous weed—a sad ending to a saga of incomparable accomplishment.

## STATISTICAL VERIFICATION OF MENDEL'S LAWS

As we have seen, the results of experiments in genetics are often expected to fit certain predicted ratios. Sometimes the data are so close to the ratios that there is little doubt that they conform to the predictions. Frequently, however, results deviate sufficiently from predictions so that a question arises about whether the deviation is due to chance or to a faulty hypothesis. In such cases, statistical analysis is necessary.

In 1900 the statistician Karl Pearson developed a technique [*chi-square* ($\chi^2$) *analysis*] for determining the *goodness of fit* of a particular set of data with a hypothesis. For example, if a cross is made between heterozygous tall pea plants and homozygous short plants, our hypothesis would be that we should get a 1:1 ratio of tall to short phenotypes in the progeny. Deviations from a 1:1 ratio could be interpreted in one of two ways:

1.   The deviations are caused by chance alone and are not sufficient to make us suspect that the variation is significantly different from the expected 1:1 ratio. An assumption that there is no significant difference between the real and the expected results is called the *null hypothesis*.

2.  The differences are so great (chi square is high) that there is little likelihood that chance fluctuations are responsible. In such a case the null hypothesis would be rejected, and the ratio would not be regarded a 1:1 distribution.

In performing a chi-square analysis, the difference (deviation) between the observed value ($O$) and the expected value ($E$), for each component of the ratio, is determined. Each deviation is then squared and divided by this predicted value $E$, or

$$\frac{(O - E)^2}{E}$$

$\chi^2$ is derived by summating these values and then comparing the sum with a figure in a chi-square table of probabilities. (An exact fit would yield a chi square of 0.) If the calculated sum ($\chi^2$) is very high, we would probably reject the hypothesis that the ratio fits the expected distribution, since the probability for such a high value by chance alone would be very low. The probability figure chosen from the table will depend on the *level of confidence* desired; if the probability of getting a chi square that we have computed is less than 0.05, then we would reject the null hypothesis with the confidence that 95 times out of 100 instances this was not a case of the expected ratio.

**EXAMPLE 7**  For the cross between heterozygous tall pea plants and homozygous short pea plants, suppose we obtain a ratio of 30 tall plants to 38 short plants. Is the deviation from the expected 1:1 ratio significant at the 0.05 level (i.e., is the deviation sufficiently great that 95 percent of the time we would be correct in rejecting the hypothesis that the data fit a 1:1 ratio)?

Since 68 plants were produced, we would expect 34 tall ones and 34 short ones. Computing $\chi^2$, we get

$$\chi^2 = \frac{(30 - 34)^2}{34} + \frac{(30 - 38)^2}{34} = 0.47 + 0.47 = 0.94$$

Chi-square probability tables show $\chi^2_{0.95} = 3.84$. Our calculated figure is less. We can then accept the hypothesis at the 0.05 level, since our value for chi square will occur by chance alone with a greater frequency than 5 out of 100 times. We could actually expect to get a chi square as high as 0.94 from 30 to 50 percent of the time by chance alone.

This same statistical tool may be used to evaluate other ratios, such as the 3:1 phenotypic ratio of a hybrid cross. We may also test results in which we have three or more classes in our results.

### CONNECTING MENDEL'S LAWS TO THE CHROMOSOMES

Mendel's only paper of his results was published in 1866, but it received little attention from other scientists. In 1900, Mendel was "discovered" by three separate groups, each of whom had independently worked out the law of segregation. They included Hugo De Vries of Holland, Carl Correns of Germany, and Erich Tschermak von Seysenegg of Austria.

In 1901, William Sutton, a graduate student at Columbia University in New York, was studying the migration of chromosomes during meiosis, which had first been described by Theodor Boveri. Mendel's newly discovered work was being talked about at Columbia at that time. Sutton put the two ideas together and developed the concept that chromosomes are the physical basis of inheritance. It is the chromosomes that contain a pair of alleles, since each chromosome shares with its colleague an identical locus for the same gene. It is the chromosomes which segregate from one another during gamete formation, and it is the chromosomes of sperm and egg which come together randomly at fertilization to provide even further variety in the recombination of chromosomes. For Sutton it was clear that genes are chromosome parts, and yet most cytologists and geneticists for the next 17 years would contend that chromosomes were casual constructs in the nucleus and not associated with the genes themselves.

### 9.3  LINKAGE

In the early 1900s, Thomas Hunt Morgan set up a genetics laboratory at Columbia University that launched several Nobel laureates and provided the superstructure for the foundations of classical genetics. Morgan worked with the tiny fruit fly, *Drosophila melanogaster*, which was easy to use, cheap to maintain, and had a generation time of less than a month.

Working in a relatively small space dubbed the "fly room," Morgan's group found early on that Mendel's law of independent assortment was not always true. A number of traits of *Drosophila* were inherited together rather than separately.

As we have seen, when dihybrids are crossed, we expect all combinations of the traits to show up in the next generation. This is because each trait is inherited independently, according to Mendel. With two traits involved there are four phenotypic classes possible: dominant for first, recessive for second; dominant for first, dominant for second; recessive for first, recessive for second; and recessive for first, dominant for second.

In Morgan's laboratory, it was found that certain dihybrid combinations produced only two of the four possible classes. Those traits that did not segregate independently were said to be *linked*. Linkage occurs between genes that lie on the same chromosome. Independent assortment of alleles for such genes cannot occur, because during gametogenesis they migrate as a unit (i.e., they migrate together on their chromosome). Thus, an allele at locus 1 does not have a choice of entering a gamete either with the allele at locus 2 on its own chromosome or with the allele at locus 2 on the homologous chromosome. It must migrate only with the locus 2 allele on its own chromosome. (As will be discussed in the next section, such linkage can be broken by crossing over.)

**EXAMPLE 8** Assume that at the gene locus for flower color a dominant mutation occurs that causes blossoms to be blue, instead of the normal red. If, on the same chromosome, a nearby gene locus that codes for leaf shape has a dominant allele for round leaves, these two alleles, because of their linkage, will always enter gametes together (discounting crossing over), and, whenever blue blossoms are present, only round leaves (not the other, unlinked alleles for leaf shape) will accompany them.

Morgan and his collaborators found that there were four linkage groups in *Drosophila*. Each of the many traits manifested by the flies could be assigned to one of these four groups. *Drosophila* also contains four chromosomes as a haploid set. This certainly constituted a strong argument that chromosomes are the physical basis for heredity, the carriers of the genes.

## 9.4 MAPPING THE CHROMOSOMES

One would assume that linkage, like pregnancy, is an all-or-none affair. But in many of the crosses prepared by Morgan's group, linkage appeared to be "leaky." Although only two of the four possible phenotypic combinations would occur for most of the offspring studied (a clear indication of linkage), out of hundreds of these offspring a very few would show *reassortment* (separation) of the linked traits. How could traits be linked and then show evidence of being unlinked? The answer is simple in light of what we now know about crossing over (Chap. 8). Genes that are linked may, during crossing over, break that linkage through an exchange of parts between homologous chromosomes. If genes *A* and *B* are found on one homologue and *a* and *b* on the other, the two genes are said to be linked. Linkage dictates that gametes would contain the *AB* homologue or the *ab* homologue; a gamete with an *Ab* or *aB* combination would supposedly not form. However, if crossing over occurs anywhere along the length of the chromosome that lies *between* the two gene loci, the exchange of chromatids would interchange one of the loci, while leaving the other locus unmoved. This would produce these "odd" combinations of alleles and permit the unexpected outcomes.

It was actually a young associate of Morgan, A. H. Sturtevant, who realized that the frequency of recombination of linked genes (the "leakiness" of linkage) could be an indication of how far apart genes were positioned along the linear chromosomes. Since crossover sites are random occurrences along the chromosomes, the further apart two linked genes are located on their chromosome, the greater the likelihood that crossover events, which will break the linkage, will occur between them. Obviously, genes that are close together will be very tightly linked, since only a rare crossover event will occur between them to break their linkage.

By carefully determining the relative frequency of linkage leakiness between all the traits of *Drosophila*, taken two at a time, Sturtevant was able to draw hypothetical maps of the relative positions of all the genes. This involved an inordinate number of crosses to establish recombination frequencies. These maps actually reflected recombination (leakiness) rather than actual linear distances, but it is presumed that the two are closely related.

## 9.5  SEX-LINKAGE

In the crosses that Mendel performed, the results did not depend on which parent contributed a particular set of alleles to the zygote. The cross of a homozygous dominant with a homozygous recessive was always the same regardless of whether the mother or father was dominant. The neutrality of gender in most genetic crosses in the fruit fly was also apparent, but for some traits gender did influence the outcome. In those situations, first discovered by Charles Bridges in Morgan's laboratory, where traits are inherited differently in males than in females, we use the term *sex-linkage.*

Traits that are sex-linked have genes on the sex chromosomes. The best known of these traits are those associated with the X chromosome. In both *Drosophila* and humans (and other mammals) the male is usually XY and the female is XX. Since the male has only one X and a Y that has comparatively few genes, a recessive mutation occurring on the X chromosome of the male will be expressed, because there is no homologous chromosome to contain an allele that might suppress the recessive. In the female, only the homozygous recessive will produce expression of the mutation, since a dominant *wild-type* allele (the more common allele occurring in nature) on one of the X chromosomes would suppress expression of the one recessive mutation on the other X chromosome.

**EXAMPLE 9**  In *Drosophila*, white eye is a mutant recessive trait, while red eye is the wild type. White-eyed males crossed with homozygous red-eyed females produce offspring all of whom are red-eyed. This is what one would expect from any homozygous dominant and recessive cross. On the other hand, white-eyed females crossed with red-eyed males produce offspring in which the females are all red-eyed while the males are all white-eyed. Since the gender of the offspring is a factor in the pattern of inheritance, this is a classic case of sex-linkage. All female offspring must receive an X chromosome from the male parent. In the latter cross, since the dominant red-eye allele was on the males' only X chromosome, all females received the dominant allele and were red-eyed. All male offspring must receive an X chromosome from the female parent; in this cross the males received the recessive allele for white eyes and, lacking a homologous allele, expressed the recessive.

In humans a variety of karyotypic diseases involving sex-linkage have been described. *Hemophilia,* a disease in which blood-clotting mechanisms are impaired, is caused by a recessive allele for a gene lying on the X chromosome. Women with one wild-type allele and one mutant allele are of normal phenotype but are *carriers* for the disease. Half their sons will have the disease and half will be normal. Color blindness is also carried as a mutant recessive on the X chromosome, and the dynamics of its inheritance parallel those for hemophilia.

At one time it was thought that there were few functional genes on the Y chromosome. We now know that there are genes with phenotypic expression that lie on the Y chromosome. In humans, a gene for baldness probably has its locus on the Y chromosome, and only males are involved in its inheritance.

In 1948 Murray Barr and Dewart Bertram discovered a dark-staining locus in the nuclei of female mammals that was not present in the nuclei of male cells. Dubbed *Barr bodies*, these deeply stained structures were later found in the cells of men who suffered from Klinefelter's syndrome (XXY genotype).

Some years later, the British geneticist Mary Lyon provided an explanation for the appearance of the Barr body. It represents a highly condensed inactivated X chromosome. Whenever two X chromosomes are present together, only one will exert a genetic effect; the other will remain inactive as a tightly coiled mass of heterochromatin. The logic underlying the phenomenon is elegant: only one X chromosome is present and active in the cells of the male so that an equivalent gene dosage is present in the female as a result of the inactivation of one of her X chromosomes. Lyon maintained that the inactivation is a random event, so that some of the cells of the female are influenced by the paternal X and others by the maternal X. Descendents of a cell in which a particular X had been inactivated continue to have the same X inactivated.

In a variety of karyotypic diseases, multiple copies of the X chromosome may arise. The evidence is strong that only one X is ever active in these cells, and two or more Barr bodies may be found. The inactivation of the X occurs after gender has been determined, since the two X chromosomes are necessary to effect primary sexual differentiation. Because inactivation may occur late or even be incomplete, there may be an abnormal dosage of gene activity in these cells.

## 9.6   VARIATIONS IN GENE EXPRESSION

We have already seen that a dominant gene can suppress the expression of a recessive one; however, this is just one of many ways in which genes interact with each other and with their environments. These interactions all influence how the gene is ultimately expressed. In the case of the four-o'clock flower, for example, a cross between pure-bred red flowers and pure-bred white flowers produces an $f_1$ with pink flowers. This phenomenon was called "blending" at one time, but it is an unfortunate usage because there is, in reality, no dilution of gene action. If the $f_1$ generation is crossed, we get offspring which are red, white, and pink in a 1:1:2 ratio. Thus, the individual alleles for color have not actually blended; instead, the explanation is that neither allele (red or white) is completely dominant. In this case, *incomplete dominance*, both red and white alleles independently produce products which, when combined, give a phenotype of pink. *Codominance* involves equal expression of more than one allele, i.e. human blood type AB.

Still another aspect of gene expression was discovered by the English geneticist William Bateson. Termed *epistasis*, it involves the effect of the alleles of one gene on the expression of the alleles of an entirely different gene. Thus, one gene exerts a permissive or modifying action on another.

In *pleiotropy* a single gene exerts an influence on several characteristics. Quite often in pleiotropy the underlying biochemical or molecular action of the gene remains constant, but its effect is expressed differently in a variety of organs to produce a complex of symptoms.

## 9.7   CHROMOSOMES AND GENE EXPRESSION

In Morgan's laboratory it became clear early on that the hypothetical hereditary units of Mendel were actually arrayed along the chromosomes. The focusing of classical geneticists upon the chromosomes gave rise to a subdiscipline, *cytogenetics*. It was within the framework of cytogenetics that sex determination was established—the *autosomes* (nonsex chromosomes) were distinguished from the sex chromosomes. In many species XX was associated with the female, but later work showed that in many birds it was the male that was XX, and the female XY. Diseases of nondisjunction were studied by analyzing the aberrant karyotypes of such syndromes as Down's, Klinefelter's, Turner's, etc. Cytogeneticists also showed that chromosomes may break apart at two or more points and then rejoin with the middle segment inverted. In such instances of *inversion*, the linear order of the genes on the chromosome is changed, but the content of gene material remains the same. With some inversions, changes in gene expression have been detected, and this has led to the recognition of a *position effect*. Genes are affected in their actions by their position within the chromosome.

Unusual types of chromosomes in terms of length or shape have also been associated with particular genetic diseases. In certain types of leukemia, a chromosome of aberrant length has been detected. It is called the *Philadelphia chromosome* because of where it was first described. These chromosomes may arise as a result of a transfer of parts between nonhomologous chromosomes—a phenomenon known as *translocation* and similar in its mechanism to a crossover. Many of the structural alterations of chromosomes were first identified in the so-called giant salivary gland chromosomes of the larva of *Drosophila*. These chromosomes are the result of numerous replications of the chromosome with no intervening cell divisions.

## 9.8   TREATING GENETIC DISEASES

There are serious dangers associated with human intervention into the genome. The question of who has the right to decide which genes should be engendered and which eliminated can never be unequivocally answered. More subtly, many alleles that are deleterious in one environment are beneficial in another. Thus, decisions about which should be eliminated are not necessarily straightforward. Nevertheless, it is difficult to possess knowledge about genetic effects and not use it to ameliorate hereditary pathologies.

A modest approach stems from the recognition that alleles do not achieve expression unless they interact with the environment and the experiences of an individual. Professor I. Michael Lerner of Berkeley uses the term *euthenics* to refer to measures in which we alter the environment in a *general* manner to discourage the expression of harmful genetic traits in the population as a whole. As an example, the development of high blood pressure (hypertension) may have a genetic base, but if we substitute low-salt products like fruits and certain vegetables for snack products such as potato chips and pretzels in school vending machines, we may avoid the expression of hypertension until late in life. Diminishing stress by, for example, lowering noise levels in city environments would also reduce the dangers associated with genetic vulnerabilities.

A second environmental approach might be *euphenics*, an intervening in *specific* cases with biological means to overcome genetic problems. Some infants are born with the inability to handle the amino acid phenylalanine. This condition, phenylketonuria (PKU), is due to a single gene mutation and can be dealt with by providing a diet low in phenylalanine through infancy and early childhood. This treatment avoids the mental retardation that was at one time associated with the disease. Another example of euphenics is the replacement of defective organs with transplants. In each of these cases a specific prescriptive intervention occurs in which the genotype is unaltered but compensations are provided.

*Genetic counseling* offers advice to prospective parents about the likelihood that their offspring will develop various genetic diseases. Concern usually centers upon the possibility that a normal couple carries recessive alleles for conditions such as Tay-Sachs disease (a fatal lipid metabolism disorder) or sickle-cell anemia. Since these diseases occur only in the homozygous recessive condition, a mating of two carriers provides a 1/4 chance of producing a defective child. With *amniocentesis*, which is a technique for obtaining cells from the unborn child for analysis, the status of the fetus can be determined.

*Eugenics* refers to social programs in which matings of individuals with "desirable" traits are encouraged and matings of individuals with "undesirable" traits are discouraged. Eugenic measures have included monetary awards to those deemed "fit" and sterilization or even incarceration of those regarded as unfit. The opportunity for social abuse is enormous.

Perhaps the most hopeful scenario is yet to be written—the use of genetic engineering to repair damaged genes. Work is already under way to insert into the chromosomes of individuals suffering from hereditary anemias the functioning genes that are missing or defective. Bone marrow is a tissue of choice because it can be removed from the body, treated, and then reintroduced into the bone, where it may produce a healthy lineage of cells. Genetic engineering of bacteria is already used to produce insulin, growth hormone, and a number of protein products necessary for the treatment of patients with genetic diseases.

# Solved Problems

**9.1**   If flies with straight wings (wild type) are crossed with flies with wrinkled wings, all the $f_1$ progeny have straight wings. Using a Punnett square, predict the phenotype of the $f_2$ progeny and their relative proportions.

Since no flies with wrinkled wings appear as a result of the $P_1$ cross, we will assume the parents were homozygous. Since all $f_1$ progeny had straight wings, we will also assume that straight wings ($S$) are dominant over wrinkled wings ($s$). The $f_1$ progeny, being heterozygous ($Ss$), will produce both $S$ and $s$ gametes in equal numbers. Therefore the Punnett square for the $f_1$ cross would be:

♂<br>
$S$    $s$

|   | $S$ | $s$ |
|---|---|---|
| $S$ | $SS$ | $Ss$ |
| $s$ | $Ss$ | $ss$ |

♀

Since $S$ is dominant over $s$, all genotypes containing even one $S$ will have straight wings; only $ss$ genotypes will express the wrinkled-wing phenotype. Thus, the $f_2$ progeny will show a typical Mendelian dominant-recessive relationship of 3:1.

**9.2**  Of the three-fourths of the $f_2$ pea plants that were dominant *in appearance* in Mendel's experiments with height, some were homozygous dominant ($TT$), while others were heterozygous dominant ($Tt$). Using another cross, how would it be possible to show which individuals were $TT$ and which $Tt$?

Mendel hypothesized that if he could cross these dominant tall plants with a recessive partner, he could distinguish the heterozygotes from the homozygotes. This is called a *test cross* or *back cross*.

The recessive parent would produce only one class of gametes, those containing $t$. If crossed with the homozygous dominant ($TT$), all the offspring should be tall ($Tt$). (This is exactly the same as the original $P_1$ cross carried out earlier by Mendel.) The heterozygous parent, on the other hand, would produce two classes of gamete—one containing the $T$ allele and one containing the $t$ allele. In a cross with $t$ gametes, one would expect both tall and short offspring in equal numbers.

For all practical purposes, the appearance of even one recessive product of this cross of a dominant of unknown genotype with a recessive would indicate that the dominant is heterozygous rather than homozygous.

**9.3**  A mark of Mendel's care and persistence is that he performed a test cross on each of the tall plants that arose from the $f_2$ of his original cross with tall and short plants. What ratio of tall to short plants would you expect him to have obtained from his test crosses?

Mendel's $f_2$ generation had genotypes in the ratios of 1 $TT$:2 $Tt$:1 $tt$. Looking only at the tall plants produced in this generation (the only plants involved in the test cross), we can see that there are twice as many heterozygotes as homozygotes. Another way of viewing this fact is that, relative to the homozygotes, the heterozygotes produce twice as many gametes and therefore twice as many alleles as the tall homozygotes. Now, consider how many $T$ alleles versus $t$ alleles there are in the tall population. The homozygous tall plant contributes two $T$ alleles, so there are $2x$ $T$ alleles contributed by the homozygotes, where $x$ is the total number of homozygous plants. The heterozygous tall plant contributes only one $T$ allele; however, since there are twice as many heterozygotes, these plants also contribute $2x$ $T$ alleles. The heterozygote also produces a $t$ gamete, and the whole population of heterozygotes produces $2x$ $t$ alleles. Comparing $T$ and $t$ alleles, there are $4x$ $T$ alleles ($2x + 2x$) and $2x$ $t$ alleles in the population, or a ratio of 2:1. Each of these alleles will combine with a $t$ allele from the homozygous short plant in the test cross; a $T$ allele from the population being tested will produce a tall plant ($Tt$), and a $t$ allele a short plant ($tt$). Since the ratio of $T$ alleles to $t$ alleles is 2:1, this will also be the ratio of tall to short plants we would expect from the test crosses. This is, in fact, what Mendel obtained and provides clear verification of the law of segregation—a law with its roots in the operation of chance for the distribution of the discrete particulate alleles.

**9.4**  Describe Mendel's second law in terms of the alleles that segregate independently.

To illustrate the independent assortment of alleles, we will deal with two traits at a time—the form (round or wrinkled) and color (yellow or green) of the seed. Let $R$ be the round allele, $r$ the wrinkled allele, $Y$ the yellow allele, and $y$ the green allele. Note: $R$ is dominant over $r$, and $Y$ is dominant over $y$. Let us cross the homozygous dominant for both traits with the homozygous recessive in $P_1$. Only one class of egg or sperm is possible (the gametes are circled):

$$P_1 \quad RRYY \quad \times \quad rryy$$
$$\widehat{RY}\, ♀ \qquad\qquad \widehat{ry}\, ♂$$
$$f_1 \qquad\qquad RrYy$$

The $f_1$ progeny will all be round and yellow and heterozygous for each trait. This kind of genotype is called a dihybrid.

Let us now cross the $f_1$. If independent assortment occurs, the dihybrid will produce gametes with all the possible combinations ($RY$, $Ry$, $rY$, and $ry$); if assortment does not occur, then one scenario could be that the gamete containing an $R$ will also always contain a $Y$, while a second class of gamete will contain $ry$. When Mendel performed the $f_1$ cross, he found that the phenotypic ratios of the $f_2$ corresponded to a situation in which four classes of sperm were able to unite with four classes of egg:

|   |   | | ♀ | | |
|---|---|---|---|---|---|
| | | $RY$ | $Ry$ | $rY$ | $ry$ |
| ♂ | $RY$ | $RRYY$ | $RRYy$ | $RrYY$ | $RrYy$ |
| | $Ry$ | $RRYy$ | $RRyy$ | $RrYy$ | $Rryy$ |
| | $rY$ | $RrYY$ | $RrYy$ | $rrYY$ | $rrYy$ |
| | $ry$ | $RrYy$ | $Rryy$ | $rrYy$ | $rryy$ |

Mendel actually obtained an $f_2$ ratio of 9 round yellow:3 round green:3 wrinkled yellow:1 wrinkled green seed. If we examine the Punnett square above, we note that those ratios are precisely what one would expect from a cross in which each pair of contrasting alleles is free to sort out independently.

Suppose the alleles did not segregate independently and the gametes produced by the $f_1$ were of only two classes: $RY$ and $ry$. Then we would have a cross

| | $RY$ | $ry$ |
|---|---|---|
| $RY$ | $RRYY$ | $RrYy$ |
| $ry$ | $RrYy$ | $rryy$ |

3 round yellow: 1 wrinkled green

Only two phenotypic classes would be produced in the $f_2$, in a 3:1 ratio. Since Mendel's $f_2$ consisted of four phenotypic classes in a 9:3:3:1 ratio, an independent assortment of different gene categories was clear.

It is also of considerable interest that Mendel chose seven traits for which the genes were all on different chromosomes and therefore were not linked. The haploid set in the garden pea consists of seven chromosomes, and the odds are astronomically high that the genes of seven traits chosen at random should lie on the seven different chromosomes. Some geneticists feel that Mendel may have ignored data for some traits that did not follow the second law.

**9.5**  Yellow-haired house mice interbreed and produce progeny with a 2:1 ratio of yellow to nonyellow. When yellow is crossed with nonyellow, a 1:1 ratio of the two classes is obtained. Nonyellows interbreed to produce all nonyellow offspring. How can this be explained?

This would appear to be a case in which yellow is dominant and nonyellow is recessive. The nonyellows always produce nonyellow offspring; this is consistent with its recessive character. The only anomaly is the behavior of the yellows. They seem to be heterozygous but fail to yield the usual 3:1 ratio of a hybrid cross.

The explanation is that the homozygous dominant, *YY*, is a lethal combination and all such individuals die before birth. All surviving yellow mice are therefore hybrid. The hybrid cross, which usually gives a 3:1 ratio, is characterized here by a 2:1 ratio because the homozygous dominant, making up one-third of the dominant phenotype in the cross, does not show up in the final accounting.

**9.6** In domestic poultry, the character of the comb is controlled by two genes, rose and pea. If the dominant allele *R* is present with a dominant *P*, then a "walnut comb" is produced. If an individual is homozygous recessive (*rrpp*) the comb will be "single." If an *R* is present without a *P*, the comb will be rose, whereas a *P* present without an *R* produces a pea comb. Determine the phenotypes of the cross *RrPp* × *Rrpp*.

First determine the gametes; then perform the cross.

|  | ⓇⓅ *RP* | Ⓡⓟ *Rp* | ⓇⓅ *rP* | Ⓡⓟ *rp* |
|---|---|---|---|---|
| Ⓡⓟ *Rp* | *RRPp* | *RRpp* | *RrPp* | *Rrpp* |
| Ⓡⓟ *rp* | *RrPp* | *Rrpp* | *rrPp* | *rrpp* |

↓

| Walnut | Rose | Walnut | Rose |
|---|---|---|---|
| Walnut | Rose | Pea | Single |

Phenotypes:

**9.7** A walnut crossed with a single produced among the progeny only one single-combed offspring. What were the genotypes of the parents?

The single parent could only be homozygous recessive (*rrpp*). Since a single-combed offspring was produced, each parent would have to provide at least one class of gamete that was *rp*. Thus, the walnut parent must have had the genotype *RrPp*. A homozygous dominant genotype for either gene could not yield a gamete with two recessive alleles.

**9.8** The following table lists the traits in pea plants that Mendel studied and shows the results of his $P_1$ and $f_1$ crosses. For example, the data show that of the 8023 $f_2$ progeny, 6022 showed the dominant phenotype of yellow seed color, while 2001 showed the recessive green seed color. Of the 6022 $f_2$ seeds showing yellow color, how many would be expected to be round and produce short-stemmed plants with purple axial flowers and constricted green pods?

| Trait | $P_1$ Cross (Dominant × Recessive) | All $f_1$ Progeny | $f_2$ Progeny Dominant | $f_2$ Progeny Recessive | Total $f_2$ Progeny | Ratio, Dominant to Recessive |
|---|---|---|---|---|---|---|
| Stem length | Tall × short | Tall | 787 | 277 | 1064 | 2.84:1 |
| Seed form | Round × wrinkled | Round | 5474 | 1850 | 7324 | 2.96:1 |
| Seed color | Yellow × green | Yellow | 6022 | 2001 | 8023 | 3.01:1 |
| Flower position | Axial × terminal | Axial | 651 | 207 | 858 | 3.14:1 |
| Flower color | Purple × white | Purple | 705 | 224 | 929 | 3.15:1 |
| Pod form | Inflated × constricted | Inflated | 882 | 299 | 1181 | 2.95:1 |
| Pod color | Green × yellow | Green | 428 | 152 | 580 | 2.82:1 |

The probability of several events occurring simultaneously is equal to the product of their individual probabilities of occurrence. Mendel's data show that an $f_1$ cross yields a 3:1 phenotypic ratio; i.e., for every four progeny, three show the dominant phenotype, and one shows the recessive phenotype. Therefore, the probability is 3/4 that one of Mendel's plants would show a dominant characteristic and 1/4 that it would show a recessive trait. The individual probabilities of occurrence for the various phenotypes are as follows: short stem, 1/4; round seed, 3/4; purple flower, 3/4; axial position, 3/4; constricted pod, 1/4; and green pod, 3/4.

The probability of all these phenotypes occurring at the same time (i.e., in the same plant) is:

$$1/4 \times 3/4 \times 3/4 \times 3/4 \times 1/4 \times 3/4 = 81/4096$$

Therefore, of the 6022 yellow-seeded progeny, 119 (i.e., $6022 \times 81/4096$) would be expected to have all these traits at the same time.

**9.9**    In the preceding problem, the probability of a multihybrid phenotype's occurring was determined from the individual probabilities of each of the *phenotypic* traits. The laws of probability can similarly be invoked in predicting genotypes. For a dihybrid $f_1$ cross for height and seed color, use the probabilities of the various *alleles* (instead of the phenotypic probabilities) to determine what proportion of the $f_2$ progeny would be expected to be short-stemmed plants produced from yellow seeds.

Such plants would have the genotype *ttYY* or *ttYy*. There is only one way that the short-stemmed phenotype can be produced: each parent must contribute a *t* allele to create a homozygous *tt* individual. Since the probability of a *t* allele's being produced in a *Tt* heterozygote is 1/2, the probability of two such events occurring together (i.e., of two gametes that contain the *t* allele merging into a *tt* zygote) is $(1/2)(1/2) = 1/4$.

The yellow seed coat can be produced in three different ways: (1) each parent's contributing a *Y* allele (probability of $1/2 \times 1/2 = 1/4$); (2) the male's contributing the *Y* allele and the female's the *y* allele (probability of $1/2 \times 1/2 = 1/4$); or (3) the male's contributing the *y* allele and the female's the *Y* allele (probability of $1/2 \times 1/2 = 1/4$). Each of these three mechanisms contributes to the *total* probability of producing a yellow seed and so must be *added* to determine this total probability: $1/4 + 1/4 + 1/4 = 3/4$. [An alternative way of seeing this is to recognize that the only genotype that will not yield a yellow coat is *yy* (probability of $1/2 \times 1/2 = 1/4$); therefore, the probability of all other genotypes (and, thus, yellow seeds) is $1 - 1/4 = 3/4$.]

Since the probability of short stems is 1/4 and the probability of yellow seeds is 3/4, the probability of these two traits occurring together is $(1/4)(3/4) = 3/16$. (Note: this is the same calculation we would have used if we had started with phenotypic probabilities.)

**9.10**    Suppose that $f_1$ crosses between mice heterozygous for normal ear shape ($T$) and a mutant allele ($t$) for twisted ears produce 735 mice with normal ears and 265 mice with twisted ears. Determine whether, within a 0.05 level of significance, these data conform to the Mendelian law of segregation for dominant and recessive alleles. ($\chi^2$ for 0.05 significance is 3.84.)

According to the law of segregation, we would expect a 3:1 ratio of normal mice to mice with twisted ears. For the sample size of 1000 mice, this would be 750 normal mice and 250 mice with twisted ears. Using chi-square analysis

$$\chi^2 = \frac{(735 - 750)^2}{750} + \frac{(235 - 250)^2}{250} = 0.3 + 0.9 = 1.2$$

Since 1.2 is less than the 3.84, the data conform to normal Mendelian segregation. We would attribute the deviation from the expected ratio to chance alone.

**9.11**    In *Drosophila*, gray body (wild-type allele) is dominant over black body (mutant allele) and is non-sex-linked, since this body color gene lies on an autosomal chromosome. Red eye (wild-type allele) is dominant over white eye (mutant allele) and is sex-linked, since the gene for eye color lies on the X chromosome (there is no homologous allele on the Y chromosome). In *Drosophila*, the male is generally XY and the female XX. Reconstruct the genotype and phenotype of the parents for the following described progenies. *G* will represent the gray allele, *g* the black; *R* will stand for the red-eye allele, *r* for the white-eye.

Males:      3/8 gray, white; 3/8 gray, red; 1/8 black, white; 1/8 black, red
Females:    3/8 gray, white; 3/8 gray, red; 1/8 black, white; 1/8 black, red

At first glance this seems to be a rather formidable task. However, if we take it piece by piece, it can be analyzed and solved. Let us examine the body color first. For both males and females a 3:1 ratio exists, if we add up all the gray categories and compare with the summated black group. This can only be the result of a hybrid cross: *Gg* × *Gg*. Each parent then is a heterozygote, and the phenotype will be gray body color.

Now we examine the situation for the sex-linked eye color trait. Note that the males are both white-eyed and red-eyed (1:1). Since the male receives its one X from the mother, it follows that the mother must have one $X^R$ and one $X^r$ and that she is therefore red-eyed in appearance.

The daughters also show a 1:1 ratio of red eye to white eye. If the father were $X^R$, all the females would be red-eyed, since they all receive one of their two X's from the father and would thus always receive at least one dominant allele for eye color. Thus, the father must be $X^r$ and would appear white-eyed.

To summarize:

$$♂ \qquad\qquad ♀$$
$$GgX^r \quad × \quad GgX^RX^r$$
gray, white eye    gray, red eye

Males:      3/4 gray, white; 1/4 black, white
Females:    3/4 gray, red; 1/4 black, red

Plainly, this is a hybrid cross in terms of body color (3:1 ratio).

Since the males are all white-eyed, the mother must be $X^rX^r$. Since the females are all red-eyed, the father must be $X^R$.

The final result is therefore:

$$♀ \qquad\qquad ♂$$
$$GgX^rX^r \quad × \quad GgX^R$$
gray, white eye    gray, red eye

It should be noted that coat-color ratios are the same for male and female progeny since it is not a sex-linked trait. Eye-color inheritance is different for male and female progeny since its gene is present on the X chromosome.

**9.12**   The problem of men competing as women has arisen in recent years in the Olympic games and in other major sports events. How might the work of Mary Lyon have been helpful in dealing with this problem?

Since individuals free of karyotypic disease will demonstrate Barr bodies only if they are genotypically female, Barr bodies have been used to make sure that athletes competing as women are indeed members of that gender. A buccal smear is made by scraping the inside of the cheek and staining the cells obtained. The presence of the Barr body indicates a female; its absence indicates a male. Lyon explained that the Barr body arises from the inactivation of one of the X chromosomes in cells containing two X chromosomes.

**9.13**   Quite often a population contains a larger number of heterozygotes than would be expected from the normal laws of probability. What might be responsible for this phenomenon?

When homozygotes are selectively less viable than the corresponding heterozygotes, natural selection may favor the heterozygotes. This is known as hybrid vigor. Mutations such as those for sickle-cell anemia confer a high degree of immunity to malaria. Individuals in malarial zones may benefit from carrying one mutant allele for the sickle trait, since they will be relatively resistant to malaria. However, homozygotes for the recessive mutation demonstrate full-blown, often deadly, sickle-cell anemia. On the other hand, homozygotes for the normal allele are culled because of malaria. Thus heterozygotes prevail.

# Supplementary Problems

**9.14** Mendel did not deal with   (a) segregation.   (b) incomplete dominance.   (c) linkage.   (d) both a and b.   (e) both b and c.

**9.15** Sexual reproduction is encountered   (a) only in animals.   (b) only in plants.   (c) in viruses but not in bacteria.   (d) in most organisms throughout the living world.   (e) never in marine forms.

**9.16** The alternative forms of a gene are known as   (a) isomers.   (b) crossovers.   (c) translocations.   (d) alleles.   (e) none of these.

**9.17** Assuming complete dominance, which is greater   (a) the number of genotypes in a hybrid cross or   (b) the number of phenotypes in a hybrid cross?

**9.18** Which is greater   (a) the classes of gametes produced by a homozygous individual or   (b) the classes of gametes produced by a heterozygote?

**9.19** Which is greater   (a) the number of phenotypic classes in a hybrid cross with complete dominance or   (b) the number of phenotypic classes in a hybrid cross with incomplete dominance?

**9.20** Which is greater   (a) the number of linkage groups in *Drosophila* or   (b) the number of linkage groups in humans?

**9.21** Which is greater   (a) the number of Barr bodies in a person with Klinefelter's syndrome or (b) the number of Barr bodies in a normal male?

**9.22** It is not possible to be a carrier (carry the allele for a disease but not have the disease) for traits such as Huntington's chorea that are caused by the dominant allele.
(a) True   (b) False

**9.23** Abrupt hereditary changes (first described by Hugo De Vries) are called mutations.
(a) True   (b) False

**9.24** X-rays increase the rate of mutation.
(a) True   (b) False

**9.25** The existence of a range of differences for certain traits, rather than two discrete classes, is probably due to the fact that a number of genes are involved in the trait (*polygenic inheritance*).
(a) True   (b) False

**9.26** Where multiple alleles exist for a trait, any one individual will have more than two alleles for that trait.
(a) True   (b) False

**9.27** Sex-linked traits may be defined as those traits that affect the sex organs.
(a) True   (b) False

**9.28**  The banding patterns of the giant chromosomes in the salivary glands of *Drosophila* larvae permit the association of genes with specific regions of the chromosome.
(*a*) True    (*b*) False

**9.29**  A normal woman who is a carrier for hemophilia could expect to have half her sons suffer from the disease.
(*a*) True    (*b*) False

**9.30**  Since a tossed coin has an equal probability of landing heads or tails, if we toss a coin in the air and it lands heads, then a second toss must be tails.
(*a*) True    (*b*) False

# Answers

| | | | | | | | |
|---|---|---|---|---|---|---|---|
| **9.14** | (*e*) | **9.19** | (*b*) | **9.23** | (*a*) | **9.27** | (*b*) |
| **9.15** | (*d*) | **9.20** | (*b*) | **9.24** | (*a*) | **9.28** | (*a*) |
| **9.16** | (*d*) | **9.21** | (*a*) | **9.25** | (*a*) | **9.29** | (*a*) |
| **9.17** | (*a*) | **9.22** | (*a*) | **9.26** | (*b*) | **9.30** | (*b*) |
| **9.18** | (*b*) | | | | | | |

# Chapter 10

## Control Mechanisms in Genetics

The chromosome of a bacterium like *Escherichia coli* (*E. coli*) is almost 1 mm in total length and is composed almost entirely of DNA. Virtually all this DNA is active or potential gene material that has the capacity to code for as many as 4000 different polypeptide chains. Perhaps as many as 700 functional enzymes are present in the bacterial cell at any one time, and there is considerable variation in the number of molecules of each protein present at a particular time. Some proteins may be absent or present in just a few copies.

This scenario suggests that there are elaborate mechanisms for the regulation of gene expression in bacteria. The existence of a particular gene on the bacterial chromosome does not guarantee that it will be expressed. Further, the quantitative degree of expression is subject to considerable modification. These modifications of gene expression may be asserted at the level of transcription or translation. The stability of protein synthesizing components may play a key role, as well as the rate at which these components may be fashioned and organized. Enzymic mechanisms for degradation of protein must also be taken into account, since synthesis is significant only when balanced against the opposing tendency of destruction.

Although our understanding of gene control in prokaryotes is far from complete, its details are much more amenable to experimental probing than the similar genetic control mechanisms of the complex eukaryotic cells. In eukaryotic cells a membrane separates the gene material from the cytoplasmic compartment and thus effectively segregates transcription from translation of the code. The vastly increased store of DNA in a eukaryotic cell is intimately associated with and folded around the cores of histone protein that underlie the nucleosome structure. Since it is only the exon regions of eukaryotic DNA that will ultimately be translated into protein, much of the DNA possesses a different class of functions, and this complicates the delineation of a single mechanism for controlling gene expression.

Some understanding of gene control in eukaryotic cells has been achieved with yeast, which is a relatively simple eukaryote. However, a typical vertebrate cell may be up to 40 times larger than a yeast cell, and questions arise about the relevance of control mechanisms in unicellular yeast to those of the very different and far more complex cells of multicellular organisms.

### 10.1 THE OPERON HYPOTHESIS

In 1961, Francois Jacob and Jacques Monod of the Pasteur Institute in Paris proposed a novel theory regarding prokaryotic gene regulation. They theorized that the genes coding for functionally related enzymes lie close together on the bacterial chromosome and are turned on or off as a unit. The assembled unit is controlled by a switchlike gene called the *operator* (see Fig. 10.1). The operator and its adjacent structural genes make up the *operon*. Normally, the operon is turned off because a second type of control gene, called a *regulator gene*, produces a small protein called a *repressor*. This repressor unites with the operator to keep the operator gene in the off position. When certain substances are present in the environment, they react with the repressor and effectively prevent it from tying up the operator. This results in the operon's becoming derepressed and turning on; RNA polymerase, which binds to a *promoter* site, is then free to begin transcription. The substance which ties up the repressor is called an *inducer*, since it influences the operon to turn on and begin to produce enzyme products.

**EXAMPLE 1** The first operon discovered by Jacob and Monod was the *lac operon*, which produces enzymes necessary for the degradation of lactose (milk sugar). The *lac* operon consists of three structural genes which direct the synthesis of three enzymes essential for processing the disaccharide milk sugar lactose. These three genes make up the key functional portion of the operon, and they are switched on or off as a unit. The first of these structural genes codes for the synthesis of β-galactosidase, the enzyme that actually cleaves lactose into glucose

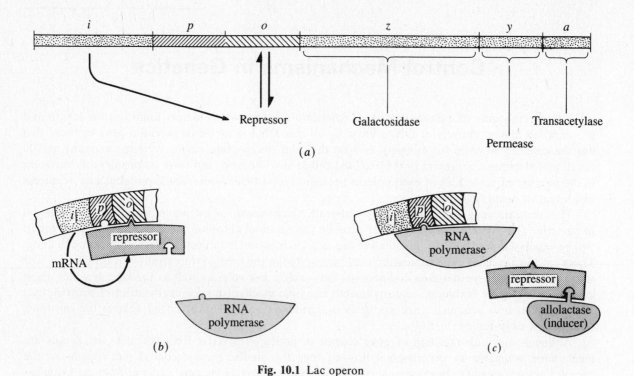

**Fig. 10.1** Lac operon

and galactose, the monosaccharide units of milk sugar. The enzyme is made up of four polypeptide chains (i.e., is tetrameric). The second structural gene codes for *permease*, the enzyme that facilitates the passage of lactose into the bacterial cell, where it may then be processed. (As its name implies, a permease enhances the permeability characteristics of specific substances.) A third structural gene coding for transacetylase completes the operon's

triad of structural genes. *Transacetylase* promotes the transfer of an acetyl group $(CH_3\overset{\overset{O}{\|}}{-}C-)$ from acetyl-CoA

to galactose but is not central to the processing of lactose itself. Usually these enzymes are not produced. When lactose is present in the medium, it acts as an inducer (or *derepressor*) for the operon by attaching to the *lac* repressor and preventing it from inhibiting the operator. Thus, lactose becomes the instrument for bringing about the production of enzymes required for its degradation and use by the cell—a very elegant mechanism for eliciting enzyme production only when these enzymes are needed.

Other operons have been discovered, and the principle of their operation is much the same as for the *lac* operon. In some cases, the operon is normally turned on. Molecules arising in the environment will then turn off the operon. An inactive repressor is first produced; this repressor is activated by the intrusive environmental molecule (*corepressor*) and turns the operon off.

**EXAMPLE 2**   The amino acid tryptophan is synthesized by a pathway that involves five interacting enzymes associated with an operon. Normally, the pathway is active, and the operon is turned on so that tryptophan may be synthesized. Should tryptophan be introduced, it acts as a corepressor—combining with the inactive repressor to activate it and turn off the operon and the subsequent synthesis of the enzymes required for making tryptophan. This illustrates the phenomenon of *end-product inhibition*, the turning off of a process by the end product of that process. Such cellular "talent" affords a trim kind of economy for the bacterial cell.

The role of repressors, both active and inactive, demonstrates negative control mechanisms for gene expression; however, positive controls also exist.

**EXAMPLE 3**   One such positive mechanism for the *lac* operon involves a protein capable of combining with cyclic AMP, a messenger substance known to mediate the effects of many hormones on cellular activity. The protein is called *cyclic AMP binding protein*, or *CAP*. It is localized on the promoter site of DNA when cyclic AMP is present. When CAP binds to cyclic AMP and attaches to the promoter, it alters the conformation of the promoter, which increases the promoter's capacity to bind with RNA polymerase.

Cyclic AMP levels are low when glucose or some similar carbon sources are present. (Glucose and a number of other monosaccharides are more efficient sources of energy than lactose, so it is of benefit to the cell to avoid processing lactose when these alternative sources are also present.) With low levels of cyclic AMP, CAP does not attach to the promoter site, and the consequent activation of the RNA polymerase does not occur. In the absence of carbon sources other than lactose, cyclic AMP levels rise. The cyclic AMP forms a complex with CAP, which then attaches to the promoter locus of the *lac* operon and brings about activation. With lactose in the medium, high activity of the *lac* operon will occur. The action of CAP in this instance is a positive control mechanism which is superimposed upon the more prominent repressor mechanism.

Operons appear to play a role in the integration of virus (*phage*) material within bacterial chromosomes. When viruses enter a cell, their nucleic acid is injected into the cell while the protein coat remains outside the cell. The viral DNA captures the genetic machinery of the cell in a biochemical seduction process, and the cell proceeds to make viral particles. The phage virus which attacks bacteria usually elicits hundreds of copies of the viral particle (chromosome + protein capsule), which *lyse* (rupture) the bacterial cell and spread to other cells.

In some cases the viral genome integrates with the host chromosome. The viral genes are replicated along with the bacterial genes through a number of division cycles. At some point the virus plays the Trojan horse and initiates a cycle of multiplication and infection through the bacterial colony. This type of virus is called a *temperate virus*, and the phenomenon in which a "low lying" virus produces periodic bursts of infection is termed *lysogeny*. The integral viral particles are known as *prophages*.

It appears that the initiation of the lytic state, during which the phage separates from the bacterial genome and multiplies, comes about through the turning on of an operon. Several operons work sequentially to initiate the cascade of effects associated with lysis. During the prophage stage, the virus itself produces the repressor that keeps the operons of the lytic cycle in check. It is only when a DNA repair enzyme is produced in large amounts as a result of injury to the cell that the repressor is partially broken down and the virus begins the production of enzymes required for viral multiplication and lysis of the bacterial cell. In one of the more actively studied temperate viruses, the *lambda virus*, ultraviolet light readily initiates the lytic cycle, which includes a suppression of the formation of the repressor.

## 10.2   CISTRON, RECON, AND MUTON

A *gene*, in any of its specific allelic forms, has been defined as a linear array of bases sufficient to code for a single protein. In some cases, the gene contains codons (three successive bases) coding for a single polypeptide chain rather than the more complex protein. This functional unit for polypeptide, or protein, coding has been named a *cistron*. It has been differentiated from the much more limited stretch of DNA involved in mutations. The smallest number of bases that when mutated produce a phenotypic change is called a *muton*; each cistron may have hundreds of mutons, since alteration of even a single base may be sufficient to produce a phenotypic alteration. The same reasoning has been applied to the unit involved in effecting a transfer of gene material between two homologous chromosomes. The *recon* is the part of the cistron that is involved in an exchange of gene information (recombination) between two chromosomes. The specific number of base pairs involved in this recombination phenomenon is not as well understood as the DNA region associated with the cistron and the muton.

## 10.3   GENE REGULATION IN EUKARYOTES

Theoretically, the mechanisms that exist for gene regulation in prokaryotes, such as operon function, could exert a control function in the eukaryotic genome. However, the likelihood is that qualitative and quantitative differences in gene control exist within eukaryotic chromosomes.

The number of genes within the total genome of eukaryotes is up to 800 times greater than that of prokaryotes. In addition, for any single eukaryotic cell, as many as 99 percent of all potential genes are shut off. Finally, the translatable gene material of the eukaryote is interrupted by intervening sequences that do not get expressed. The latter is an intriguing finding we shall now consider further.

**EXAMPLE 4**  In 1977, an analysis of the gene for the synthesis of ovalbumin in chickens revealed that several of the sections involved in coding for the protein were interrupted by segments of nucleotides with no coding information. The coding bands are now called *exons*, while the noncoding regions that split the gene are known as *introns*. Further analysis of a variety of advanced eukaryotes revealed that most genes are split in this fashion. In the case of the ovalbumin gene, eight exons are distributed over the gene and seven introns are present. Much more of the gene consists of intervening sequences of introns than of protein-coding exons. Mammalian $\beta$-globin genes usually contain only two introns. Approximately equal division of the gene into intron and exon is found, but in other genes the intron regions far exceed the exons.

Such *split genes* are not found in all situations. The various genes coding for histones integral to nucleosome structure are not split, and most of the interferons are coded for by continuous genes. Many of the genes that code for both tRNA and rRNA however do have introns. Generally, a greater tendency for intron interruption occurs as we move from more primitive to more advanced eukaryotes.

All these findings are consistent with a different set of control features for the eukaryotic genome.

**EXAMPLE 5**  There appear to be three basic classes of DNA within the eukaryotic chromosomes. The surprise is that in many eukaryotes only about 1 percent of that DNA may actually code for translated proteins. Nor is there a simple relationship between the total amount of DNA in a cell and the complexity of that cell. *Drosophila* has approximately 75 times more DNA than *E. coli*, but humans have only one-thirtieth the DNA content of a salamander.

One type of DNA is known as *highly repetitive DNA* because its base sequences are found many thousands of times within the genome. These sequences are often short and repeated one after the other in various locales along the chromosome. Because these tandem sequences usually separate from other kinds of DNA when density gradient centrifugation is applied to DNA extracts, this DNA has also been called *satellite DNA*. Particularly short types with only moderate variation are found with the greatest frequency; they are associated with the centromere of the chromosome as well as other heterochromatic regions within the chromosomal arms. They do not seem to be transcribed. Longer units of highly repetitive DNA appear to code for the lightest of the RNA subunits of the ribosome, but do not produce protein. Other highly repetitive sequences have been identified, but no function has yet been discovered for them.

*Moderately repetitive DNA* constitutes a second type of DNA, consisting of almost 25 percent of the chromosome. Its sequences are found hundreds of times throughout the chromosome. It has been identified with the genes coding for histone proteins. Its repetitive character can be attributed to the need in most eukaryotic cells to form many molecules of the all-important histones.

Almost three-fourths of the eukaryotic genome consists of *single-copy DNA*, the source of the protein-coding (structural) genes within the chromosome. Much of this DNA is never transcribed. Some of the nontranscribed sequences are almost identical with true structural genes, and such genes are called *pseudogenes*. An intriguing characteristic of some pseudogenes is their apparent derivation from RNA in a reverse transcription process. Since pseudogenes may represent the result of significant microevolutionary processes, their origin is of considerable interest. If they are merely nonfunctional because of maladaptive accumulations of mutations, they represent a considerable burden for the genome.

The prime suspects as unique shapers of gene expression in eukaryotes are the proteins of the individual chromosomes. DNA is so wound around the histone cores of the nucleosome that there are sites of sharp bending in the polynucleotide. It is at these stress points that regulatory factors may exert their effects. Alternatively, the histones may alter the tightness of their interaction with DNA. DNA that is bound more loosely to histone may be more accessible to the factors that initiate transcription. Larger and more acidic proteins have also been found in the nucleus and have been considered as possible participants in gene expression. The nonhistone proteins are not a vital component of chromatin structure and tend to be variable components of the nucleus—facts that argue for a regulatory role rather than for that of a constant component of an invariant process.

Genes that are actively involved in transcription and translation are found along less deeply staining portions of the chromosome, the *euchromatin*. Deeply stained regions, the *heterochromatin*, generally contain genes that were never or are no longer active.

**EXAMPLE 6**  In *polytene* chromosomes, which contain many strands because of repeated replications, specific regions with clearly marked bands can readily be studied and identified. In such chromosomes euchromatic regions present in early development are superseded by heterochromatic regions later in development. This suggests active gene activity at one stage followed by a switching off at a later stage. In the case of cells with two X chromosomes, one entire X will be converted to a heterochromatic Barr body in random fashion in different cells.

*Drosophila* polytene chromosomes have demonstrated the importance of structural changes in chromatin as a corollary of gene transcription. Within regions of euchromatin that are being transcribed, a heightened sensitivity to DNase I occurs. This is an enzyme that digests DNA. This increased susceptibility to digestion suggests a marked unfolding and disruption of the regular chromatin structure. Hypersensitive sites may exist within a stretch of sensitive DNA, and these sites are associated with short lengths of DNA that are completely lacking in nucleosomes. Specific protein signals for transcription may be localized in these hypersensitive sites.

Any of the nontranscribable material of the chromosome may be a potential switch to turn on the genes in the euchromatic region. Exons may play a role as control centers, and the same is true of all the protein and RNA associated with the chromosome. It is clear that genes may be highly active at one point in time and may be suppressed in their protein synthesizing function at other times. The increase in gene activity in the polytenes of *Drosophila* larvae has actually been visualized in the form of chromosome *puffs*, clusters of DNA loops with large amounts of enmeshed RNA that burgeon out from regions of activated exons in the chromosome. In another form of increased gene activity, entire sets of genes may be duplicated many times to provide needed RNA or protein species (the process of *amplification*). This is exemplified by the formation of a nucleolus containing multiple copies of the genes that transcribe ribosomal RNA. The nucleolus becomes a site for both the production and the storage of rRNA.

A fundamental molecular mechanism for gene activation is strongly suggested by the correlation of gene activity with *undermethylation*. As many as 5 to 7 percent of the carbon residues in cytosine are attached to methyl groups in mammalian DNA. In gene sites where transcription is (or will be) occurring, a marked demethylation has been detected. The basis for the involvement of demethylation in gene activation lies in the influence of methyl groups on the conformation of the DNA molecule. Conformational changes may alter the capacity of effectors and regulators to bind the DNA.

A great deal of control may be exerted at levels beyond transcription, such as posttranscription processing of mRNA and modification of the several steps of translation. The synthesis of hemoglobin in mammals, for example, is significantly controlled at the translational level. The mode of control is the phosphorylation of factors in translation that are sensitive to the action of kinase proteins. The numerous factors required to begin translation in eukaryotes are known collectively as *eukaryotic initiation factors* (eIFs).

Differences in control programs of prokaryotes and eukaryotes may stem from the vastly different tasks imposed on the two groups. In prokaryotes, protein synthesis must respond exactly and quickly to changes in the environment. More than 90 percent of the blueprint (genome) of prokaryotes may be expressed. In eukaryotes, only a small portion of the total genome will be expressed in any single cell; the major function of that expression is the creation of a highly specialized cell.

## 10.4  CANCER AS A GENETIC ABERRATION

The malignant (deadly) tumors that are classified as cancers represent a broad variety of pathologies. The word *cancer* derives from the Latin word for "crab" and was chosen to represent the creeping pattern of growth produced by solid cancers. But cancers also include the unrestrained growth of fluid tissues like blood. Cancers are characteristically cell populations that show unrestrained growth. They tend to invade adjacent structures and may disrupt normal function in surrounding tissues. A more serious phenomenon is the widespread colonization of distant regions of the body by cancer cells that are carried by blood and lymph from the primary cancer growth. This general spread of the cancer is termed *metastasis.*

A great deal has been learned about the nature of cancer growth from studies of cancer cells grown in tissue culture outside the body. One of the earliest successfully grown lines of such cancer cells is the HeLa cells, which have been maintained in culture in many laboratories since 1950. They were originally scraped from a cervical carcinoma of a black woman for whom the pseudonym *Helen Lane* was provided (her actual name was Henrietta Lacks). A striking finding in these cultures is the absence of *contact inhibition,* the tendency of cells to stop growing when they come into contact with other cells. Receptors in normal cells are stimulated to signal cessation of growth when crowding threatens, but this mechanism does not exist for cancer cells. Growth in normal cell stocks is also limited by the tendency of these cells to undergo a specific number of divisions and then cease further growth. Though not as precise a growth control mechanism as contact inhibition, its absence in cancer tissue also contributes to the unrestrained growth of malignant tumors.

Convincing evidence has accumulated that cancer may be a disease rooted in the genes. Hereditary predispositions exist for many cancers. In many cancer cells an abnormal number of chromosomes occur, while other cancers are associated with such chromosome abnormalities as translocations, inversions, and deletions. An impaired ability of cells to repair damaged DNA may be associated with a greater susceptibility to some types of cancer. Also, the *carcinogens,* agents that cause cancer, are themselves *mutagenic* (produce mutations). So intimately are the two processes related that the standard test for the evaluation of the carcinogenicity of various chemical reagents or drugs (Ames test) involves a determination of the mutagenicity in bacteria of the substance in question.

Perhaps the strongest line of evidence for a genetic basis for cancer comes from the discovery of *oncogenes.* These are genes whose presence and activity appear to be involved in the transformation of normal cells to cancerous cells. Oncogenes usually arise from a number of ordinary genes within the host, which are known as *protooncogenes.* The protooncogene may play an ordinary housekeeping role within the genome until it is altered to an oncogene by a particular event that changes its structure or rate of activity. One kind of event known to convert a protooncogene to an oncogene is the incorporation of a retrovirus by a normal cell. A retrovirus contains RNA as its core blueprint but also possesses a reverse transcriptase that permits the virus to synthesize a DNA transcript from its RNA. This DNA transcript may then be incorporated into the genome of the cell. Where viral DNA is incorporated into the host genome, a possibility exists that damage will occur directly to adjacent cellular genes or that they will come under the control of viral regulatory factors that will alter their customary modes of expression. These two events, known collectively as *insertional tumorigenesis,* have been implicated in tumor induction.

A number of lines of evidence implicate oncogenes as sources of cancerous growth. In some cases the protein product of a cellular gene has been implicated in *neoplastic* change (i.e., tumor formation). Later it was shown that insertional mutagenesis was responsible for altering the cellular gene. In cultured cells, phenotypic changes occur when mutant protooncogenes are brought in by viral vectors; this attests to the biological activity of these putative inducers.

Perhaps most telling is the detailed study of chromosome translocations (exchanges between nonhomologous chromosomes). The damaged chromosomes encountered in cancer cells were long recognized, but some investigators argued against a specific significance of such damage. In precise analyses of translocations it was found that a new segment would lie next to a protooncogene; this juxtaposition, owing to an exchange of chromosome parts, produces a new neighbor for the proto-oncogene that may change its expression. This phenomenon may underlie the cancer known as *Burkitt's lymphoma.* In the case of chronic myelogenous (bone marrow and spleen tissue) leukemia a different mechanism operates. A translocation occurs between chromosomes 9 and 22. In the process, part of the protooncogene *c-able* is fused with another genetic locus called *bcr.* The novel gene resulting from this welding of distinct gene material produces an unusually active enzyme product which is biochemically distinct from the usual cell product produced by the protooncogene. Presumably, this may be associated with the tumorigenic process.

In 1982, experiments were conducted in which DNA was transferred from tumor cells to normal cells kept in culture. In a significant number of cases the normal cells were transformed to neoplasms (tumors). A number of oncogenes were identified in the DNA that were essentially mutant alleles of normal cellular genes. The fact that the tumor DNA contains identifiable oncogenes supports the role of these altered genes in tumorigenesis. In three different studies it has also been shown that the amplification of an oncogene is correlated with a poorer prognosis for survival; this correlation establishes a quantitative link between the oncogene and virulence of the cancer.

More than 20 retroviral oncogenes have been identified. The pioneering work was actually carried out by Peyton Rous who showed, in 1910, that a virus could produce cancer in chickens. More than 50 years later Rous was awarded a Nobel prize for this implication of viruses with the induction of cancer. An oncogene, known as *src*, is actually brought into chicken cells by a retrovirus, which is now called the *Rous sarcoma virus*.

It must be emphasized that few clear-cut associations between viruses and human cancers have yet been established, although such benign tumors as warts do seem to have a viral basis. Two human cancers, Burkitt's lymphoma and nasopharyngeal carcinoma, have been linked to DNA viruses related to the herpes family. A rare T cell leukemia is caused by a retroviral relative of the AIDS virus.

# Solved Problems

**10.1**    How is the genetic blueprint (DNA) distributed in the bacterial cell?

The best-studied bacterial cell is *E. coli*, and it will serve as our model for bacteria in general. All the genes that govern metabolism, growth, and reproduction are found on the single chromosome, which consists of a long thread of double-stranded DNA containing more than 2 million base pairs. When the chromosome is separated from the bacterial cell, it assumes a circular shape; within the cell, however, it is highly folded. The folding is accomplished with the aid of RNA and protein, but the precise details are not yet known. Histone proteins are not found within the bacterial chromosome, so that folding is probably brought about by other protein types.

A group of genes may also be found in small circular pieces of DNA lying outside the chromosome. These structures, known as *plasmids*, are encountered in both bacteria and yeast. They may contain just a few genes or as many as 25 different genes—usually those conferring resistance to specific antibiotics. Plasmids are capable of self-replication and usually reproduce in synchrony with the cell itself. They may easily be isolated from their cells and concentrated for experiments in genetic engineering. When plasmids are opened with restriction enzymes, they may incorporate foreign genes within their circular structure and bring these genes into new host bacterial cells where they can be expressed. Both plasmids and viruses have thus been used as vectors for introducing foreign DNA into bacterial genomes.

*Episomes* are plasmids that at one time may exist as tiny independent ringlets of DNA and at other times be integrated into the chromosome itself. Episomes generally carry the sex factors that separate bacteria into various mating types. The sex factor, F, is present as an autonomous circlet of DNA in strains of *E. coli* designated as $F^+$. The factor may replicate and pass into a cell ($F^-$) lacking the factor. The integration of F within the main chromosome results in the creation of a strain known as *Hfr* (for "high frequency recombination"). Cells of this strain tend to conjugate with cells of the $F^-$ strain in a characteristic fashion. The chromosome of the Hfr strain begins to replicate, and this replicate strand is inserted across a conjugation bridge into the $F^-$ cell. The $F^-$ cell undergoes recombination with the inserted strand in a time-dependent fashion, so that the longer the conjugation is allowed to continue, the greater the amount of gene exchange. An approximate mapping of chromosomes may be carried out on the basis of the time required for gene exchanges for any particular trait. Clearly, the greater the time required for recombination, the farther away from the leading end of the inserted donor chromosome the trait is located.

**10.2**   Suggest the mechanism by which operons control transcription.

In order for DNA to initiate the formation of RNA for subsequent translation, it must first unite with an RNA polymerase. The site for RNA polymerase attachment on the DNA is called the *promoter*. The operator lies within the promoter region (or overlaps it) so that attachment of the RNA polymerase is blocked when the repressor is present. In the presence of an inducer (lactate or a closely related isomer, in the case of the *lac* operon), the repressor is inactivated (and thus detaches from or fails to attach to the operator), and the operon turns on in the typical inducible fashion. RNA polymerase may then attach to the promoter site, and transcription can be carried out.

Transcription occurs along the DNA template in a 3' to 5' direction. In the case of the *lac* operon, for example, a single RNA transcript is synthesized for the three structural genes of the *lac* operon. Before translation, processing of this single RNA molecule will occur to permit the generation of three separate enzymes.

**10.3**   What are the advantages of the operon organization within the bacterial chromosome?

A major advantage is that it leads to the synthesis of groups of functionally related enzymes, usually from a single mRNA transcript. Since a functional pathway must be activated in terms of all of its components, the operon affords an all-or-none response that serves efficiency.

The control mechanisms existing for turning on or turning off the operon make the operon sensitive to those changes in the environment to which the bacterial cell must adjust. The ability of the cell to fashion repressors that are functional as well as repressors that are nonfunctional permits a broad variety of environmental triggers to exert their effects. Inducible systems may be activated by agents that tie up the repressor, and constitutive systems may be shut off by corepressors that activate initially "blank" repressors. Further fine tuning is achieved by the participation of CAP protein and cyclic AMP in the activation of the promoter site.

**10.4**   Why is the concept of a single gene as the ultimate unit of inheritance inadequate to provide a unitary explanation for protein synthesis, recombination, and mutation?

The primary function of the gene is to code for a protein product. Sufficient DNA must be present to account for each of the amino acids making up the primary structure of the protein. It is this length of DNA that is designated the cistron. This is the basic unit of function of the gene; however, there are units of function below this primary level. A mutation involves a change in the original message contained within the cistron. Such a change may lead to the manufacture of no protein or an altered protein. Since, in some cases, change in even a single base may produce a mutation, only tiny lengths of DNA may represent a unit for mutation—the muton. The cistron may thus contain hundreds of mutons.

The minimum length of DNA participating in exchanges of genetic information is not clearly apparent. Exchanges of relatively few bases between chromosomes would probably not result in the mutual transfer of intact message material. The recon, the unit involved in recombination, may be only slightly shorter than the full cistron.

**10.5**   Describe two mechanisms found in eukaryotic cells for mass producing proteins and RNA.

The existence of multiple copies of many genes, particularly those that code for ribosomal RNA units, provides an amplified pathway for manufacturing materials that are required in large numbers. Many of these repetitive genes occur as tandem (linear-succession) repeats, but some repetitive sequences achieve amplification through a localized replication of the DNA strand to produce a polytene segment within the larger chromosome. The nucleolus is one such region, which is associated with the production and storage of rRNA.

**10.6**   Thus far, no operon mechanism has been found in eukaryotes. How, then, do you suppose control of the genome is accomplished?

Highly suggestive of large-scale switch mechanisms is the tendency of active gene material to be localized in one area of the chromosome, while nontranscribing heterochromatic regions are found at other sites. Perhaps these regions are handled as functional units. Although no operon mechanism has been found in eukaryotes, the patterns of distribution of the different classes of DNA within the genome suggest a corresponding organization of control systems. The sheer bulk of DNA that is not involved in making protein or even RNA suggests that much of the DNA of the genome is involved in control of gene expression. Since so much of this seemingly nonfunctional DNA is scattered throughout the genome, the control systems may form an interlocking molecular network analogous to the nerve nets of primitive invertebrates.

**10.7**    What advantages might arise from the existence of split genes?

Exons may represent subunits of a large polypeptide chain. The existence of a number of separable exons within the gene affords an opportunity to "mix and match" subunits to create new kinds of proteins. Such mixing of subunits could not easily occur in prokaryotes.

The extending of the gene over a longer length enhances the possibility of recombination between different genes. Further, the point at which chromosome material will actually transfer from one chromosome unit to another will more likely be a noncoding region than an exon. This lowers the possibility of damaging a protein product as a result of DNA exchange.

Since the excision of noncoding regions is part of the processing of premessenger transcripts, the alteration of splice sequences may permit the fashioning of different kinds of protein at one and the same time. Thus, a cell might experiment with a new combination of exons while still holding on to older combinations.

In many of the globular proteins, the boundary between exon and intron corresponds to a protein folding site. This provides a functional connection between the blueprint material and the substructure of the final protein product. Such mirroring of structural domains in proteins by intron-exon boundaries in mRNA is seen in several dehydrogenase enzymes.

**10.8**    Describe two possible roles for histones in the activation of genes during transcription.

The tight winding of DNA around histone cores in the nucleosome may produce stress points that are susceptible to regulatory factors. Alternatively, the histones may alter the tightness of their interactions with the DNA and thus make some sections of DNA more accessible to the factors initiating transcription.

**10.9**    How has the methylation of cytosine been implicated in the initiation of transcription?

Marked demethylation occurs in areas of transcription. It is believed that methyl groups influence the conformation of DNA and, thus, that changes in methylation produce corresponding changes in DNA conformation that affect the ability of regulators to bind the DNA.

**10.10**   The alteration of a normal cell into a renegade cancer cell may be the result of a single inductive event or of two or more separate events. Since induction, once it has occurred, appears to alter the cell in an irreversible and profound manner, we can conclude that the induction process involves the genetic machinery. If only a single step were involved, we might expect that the probability of that step occurring would increase proportionally with time. Thus, if the probability for developing cancer because of that single event were 0.1 percent each year, then by the end of 10 years there should have been a 1 percent incidence of cancer within a starting group. However, the frequency of most cancers does not show this slow rise with time. Rather, cancer incidence seems to be relatively low early in life and then rises sharply with increasing age—an exponential rather than a linear model. This suggests that cancer is actually the result of two or more steps—the *multistep hypothesis* for cancer induction. The chances of developing cancer then would be the chance of the first step's occurring multiplied by the probability of the second step's occurring. Such a curve would indeed be very flat in early life but would then rise sharply with advancing years, since each of the factors would be increasing with time. Figure 10.2 shows plots of cumulative cases diagnosed versus age in years for prostate cancer and retinoblastoma. What do these two curves suggest about the respective underlying mechanisms for each of these cancers?

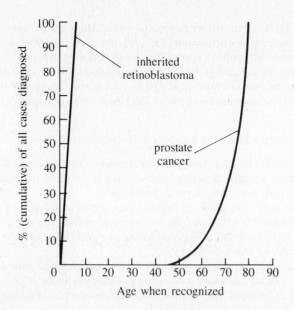

**Fig. 10.2**

It is difficult to prove a precise number of causal events using only demographic studies and statistics, nor do statistics explain the nature of these events. However, the curve for prostate cancer demonstrates an exponential relationship between age and incidence and thus suggests a multistep process, possibly including up to five events. On the other hand, persons afflicted with retinoblastoma, a cancer of the eye, show a proportional increase with age; this suggests that a single event precipitates this cancer in susceptible individuals. The tendency toward this type of cancer is inherited; we may thus assume a primary genetic event. (Evidence indicates that recessive mutations on chromosome 13 are associated with retinoblastoma in humans, and in 1986 attempts were begun to overcome the disease through the direct replacement of damaged genes.)

**10.11** Given our limited knowledge of carcinogenesis, what practical steps can be taken to minimize the incidence of cancer?

Clearly the highest priority must be assigned to increasing our knowledge about the induction of cancer. The work on oncogenes and protooncogenes shows great promise in terms of identifying the sequelae of events involved in the production of cancer. With greater understanding may come the possibility of blocking key steps in tumorigenesis. The array of molecular tools we possess may one day permit precise gene manipulations to halt the process.

We already know that radiation and a variety of chemical mutagens are associated with the induction of cancer. Clearly, the reduction of such environmental pollutants as automobile exhaust, coal-tar derivatives, heavy-metal compounds, and manufactured radioactive by-products could reduce the incidence of cancer. The advantages of industrial processes which spew these pollutants into the environment must be measured against the health hazards that ensue. Information that is objective and detailed must be available if decisions are to be responsible and reasonable.

Another practical approach could be the development of "antidotes" to the mutagens in the environment. This could take the form of agents which precipitate, alter, or neutralize the offending species. Ingested carcinogens may be neutralized by compounds that react at the cellular level. Antioxidants have already been prescribed to prevent damage from peroxides. Chelating agents, which sequester heavy metals, may also serve as models for reducing risk. Changes in susceptibility to cancer may also occur in a population, but the reasons for this at present are not understood, and thus these changes cannot be produced experimentally.

# Supplementary Problems

**10.12** The number of genes in humans is about ____ the number in bacteria. (*a*) 1000 times  (*b*) 10 times  (*c*) the same as  (*d*) 1/10  (*e*) 1/100

**10.13** The operator in an operon is kept in the "off" position by a small protein called a  (*a*) promoter.  (*b*) modulator.  (*c*) repressor.  (*d*) histone.  (*e*) globin.

**10.14** When a repressor is attached to an operator site, it blocks the  (*a*) promoter.  (*b*) cyclic AMP.  (*c*) mRNA.  (*d*) recon.  (*e*) regulator.

**10.15** The attachment site for RNA polymerase on the DNA template is called the  (*a*) cistron.  (*b*) regulator.  (*c*) operator.  (*d*) promoter.  (*e*) none of these.

**10.16** All the structural genes of the operon are transcribed on a single long mRNA.
(*a*) True  (*b*) False

**10.17** In repressible operons, an inactive repressor is normally produced that is activated by the metabolic product (corepressor) of the reaction controlled by the operon enzymes.
(*a*) True  (*b*) False

**10.18** Lactose is a corepressor for the *lac* operon.
(*a*) True  (*b*) False

**10.19** A muton may consist of up to 100 cistrons.
(*a*) True  (*b*) False

**10.20** The transcribable genes are found within the euchromatic chromosome region.
(*a*) True  (*b*) False

**10.21** Enzymes necessary for nuclear functions are actually made in the cytoplasm of eukaryotes.
(*a*) True  (*b*) False

**10.22** A protooncogene can arise only from its introduction into a eukaryotic cell by a retrovirus.
(*a*) True  (*b*) False

**10.23** Peyton Rous first showed the association between viruses and the induction of cancer in 1910.
(*a*) True  (*b*) False

In Problems 10.24 through 10.28 match each item in column **A** with one from column **B**.

| | **A** | **B** |
|---|---|---|
| **10.24** | Synthesis and storage of rRNA | (*a*) CAP |
| **10.25** | Positive control of promoter | (*b*) Plasmids |
| **10.26** | Extrachromosomal DNA | (*c*) Nucleolus |
| **10.27** | Temperate viruses | (*d*) Transcription |
| **10.28** | 3' to 5' direction on DNA template | (*e*) Prophage |

# Answers

| | | | | | | | |
|---|---|---|---|---|---|---|---|
| **10.12** | (b) | **10.17** | (a) | **10.21** | (a) | **10.25** | (a) |
| **10.13** | (c) | **10.18** | (b) | **10.22** | (b) | **10.26** | (b) |
| **10.14** | (a) | **10.19** | (b) | **10.23** | (a) | **10.27** | (e) |
| **10.15** | (d) | **10.20** | (a) | **10.24** | (c) | **10.28** | (d) |
| **10.16** | (a) | | | | | | |

# Chapter 11

# Development

In all advanced multicellular forms development involves a program of:

1. Cell division, or growth
2. Specialization of cells into highly differentiated tissues
3. Formation of the adult form with its characteristic shape and functioning organ systems or corresponding plant parts

These complex processes are directed by the tiny speck of DNA present in the nucleus of the cell undergoing development. Proteins produced by developing cells under the direction of DNA provide the structural building blocks and enzymatic tools to fashion the adult organism. The common pool of DNA produced in each of the cells of the developing organism assures coordination and unity in developmental patterns.

## 11.1 ANIMAL DEVELOPMENT

### FERTILIZATION

In multicellular animals life generally begins with the union of two *heterogametes*, the *sperm* and *egg*. The single cell resulting from that fusion is the fertilized egg, or *zygote*. Fertilization is a complex process involving the penetration of protective coats around the egg by the motile sperm, incorporation of the sperm nucleus into the egg cytoplasm, and ultimately fusion of the two *pronuclei* (each gamete nucleus is called a *pronucleus* before it combines) to yield a single diploid nucleus.

Most of our understanding of fertilization derives from detailed studies of the process in sea urchin eggs. However, an extensive literature has developed in the area of mammalian fertilization, and many of its aspects have practical relevance to problems of contraception, in vitro fertilization, etc. Striking parallels exist between what has long been known of the union of sea urchin gametes and mammalian fertilization in terms of the general steps; differences exist largely in specific details.

The mammalian egg (see Fig. 11.1) is surrounded by a thick extracellular coat called the *zona pellucida*. The first step in fertilization is a loose *attachment* of sperm to the surface of the zona pellucida.

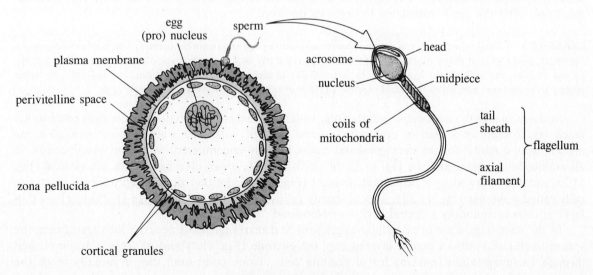

**Fig. 11.1** Fertilization

153

This is followed by the *binding* of sperm to the zona pellucida. Binding is both highly specific between classes of egg and sperm and extremely tight, as compared with the loose attachment phase. Receptors for sperm binding are present in the zona pellucida, and proteins specific for egg binding are contained in the plasma membrane of the sperm. Many thousands of sperm may attach to an egg, but little more than a thousand reach the stage of intimate binding. The actual sperm receptor in the zona pellucida of the mouse has been isolated and found to be a glycoprotein with a molecular weight of 83,000 daltons. The recognition signals within the glycoprotein appear to be associated with small carbohydrates known as *oligosaccharides*. Similarly, small carbohydrates are believed to mediate the binding of sperm and egg in the sea urchin as well.

The attached sperm then complete the *acrosome reaction*, which is a preparation for the fusion of the sperm and the egg proper. The outer membrane of the two-layered acrosomal structure attaches to and fuses with the sperm plasma membrane at many places along the periphery of the sperm head. This is associated with (1) the formation of small vesicles that reach the surface of the sperm and (2) exposure of the inner acrosomal membrane.

The acrosomal reaction releases hydrolytic enzymes that enable the sperm to move through the zona pellucida to the egg. A very narrow bore pathway is produced by sperm during its journey through the zona, which suggests a rapid and specific reaction at the leading end of the migrating sperm. This is mediated by an enzyme called *acrosin*.

Once across the zone pellucida the sperm reach the *perivitelline space*, which separates the egg from the zona pellucida. One sperm undergoes *fusion* with the egg by a union of the posterior acrosomal membrane of the sperm and the plasma membrane of the egg. An immediate block to *polyspermy* (fertilization of an egg by more than one sperm) probably occurs through changes in the electrical potential of the egg membrane following sperm entry.

Sperm entry activates the egg and its nucleus. Meiosis is completed (see Chap. 8), and sperm and egg pronuclei come together. At the same time, cortical granules in the periphery of the egg cytoplasm fuse with the overlying plasma membrane, and a variety of enzymes are released into the perivitelline space. It is these enzymes that cause a stiffening of the zona pellucida and an associated loss of its ability to bind sperm. The stiffened zona provides a long-term block to polyspermy, but the changes in the zona in mammals are more modest than those in sea urchins, in which a patent fertilization membrane forms above the egg's plasma membrane and actually pushes the extraneous sperm away.

## CLEAVAGE PATTERNS AND MORPHOGENESIS

Following fertilization the single zygotic cell begins its destined odyssey to adulthood by undergoing a regulated series of divisions. The early cell divisions are called *cleavage*. It is now clear that cleavage starts only after the egg is stimulated to begin its division.

**EXAMPLE 1**   Usually the entrance of sperm serves to activate the egg and begin the formal process of *embryogenesis*. However, in the case of many insects an *unfertilized* egg regularly begins its cleavage divisions to produce a male. In sea urchins, the unfertilized egg can be experimentally induced to begin its divisions, for example, by being placed in hypertonic salt solutions. Mild electric shock will initiate divisions in amphibian eggs.

In those eggs with comparatively little yolk, which is evenly distributed, early division patterns are much less complicated than in eggs with a great deal of yolk. The starfish is often used as an uncomplicated model for the developmental stages that are encountered, with some modification, in all multicellular organisms (see Fig. 11.2). In starfish, the first two cleavage planes are vertical (Fig. 11.2a and b), followed by a horizontal division (Fig. 11.2c). Further divisions produce a cluster of cells called a *morula* (Fig. 11.2d). This is shortly followed by the *blastula* stage (Fig. 11.2e)—a thin layer of cells surrounding a central cavity, or *blastocoel*.

In the next stage a series of cellular migrations and rearrangements occurs which transforms the sphere-like blastula into a double-layered cup, the *gastrula* (Fig. 11.2f and g). This is accomplished through an invagination (pushing in) of blastula cells at one point until they eventually reach the opposite side and thus form a hollow tube from one end of the embryo to the other. The outer layer

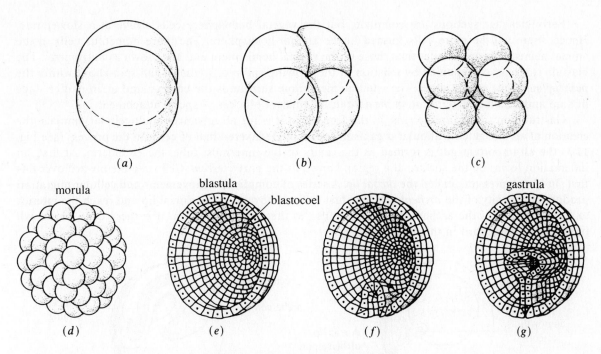

**Fig. 11.2** Development in starfish

of the gastrula will become the *ectoderm*, which later gives rise to the skin and nervous system. The inner layer of cells (those forming the hollow tube) will become the *endoderm*, which will produce the inner lining of the gut and a variety of accessory structures which are formed as outpockets of the primary gut. The new cavity enclosed by the endoderm is called the *archenteron*, or primitive gut. The initial opening into the archenteron is the *blastopore*, which is the original site of invagination. Later, an opening will form opposite the blastopore site so that the archenteron is open at each end. In starfish and other echinoderms (spiny-skinned creatures) as well as in vertebrates, the blastopore becomes the anus, and the opposite opening gives rise to the mouth. In other multicellular animals the blastopore becomes the mouth.

Concurrent with or slightly after the complex early movements of gastrulation, two evaginations, or outpocketings, of the endoderm pinch off into the blastocoel and give rise to a third layer of cells that becomes established between the ectoderm and endoderm and is known as *mesoderm*. Gastrulation thus results in the formation of a three-layered embryonic structure with an established longitudinal axis and a general outline of the broad features of the adult. Each of the three *primary germ layers* (ectoderm, endoderm, and mesoderm) will soon produce a variety of differentiated structures.

In the case of the frog and most amphibians, the development of the embryo is complicated by the fact that the egg contains an appreciable amount of yolk material that constrains the divisions. Eggs with a small amount of yolk material evenly distributed throughout, as in starfish eggs, are termed *isolecithal* eggs (from the Greek words *isos* meaning "equal, uniform" plus *lekithos* meaning "egg yolk"). Amphibians, which have an appreciable amount of yolk which is unevenly distributed, produce eggs referred to as moderately *telolecithal* (from the Greek word *telos* meaning "consummate, complete"). In the extreme telolecithal egg of birds, the mass of the yolk is so great that embryogenesis involves only a small disk of cytoplasm at one end of the original egg cell.

In moderately telolecithal organisms such as frogs, a considerable degree of differentiation marks the unfertilized egg. The upper, *animal* hemisphere is heavily pigmented, but the lower, *vegetal* hemisphere is much denser and yolk-laden. Shortly after the attachment of the sperm, the eggs orient themselves with the animal poles uppermost. Meiosis, arrested in metaphase II, is activated by the sperm entry and proceeds to its completion. *Syngamy* (union of the nuclei) will occur, and the first cleavage will be completed about two to three hours after sperm entry.

Early cleavage sections are complete, but the vegetal hemisphere cells cleave at a slower rate. Hence, there will be more cells formed in the animal hemisphere. The more numerous cells in the animal hemisphere are smaller than those of the vegetal hemisphere and are known as *micromeres*. The blastula stage is accompanied by ciliation of the growing embryo, so that it can spin about within the *perivitelline space* at this stage. The vitelline membrane surrounds the embryo and at an earlier stage lifts up and becomes a *fertilization membrane*, an efficient blocker of sperm attachment.

Gastrulation (gut formation) is, in the frog, what it is for all advanced multicellular animals, the creation of an elongated tritubular organism from a single-layered ball of cells. In the process (see Fig. 11.3) the all-important gut is formed as the cavity of the innermost tube, the endoderm. At first, an indentation forms on the surface at a region known as the *gray crescent*. Cells migrate inward over this first slit-like depression, called the *dorsal lip*. A series of complicated movements and cellular migration produces the cavity of the *archenteron*. The first cells to move over the dorsal lip and reach the interior to form the roof of the archenteron will give rise to the *chordamesoderm*. It is these cells which will produce the notochord in the neurula stage.

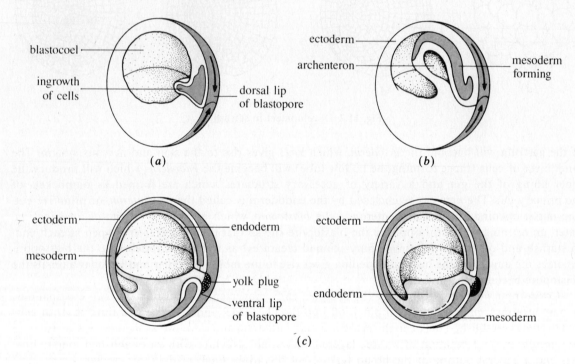

**Fig. 11.3** Development in amphibians

The dorsal lip soon bends down on each side to form lateral lips, and eventually these lateral slits bend around underneath and meet to form the ventral lip and thus close the circle and provide a complete blastopore for the continually growing archenteron. A considerable population of yolk-filled cells protrudes through the blastopore to form a *yolk plug*.

Chordamesoderm gives rise to both notochord and mesoderm. The original cells of the vegetal hemisphere move into the interior of the embryo and become the endodermal layer. The archenteron will form the gut and provide a complete inner tube of endoderm. Mesoderm is intruded between this tube of endoderm and the ectoderm, the outermost tube made up of animal hemisphere material that never moved to the interior.

In chickens (and other birds) the egg proper consists only of the yolk and, on one side, a thin region of cytoplasm and a nucleus. Fertilization occurs within the oviduct, and the albumin and shell are secreted as extraneous coats by special glands as the egg moves down the oviduct. Blastula and gastrula stages occur while the egg is still located within the oviduct. The *blastodisk* (see Fig. 11.4), a

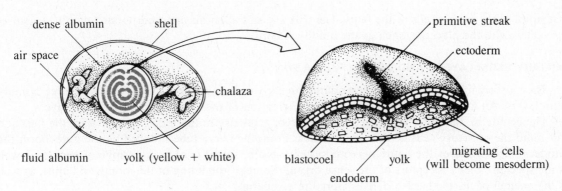

**Fig. 11.4** Development in birds

layer of cells derived from the nucleus and cytoplasm of the fertilized egg, undergoes delamination to produce a two-layered disk surrounding a blastocoel. The lower layer is largely made up of yolk-laden cells and is termed the *hypoblast*. The smaller cells of the upper layer constitute the *epiblast*. A long constriction, called the *primitive streak*, develops along the midline of the epiblast. Epiblast cells will migrate downward along this line to form mesoderm and contribute to endoderm formation together with hypoblast cells. The hypoblast will also form the *yolk sac*, an extra-embryonic compartment containing food material.

## EXTRAEMBRYONIC MEMBRANES

Extraembryonic membranes are membranous extensions of the tissues of the embryo. They are essentially folds which eventually grow around and encase the embryo to create four sacs which service the growing embryo. Each of the membranes is formed from cells of two different primary germ layers (see Fig. 11.5).

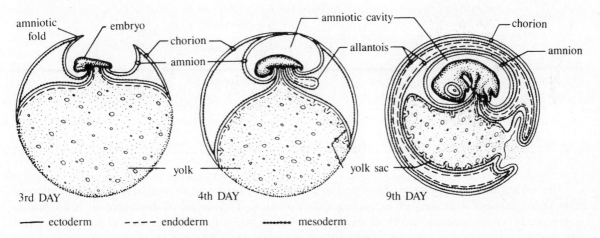

——— ectoderm    - - - - endoderm    •••••• mesoderm

**Fig. 11.5** Extraembryonic membranes

Two of the membranes are formed as the result of the folding and then meeting of a membrane over the embryo. The inner membrane (the *amnion*) encloses the *amniotic cavity*, while the more extensive outer membrane is called the *chorion*. From the anterior end of the gut and continuous with that tube, a membrane grows around the yolk mass to form the *yolk sac*. A posterior outpocket of the gut forms the *allantois*, a sac involved in both excretory and respiratory functions within the enclosed egg.

In reptiles and birds all four membranous sacs are prominent throughout development. In mammals the chorion participates in the formation of the *placenta*, the spongy mass of maternal and fetal tissue that supplies the basic needs of the fetus. The yolk sac and allantois are incorporated within structures associated with the placenta, such as the umbilical cord.

## PRIMARY GERM LAYERS AND ORGAN FORMATION

The most significant aspect of differentiation in early embryogenesis is the creation of three primary germ layers. All later development is cued by the creation of these three "skins," or cell layers.

The ectoderm, the outer covering of the embryo, gives rise to the outer layer of skin, the hair and nails, and the secretory cells of the sweat glands. From the nerve tube induced by the notochord, the ectoderm also produces the entire nervous system—brain, spinal cord, and peripheral nerves, as well as the specialized end receptors of the sense organs. Some of the lining of the mouth and anus, as well as the enamel of the teeth, also derives from the ectoderm.

The endoderm provides the linings of the digestive tract and the major respiratory ducts, most of the liver and pancreatic cells, the lining of the urinary bladder and the inner layer of the urethra, and the thyroid and parathyroid glands.

The mesoderm is the third of the germ layers to be formed, but it is the source (see Fig. 11.6) of the greatest bulk of living material within the organism. All muscles, solid connective tissues (bone, cartilage, and fiber), blood and its vessels, and thin mesenteries that connect most visceral organs to the body wall are derived from the mesoderm. The dermis, the primary functional stratum of the skin, also arises from the mesoderm, as do the kidneys and the reproductive organs. The segmentation that characterizes many multicellular groups is a result of the division of mesodermal tissue into a sequence of repeating blocks early in development. Later structures, such as the vertebrae and major muscle systems of the body, are arranged into *somites*, or regular repeating units running from front to back, because of this prior segmentation of their precursor mesodermal structures.

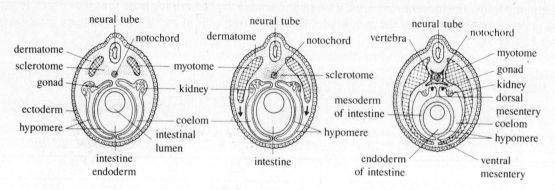

**Fig. 11.6** Organ formation

The organs and organ systems of multicellular forms arise from undifferentiated cells of the primary germ layers through a process of increasing differentiation. This mode of development is known as *epigenesis*. It can be contrasted with an earlier concept, *preformation*, which assumed that a tiny adult was already present in the egg and merely unfolded and increased in size to produce the adult.

**EXAMPLE 2**   The earliest organ formed in vertebrates is the *nerve tube*, which arises from that portion of ectoderm overlying the notochord. The notochord first induces a thickening of the ectodermal layer above it, the *neural plate*. Shortly an indented region can be discerned running lengthwise along the middle of the plate, the *neural groove*. This depression is formed from the elevation of two ridges that flank the groove on each side, the *neural folds*. The apex of the neural folds is called the *neural crest*, a region that will later give rise to groups of nerve cells lying just outside the central nerve tube. Eventually the neural folds join to form the tube. Both mesoderm and ectoderm cells now surround the tube as it sinks below the surface. These surrounding cells produce the spinal column and cranium that support and protect the vulnerable nerve tube, a structure which will form the brain and spinal cord.

The formation of one structure triggers, in turn, the development of another. This pattern of increasing differentiation is apparent in the development of a variety of organs found in the anterior end of the organism.

**EXAMPLE 3** The anterior portion of the nerve tube produces three distinct bulges that will give rise to the three regions of the adult brain—the *forebrain* (most anterior), the *midbrain*, and the longer, posterior *hindbrain*. As the brain begins its formation, two vesicular protrusions just posterior to the forebrain extend toward each side of the embryo. These are the optic vesicles. At their leading end a partly invaginated cuplike structure will eventually develop into the retina. Before this occurs, however, the optic vesicle will reach the underside of the surface ectoderm and induce the formation there of lens and cornea. Thus, an orderly sequence of events eventually produces end structures from earlier triggering processes.

## 11.2 HUMAN DEVELOPMENT

The human egg is fertilized in the fallopian tube of the female. Early divisions are rather slow, and blastomeres of equal size are produced. These blastomeres may separate to produce multiple embryos. About 5 days after fertilization, a blastula stage known as a *blastocyst* is formed. This consists of an outer circle of cells known as the *trophoblast* and an inner round mass of cells hanging down at one end of the trophoblast. The inner mass of cells will produce the embryo, while the trophoblast will give rise to the chorion and the fetal component of the spongy *placenta*, the structure within which exchanges of gases and fluids between mother and child will occur.

At about the seventh day the trophoblast buries itself into the inner lining of the uterus (the *endometrium*), and a network of projections from embryo to uterus forms. This process is called *implantation*. Soon after implantation the extra-embryonic membranes form.

The yolk sac does not surround yolk material, because the mammalian egg is generally tiny and relatively free of yolk. This accords with an internal developmental pattern in which the mother will be supplying nutrition. Instead, the yolk sac becomes a repository for the future gametes of the embryo, the *germ cells*. The allantois develops into both the *anlage* (precursor) for a urinary bladder and the umbilical cord, which connects the embryo to the placenta. *Chorionic villi* form as the embryonic contribution to the placenta, while the amnion develops as a fluid-filled chamber within which the embryo proper is safely nestled. The chorion also produces a hormone, human chorionic gonadotropin (HCG), which initiates in maternal glands the production of hormones supporting pregnancy. These hormones include estradiol and progesterone. By the second trimester of pregnancy, hormone production is a function of the placenta.

Pregnancy, or *gestation*, is divided into three parts. During the first trimester all the major systems have been formed. By the second week mesoderm is formed, and the segmentation into somites is clearly apparent. The heart and nerve tube have formed by the end of the third week. After two months the embryo takes on a human appearance and graduates to the status of *fetus*.

During the second trimester, movements and reflex actions are quite pronounced and refinements in organ development occur. A temporary covering of hair called *lanugo* is formed.

The major events of the third trimester are increases in body mass and size. Migration and extensions of nerve processes continue through *parturition* (birth). Late in the third trimester all nerve cell division ceases. Muscle fibers also cease to multiply shortly after birth. After about 270 days of gestation, birth occurs.

## 11.3 PLANT DEVELOPMENT

Recall (from Chap. 8) that in plants two life forms, a haploid gametophyte and a diploid sporophyte generation, alternate with one another through time.

## ALGAL REPRODUCTION AND DEVELOPMENT

In algal forms the gametophyte generation is usually the prominent form. In many of the green algae the sporophyte stage is completely absent, and no alternation of generations can be found.

**EXAMPLE 4**   The filamentous alga *Spirogyra* exists as a longitudinal set of attached haploid cells through most of its life cycle. During the reproductive phase, adjacent filaments lie next to one another, conjugation bridges are formed in a uniform fashion between neighboring cells, and the cellular contents (*protoplasts*) of the cells in one filament migrate across the bridges to fuse with the protoplasts of the recipient strand. The zygotes thus formed are the only representatives of a diploid sporophyte stage, for shortly after their formation the cellular partitions break down and the encysted zygotes, called *zygospores*, settle to the bottom of the pond or stream and undergo meiosis to produce a new haploid gametophyte. Each haploid cell may form (through mitosis) a new filament. It is of considerable interest that in the meiotic process here only one functional cell is eventually produced. This resembles the meiotic process in the oogonia of multicellular animals, in which a single ovum and several polar bodies are produced.

## REPRODUCTION AND DEVELOPMENT IN BRYOPHYTES

The bryophytes are nonvascular plants, including liverworts, hornworts, and mosses. They are found on land, but they are not completely independent of an aquatic environment. In all cases, the gametophyte generation is dominant. Usually, the sporophyte is attached to and dependent on the gametophyte. Bryophytes require a film of moisture because the sperm are motile and must swim to the eggs of the gametophyte.

Bryophytes are considerably more advanced than the algae and belong to a group of plants known as the *embryophyta*, which includes all the more advanced forms such as ferns and seed plants. Among the characteristics of the embryophyta are multicellular sexual organs that enclose the gametes and protect them from the vicissitudes of the harsh terrestrial environment. In bryophytes, the sperm are encased within the *antheridia*, while the eggs are found within the *archegonia*. Motile sperm swim to the narrow opening of the archegonium, where fertilization will occur. The zygote develops within the archegonial receptacle, and the sporophyte usually remains attached to the tissues of the gametophyte from which it arose. A basal *foot* of the sporophyte attaches to and derives nourishment from the gametophyte tissue, while a stalk extends from the foot to a capsule within which haploid spores will be produced.

**EXAMPLE 5**   *Marchantia*, a representative bryophyte (see Fig. 11.7), is a liverwort with a prominent gametophyte state and a sporophyte that exists only as a parasitic structure within the archegonium. The gametophyte consists of a fairly tough elongate structure resembling a segmented leaf. The sexes are separate (*dioecious*). The eggs are formed within the flask-shaped archegonia located on the underside of the *archegonial receptacles*. The sperm are produced within the antheridia that lie atop the *antheridial receptacles* of the male gametophyte. The sperm released in the antheridia swim through water films to the eggs found in the archegonia. Special cells called *gemmae* are produced within cuplike structures on the surface of the *thallus*, the extensive flat portion of the gametophyte. These gemmae represent asexual cells that are capable of producing a new thallus during fragmentation of the original thallus.

A fertilized egg develops into a sporophyte within the archegonium. The *foot* remains attached to the cells of the archegonium, and the *seta* extends as a stalk bearing the *capsule* at its distal end. Within the capsule, spore mother cells undergo meiosis to form a tetrad of haploid spores. At that juncture the seta elongates, and the capsule is pushed into the air. Irregular drying of the capsular cover produces a rupture, and the spores are scattered by the twisting movements of the *elaters*. Spores grow into new gametophyte plants.

## REPRODUCTION AND DEVELOPMENT IN FERNS

In the fern (see Fig. 11.8), the dominant phase is the sporophyte generation. The gametophyte exists as an inconspicuous heart-shaped *prothallus*, extremely thin and notched at its broad end. It is *monoecious* (both sexes on one plant); the archegonia develop near the notch, and antheridia form further back among the *rhizoids*, the delicate tendrils that tend to hold the gametophyte to the substratum of soil. There are no gemmae cups or parallel asexual structures.

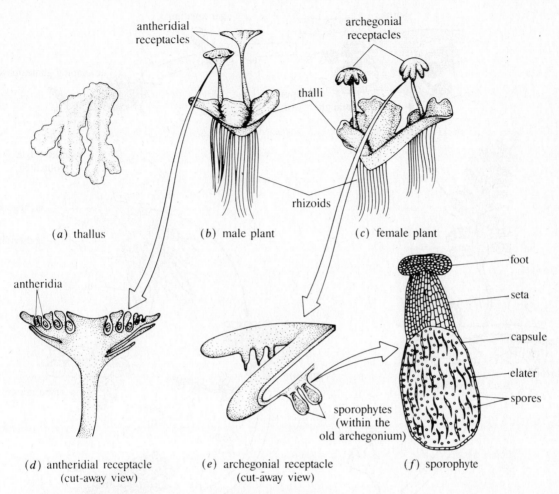

antheridial
receptacles

archegonial
receptacles

thalli

rhizoids

(a) thallus              (b) male plant              (c) female plant

antheridia

foot

seta

capsule

elater

spores

sporophytes
(within the
old archegonium)

(d) antheridial receptacle        (e) archegonial receptacle        (f) sporophyte
(cut-away view)                    (cut-away view)

**Fig. 11.7** Marchantia

Sperm released within the antheridia must swim to the egg cells, which lie within the archegonia. The fertilized egg develops into a sporophyte that, like *Marchantia*, is parasitic upon the tissues of the archegonium at first. In a short while the sporophyte develops a true root that extends beyond the parent plant, stem and leaves push up to overshadow the gametophyte, and the adult fern sporophyte is on its way.

On the underside of most fern leaves, or on modified *sporophylls*, diploid spore mother cells undergo meiosis to produce tetrads of spores arranged within *sporangia* (spore cases). The sporangia are accumulated in round clusters known as *sori*, which are prominent near the end of the growing season in temperate zones. As the sporangia dry, a thick-walled belt of cells (*annulus*) that partially encases each sporangium flicks out and then back and releases the mass of spores. The cycle is complete when each spore germinates into a new haploid prothallus.

## REPRODUCTION AND DEVELOPMENT IN FLOWERING PLANTS

The flowering plants (*angiosperms*) represent the most recently evolved of plant forms and are best equipped to withstand the challenges of a terrestrial lifestyle. In flowering plants the sporophyte generation is the only form that is visible. The haploid gametophyte has been so reduced that it exists only as a parasite within the tissues of the flower of the host sporophyte (see Fig. 11.9). In many angiosperms the sexes are separate. The significant feature of the group is the flower, a complex reproductive organ that probably arose through the modification of leaves.

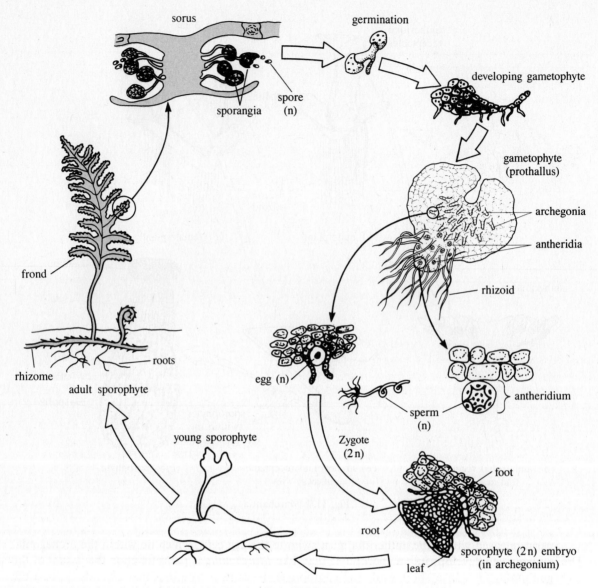

**Fig. 11.8** Fern

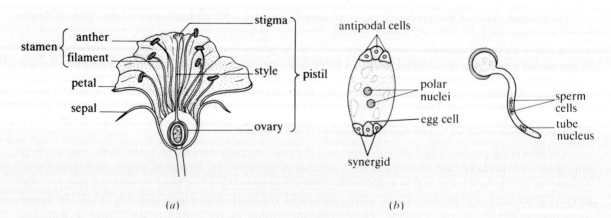

**Fig. 11.9** Flowering plant

The parts of a flower are arranged in four *whorls*, or concentric layers nestled one within the other. The innermost whorl comprises the female organ, or *pistil*, which is made up of a sticky *stigma* resting on a long, narrow *style* leading to the *ovary* and *ovules*. Surrounding the pistil are the *stamens*, the male organs of the flower. External to the stamens are the *petals*, the often colored, aromatic structures that collectively make up the *corolla*. Finally, the outermost whorl, the *calyx*, is composed of the tough *sepals*, which completely enclose the unopened bud of a developing flower. The unfolding of the flower begins with the calyx and extends inward.

A mature *embryo sac*, the female gametophyte of a flowering plant, arises within the ovule by the mitotic division of a single haploid megaspore. Essentially, seven cells are produced by cytoplasmic compartmentalization of the megaspore, but eight nuclei are scattered among these cells. Near the *micropyle*, or opening into the ovule, are three cells—the egg cell and two accompanying *synergid* cells. (A rather fanciful explanation of the existence of these seemingly nonfunctional synergids is that they represent a remnant of the ancestral archegonium, which is no longer found in these seed plants.) At the opposite end are three *antipodal* cells, which some say are analogous to the vegetative (non-reproductive) tissue of the gametophyte. Such vestiges establish an evolutionary link to ancestral forms. In the center of the embryo sac are the last of the nuclei, the two polar nuclei. They are called *polar nuclei* because each has arisen at an opposite pole and then migrated to the center. The polar nuclei quickly fuse to produce a central, diploid *endosperm* nucleus. At this point the ovule, or more specifically, the embryo sac is ready for fertilization.

Now we turn to the pollen grain, the haploid *microspore* that develops within the anther sac of the stamen. Once carried to the stigma, the pollen grain, whose nucleus has divided, now elongates into a *pollen tube* (the male gametophyte) and grows down the style (pistil stalk) toward the micropyle. The leading nucleus is called the *tube nucleus* and appears to guide the extending tube. The trailing nucleus is the *generative nucleus*, which divides into two sperm nuclei. Once the pollen tube reaches the embryo sac, a *double fertilization* occurs: one sperm nucleus joins the egg nucleus to produce a zygote. The second sperm nucleus unites with the diploid endosperm nucleus to produce a triploid cell, which grows into the endosperm, or seed food.

Although an endosperm develops in the *gymnosperms* (nonflowering seed plants), double fertilization is found only in angiosperms. It should be stressed that the reproductive pattern of the seed plants involves a total independence of environmental water in the process of fertilization and in the further development of the embryo within the sealed seed.

Following fertilization, the ovule matures into a seed. The seed is particularly adapted for colonization of the land. It consists of (1) an embryo sporophyte which is in an arrested state of development, (2) a relatively large supply of food, and (3) a resistant seed coat which protects the embryo and its food supply until conditions are ripe for resumption of growth of the embryo. Seeds develop within the ovary of the flower. A ripened ovary containing seeds is called a *fruit*. When fruits are eaten or decay, the seeds are released for new growth.

There are two major groups of flowering plants: the *monocotyledons* and the *dicotyledons*. The monocots have leaves with parallel venation, the embryos in the seed are small and flat and protected by a sheath called a *coleoptile*, and the first seed leaf (*cotyledon*) is single. Most of the stored food of the seed is associated with the endosperm. The grasses, including wheat, corn, and rice, are all monocots.

The dicots demonstrate netted venation in the leaves. The seeds contain embryos that tend to be large, and two cotyledons are always present. Most of the food is contained within the cotyledons, which often occupy most of the space of the seed. Dicots do not have a coleoptile.

## 11.4  CONTROL OF DIFFERENTIATION

The cells that make up a differentiating embryo are, for the most part, genetically identical. The specialization process, then, must involve the differential expression of genes. This may include transcription of some genes while other are suppressed, an increase in the activity of a particular gene (amplification), or the modification of post-transcriptional products. Cytoplasmic factors, already present in the egg and passed along to the blastomeres during cleavage, may also be involved in embryogenesis.

The egg cell is capable of forming a complete organism. Such a capacity, which is shared with the blastomeres produced by at least the first three cleavages in mammals and amphibians, is termed *totipotency*. Beyond the early cleavage stages, the cells produced lose their plasticity and become *determined*; i.e., they are destined to become specific structures within the total embryo. If isolated, they no longer form a complete embryo. The totipotent cell can be characterized as existing in a state of "innocence," with all options open to it. Its loss of innocence corresponds to the narrowing of options and the assumption of specialized responsibilities better-suited for functioning within a complex network of coordinated parts.

Older theories of cellular differentiation involved the notion of "developmental fields." An embryo developed within an energetic field of its own creation. One end tended to be more active than the other, leading to a head-tail orientation. Generally, greater metabolic activity was associated with the head end. Structures were molded from cellular precursors based on these metabolic gradients.

In some organisms, environmental influences are believed to be responsible for the creation of metabolic fields. The effect of light or pressure on the components of a cell may cause an orientation of these internal entities that establishes a developmental field and influences the course of embryonic growth patterns.

Should plant parts be excised from a main stock, a field is set up that influences the development of the free portion. The formerly proximal (attached) end will tend to develop a root system, while the formerly distal (free) end will grow leaf or flower buds. The polarity established in an isolated twig, for example, is quite similar to the polarity established within a frog egg shortly after fertilization.

The nature of the differentiation process in embryos was further elucidated in the 1920s. Dr. Hans Spemann and Dr. Hilde Mangold established the role of primitive structures arising early in embryogenesis in directing the creation of later, more elaborate structures in the embryo. The early primitive structure is called an *organizer*, and it is thought to orchestrate development by forming and then secreting a chemical *inducer* that diffuses into a specific region of the embryo and effects the synthesis of a particular structure. The inducer is believed to be a small, diffusive molecule, but the search for the substance has been marked by disappointment.

In the case of the frog egg, the earliest indication of a developmental control structure is the gray crescent area, which arises exactly opposite the point of sperm entry. During the ensuing cell divisions and morphogenetic arrangements, a tadpole develops whose orientation and major developmental axis is based on the location of the original gray crescent. While each of the blastomeres formed during the early cleavages can actually give rise to an intact tadpole if separated from one another, this will not be true if the first cleavage is artificially induced to produce one blastomere with all the gray crescent and a second with no gray crescent.

The dorsal lip of the gastrula was extensively studied as an organizer in frog development by Mangold and Spemann. Under normal conditions of development the cells of the dorsal lip derive from the gray crescent. The first cells that move across the dorsal lip to produce the original roof of the newly formed archenteron will play a key role in inducing the formation of a neural plate. So important is the dorsal lip to developmental patterns that experimental implantation of a second dorsal lip in a normally developing embryo causes the host embryo to develop two regions of gastrulation and two nerve tubes.

One area of development actively being studied is the reversibility of differentiation once it occurs. Recent work suggests that the nucleus of even a determined cell can be brought back to a state of totipotency under appropriate conditions. The role of cytoplasmic inclusions in the egg in developmental processes is also being explored, and research has shown that in snail development only the blastomere containing a cytoplasmic structure known as a *polar lobe* is capable of forming a complete embryo. Perhaps most exciting of the new developments in embryology is the recognition of highly conserved groups of genes, found in many different species, that are turned on or off as a unit. The term *homeotic box* has been applied to these genetic aggregates, which appear to be involved in forming structures arising from the segmented blocks, or somites, of organisms as diverse as insects and mammals.

## 11.5  MAJOR CONTRIBUTIONS TO THE FIELD OF EMBRYOLOGY

Embryology as an organized discipline within biology began with the seminal work of William Harvey and Marcello Malpighi in the seventeenth century. Harvey's classic embryological observations were contained in *Exercitationes de Generatione Animalium* (1651) in which development of the chick and even of a mammal (deer) were described. He stressed the view that all living things come from an egg. He also anticipated the epigenetic view that new structures arise de novo from the raw materials of cells.

Malpighi was the first investigator to describe and illustrate the stages in the development of the chick. Although Malpighi was a preformationist, his splendid drawings and descriptions of neurulation are regarded as a major contribution to embryology.

In the 1700s a preoccupation with observation of sperm and egg led to the *preformation doctrine* that a little human being (homunculus) was encased within either sperm or egg. In the *Theoria Generationis* (1759) Friedrich Wolff attacked this notion and defended instead the epigenetic view, held to this day, that new parts arise gradually over time.

The towering figure of the premodern era of embryology is Karl Ernst van Baer (1792–1876). He is celebrated for having raised the standards for experimental work in embryology as much as for his considerable conceptual contributions. He established the germ layer theory, which represents a basis for the development of almost all multicellular organisms. He also made embryology a comparative study. He discovered the existence of the mammalian egg and showed that a notochord is present in all vertebrate embryos. These contributions have earned him the sobriquet of "the father of modern embryology."

Oskar Hertwig was a nineteenth-century embryologist who initiated studies in fertilization. In that same era, Wilhelm His explored the development of the nervous system and the specific patterns in human embryology. In the twentieth century, genetic analyses and experimental manipulation of embryos have dominated the field of embryology.

# Solved Problems

**11.1**  Describe development in the isolecithal, moderately telolecithal, and extremely telolecithal egg, emphasizing differences necessitated by the varying amounts of yolk.

It should be stressed that a gradation in the amount and distribution of yolk exists in a continuum from the very little, evenly distributed yolks in starfish eggs to the extreme telolecithal situation in bird eggs. To focus on examples from each end of that spectrum and a representative of the middle is not intended to suggest sharply defined categories that may not exist in nature.

The starfish or sea urchin is often selected as an example of development from the isolecithal egg. The egg tends to homogeneity, but there is some slight polarization. Slightly more yolk is found at one pole (vegetal), while a higher metabolic rate is associated with the cytoplasm of the other animal pole. Following syngamy (union of sperm and egg pronuclei) cleavage begins. Planes of cleavage division extend completely through the egg, and the early blastomeres are fairly equivalent in size. The blastula is a hollow ball with a thin, one-cell-thick layer surrounding the blastocoel. Gastrulation involves mainly an invagination of one wall of the blastula. The region of indentation eventually reaches the opposite wall. The indented material forms an inner tube of endoderm with an enclosed cavity known as the archenteron. Gradually the blastocoel is obliterated. Evaginations, or pouches, arise from the endoderm at the end farthest from the blastopore. These pouches give rise to the mesoderm, the third germ layer, between the external ectoderm and the internal endoderm.

The formation of a blastula and the onset of gastrulation are similar in frogs and starfish, but complicated in the frog by the increased amount of yolk. The region of the egg that contains the greater amount of yolk becomes the vegetal pole. Later cleavages occur more frequently in the region of the opposite pole, the animal pole. The cells here are smaller and more numerous and possess a higher metabolic rate. It is at the animal pole that the polar bodies were originally extruded during oogenesis (see Chap. 8, Example 5).

The blastula of the frog is composed of an irregular array of cells that enclose a somewhat smaller blastocoel than that in the starfish. Gastrulation begins with the formation of a crescent-shaped indentation at the border of the unpigmented vegetal hemisphere, the blastopore. Cells move over the dorsal boundary of the blastopore and migrate inward to create the cavity of the archenteron. The dorsal lip of the blastopore eventually curves to form a complete circle, which gives rise to infolding cells which completely encompass the yolk within the growing archenteron. At this stage the infolding and migration of cells have created an elongated embryo in which three primary germ layers can be distinguished. The middle layer, which is located centrally and dorsally, is called chordamesoderm, and it will give rise to the notochord, a stiff cartilaginous rod found in all vertebrates. The lateral and ventral portions of the middle layer become the mesoderm.

In birds the formation of blastula and gastrula involves only a thin stratum of cells, the blastodisk, atop the yolk. The blastodisk in chickens consists of two layers of cells separated by a thin space, the blastocoel. The upper layer will give rise to ectoderm, and the lower layer will form endoderm. Gastrulation begins with the appearance of a thin line of indentation on the surface of the blastodisk, the primitive streak. The primitive streak, analogous to the blastopore of amphibians, acts as an organizing conduit through which cells migrate to form mainly the middle mesoderm layer. As in the frog and all other vertebrates, a notochord will develop, which influences the formation of the nerve tube in the overlying ectoderm. The chick differs from the starfish and frog in that the early cleavage planes do not transect the entire egg, but cut only through the blastodisk region. The yolk prevents more extensive divisions. This is known as *meroblastic* cleavage. Also, the chick will develop four extra embryonic membranes that characterize the embryology of reptiles and mammals as well.

**11.2**　What sorts of cellular movements would you suppose are involved in embryogenesis?

1. There is a downward spreading movement (*epiboly*) in which cells from the animal hemisphere cover the lower vegetal surface of an embryo. This is especially significant in the gastrulation patterns of the frog. The term *epiboly* has also been applied to the lateral movements of cells from a central axis in the flattened disks of birds or teleost (i.e., bony) fishes. At first it was reasoned that epiboly was associated with localized rapid growth, but this is not actually the case. It is obvious that movements in one region of an embryo must be elegantly coordinated with complementary movements in other areas to prevent stretching or tearing of the embryo.
2. *Migration* refers to cell movements from the surface to the interior. If the surface cells simply begin to project into the interior through "inpocketing," the movement is termed *invagination*. *Delamination* involves the splitting of a surface layer to produce two separate sheets.
3. *Convergence* involves the movement of cells toward the central axis of an embryo. Divergence is the opposite.
4. *Extension* involves the movement of cells toward the ends of an embryo to increase total length. The same purpose may be accomplished by a lengthening of the longitudinal axis of linearly arrayed cells.

Most of these classes of movements were delineated in the classic work of Wilhelm Vogt in the 1920s. By using vital stains, which do not harm the embryo, Vogt was able to construct *fate maps*; i.e., he could follow the eventual placement of some early part of a blastula or even egg portion in the later embryo by tracing the movement of stained cytoplasm.

**11.3**　Many organisms experience two separate episodes of morphogenesis during their lifetime. Such organisms develop partially to reach a *larval stage.* They then enter a transient inactive stage, called the *pupa*, in which a new series of morphogenetic changes occur, after which they emerge as adults, usually totally different in appearance from the larval form. The change from larva to adult is called *metamorphosis.*

In the case of some organisms, metamorphosis is extreme. There is no resemblance between larval and adult forms. In other organisms the transition occurs in phases during which no sharp demarcations are discernible. Give an example of metamorphosis, and explain the advantage this imparts.

In the butterfly, the egg hatches into a most mundane, earthbound caterpillar whose entire life is spent eating. This caterpillar then weaves a cocoon around itself and breaks down, only to be reconstituted as a flying creature of great beauty, with a sexual mission and often no appetite for food whatsoever. The caterpillar or grub is ideally suited for laying in huge caloric stores, while the flying adult is eminently equipped for its reproductive mission.

**11.4** What is the relationship between the extraembryonic membranes and a terrestrial lifestyle?

Reptiles, birds, and mammals have all successfully invaded dry land, and each of these classes possesses extraembryonic membranes. The fundamental advantage conferred by the membranes is the independence from water they provide for the embryonic phase of the life cycle.

In the case of birds and reptiles the embryo develops within the protective confines of the *cleidoic* (enclosed) egg, safe from desiccation and other hazards of dry land. The amnion affords a moist, relatively shock-free milieu for development. The yolk sac offers a continuous supply of food through the developmental process. Early on the allantois is a repository for waste materials generated by the embryo. Later it fuses with the chorion to provide a broad respiratory surface for the metabolically active embryo.

In mammals, the embryo and the later fetus develop within the body of the mother. Modification of the four sacs occurs to accommodate the placental style, which involves a parasitizing of maternal tissues and resources by the fetus. Here, too, the sacs (membranes) are associated with accommodation to a terrestrial lifestyle by the creation of a minienvironment for development isolated from the larger natural environment.

In amphibians, the development of the young requires a return to an aqueous environment. For many amphibians, metamorphosis involves the transition from a water-dependent organism to a form less dependent on ponds and streams. Amphibians are clearly not to be regarded as conquerors of dry land.

**11.5** Generally speaking, during embryogenesis the ectoderm surrounds the mesoderm, which in turn surrounds the endoderm. In one of the extraembryonic membranes, however, this relationship does not hold. Which membrane departs from this pattern, and why?

Each of the extraembryonic membranes is formed from the mesoderm and either the ectoderm or the endoderm. During the formation of the chorion and amnion, a bilayer of ectoderm on top of the mesoderm grows up over the embryo and forms the amniotic folds, as shown on the left in Fig 11.5. In this figure it can be seen that because of the folding action, in the bottom half of the fold the mesoderm is inverted and lies on top of the ectoderm. Since this bottom half forms the amnion when the folds from the two ends of the embryo meet, the amnion consists of ectoderm surrounded by mesoderm.

**11.6** List the basic differences between the bryophytes and the ferns.

| Bryophyte | Fern |
|---|---|
| Gametophyte dominant | Sporophyte dominant |
| Nonvascular | Tracheophytic (vascular) |
| Dioecious | Monoecious |
| Asexually reproducing gemmae | No asexually reproducing structures |
| No true roots, stems, or leaves | True roots, stems, and leaves |

**11.7** What is unusual about the endosperm nucleus in the embryo sac of angiosperms?

The embryo sac in angiosperms constitutes the female gametophyte. Its cells are correspondingly haploid; however, the endosperm nucleus, though part of the gametophyte, is formed by the fusion of the two polar nuclei and is thus diploid. When it later unites with one of the sperm nuclei to form the endosperm during double fertilization, it becomes triploid.

**11.8** In experiments by Briggs and King in the early 1950s nuclei from embryonic frog cells were transplanted into frog eggs that had been enucleated. They found that nuclei from early embryonic cells could produce normal frogs, whereas nuclei from cells of older embryos could not produce normal and complete development. (The production of an individual that is a replica of the nucleus donor is called *cloning*, an asexual kind of replication.) The results suggested a critical period during which the nucleus was fixed in its outcome. Later, however, very painstaking experiments by John Gurdon indicated that the nuclei from cells of late-stage tadpoles could, in rare cases, act in a totipotent manner. What are the implications of Gurdon's work?

Gurdon showed that the changes that occur during cell differentiation do not involve the loss of genetic material, since complete organisms could not have been grown from fully differentiated cells if the genome were not complete. Therefore, differentiation more likely involves a regulatory interplay between cytoplasmic factors and the nucleus.

# Supplementary Problems

**11.9**   Egg pronucleus   (*a*) haploid or   (*b*) diploid?

**11.10**   Archegonial cell   (*a*) haploid or   (*b*) diploid?

**11.11**   Germinal vesicles in starfish egg   (*a*) haploid or   (*b*) diploid?

**11.12**   Endosperm nucleus   (*a*) haploid or   (*b*) diploid?

**11.13**   Tube nucleus of pollen tube   (*a*) haploid or   (*b*) diploid?

**11.14**   Gemmae of a bryophyte   (*a*) haploid or   (*b*) diploid?

**11.15**   Cell from true root of a fern   (*a*) haploid or   (*b*) diploid?

**11.16**   Which are larger   (*a*) cells at the animal pole or   (*b*) cells at the vegetal pole?

**11.17**   Which is greater   (*a*) the amount of yolk in a frog egg or   (*b*) the amount of yolk in a human egg?

**11.18**   Which is larger   (*a*) the sporophyte in a bryophyte or   (*b*) the gametophyte in a bryophyte?

**11.19**   Which is larger   (*a*) the sporophyte in an angiosperm or   (*b*) the gametophyte in an angiosperm?

**11.20**   Which is greater   (*a*) the number of sexes in a prothallus or   (*b*) the number of sexes in a thallus?

**11.21**   The nerve tube arises from the ectoderm.
(*a*) True   (*b*) False

**11.22**   In gymnosperms, e.g., pine trees, there are seeds but no flowers.
(*a*) True   (*b*) False

**11.23**   Von Baer established the germ layer theory of development.
(*a*) True   (*b*) False

# Answers

| | | | | | | | |
|---|---|---|---|---|---|---|---|
| **11.9** | (a) | **11.13** | (a) | **11.17** | (a) | **11.21** | (a) |
| **11.10** | (a) | **11.14** | (a) | **11.18** | (b) | **11.22** | (a) |
| **11.11** | (b) | **11.15** | (b) | **11.19** | (a) | **11.23** | (a) |
| **11.12** | (b) | **11.16** | (b) | **11.20** | (a) | | |

# Chapter 12

# Animal Reproduction

## 12.1 AN EVOLUTIONARY SURVEY

As discussed in Chaps. 8, 10, and 11 *sexuality*, a union of parts producing new genetic combinations, is virtually universal in the living world. Even viruses, existing at the borderline of life, demonstrate recombinations of their nucleic acid cores.

Specialized cells, the gametes, unite in sexual union to produce the zygote. In a variety of algal forms the gametes are alike and are called *isogametes*. In other primitive plants, such as spirogyra, the gametes are structurally alike but function differently. Such a situation is termed *functional heterogamy*. In most advanced plants and animals the gametes have achieved a high degree of specialization as a motile sperm and a usually passive, food-bearing egg. One older idea dealing with the origin of sexuality was the *hunger theory of sex*—the concept that sporelike cells originally united as gametes in order to share their food supplies.

**EXAMPLE 1**   There is considerable variation in the shape and activity of sperm. Nematode worms like the parasite *Ascaris*, arachnids (spiders), mites, centipedes, and some species of crab have amoeboid sperm, which move by means of undulations. In most of the higher animals (vertebrates), sperm cells are elongated structures with long flagella emanating from and continuous with a midpiece. The midpiece is rich in mitochondria which provide the ATP necessary for the sculling movements of the flagella, the basic movements which propel the nuclear head eggward.

In mammals, individual mitochondria are observed in the midpiece, but in a variety of other vertebrates and most invertebrates the mitochondria are fused to form a single large metabolic structure. Cristae may also be modified in many invertebrates. Where a flagellum is present, it usually presents the $9 + 2$ microtubular structure characteristic of motile appendages. In many species the Golgi apparatus, endoplasmic reticulum, ribosomes, and other cytoplasmic organelles are sequestered into a bud, which then breaks away from the streamlined sperm.

The nuclei of most sperm cells, which constitute the payload for fertilization, are relatively small. Both nucleus and surrounding cytoplasm present a microtubular structure; even the chromatin may be present in a tubular array. Histones are usually present in large amounts, but in many plant and animal species the histones are replaced by highly basic proteins called *protamines*. Chromosomes may actually be oriented as longitudinal fibers along the length of the head. In primate sperm, fructose replaces glucose as the primary monosaccharide fuel.

The life span of sperm varies greatly. In insects, sperm from a drone may live in the body of the queen bee for more than a year. In humans, activity may continue up to a week in the female reproductive tract and for 3 days in the body of a dead male.

The overwhelming advantage of sexuality lies in its capacity to provide variation in the progeny for each generation. Such variation permits a selection of better adapted individuals in the inexorable move toward enhancement of fitness for extant populations. The obvious disadvantages of sexual reproduction are that, at least in dioecious forms, it requires a meeting of two different organisms. Where gametes are shed in the water, perhaps only a rendezvous of gametes is required, with the gamete makers free to pursue their separate paths.

**EXAMPLE 2**   Among some salamanders the male deposits packets of sperm on a moist stretch of forest floor. A female comes along and picks up the sperm with her cloacal lips. Fertilization will occur within the reproductive tract of the female, but the eggs will be deposited along the banks of a stream or other moist environment for development. In this case, parents never meet but produce offspring.

### INTERNAL FERTILIZATION

So long as mating occurred in aqueous surroundings, in which transport and hydration of the gametes were guaranteed, external fertilization was a suitable method for bringing sperm and egg together. Consequently it is seen most frequently in aquatic animals and in amphibians.

**EXAMPLE 3**   In the frog, external fertilization of the eggs occurs in a pond or stream. The male mounts the female and proceeds to squeeze her sides with his thickened thumb pads, which are a secondary sexual characteristic of the male. This embrace is called *amplexus*. As a string of eggs is extruded to the outside in response to the squeezing actions, the male proceeds to urinate a sperm-rich fluid directly on the eggs and fertilization occurs in the vicinity of the clasped couple.

With a movement into terrestrial habitats, animals (and plants as well) required adaptation to maintain an aqueous environment for gametes as well as for the developing embryo. In some cases, as we have seen, a return to ponds and streams occurred during breeding activity. This reflects a continued dependency on environmental water. Other organisms, such as mammals, overcame this problem by providing an internal, rather than an external, medium for fertilization. In reptiles and birds, sperm are deposited within the reproductive tract of the female, and fertilization results in a zygote that is covered by albumin and a shell and is soon deposited externally to complete its development under the care of the mother or even both parents.

In coitus (sexual intercourse) an introjecting organ, the penis, is actually inserted into the reproductive tract of the female, and sperm are deposited internally. In the bird, internal fertilization is accomplished when male and female place their *cloacae* (collecting chamber to outside) close together in a modified form of coitus called a *cloacal kiss*. No penis is present, but in some birds the cloaca of the male actually everts into the cloaca of the female. In reptiles, an imperfect *hemipenis* introduces sperm within the vagina of the female. In most insects in which the male supplies sperm, fertilization is accomplished through copulation.

**EXAMPLE 4**   An almost heroic commitment to internal fertilization is shown by the bedbug. In the course of evolution the male's penis enlarged and could no longer be inserted into the female's oviduct. So vital is sexual recombination to that species that specialized pouches have developed along the sides of the female's abdomen into which the male deposits packets of sperm. These sperm then burrow through the female's tissues to unite with eggs in the ovary.

### INTERNAL DEVELOPMENT

In mammals, an ingenious strategy has developed to cope with the exigency of a terrestrial habitat. Not only is fertilization internal, but the embryo itself develops within the body of the mother in parasitic fashion. This developmental lifestyle completely liberates mammals from dependency on water for reproduction. In some special cases, insects, spiders, and fish also show modifications for the development of the young within the body of the female, although usually this involves brooding fertilized eggs in the reproductive tract until they hatch.

Major changes in the reproductive tract of the female are a prerequisite for successfully carrying the young. These involve minimally a *uterus*, or *womb*, within which the embryo can develop; a *placenta* in which exchanges of food, oxygen, and waste material can occur between mother and child; and an elaborate repertoire of hormones which can regulate periodic changes in the uterus and other structures within the female reproductive tract. Clearly, a uterus which is carrying an embryo will be different from one which is not so challenged.

## 12.2   MALE REPRODUCTIVE SYSTEM IN HUMANS

The male reproductive system:

1.  Produces sperm (male gamete).
2.  Delivers sperm to an appropriate region within the female reproductive tract.
3.  Produces the male sex hormones, or androgens, which prepare the glands and tubular conduits of the male reproductive tract for functioning and produce such secondary sexual characteristics as facial hair, increased muscular strength, and libido (sex drive).

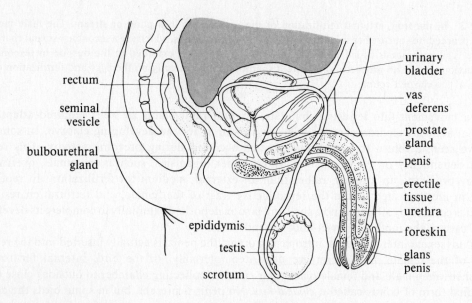

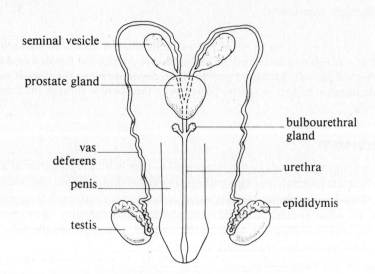

**Fig. 12.1** Male reproductive system

The *gonads* of the male, within which the sperm are produced and androgens synthesized, are the *testes* (see Fig. 12.1). In humans, the term *testicles* is often used. In many mammals the testes descend from their origins in the kidney region of the abdomen to the loose *sacs* of the *scrotum, a* pouch that hangs down just behind the penis. Both penis and scrotal sacs are external in these mammals and lie between the legs.

Sperm are formed through a set of meiotic divisions (see Chap. 8) within extended, but extensively coiled, *spermatic tubules.* Scattered among the tubules in the testis are the *interstitial cells*, which continually secrete *testosterone*, the major androgen in the male. Androgens are synthesized and secreted at high rates following *puberty*, the time at which sexual maturity occurs.

Sperm produced at a steady rate in the spermatic tubules of the testicle are stored in the epididymis, a highly coiled tubular structure lying atop each testicle. The sperm are carried through the abdominal cavity by the *vas deferens*, a long tube that joins the *urethra* in the region of the *prostate gland.* The urethra, a tube that exits through the penis, carries urine to the exterior and, alternatively, serves as a conduit for *seminal fluid*. Sperm is carried in a fluid formed along the male reproductive tract by the *seminal vesicles, prostate gland*, and *Cowper's glands*, accessory glands of the male reproductive tract.

The penis is an efficient tool for delivering sperm (semen) to the female reproductive tract, but it must first become erect. This is accomplished during sexual excitation in humans by the triggering of impulses in a nerve from the parasympathetic nervous system, which causes blood to engorge the spongy erectile tissues within the shaft and glans of the penis. A *vasodilation* (widening of blood vessels) of the arterioles bringing blood to the tissues of the penis brings in more blood than can leave, causing distension and stiffening—an *erection*. In its stiffened state the erect penis can be thrust into the female reproductive tract. To and fro movements increase excitation and culminate in the expulsion of seminal fluid from the penis in a series of spurts at *climax*. These forceful emissions are known as *ejaculations*. They are caused by rhythmic waves of contraction along the lower two-thirds of the male genital tract, generally with a periodicity of 0.8 second.

## 12.3  FEMALE REPRODUCTIVE SYSTEM IN HUMANS

### ANATOMY

The gonads of the female are the paired ovaries that remain within the abdominal cavity throughout life (see Fig. 12.2). More than 400,000 potential eggs are present in the ovaries, but only about 400 will ever complete meiosis.

The ovaries are held in place by *ligaments* (bands of connective tissue) and thin membranous *mesenteries*. Overhanging each ovary are the *fallopian tubes* (*oviducts*), which lead to the thick, muscular uterus. It is within the uterus that development of the embryo, later the fetus, will occur. Eggs that are fertilized in the fallopian tubes (usually only one egg) are swept along by cilia into the uterus, where early development and eventual implantation will occur.

The uterus adjoins a short corrugated tube known as the *vagina*. The vagina stretches over the narrow neck of the uterus (*cervix*) at one end and extends to the outside at its distal end. The external sex organs—the labia majora and minora and the clitoris—that surround both the opening into the vagina and the separate opening into the urethra form the *vulva*; the external lips (*labia majora*) enclose a pair of internal lips (*labia minora*). The labia are homologous to the scrotum of the male. The *clitoris*, which is a knob of erectile tissue lying just above the vaginal opening, is homologous to the penis of the male.

### MENSTRUAL CYCLE

The *menstrual cycle* refers to the series of events that periodically modifies the female reproductive tract of humans and advanced primates. In other mammals a cycle of receptivity to sexual activity occurs that is probably less complex than the menstrual cycle and is known as the *estrous cycle*.

**EXAMPLE 5**  *Estrus* comes from a Greek root meaning "gadfly," probably because the goading influence of the sexual libido was felt to be similar to the effect of a stinging fly. As the name implies, animals at the height of estrus experience a strong, short-lived compulsion to mate; they are said to be "in heat" or "in estrus." On either side of this brief period, their sex drive is quiescent. At the physical level, the estrous cycle prepares the female reproductive tract for copulation. There is not the elaborate development of the uterine lining seen with the menstrual cycle. If fertilization does not occur, any preparatory thickening of the uterine wall is simply reabsorbed into the body. The events of the cycle are subject to environmental influence; in some animals, the release of the ovum is dependent on copulation.

The menstrual cycle is regulated by hormones produced by the hypothalamus of the brain, pituitary gland, and endocrine structures present in the ovary. The gonadotropin-releasing hormone (GnRH) from the hypothalamus of the brain probably moves across a tract of blood vessels called the *median eminence* directly to the pituitary. The pituitary is then stimulated to produce two hormones, the

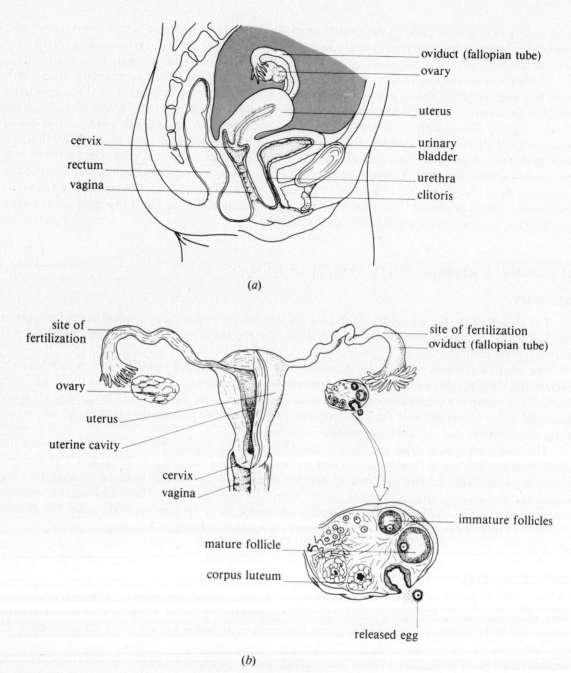

Fig. 12.2 Female reproductive system

follicle-stimulating hormone (FSH) and luteinizing hormone (LH), which stimulate the ovaries to produce the female sex hormones *estradiol* (from the graafian follicle encasing an egg) and *progesterone* (from the *corpus luteum* which is formed from a ruptured follicle). The pituitary also produces *prolactin*, a hormone that acts directly upon the breast and its milk glands and also influences mineral and fluid balance.

It is estradiol (and perhaps other species of estrogen) and progesterone which produce menstrual changes in the ovaries and uterus and throughout the body. The hormones involved in the menstrual cycle may also interact with the nervous system to modify the menstrual response. Under fasting conditions, as well as during rigorous training for athletic competition, menstruation often ceases. Women who were incarcerated in Nazi concentration camps often experienced loss of the menstrual cycle.

The total cycle may be divided into three parts: the *menses* (flow of blood), which lasts about 5 days on average; the *follicular* (or proliferative) phase, which follows the cessation of flow and lasts about 9 days; and the *luteal* phase, which comprises about the last 2 weeks of the cycle.

During the 5-day period of flow, the levels of GnRH are low to moderate. FSH is low but will go lower still at the time of ovulation. LH is also at a low plateau. Both estradiol and progesterone levels are similarly low, and the *endometrium*, or inner lining of the uterus, is being shed along with a modest amount of blood and deteriorated vascular tissue. Despite higher estimates by those undergoing menstruation, the total volume of blood lost rarely exceeds 1.5 cupsful. The question has been raised about why the menstrual flow of blood is not staunched by the clotting mechanism. Actually, menstrual blood arises from clots that have formed but then have been dissolved by an enzyme called *fibrinolysin* or *plasmin*. Thus, menstrual blood is declotted blood.

During the proliferative phase, follicles begin to develop in the ovaries under the influence of FSH and LH from the pituitary. Hypothalamic production of GnRH is appreciable. A single growing follicle (others are suppressed late in this follicular stage) continues to produce increasing amounts of estradiol, which influences growth and vascularization of the endometrium. At the time of ovulation (release of the ovum) there is a sharp drop in estradiol and a surge in LH levels. This marks the transition from the follicular to the luteal phase. One of the effects of the short surge of LH is the conversion of the ruptured follicle through a series of changes to the yellowish *corpus luteum*. A small amount of estradiol is produced by the corpus luteum, but more characteristically it secretes high levels of progesterone, the second female sex hormone, which is largely responsible for the changes associated with the luteal phase. The thickened endometrium becomes highly glandular, and an increase in glycogen occurs. These maturational changes of the uterus prepare it for implantation and the sequelae of pregnancy. Very little GnRH is produced during this third phase of the cycle. One important effect of the high levels of progesterone is the suppression of follicular development in the ovary. It is progesterone that acts as the major ingredient of the birth control pill, since a woman is maintained in the postovulatory condition under its influence, with no follicle formation.

Should pregnancy fail to occur, the corpus luteum will break down toward the latter part of the luteal phase (about day 21) and soon the endometrial lining will degenerate and a new flow will begin. The boundary between luteal and flow phases is marked by a sharp drop in progesterone. All other hormones are also low at that juncture. However, with the loss of the corpus luteum a new buildup of GnRH occurs, and by the end of flow a buildup of estrogen will be renewed by newly developing follicles.

*Menstruation* derives from the root for "moon," because in many women the menstrual period approximates the duration of the lunar month, 28 days. Since the tides, influenced as they are by the moon, also have a 28-day cycle, some biologists contend that menstruation is a vestige of a cycle established when our ancestors lived by the shore and were influenced by the tides. Menstruation usually ceases during pregnancy and trails off when older women enter *menopause*, the period of slackening of hormone secretions that control the menstrual cycle. The cessation of the menstrual cycle at menopause is not necessarily correlated with a decrease in sexual desire.

## 12.4  SEXUAL RESPONSE IN HUMANS

Both genders have an identical repertoire of physiological responses in their sexual activity: *vasocongestion* and *myotonia*. Vasocongestion, the movement and pooling of blood, is responsible for a ventral sex flush in both sexes, primary erection of both the penis and the clitoris, secondary erection of the nipples, and even the *vaginal sweating phenomenon*, in which the pressure from pooling blood forces blood fluids across the wall of the vagina, where it collects on the internal surface. Myotonia comprises the often spasmodic contraction of muscle which produces waves of contractions along the male and female reproductive tracts and anus during orgasm; the tremors of the buttocks, arms, and legs during climax; and the minor contractions of a more superficial nature which occur as the "aftershock" phenomena during resolution in the sexual response.

In both males and females the sexual response consists of four phases. Initially erotic stimuli induce an *excitement phase* in which engorgement produces an increased level of tumescence. A high degree of excitation is sustained at peak levels in the *plateau phase*, which is followed, usually with an accompanying sense of physiologic inevitability, by the *orgasmic phase*, in which the exquisite feelings of release are accompanied by the myotonia of orgasm. In the *resolution phase*, a return to ground level occurs, after which arousal is again possible.

Despite a basic similarity, differences do exist in the way the sexes respond. In women, each phase of the sexual response tends to be drawn out, so that the qualitative similarity in response of men and women is modified by a quantitative difference. The differences in duration may underlie difficulties in sexual adjustment for heterosexual couples, particularly the possibility that men may reach resolution before their partners are past plateau.

A second difference is that women have a much greater variety of postplateau responses including multiple orgasm, which is a movement into and out of plateau and orgasm without intervening resolution. Most men must move along in lockstep fashion with single consecutive stages. Orgasm is usually followed by resolution, and in older men the resolution phase may be inordinately long.

The first sign of sexual excitation in a male is a phallic erection. The first sign of a female's sexual arousal is the vaginal sweating phenomenon. In the female, orgasm is associated with a bright red coloration of the vulval region, reminiscent of some of the lower primates.

The menstrual cycle influences fluctuations in the physical and emotional parameters of sexuality in females. Fluctuations in the male tend to be less well defined.

## 12.5  CONTRACEPTION

Unlike almost all other mammals, humans have no season for sexual activity. Coitus may occur at any time. Although coitus is generally associated with reproductive functions, in humans it may serve to maintain a permanent bonding between a man and a woman.

**EXAMPLE 6**  The love or attachment between husband and wife is often expressed in some form of sexual activity, certainly kissing and hugging if not sexual intercourse itself. By maintaining sexual interest throughout the year, the family constellation may be maintained on a permanent basis. Since family structure, with its cooperation and specialization of cultural tasks, may have enhanced the chances for survival of early human societies, a trend toward the maintenance of sexual expression may have had evolutionary advantages for humans.

If we view sexuality from a social rather than reproductive perspective, preventing pregnancy while still maintaining sexuality (i.e., using *contraception*) may be seen to fulfill certain social functions. For many individuals sexual activity may reduce tensions that arise from other areas of life and also provide intense gratification and enhanced feelings of self-worth. For these reasons, as well as for the pragmatic necessity of controlling family size, contraceptive techniques have been of increasing interest.

A variety of relatively recent technologies for contraception have been added to the older, more traditional methods.

From the point of view of effectiveness, the best methods for contraception are the *vasectomy*, a surgically simple cutting and tying of the vas deferens near the testicles of the male, and *tubal ligation*, a parallel procedure on the fallopian tubes of the female. Vasectomized males may copulate, have orgasm, and ejaculate, but the seminal fluid is free of sperm because of the surgical block between the epididymis and vas deferens. The comparable tying of the female tubes prevents sperm from reaching the eggs. Once performed, it requires no further preparation from the sexual partners; however, a disadvantage is its usual irreversibility. Extensive counseling should precede this course of action.

The *birth control pill* is relatively effective; careful application of this method would probably produce even higher effectiveness. Since it must be taken on a regular basis, some degree of sophistication is required for its use. Its expense, while not inordinate, militates against its large-scale use in developing countries. Some increased risk of stroke has been reported with its use.

The *intrauterine device* (coil, loop) was widely used until a high rate of infection and discomfort was noted, especially for the type called the Dalkon shield.

The *condom* has become much more popular because it is particularly effective in preventing venereal disease in addition to its contraceptive role. The concern with the spread of AIDS (acquired immune deficiency syndrome) has markedly enhanced the popularity of this technique, which prevents contact with blood or other bodily fluids during intercourse. Condoms have specifically been recommended for use during homosexual activity, especially in anal intercourse.

The *diaphragm* used with spermicidal jelly can be highly effective if fittings by gynecologists are carefully done each time so that complete blockage of the cervical opening is assured. No side effects are associated with this approach.

Statistical evidence argues against the use of vaginal foams, withdrawal, rhythm (avoiding sex around the time of ovulation when pregnancy is likely to occur), and the douche. However, the Natural Family Planning method, which is a newer application of the rhythm method based on more reliable examinations of cervical mucous, seems to be a more effective method for controlling family size.

Promising research continues on the development of a male contraceptive which involves the injection of testosterone derivatives. Such a regimen markedly reduces the production of sperm. It is an approach that does not require mechanical interference with intercourse.

In some communist countries a program of *induced abortion*, an active intervention to terminate a pregnancy, is used to limit population size. Abortion is also quite common in the west, but increasing opposition has led to a reappraisal of legalized abortion in the United States.

**EXAMPLE 7** Such issues as the abortion controversy and the debate over the permissible extent of heroic measures to stave off death in terminal patients have kindled partisan interest in when life actually begins and when it can be thought of as having ended. In some religious denominations, life begins when the sperm enters the egg, but for other religious groups life starts at the moment of birth. This obviously bears some relevance to the abortion controversy. The quality of life has also been explored, and some religious experts have testified that the mere possession of vegetative functions without a manifestation of feelings and thoughts does not justify artificial maintenance of breathing, circulation, etc.

## 12.6 SEXUAL PREFERENCE

The libido, or sex urge, is a generalized propensity for tension-reducing activities, which may be expressed in a variety of ways. Both genetic makeup and early training and experience probably determine a person's eventual sexual choices. Heterosexual intercourse is often cited as the natural expression of sexual energies, but a variety of alternatives exist. Among them are *masturbation* (self-stimulation to sexual climax), *homosexual intercourse* which involves a sex partner of the same gender, *bestiality* (sexual activity with nonhuman creatures), and even *abstinence*, or noninvolvement in the sexual arena. In many societies strong condemnation is directed against some of or all these practices. Masturbation and homosexual activity can be found in other mammals, especially under conditions of stress.

In an early study of the sexual activity of human males (late 1940s), Alfred Kinsey and his research group discovered that there is tremendous variation in both the nature of the sexual outlet chosen and the amount of sexual activity that occurs. Whereas certain physiological functions, such as blood pressure and blood sugar, exist within definite parameters in a healthy individual, sexual activity eludes the establishment of a norm. At least six different kinds of outlets were repeatedly cited by the experimental population, a diversity that was not readily condoned during that era of intolerance of the sexually adventurous.

## 12.7  SEXUAL DYSFUNCTION

Sexual fulfillment, as well as reproductive success, is dependent on several functional levels. The ability of the aroused male to carry out sexual intercourse depends on his achieving an erection. Failure to achieve or maintain an erection is known as *impotency.* While all males may experience temporary periods of impotency, a long-term failure of potency in the absence of debilitating disease is a cause for medical concern. Most often impotency has a *psychogenic* cause, a response to emotional conflict or psychic confusion. Since sex is one of a group of vegetative functions (such as eating and sleeping) that render the organism vulnerable to external dangers, any indication of a threat to safety tends to suppress these "luxury" functions as the organism prepares for "fight or flight." This mobilization for danger is under the control of the sympathetic nervous system (see Chap. 15) and tends to override the vegetative functions. Thus, in the subtle complex of reactions embodied in the sexual response even an imagined or projected threat, whether brought to consciousness or not, might interfere with performance. At the same time, the possibility of an underlying organic dysfunction should not be ignored.

The inability to become a biological parent is called *sterility.* A sterile male will not produce sperm or will produce too few to assure success in reaching a viable egg in the fallopian tube. Sterility may also be associated with defective sperm, particularly if such a defect impairs sperm mobility. Impotency and sterility are separable phenomena; that is, an impotent male may be highly fertile. In such males a semen sample could be obtained and a pregnancy produced in a mate through *artificial insemination,* in which semen is introduced into the reproductive tract by a syringe.

A sterile (infertile) female may fail to produce eggs, or her eggs may not discharge from the surface of the ovary to reach the fallopian tubes. In some sterile females, a block to fertilization occurs in the tubes, usually associated with chronic inflammation. Indirect evidence indicates that some women may produce lethal factors which destroy or inactivate sperm which are deposited in the region of the cervix. In women, a failure to respond to sexual stimuli does not preclude copulation as does the failure to achieve erection in the male. However a lack of receptivity may make coitus less pleasurable for either partner.

The height of sexual tension is relieved by a sudden relaxation that is intensely pleasurable—the *orgasm.* In the male, orgasm is usually accompanied by the ejaculation of semen. The failure to achieve orgasm has been the subject of intense scrutiny, especially in the case of women, and it is currently regarded as a sexual dysfunction. A common problem in many men is *premature ejaculation,* in which orgasm and expulsion of semen occur within an inappropriately short span of time.

In humans, sexual expression is a complex of emotional and physical responses, a combination of plumbing and poetry. Difficulties may arise in any aspect. The increasing availability of skilled professionals in this area and the growing willingness of most societies to deal with sexuality and to recognize it as a significant area for self-actualization promise hope for people suffering sexual dysfunction.

**EXAMPLE 8**   William Masters and Virginia Johnson began a series of studies of human sexual response in 1954 at Washington University Medical School in St. Louis, Missouri. They developed a conceptual framework for understanding human sexuality and a series of clinical techniques for dealing with sexual inadequacy.

Even before their collaborative explorations, Masters had discovered the existence of the vaginal sweating phenomenon. Other physiological responses, mostly unknown to gynecologists, were soon discovered. A variety of concerns about sexuality as a complicating factor in the presence of other diseases were explored by the group. Their contributions have been seminal to the fields of comparative physiology, *ethology* (comparative study of animal behavior), and sexual dysfunction. They have spawned centers that have been highly successful in curing dysfunctions such as impotency and premature ejaculation through specific clinical approaches in a relatively short time.

Perhaps most important of all, they have brought a profound measure of respect for the scientific study of human sexuality and have made it possible for troubled patients everywhere to arrive at a satisfactory resolution of their sexual problems.

# Solved Problems

**12.1**  Describe some asexual processes occurring in animal reproduction. What is the chief disadvantage of this form of reproduction?

We have already referred to parthenogenetic development of unfertilized eggs as a regular procedure in the production of male ants and bees (Chap. 11), and we also saw that artificial stimulation of sea urchin eggs induces cleavage. Although the egg is a sex cell, the development of unfertilized eggs is an asexual process.

Among sponges and hydra *budding* is a common occurrence. This involves an outgrowth of a portion of the parent's body to produce a new individual. Occasionally the new growth will remain attached to the parent stock, but usually the bud breaks away to produce an individual organism.

In the phylum Platyhelminthes a process called *fragmentation* occurs in which some of these flatworms spontaneously separate into separate lengths. Each of the fragments produces a new flatworm. Related to fragmentation is *regeneration*, or the restoration of lost parts. Seemingly, this is a mechanism for compensating for the accidental removal of organs or extended structures. However, among the echinoderms (starfish, brittle stars, etc.) the removal of an arm and part of the central disk leads to the formation of a new organism from that arm, so that the process may be considered both fragmentation and regeneration.

Asexual reproduction is a simple procedure for producing progeny, but it tends to minimize the variation which is grist for the mill of evolution. It also tends to eliminate the existence of parents, except in the instance of budding. In almost all animal organisms, asexual reproduction is only a supplement to sexual reproduction.

**12.2**  What do insects and mammals demonstrate in common in the reproductive area?

Despite their clearly disparate lineages, both insects and mammals represent marked successes in dealing with the challenges of a terrestrial environment. Both classes have independently evolved mechanisms to prevent water loss; the insects by means of a virtually impervious chitinous exoskeleton and mammals through an elaborate integument covered with a layer of down or fur. But it is in terms of reproductive adaptations that the strategies of insects and mammals show a remarkable convergent similarity.

In both groups the penis, consisting of a closed tube for discharging male gametes within the reproductive tract of the female, has become efficient as a conduit for seminal fluid. Copulation occurs in all insects in which sperm is supplied by the male. There are a number of species in which the males are no longer present and eggs develop parthenogenetically. (This is also true for some groups of lizards and fish among the vertebrates.)

A major modification of the mammalian female is the internal development of the young through the evolution of a placenta. In a number of species of insects, particularly among roaches and flies, live young are "birthed" by the mother. This is not a strict parallel with the mammals, since the insect mother merely retains eggs within her reproductive tract until the young are ready to hatch. They are usually expelled from the reproductive tract shortly after hatching.

**12.3**  In fish and most other nonmammalian vertebrates, the testes remain within the abdominal cavity. However, in elephants they descend during the breeding season; in rodents, bats, camels, and a few apes they descend only during copulation. In marsupials (pouched animals), most carnivores, and almost all primates, the testes are permanently descended. What is the likely explanation for this variation?

Mammals tend to maintain a body temperature of 37 °C. Such high temperatures interfere with sperm production, as well as sperm release and motility. Consequently, most warm-blooded animals have testes positioned outside the body core. Cold-blooded vertebrates are not faced with this problem and can maintain their testes within the abdominal cavity.

In humans, the descent of the testes is associated with certain potential difficulties. Shortly after birth the testes leave the abdominal cavity and move along the *inguinal canal*, a preformed channel cutting across the heavy muscular body wall to the scrotal sacs. Once the testes have descended, the rift in the body wall heals over but it remains a permanently weakened spot. Under the pressure of sudden movement, severe coughing, or an incorrect lifting of weights, it may tear and produce an *inguinal hernia.* If a loop of intestine becomes trapped in the crevice, it may lead to a cutting off of blood supply and consequent gangrene, a serious condition called *strangulated inguinal hernia.*

Should the testes fail to descend, a number of serious consequences may follow. These include failure of the male to develop sexually, generalized inhibition of body development, and sterility. The medical term for this condition is *cryptorchism* (or *cryptorchidism*), which literally means "hidden testis." The term *orchid* (Greek) refers to the testis; the flower is so named because of the oval swellings along its "roots," which resemble testicles.

**12.4** The menstrual cycle comprises an intricate interplay of hormones. We have seen that gonadotropin-releasing hormone stimulates the pituitary to release follicle–stimulating hormone and luteinizing hormone, which engender growth of the follicle and its transformation into the corpus luteum. The follicle in turn produces large amounts of estradiol, and the corpus luteum is responsible for a sharp rise in progesterone levels. Negative feedback, however, plays an equally important role. High levels of estradiol inhibit the FSH-stimulating center of the hypothalamus, and high levels of progesterone inhibit the FSH-stimulating centers and, in the absence of fertilization, the LH-stimulating centers. Given the timing for the release of estradiol and progesterone, what roles do you suppose such inhibition plays in the menstrual cycle?

The production of estradiol rises significantly as one of the follicles begins to mature faster than the others. Since one of the follicles has been "chosen" and developed, the role of FSH has been fulfilled; its inhibition by estradiol thus makes metabolic sense. The inhibition of the FSH-stimulating center by progesterone ensures that FSH levels do not rise and trigger the beginning of another cycle. If fertilization does not occur, progesterone begins to inhibit the LH-stimulating center. Since LH is necessary for the maintenance of the corpus luteum, this body begins to deteriorate, with a resultant decrease in progesterone. The diminished levels of progesterone cause atrophy and sloughing of the uterine wall. The decrease in estradiol and progesterone frees the inhibited hypothalamus to produce a gonadotropin-releasing hormone, and start a new cycle.

**12.5** How do the estrous and menstrual cycles differ?

Animals in heat experience a brief, strong sexual urge during the middle of the estrous cycle but are not sexually receptive at any other time; sexual receptivity occurs throughout the menstrual cycle. Physically, estrus prepares the female reproductive tract for copulation, whereas the menstrual cycle involves an elaborate preparation of the endometrium for implantation of a fertilized egg. Consequently, if fertilization does not occur, any preparatory thickening of the uterine wall in estrous animals is merely reabsorbed; in menstruating animals the hypertrophic lining is sloughed off as menstrual flow. Finally, the events of the estrous cycle are much more subject to environmental influence than those of the menstrual cycle.

**12.6** What would you guess are some common deterrents to the effective use of contraceptives?

Religion is probably the greatest deterrent. In Catholicism, Orthodox Judaism, and certain sects of the Moslem faith contraception is formally forbidden. In all three cases the basis for the proscription is rooted in the biblical mandate to "be fruitful and multiply," which discourages the separation of sexual activity and reproduction. The effectiveness of large-scale birth control programs in underdeveloped countries is also diminished by a lack of understanding on the part of the populace in the proper use of various contraception techniques.

Inconvenience and expense are two more factors.

**12.7** Sterility and impotence are both sexual dysfunctions in men; however, only one of these is associated with women. Explain.

Sterility is a dysfunction that affects the gametes directly. In females this can involve an inability to produce ova or to expel ova successfully into the fallopian tubes. It can also involve impediments to fertilization such as inhospitable environments in the reproductive tract that are lethal to sperm and blockages in the fallopian tubes caused by inflammation. However, impotence involves an inability to perform sexually. Unlike the situation for males, who must achieve erection, the physical changes associated with a healthy sexual response in females are not such that their absence would preclude intercourse.

# Supplementary Problems

**12.8**  Functional heterogamy and structural isogamy are exhibited by  (*a*) humans.  (*b*) ulothrix.  (*c*) spirogyra.  (*d*) hydra.  (*e*) all of these.

**12.9**  Amplexus in frogs is an example of internal fertilization.
(*a*) True  (*b*) False

**12.10**  Gametes are usually produced in specialized structures known as gonads.
(*a*) True  (*b*) False

**12.11**  The most complete and competent of penes (pl. of *penis*) is found in successful terrestrial forms.
(*a*) True  (*b*) False

**12.12**  Accessory glands like the seminal vesicles contribute to the fluid substance of the seminal fluid.
(*a*) True  (*b*) False

**12.13**  Ejaculation of seminal fluid occurs with a periodicity of 0.8 second in both males and females.
(*a*) True  (*b*) False

**12.14**  The initiating hormones for the menstrual cycle arise in the hypothalamus.
(*a*) True  (*b*) False

**12.15**  Match the following events of the menstrual cycle in column **A** with the appropriate phrase from column **B**.

| **A** | **B** |
|---|---|
| 1.  Sudden surge of LH | (*a*)  Menses (flow) |
| 2.  Shedding of endometrium | (*b*)  Proliferative phase |
| 3.  High levels of progesterone | (*c*)  Secretory (luteal) phase |
| 4.  Development of corpus luteum | (*d*)  Ovulation |
| 5.  Production of estradiol by graafian follicle | |

**12.16**  High levels of progesterone will result in a diminution of the trophic hormone LH. This is called  (*a*) the cascade effect.  (*b*) negative feedback.  (*c*) hermaphroditism.  (*d*) positive feedback.  (*e*) covariance.

**12.17**  The existence of sperm banks is an important consideration in which method of contraception?  (*a*) condom.  (*b*) vasectomy.  (*c*) tubal ligation.  (*d*) withdrawal.  (*e*) abstinence.

# Answers

| | | | | | |
|---|---|---|---|---|---|
| **12.8** | (*c*) | **12.12** | (*a*) | **12.15** | 1—*d*;  2—*a*;  3—*c*;  4—*d*;  5—*b* |
| **12.9** | (*b*) | **12.13** | (*b*) | **12.16** | (*b*) |
| **12.10** | (*a*) | **12.14** | (*a*) | **12.17** | (*b*) |
| **12.11** | (*a*) | | | | |

# Chapter 13

# Basic Structure and Function in Vascular Plants

In Chap. 31 we will survey the entire plant kingdom from an evolutionary perspective. In this chapter we will concentrate on the structures of those plants that have made a successful transition to a terrestrial habitat. These are the *tracheophytes*, or vascular plants, which have evolved an apparatus of conducting tubules. This network of tubular vessels permits water and nutrients to be transported from one part of the plant to another, an arrangement that frees most of the plant from the necessity of remaining in contact with water.

Among the primitive *divisions* (groups equivalent to phyla making up the kingdom Plantae) of tracheophytes are the *Psilophyta*, with no differentiation of cells into true roots and leaves and a very simple kind of conducting system; the *Lycophyta*, or club mosses, with cells differentiated into roots, stems, and leaves; and the *Sphenophyta*, or horsetails. All three contain a simple strand of vascular tissue, possess motile sperm, and are regarded by many botanists as relics of ancient forms that have largely disappeared.

The ferns (Pterophyta), consisting of more than 12,000 species, are a highly successful tracheophytic division containing flattened leaves with a broad expanse of photosynthetic surface. The success of the fern probably lies in this trend toward an increased leaf surface.

**EXAMPLE 1** Typical fern leaves (fronds) are large, extensively branched structures. The pinnate (featherlike) arrangements of leaf parts maximize the surface available for photosynthesis. Some botanists regard the fern leaf as a modified stem, still evolving toward the flat leaf of the seed plants.

The internal structure of the fern is quite similar to the organs of a typical flowering plant. It consists of water-conducting, thick-walled tubes known as *xylem* and thinner-walled tubes known as *phloem*, which carry organic material from the leaves downward to the rest of the plant.

Although xylem and phloem are encountered in the fern, the cells of these tissues are not as specialized or diverse as those in either the *gymnosperms* (seed plants with no flower) or the *angiosperms* (flowering plants). Also, true roots are not nearly as extensive in the fern as in higher plants, and the stem is a rather inconspicuous structure, usually an underground, horizontal extension known as a *rhizome*. Most ferns are rather small plants whose more prominent sporophyte lasts but a single season in temperate zones.

The seed plants, which include gymnosperms and angiosperms, differ from the ferns in two important aspects: they produce seeds, and they possess, in most instances, a *cambium* layer, which is a meristem structure (layer of undifferentiated cells) capable of continuously producing secondary tissues. These secondary tissues may be xylem or phloem or even the cork cells that protect the trunks of trees. These seed plants are so successful they make up the bulk of the botanical world today.

**EXAMPLE 2** In gymnosperms the seeds are nestled within open cones (*gymnosperm* means "naked seed"), while in flowering plants (the most abundant of all plants) the seeds are found within an enclosed structure, the ovary of the flower. *Angio* comes from a Greek root meaning "vase," and angiosperms are so named because the base of many flowers does resemble a vase containing seeds. A major difference between gymnosperms and angiosperms is that the latter group demonstrates a double fertilization (see Chap. 11).

## 13.1 PLANT NUTRITION

We saw in Chap. 6 that during photosynthesis atmospheric carbon dioxide supplies the carbon and oxygen used in producing carbohydrate. This implies that most of the mass of even the tallest tree is produced quite literally from air. Nevertheless, the soil provides many essential nutrients required in the biochemical structures and processes of the plant.

182

## MACRONUTRIENTS AND MICRONUTRIENTS

Those mineral elements (inorganic elements) required in relatively large amounts are called *macronutrients*. The element nitrogen is drawn from the soil in greatest quantity: it comprises almost 4 percent of the dry weight of many plants. Potassium has been found to constitute more than 5 percent of the dry weight of several plants. (People who must supplement their diets with potassium are often advised to eat such fruits as bananas, apricots, and oranges.) Calcium makes up from 0.5 to 3.5 percent of the dry weight of a plant. Other macronutrients, including phosphorus, magnesium and sulfur, are often found in essential macromolecules of the cell, such as chlorophyll and a variety of enzymes.

Minerals found in relatively smaller amounts are called *micronutrients*. These include iron, boron, chlorine, manganese, and sodium. Other elements occur in such small quantities that they are referred to as *trace elements*. These include zinc, copper, and molybdenum. Though not present in large quantities, the micronutrients and trace elements are essential for such processes as enzyme activation, chloroplast development, and metabolism of other minerals.

## THE NITROGEN CYCLE

Plants, like all life forms, require nitrogen. Although about 79 percent of the atmosphere consists of elemental nitrogen, plants cannot incorporate it directly. Intermediate processes must first convert the nitrogen to usable forms.

*Nitrogen-fixing bacteria* are able to combine atmospheric nitrogen with hydrogen to form ammonium ions ($NH_4^+$). The *cyanobacteria* (blue-green algae) also have this ability. The ammonia is then released into the environment, where it can be taken up and used by plants. Alternatively, nitrogen-fixing bacteria may live mutualistically within special nodules in plants such as clover, alfalfa, and legumes (beans and peas). These bacteria fix nitrogen and supply the resulting ammonia *directly* to the plant. Ammonia absorbed by plants is converted to amino acids and other nitrogen compounds through the mediation of the enzyme *nitrogenase*.

It should be noted that some ammonia is created through physical processes such as volcanic activity.

Ammonia created by nitrogen fixation may undergo further processing, called *nitrification*. In the first step, bacteria such as *Nitrosomonas* convert the ammonia (or ammonium ions) to nitrites ($NO_2^-$). Nitrites are very toxic to plants, so a second conversion is necessary. This is accomplished by bacteria, such as *Nitrobacter*, that convert the nitrites to nitrates ($NO_3^-$). Since flowering plants use nitrates more readily than ammonia as a nitrogen source, nitrification is important to the maintenance of the earth's flora.

Nitrates may also be produced from atmospheric nitrogen directly through various physical processes such as lightning and reaction with photochemical fog.

Nitrogen fixation introduces *new* nitrogen to the cycle; however, *decomposition* acts on organic sources of nitrogen, that is, nitrogen that has already been incorporated into living organisms. Nitrogenous wastes, such as urine, protein, and other nitrogenous compounds freed by bacteria and fungi during the decay of dead plant and animal matter provide this second source of nitrogen. During the process of *ammonification*, these organic compounds are converted to ammonia; the two-step process of nitrification then produces nitrate.

Some elemental nitrogen is returned to the atmosphere through the action of *denitrifying bacteria*, which act on $NO_3^-$ and $NO_2^-$ to produce $N_2$.

All these processes—the introduction of elemental nitrogen through nitrogen fixation and nitrification; the restoration of $N_2$ to the atmosphere through denitrification; the recycling of organic nitrogen through decay, ammonification, and nitrification—are all elements of the *nitrogen cycle* (see Fig. 13.1).

Any close living relationship between two different species, such as that seen between nitrogen-fixing bacteria and plants, is termed *symbiosis*. When both species benefit, the relationship is called *mutualism*, a specific kind of symbiosis.

The *mycorrhizae*, intimate associations of certain fungi and the root systems of vascular plants, are another example of mutualism. In most instances the fungi penetrate the roots and make a variety of

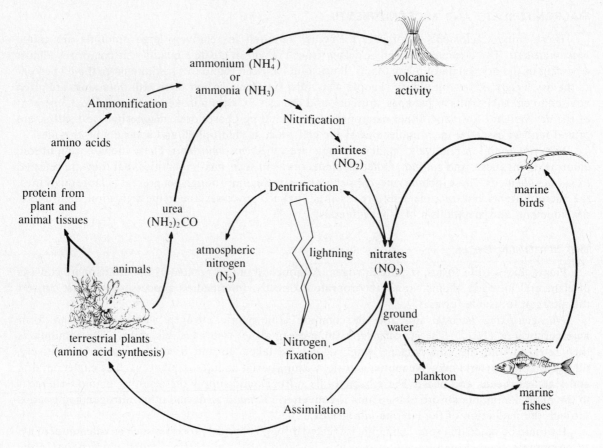

**Fig. 13.1** Nitrogen cycle

nutrients from the soil readily available to the root system of the plants. Even where no penetration occurs, the fungi apparently increase the accessibility of nutrients to the plant's root system. This is the case for pines and willows. Evidently, the association of fungus and root system is very widespread, involving at least three-fourths of all seed plants. Many fossil plants also demonstrate a fungal association. Despite the virtual ubiquity of the phenomenon, the precise way in which the fungus enhances the availability of mineral nutrients in the soil is not known.

**EXAMPLE 3**   *Horticulture* involves growing plants under carefully controlled conditions for commercial purposes. Many of the plants grown in greenhouse nurseries by the horticulturist are raised for decorative purposes. Among these plants it has been found that unless infection of the plant by fungi occurs at an early stage, the plant will fail to develop, even in soils that are high in essential nutrients. Even young trees in outdoor or indoor nurseries require a "fungal fix" in order to grow.

## ROOTS

Water and dissolved nutrients move into a plant through the roots. Thin extensions of the cells at the surface of the root, *root hairs*, actually absorb the water and bring it to the conducting tissue within the root. The factors regulating the movement of materials into the root are diffusion, osmosis, and even active transport. The nature of the soil surrounding the root system affects the supply of water available to the roots. Too compact or too loose a soil impedes the absorption of fluid by soils that have lost water to absorbing root systems.

Root systems that tend to be diffuse and highly branched, like those of the grasses, are known as *fibrous root systems*. They are eminently suited both for anchoring the plant in the soil and for extracting water from a broad expanse of soil. A less common root type is the *taproot*, a fleshy, deeply penetrating main root with only modest lateral roots arising from it. It is found in plants like the tapioca, carrot, and dandelion. Often the taproot becomes a major food storage organ. *Primary roots* are the direct continuations of the radicle (original root) of the seedling, while *secondary roots* are branches of this original root. Occasionally, roots may develop from nonroot sources, such as the stem or even the leaves. Such roots are called *adventitious* and usually serve to anchor parts of the plant.

**EXAMPLE 4**   A prominent system of adventitious roots grows down from the stem of the corn plant. These roots keep the plant upright—a position that would otherwise be difficult because of the plant's long shape. Adventitious roots may also aid the horizontal spread of such plants as the banyan tree of India.

Roots grow lengthwise (*primary growth*) through the cell divisions of the *apical meristem*, the growing region at the tip of the root. A *root cap* protects this soft tip as it is projected through the hard edges of soil particles (see Fig. 13.2). A region of cell elongation lies just behind the meristem and slowly changes to a region of cell differentiation nearer the main body of the plant. It is in the region of the zone of cell differentiation that root hairs are formed by the surface epidermal cells. These root hairs are part of an active system in which the older hairs erode and new ones are continually added. The major absorptive surface of the root system derives from the root hairs lying just beyond the growing tip of the root.

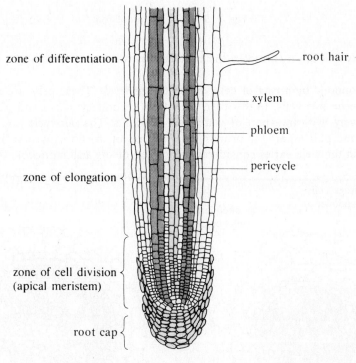

**Fig. 13.2** The root

In cross section the root appears to be formed of a series of concentric cylinders (see Fig. 13.3). The outermost ring is a single-celled layer of *epidermis*. Next is a thick ring of parenchyma called the *cortex*. The cortex may also contain some sclerenchyma near its outer boundary. The round, loosely packed cells of the cortex have thin walls, facilitating the passage of water and minerals through the cells, toward the center of the root. Water also easily passes between the cells. In older cortex, these cells are used for food storage. The innermost layer of the cortex is a single ring of tightly packed cells, the *endoderm*. Some of the cells of the endoderm have noticeably thinner walls than others; these are

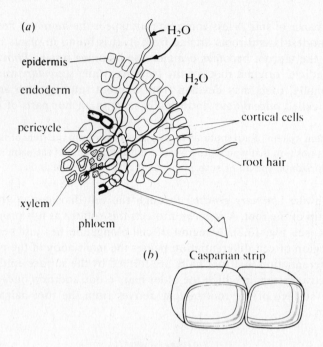

**Fig. 13.3** Cross section of root

called *passage cells* and are involved in the transport of water and minerals to the inner core of the root, the *stele.*

The stele is bounded by a ring of cells called the *pericycle.* These cells are meristematic; that is, they give rise to secondary cells. In this case, they form secondary *xylem* and *phloem*, the two tissues involved, respectively, with transport of water and nutrients. The pericycle also gives rise to branch roots. The xylem and phloem are the primary components of the fibrovascular bundle, or stele. Their arrangement within the stele varies considerably between dicots and monocots (see Fig. 13.4).

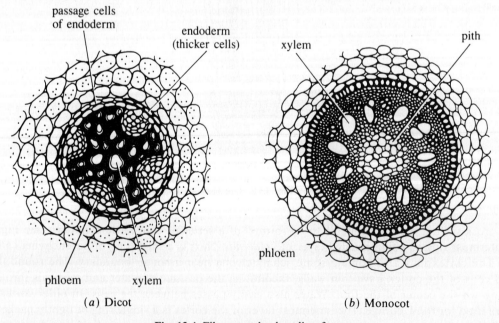

**Fig. 13.4** Fibrovascular bundle of root

In a cross section of the typical herbaceous dicot root the thick-walled xylem lies crosslike in the center of the fibrovascular cylinder, while the phloem cells are nestled within the arms of the cross. In a typical herbaceous monocot, such as *Smilax*, the vascular cylinder is made up of a central core of thin-walled cells, the *pith*. Encircling the pith is a vascular layer of alternating xylem and phloem. The xylem cells are easily distinguished by their larger diameters and thickened, *suberized* (containing a waterproofing pitch) walls.

## STEMS

A simplified view of plant function is that the roots absorb water, minerals, and even some carbon dioxide (in the form of $HCO_3^-$) from the soil, while the leaf uses these materials to weave the organic nutrients required by the plant. The stem, then, may be viewed as connecting roots to leaves. More than this, however, the stem often comprises most of the plant and may be involved in supporting the positioning of the leaves, carrying out photosynthesis (especially in herbaceous plants in which the stem does not become woody), transporting raw materials and finished primary and secondary photosynthetic products, and storing food materials.

The stem and its branches, along with the leaf system, make up the *shoot*. In annuals, such as herbs, the entire plant may die after a single growing season, or in *biennials*, such as carrots, the whole shoot may die and be resurrected for a second growing season. In most *perennials*, the stem persists throughout the life cycle and grows thicker with each growing season as a result of the secondary growth of xylem and phloem. If the stem remains relatively short with extensive branching throughout its length, it is classified as a *shrub*. Taller perennial plants with thick trunks showing little branching at the base are called *trees*. In both shrubs and trees the fibrovascular cylinder is always well-developed and contributes to the longitudinal strength of the stem. For some plants, the environment determines the lifestyle: the castor bean is herbaceous in the temperate zone but is a woody shrub in the tropics.

All young stems tend to carry out photosynthesis, as evidenced by their green coloration. The typical surface layer, or epidermis, is usually present in herbs, but in woody plants an outer layer of cork-laden cells (*bark*) forms as a watertight shield around the stem. Openings (pores) of the bark called *lenticels* afford exchanges of gases between the internal stem cells and the atmosphere.

Leaves are attached to the stem in a characteristic fashion. The point of attachment is called a *node*. A length of stem between two nodes is an *internode*. Usually, a leaf is attached to the stem through a thin stalk called a *petiole*.

Stems grow through specialized growing structures called *buds*, which are sites of apical meristem activity. *Terminal buds* at the tip of the stem allow for the elongation of the stem; *lateral buds* at the sides produce branches. These lateral buds usually arise at the acute angle (*axil*) between leaf attachment and stem and are thus also called *axillary buds*. Buds may give rise to branches of the stem, or they may be specialized to produce flowers.

Stems may broadly be divided into woody and herbaceous. Woody stems are characteristic of trees and are usually found in dicots. Herbaceous stems remain soft and frequently carry on photosynthesis. They are characteristic of most monocots and many dicots.

In cross section, the outer layer of a herbaceous stem is composed of epidermis (see Fig. 13.5*a* and *b*). Next is a thin layer of cortex, often divided into collenchyma cells near the epidermis and parenchyma more internally. There may be a layer of endoderm internal to the cortex, but this is rare. The vascular tissue is next. There are several differences between vascular tissue in dicots and monocots. In the herbaceous dicot, there may be a ring of vascular bundles, with each bundle containing a mass of phloem toward the outside and a larger mass of xylem toward the center (see Fig. 13.5*a*). Between the xylem and phloem lies meristematic tissue called the *vascular cambium*. The center of the stem is filled with pith. Alternatively, instead of discrete bundles, there may be a solid ring of phloem outside a solid ring of xylem, with a ring of vascular cambium between the two (Fig. 13.5*c*). In most monocots, the vascular bundles lack an intermediary layer of cambium and are usually scattered through the stem and buried in a sea of parenchyma cells that may be considered cortex or pith (see Fig. 13.5*b*).

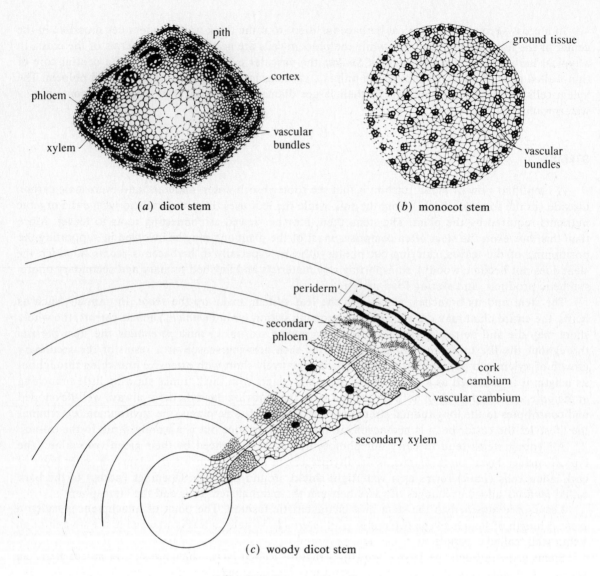

(a) dicot stem

(b) monocot stem

(c) woody dicot stem

**Fig. 13.5** Cross section of stem

Although the vascular cambium of some dicots may never become active, in other dicots, most notably ones with woody stems, the cambium produces cells on either side of itself. Those produced toward the center become xylem (termed *secondary xylem*), while cells produced on the outside of the cambium become *secondary phloem*. Just as the apical meristem of the buds is responsible for the longitudinal (primary) growth of the stem, the vascular cambium is responsible for increases in the width of the stem (termed *secondary growth*). The bulk of the stem in trees is secondary xylem (wood). In addition to secondary xylem and phloem, the vascular cambium produces *vascular rays*, which are sheets of thin-walled parenchyma and thick-walled collenchyma which form channels radiating like spokes through the wood and phloem. In the young stem, these provide for the lateral flow of nutrients into the stem. In the older tree, the rays relieve pressures created by the expanding cylinder of xylem.

As the tree matures, and its diameter widens, the epidermis and cortex split and slough off. In their place, a layer of cortical cells that have become meristematic (the *cork cambium*) begins producing *cork cells*. The dead cork cells form the outer bark (*periderm*); the thin, but functionally essential, phloem constitutes the inner bark.

In older trees, much of the xylem becomes plugged with resins, gums, tannins, and pigment and thus loses its conducting capacity, although it still provides support to the stem. This older wood is called *heartwood*. The newer xylem, which still carries water and nutrients, is called *sapwood*.

Many different types of stems have evolved, from the usual erect form to a variety of forms that range from underground thickenings to slender tendrils that entwine around sturdier structures in the environment. A number of these modifications constitute a strategy for the *vegetative propagation* of the plant, i.e., the spread and division of the parent plant by asexual means.

Many perennial herbs have stems that grow laterally underground. These are called *rhizomes*. They send shoots up afresh at each growing season from their terminal and lateral buds. Thus, a long network of underground stems may be established from each parent plant. The white potato, originally developed in Peru and later brought to Europe, is an example of the end result of this tendency, the *tuber*. Tubers are thickened underground stems that are important repositories of stored food; they no longer produce leaves directly. The *eyes* of the potato are buds which arise from a *slit*, which is a reduced modification of a leaf.

In strawberries a slender stem grows along the ground and sends new shoots up along its length. These stems are called *stolons*. In the grape the stem has been modified into tendrils that attach to trees or trellises and permit propagation of the vine.

*Bulbs* and *corms* are other stem modifications. In bulbs (the tulip, for example) a considerably shortened stem surrounded by thick fleshy leaves forms and gives rise to new plants each growing season. Corms (the crocus, for example) consist of shortened stubby stems containing considerable amounts of food but virtually no internal leaf structures. The corm is also associated with the growth pattern of certain orchids.

In some plants the stem has taken on the function of leaves. Thus, in several species of cactus a portion of the stem assumes the flattened shape of a leaf blade. Such modified stems are called *cladodes*. Some side branches of the stems of several plants have evolved into protective spines.

## LEAVES

Leaves are generally thin, flattened structures that amplify the photosynthetic capacity of the plant. They come in a variety of shapes and sizes that suggest functions in addition to photosynthesis.

**EXAMPLE 5**   Some leaves have evolved into tough scales that guard the delicate underlying tissues of the bud. In some berries the leaves have been modified into protective spines. In the onion, the thick concentric lobes in which sugars are stored are actually scale leaves. In some cases, water rather than food is stored in modified leaves. Perhaps most adventurous of all, the leaves of some forms like the pitcher plant have become an apparatus for the capture of insects.

In cross section (see Fig. 13.6), the leaf almost invariably consists of an upper and lower epidermis. Beneath the upper epidermis is a layer of column-like *palisade mesophyll*, followed by a layer of *spongy*

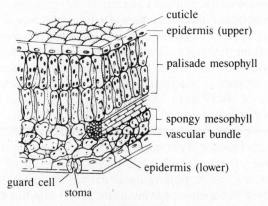

**Fig. 13.6** Leaf

*mesophyll*. Both mesophyll layers consist of parenchymal cells that are thin-walled, rich in chloroplasts, and capable of highly intense photosynthetic activity. The lower epidermis is generally spotted with a rich array of *stomata*, slitlike openings to the outside. Each stoma is surrounded by a pair of epidermal *guard cells*. The guard cells differ from the surrounding epidermal cells in their possession of chloroplasts, a characteristic related to the regulation by guard cells of the opening and closing of the stomatal opening. A waxy *cuticle* generally covers the leaf, usually more thickly on the upper than the lower surface.

Leaves are of considerable economic importance. They provide the food that is stored in various regions of the plant, but in some cases, the leaf itself is rich in stored food, as in cabbage, brussel sprout, lettuce, celery, spinach, and onion. Several types of commercial fiber come from leaves: Manila hemp and sisal hemp. A variety of drugs also come from leaves: tobacco, cocaine (from the coca plant), digitalis (from foxglove), belladonna. Tea is an important beverage throughout the world.

A significant difference between the leaves of monocots and dicots is found in the distribution of the *veins* (fibrovascular bundles). In dicots a main vein, usually in the center of the leaf, breaks up into a complex crisscrossing pattern known as *netted venation*. In the monocots a regular pattern of *parallel venation*, in which veins of similar bore run longitudinally through the leaf in parallel fashion, is found.

## 13.2  MOVEMENT OF WATER AND MINERALS IN XYLEM

The xylem is made up largely of two major cell types: *tracheids* and *vessels*. The tracheids are long, thin cells with many wafer-thin *pits* dotting their entire length. Bundles of tracheids constitute a continuous conduit for water, since liquids readily pass through the pits from one tracheid to another. Vessels are of several types and are presumed to have evolved from the more primitive tracheids. Some vessels are ringed (*annular vessels*), thus enhancing the tensile strength of the tube. In others, the cells join end-to-end, and the boundary walls at each end soon disintegrate so that an elongated single tube is formed. With the death of the cell, the functional xylem loses its cytoplasm and thereby becomes best-adapted for fluid transport. Both tracheids (found in all vascular plants) and vessels (usually found only in flowering plants) contain thickened side walls which are impregnated with water-fast substances which strengthen the tubular apparatus.

Water and dissolved minerals and gases enter the plant through the root hairs. Since the osmotic pressure within these hairs is generally greater than in the surrounding soil, a continuous inflow of fluid can be expected. This creates a push in the root region, which is called *root pressure*. This pressure is part of the dynamic that results in moving the water from the roots to the stem and eventually to all parts of the plant.

Although the bundles are arranged differently in the roots and stems, a continuous flow of fluid along the xylem occurs throughout the plant. As already noted, pressure from the roots contributes to this surging. Perhaps more significant in the transport of water is the *transpirational pull* from the leaves, the negative pressure, or suction, created by the evaporation of water from leaf surfaces. Both the push from the roots and the pull from the leaves act on a continuous, often extremely long, column of fluid within the whole xylem system of the plant. If the water column were broken, there would be a breakdown in water transport, just as the interruption of a continuous column of fluid in a siphon prevents continued drawing of water from one compartment to another. The maintenance of this continuity is dependent on the properties of water, which have already been discussed in Chap. 2. Largely as a result of extensive hydrogen bonding between adjacent water molecules, an inordinately high degree of cohesion occurs between these molecules. *Adhesion*, or the attraction of water molecules to the walls of their container, is particularly high, partly as a result of the high surface tension of water. Adhesion is believed to assist in pulling the water up the stem. These properties assure that cohesive columns of water will move along continuously in any conduit from point of entry to point of departure.

## 13.3  MOVEMENT OF FOOD THROUGH THE PHLOEM

The phloem transports food materials synthesized in the leaves to all parts of the plant. At any one time the flow of materials in adjoining xylem and phloem may be opposite in direction, but this is not invariably so. Since leaves are most abundant in regions distant from the trunk or stem, phloem flow is generally toward the stem and roots.

A variety of substances move along the protoplasm of the phloem, but sucrose is usually the most abundant. Unlike the xylem, phloem cells remain alive while carrying out their transport function.

Basically, there are two types of phloem cells, *sieve cells* and *companion cells*. A long column of sieve cells, sometimes designated a *sieve tube*, is formed by the end-to-end apposition of sieve cells. The end cell walls are studded with perforations that permit protoplasmic connections from one sieve cell to its vertical neighbor. These perforated walls are called *sieve plates*. There are also perforations in the sides of the sieve cells. The arrangement of sieve cells into long sieve tubes provides a continuous protoplasmic network within the phloem.

Lying next to the sieve cells are highly specialized, thin-walled parenchymal cells, aptly designated *companion cells*. Sieve cells generally lose their nucleus and many organelles with maturity, but the conducting cytoplasm remains. The companion cell is completely intact through its lifetime and probably provides nuclear controls for the sieve cell. The ATP necessary for functioning within the sieve cell may also come from the companion cell, which can be viewed as a tender or nurse within the phloem apparatus.

There is some evidence that during injury the perforations of the sieve plate may be partially sealed. A slime made up of proteinaceous material, *phloem protein* (*P protein*), is present within sieve cells and may participate in the sealing process. This is analogous to the sealing of compartments in the hold of a ship to prevent flooding. A polysaccharide called *callose* may also function in sealing portions of the sieve plate.

Sucrose and fructose, as well as amino acids, generally move from the leaves to the stem and roots of a plant through the sieve tubes of the phloem in a process known as *translocation*. The precise mechanisms involved in this transport are not completely understood, nor is the direction of flow always the same through a particular plant region.

Those parts of the plant that have high levels of organic nutrients tend to export these materials and are considered to be *sources* of such materials. Plant organs that are low in organic nutrients tend to import them and are regarded as *sinks* for these same materials. One interpretation of the *source-to-sink* perspective in translocation focuses on the so-called *pressure flow theory*. According to this view, the high concentration of sugars or other solutes in a source compartment results in the movement of water into that compartment through osmosis. This raises the pressure within that compartment and forces fluid with dissolved solutes into an adjacent compartment that does not contain high levels of solute. As solute enters this second compartment, it too will attract water from surrounding regions and the increase in hydrostatic pressure will push water and solute into a third compartment. Thus, solutes keep inducing a rise in pressure that forces fluids and their solutes from original sources to original sinks. A sucrose gradient does indeed exist along the phloem, and water acts to move these solutes along the continuous sieve tubes. The entire process is actually quite complex, and in several instances active transport across sieve cell membranes may be involved.

# Solved Problems

**13.1**   What significant characteristic of the fern limits its role as a land plant?

The sporophyte of the fern can grow extensively on land because of its possession of vascular tissue and an adequate root, stem, and leaf development. However, since a land plant must not only guard its extant structures from water loss and function in relative dryness but also carry out its reproductive mission on dry land, the fern fails at the reproductive level. The gametophyte (prothallus) is dependent on water both for survival and for the migration of its motile sperm. This vulnerability in the life cycle of the fern is overcome in the seed plants by the incorporation of the gametophyte into the tissues of the sporophyte and the evolution of the pollen tube as a mechanism for bringing the migrating sperm to the egg.

**13.2**  Almost all the differences between plants and animals can best be understood in terms of the autotrophy (self-feeding) of plants as opposed to the heterotrophy (food dependency) of animals. Explain.

In a certain sense plants are the solid middle-class "burghers" of the living world: they plant their roots in the earth and spin and weave in their leaves the organic compounds necessary for survival. Animals must "steal" their organic requirements from plants or from animals that have ravished a plant. The easily observable fact of animal locomotion as compared with the sessile character of most plants is explained by the way each group gets its food.

At the cellular level, plants are equipped to maintain their posture against the challenges of the environment. The very walls of each cell may serve to resist shear stress from strong winds, while tough fibers running through the stem avoid breaking of the upright shoot or trunk. The root system provides a stable anchorage in the soil, and the xylem elements of stem and leaves offer strong support.

Since plants obtain their food while remaining in place, no elaborate system of muscle or nerve has evolved as compared with animals. Coordinating mechanisms generally operate slowly and are restricted to chemical signaling.

There is a certain danger that plants may not be sufficiently appreciated for their complexity and highly evolved character because of the tendency of human evaluators to set considerable store in neural coordinating mechanisms. However, an ecosystem (see Chap. 26) is primarily defined in terms of its plant species. Furthermore, the atmosphere that now sustains actively metabolizing animals could not have come into existence without green plants.

**13.3**  Zinc is found in numerous enzymes active in plant metabolism, yet plants contain only trace amounts of it. Given its large role in plant physiology, how is this possible?

Although zinc is involved in numerous chemical reactions, as part of an enzyme it acts as a catalyst, not a reagent, and is thus not consumed in the reactions. Because it can be reused repeatedly and is thus not depleted, large amounts are not necessary.

**13.4**  Nitrogen fixation is a mutualistic form of symbiosis. What is the advantage to the plant? What do you suppose the bacteria get from the relationship? (Hint: the $N_2$ molecule is very stable.)

Many factors militate against the plant's being supplied with ammonia and nitrates from the environment. If there is drought, there may not be any water to transport these compounds to the roots of the plant; if the soil is too loose or is of the wrong pH, the nitrogen compounds may leach out and not be available. Consequently, housing bacteria in the roots that are willing to share their nitrogen is of obvious benefit to the plant. In mutualism both the organisms involved benefit, so the bacteria must gain as well. In this case the host supplies the bacteria with the ATP they need for their own maintenance and for fixing nitrogen. This is necessary, since $N_2$ is a very stable molecule and requires large amounts of energy to be split. Roughly one-fifth of the ATP produced by pea plants is contributed to mutualistic bacteria in the nodules.

**13.5**  There are really two cycles involved in the nitrogen cycle. Explain.

In one cycle elemental nitrogen is removed from the atmosphere by nitrogen-fixing bacteria and is subsequently converted to nitrates or ammonia and incorporated into organic molecules. However, nitrates and nitrites are acted on by denitrifying bacteria to produce elemental nitrogen and thus return $N_2$ to the atmosphere and complete a cycle.

In the second cycle, ammonia and nitrates are incorporated into organic molecules. In the process of decomposition, these molecules are ammonified and then nitrified to nitrate. If they are reabsorbed by plants and incorporated into organic molecules, instead of being denitrified, they complete a second, different cycle.

**13.6**   Clover is sometimes planted to improve nitrogen-poor soil. Since clover requires nitrogen, what would be the purpose of this procedure?

Clover carries on a symbiotic relationship with nitrogen-fixing bacteria. Consequently, the bulk of its nitrogen comes from the atmosphere rather than the soil. If it is plowed under, it adds this atmospheric source of nitrogen to the soil and produces a net increase in nitrogen soil content.

**13.7**   As we have seen, the innermost layer of cells in the root's cortex is the endodermis. The cells of this single layer are crowded together in a ring with virtually no space between adjacent cells. Each ring of endoderm also fits tightly with the rings above and below and thus forms a dense cylinder of cells. Within its cell wall, each endodermal cell is circled by a vertical band of wax running parallel to the ring; this band is continuous with the waxy bands of the two adjacent cells of the ring on either end and with the bands in the cells of the rings above and below. Because of their continuity, the bands are like a large gasket, with the cells of the endodermal cylinder stuck into holes in the gasket, making water passage between adjoining endodermal cells impossible. This waxy *Casparian strip* encircles the entire vascular cylinder. What do you suppose is its purpose?

We have seen that water, and whatever is dissolved in it, can flow freely between the cells of the loosely packed cortex. However, because of the Casparian strip, water must pass through, rather than around, the cells of the endoderm. Through the selective permeability of their plasma membranes, the endodermal cells act as filters, screening out various ions, large molecules, and toxins from the transport system that lies beyond in the vascular bundle.

**13.8**   Figure 13.4*a* shows the vascular bundle of the buttercup, a dicot. Surrounding the bundle can be seen the thick cells of the endoderm, as well as the endoderm's thinner-walled passage cells. A clear pattern can be seen between these two cell types and the components of the vascular bundle. Describe the pattern and explain why it exists.

The thick endodermal cells occur over the phloem, while the passage cells tend to be opposite the arms of the xylem. As we have seen, the Casparian strip forces water flowing into the stele to pass through the cells of the endoderm. The thin walls of the passage cells make them better suited than the thicker-walled cells to transport water and solutes to the stele. Since the xylem carries incoming water and nutrients to the rest of the plant, logically it should be positioned next to the passage cells, which provide it with that water.

**13.9**   Desert plants and those growing near the ocean occupy vastly different habitats, yet they each show similar modifications, particularly in regard to adaptations for holding onto water. Why do you suppose this is so?

The desert environment is patently short of water, and plants that grow in the desert show smaller cell size, greater thickness of cell wall, and fewer stomata, which remain closed for longer periods. These changes are associated with primary or even secondary adaptation to low supplies of water. However, the hypertonicity associated with ocean water also threatens to deprive plants of the water that is present in the soil and in the plant tissues themselves. It is not surprising then to encounter similar modifications in plants at the seashore as in the desert varieties, although changes in root insulation may be more significant in affording protection to seaside plants than alterations in other structures.

Plants that have become particularly well-adapted to conditions of drought are known as *xerophytes* (dry plants). They often possess a very extensive root system to maximize the uptake of water. Leaves are often reduced to the size of scales, and the stomata are either sunk into pits or covered with hair to protect against water loss. Some plants may actually grow in such hostile hypertonic environments as a salt marsh or other brackish environment. These plants are called *halophytes* (salt plants) and share many of the characteristics of desert and seashore plants.

**13.10**  In tabular form give the anatomical differences that would be found in a cross section of the root of a dicotyledonous plant versus its stem.

| Root | Stem |
|---|---|
| Lacks pith | Has pith |
| Xylem and phloem alternate | Phloem outside xylem |
| Xylem and phloem restricted to center of root | Secondary xylem and phloem expand toward periphery |
| Pericycle | Vascular cambium |
| No rays | Vascular rays |
| Endodermis present | Endodermis usually absent |
| Casparian strip and passage cells | No Casparian strip or passage cells |
| Thick cortex | Thin cortex |
| Sclerenchyma in outer cortex | Collenchyma in outer cortex |
| Epidermis | Epidermis absent in mature woody dicots |
| Periderm negligible or absent | Pronounced periderm |

**13.11**  With a few exceptions, monocots lack a vascular cambium. What prevalent characteristic of monocots can be explained by this fact?

Because they lack a cambium, monocots generally are incapable of secondary growth. They therefore tend to be herbaceous and tall and narrow.

**13.12**  Starting at the center, what is the order of tissues encountered in the stele of a woody dicot stem?

Pith, primary xylem, secondary xylem, vascular cambium, secondary phloem, and primary phloem.

**13.13**  What is responsible for the annual rings in the trunks of temperate zone trees?

In all dicots a thin layer of cambium develops between the xylem and phloem in the open fibrovascular bundles found in many herbaceous forms or in the continuous rings of xylem and phloem encountered in the perennial woody forms. In this latter group, in which the trees are found, new layers of xylem and phloem are laid down in each growing season. The xylem becomes the wood, a cellulose-rich, rugged material that both supports the tree or shrub and acts as the major conduit for water and minerals.

Phloem cells are delicate and slough off or become incorporated into the bark as the tree increases in girth from year to year. The xylem cells, both tracheids and vessels, remain intact and serve to mark the events of each growing season. Since the xylem laid down in the spring (springwood) has cells of large diameter with little fiber material interspersed as compared with the heavy fiber component of the summerwood, the xylem growth of one season is clearly demarcated from the next. Each ring consists of a clear two-zoned region, and the winter season is usually characterized by a noncellular gap between one annual ring and another.

The annual rings mark the age of a tree, but the characteristics of particular rings may also provide clues to environmental conditions for a particular year. Broader rings in general are indicators of greater rainfall.

**13.14**  Stripping a thin layer of bark completely around a tree trunk will kill the tree, but a deep gouge along one side will not. Why?

It must be remembered that the bark contains the phloem in its innermost region. The circumcision of even a thin layer of bark completely disrupts the movement of organic nutrients from the leaves and upper stem to the lower stem and roots. In a clear sense, the roots and lower stem would be starved and their functions suspended. Survival would be impossible.

On the other hand, a deep cut along one side of the trunk would disrupt completely both nutrient flow in the phloem and fluid movement in the xylem, but only in the narrow band of elements actually destroyed. The rest of the trunk would continue to provide a conduit for necessary fluids and the sugars and other organic foodstuffs manufactured by the leaves. Sometimes it is possible for trees to survive, even though their heartwood has rotted or burned. Although this diminishes their strength, it does not reduce their ability to transport nutrients and, so, they survive.

**13.15** The stomata are tiny lengthwise openings in the leaf, about 0.0001 mm wide at their maximum opening. They are concentrated on the lower surface of most leaves and range in frequency from 100 to 100,000 per $cm^2$. They are enclosed by a pair of *guard cells*, which resemble two bow-legged sausages. Just above the stoma and within the leaf is a large air space. Unlike the other cells of the lower leaf epidermis, the guard cells contain chloroplasts. What mechanism does this suggest for opening and closing the stomata?

The guard cells function through osmotic changes in their cytoplasm. During the daylight hours, when conditions are optimal for active photosynthesis, the guard cells pile up sugar molecules. These sugar molecules increase the osmotic pressure, which leads to an uptake of water and an increase in the turgor (internal pressure) of the cell. When the cells start to swell, they pull apart so that the stomatal opening (pore) increases in size. This permits $CO_2$ and water vapor to pass into the leaf and oxygen, a by-product of photosynthesis, to pass out. Thus, the leaf opens its interior to the atmosphere only when gas exchanges necessary for photosynthesis occur.

When conditions for photosynthesis are poor, the guard cells do not produce sugars and osmotic pressure is reduced. With the loss of turgor pressure, the guard cells become limp, the walls partially collapse, and the opening is occluded by the now flaccid cell walls. This prevents water loss at a time when photosynthesis is not occurring at appreciable levels.

**13.16** Crassulacean acid metabolism (CAM) involves the uptake of carbon dioxide at night (through open stomata) and its rapid conversion to such metabolites as malic and isocitric acids. In such plants as cactus, the CAM process is a vital component of the adaptation of the plant to a desert environment. Suggest a mechanism to explain this.

During the day, when the dry heat of the environment could make short shrift of the interior of the plant, the stomata remain closed. However, the $CO_2$ absorbed at night is now released from its metabolite storage form and used to carry on photosynthesis.

**13.17** Compare and contrast the methods of transport found in the xylem and phloem.

Movement in the xylem is generally against gravity. Two processes are believed to be at work. At the root end, hair cells, being hyperosmotic to the surrounding soil, draw water in. This increases the pressure in the cells of the root, a pressure which then pushes fluids up the hollow tubes of the xylem. At the other end, the transpiration of water through the stomata of the leaves creates a negative pressure, which causes a suction, or pulling, of the water column in the hollow tubes of the xylem. Some investigators believe that *adhesion*, the tendency of water to creep along the sides of containers, is also responsible for pulling up the water column. Cohesion caused by the hydrogen bonding of water molecules keeps the water column from breaking apart.

The movement of fluids and food in the phloem is generally with gravity; however, unlike the tubes of the xylem, the sieve tubes are filled with cytoplasm. As with transport in the xylem, osmosis plays a part in developing the pressure necessary to move the fluids; however, rather than occurring just at the beginning of the process, as in the xylem, the osmotic force is passed from cell to cell along the entire system. As one cell draws in water, the pressure builds and pushes the fluid into the next cell, where water is again drawn in osmotically and the pressure builds and pushes the fluid to yet another cell. Active transport may also be involved in developing the necessary pressures.

# Supplementary Problems

**13.18**  The greatest number of plants currently in existence are found within   (*a*) club mosses.   (*b*) ferns.   (*c*) gymnosperms.   (*d*) angiosperms.   (*e*) none of these.

**13.19**  Guard cells surround   (*a*) tyloses.   (*b*) sieve plates.   (*c*) companion cells.   (*d*) stomata.   (*e*) lenticels.

**13.20**  Excess water can never harm a plant.
(*a*) True   (*b*) False

**13.21**  The absence of iron could prevent a plant from synthesizing chlorophyll.
(*a*) True   (*b*) False

**13.22**  The cohesion-tension theory attempts to explain the way in which continuous columns of water and dissolved minerals move through the xylem elements of plants.
(*a*) True   (*b*) False

**13.23**  The apical meristem produces primary tissues, while the cambial layer produces secondary tissues.
(*a*) True   (*b*) False

**13.24**  Transpiration is most rapid when the stomata of the leaves are in a closed position.
(*a*) True   (*b*) False

**13.25**  In the cortex, pith, and wood rays, parenchymal cells make up most of the tissue.
(*a*) True   (*b*) False

**13.26**  The number of terminal bud scars of a 4-year-old twig should be equal to the number of rings found in a cross section of the twig.
(*a*) True   (*b*) False

**13.27**  No one type of soil is ideal for all crops.
(*a*) True   (*b*) False

**13.28**  The nodules on the roots of many legumes harbor bacteria that are capable of fixing nitrogen.
(*a*) True   (*b*) False

**13.29**  For the phrase in column **A** pick the best match from the choices in column **B**.

| A | B |
|---|---|
| 1.  Contain a palisade mesophyll | (*a*)  roots |
| 2.  Possess a cuticle | (*b*)  stems |
| 3.  Contain an x-shaped central core of xylem | (*c*)  leaves |
| 4.  In monocots, vascular bundles scattered in pith | |
| 5.  Modified to form a white potato | |

# Answers

| | | | | | |
|---|---|---|---|---|---|
| **13.18** | (*d*) | **13.22** | (*a*) | **13.26** | (*a*) |
| **13.19** | (*d*) | **13.23** | (*a*) | **13.27** | (*a*) |
| **13.20** | (*b*) | **13.24** | (*b*) | **13.28** | (*a*) |
| **13.21** | (*a*) | **13.25** | (*a*) | **13.29** | 1—*c*; 2—*c*; 3—*a*; 4—*b*; 5—*b* |

# Chapter 14

## Interactions of Vascular Plants with Their Environment

Rapid and complex responses to environmental changes (*stimuli*) are a characteristic of animals. They are intimately linked to the nutritional lifestyle of a heterotroph. An animal's procurement of food requires fast action as compared with the more leisurely manufacture of organic nutrients by green plants. Nevertheless, plants do detect changes in their environment and react to these changes in specific ways. Responses may involve movement (but usually not locomotion), alteration of growth patterns, developmental phenomena, or change in the state of particular plant structures. These responses generally affect the entire plant and alter its total potential for successfully coping with environmental challenges even though the response may seem to be localized. Thus, the responses of a plant to its environment are one aspect of integration in a plant, the interaction of individual components that serve the entire organism.

### 14.1 TROPISMS

A *tropism* is an invariable growth response to an environmental stimulus occurring in plants and primitive invertebrates. It is a relatively unsophisticated type of *irritable* reaction. *Irritability* is the capacity to respond in a characteristic way to changes in the environment. The complex repertoire of often subtle voluntary responses associated with irritability in primates affords a far greater range of adjustments to the environment than is the case with tropisms, which are stereotyped and limited types of behavior.

The tropisms are named for the eliciting stimulus and described as "positive" if growth is toward the eliciting stimulus and "negative" if growth is directed away from the stimulus.

**EXAMPLE 1** The roots of a plant grow toward the origin of gravitational force (center of the earth). Therefore, they are *positively* geotropic. On the other hand, the shoot grows away from the gravitational source. It is characterized as *negatively* geotropic.

Responses to gravitation (*geo*tropism), water (*hydro*tropism), light (*photo*tropism), pressure, touch (*thigmo*tropism), etc., are crucial to the survival of a plant. A number of tropisms may be demonstrated simultaneously.

**EXAMPLE 2** The shoot has been described as negatively geotropic. At the same time it is also positively phototropic in its tendency to grow toward the light. Two tropisms may reinforce one another, or they may contend with one another to produce a growth compromise.

### 14.2 PLANT HORMONES

The tropisms, as well as a number of other phenomena occurring in the plant, are probably mediated by plant hormones. A *hormone* is a substance produced in one part of a living organism that has profound metabolic effects throughout the organism as it travels through the vascular system. Hormones are chemical mediators of function and behavior and generally involve relatively slow responses as compared with the faster neural responses, which are found only in animals.

One of the first hormones to be discovered in plants was *auxin*, a substance both isolated and named by Fritz Went in 1926. His experiments were not the first to explore the influence of auxins, but they are regarded as the most conclusive.

The main auxin occurring naturally in plants is *indoleacetic acid* (IAA):

$$\text{CH}_2\text{COOH}$$

Other substances, such as hydroxyindole acetic acid, have also been associated with auxin action, but it is not clear whether these substances are converted to IAA as the active form.

Auxins play a significant role in a broad variety of plant behaviors and growth patterns. They are involved in (1) the suppression of lateral buds along the stem, (2) the development of root and shoot systems, (3) the growth of the fruit, (4) the dropping off of leaves and fruit (*abscission*), (5) the division of cells in the cambium, and (6) the development of new structures such as adventitious roots. Auxins exert their effects at different concentrations. Stimulatory auxin levels for the shoot may be quite different from stimulatory auxin levels for the root. In some cases, the absence of auxin may bring about a particular effect.

More than 40 years ago, auxin-like substances were produced as weedkillers. They acted primarily on the broad-leaved varieties of weeds and did not destroy grasses or the usual staple crops. Their use in Vietnam to destroy foliage cover became a controversial issue when some scientists maintained that the herbicides (Agent Orange, etc.) were responsible for cancer and related diseases in human populations exposed to them. At the concentrations used, these substances promote overgrowth of the weed and eventual death. Mechanisms for their action are still obscure.

*Cytokinins* are a class of hormones that stimulate cell division in plants. They interact with auxins to determine the differentiation of meristematic tissues. They are necessary for the formation of such organelles as the chloroplast and may play a role in flowering, developing fruit, and breaking dormancy in seeds.

The cytokinins were first discovered in the tissue culture laboratories of Folke Skoog et al. at the University of Wisconsin. They were isolated as breakdown products from plant nucleic acids. In 1964, a team led by Dr. D. S. Letham of New Zealand isolated *zeatin* from corn seeds, and this is now regarded as the active cytokinin. Although cytokinins work in conjunction with auxins, they appear to exert a focused effect on cell division, whereas the auxins seem to produce a lengthening of individual cells, largely through an effect on the cell wall.

The *gibberellins* are another class of hormones found in plants. Originally associated with fungus (*Gibberella fujikuroi*) that caused rice plants to achieve abnormally great heights, various gibberellins have also been found in the tissues of plants themselves. The Japanese, who first noted what they called "foolish-seedling disease" at the turn of the century, have studied the effects of gibberellins on plant growth extensively. It is of particular interest that the marked increase of stem length associated with this class of hormones is especially obvious in dwarf varieties of plants. Because the green revolution (a marked change in agricultural practices in undeveloped countries to improve yield using special strains of grain, extensive fertilization with synthetic products, and intensive application of pesticides) has stressed the use of dwarf strains of rice and other agricultural staples, the gibberellins may be of particular concern to societies dependent on these innovations. Gibberellins may increase plant height and alter both vigor and nutritional yield.

As in the case of auxins, the effects of gibberellins are not confined to their primary influence on the lengthening of cells. They influence the formation of enzymes which hydrolyze starch, they may break the dormancy of seeds, and they may bring about flowering under conditions which would, in their absence, produce no flowers at all. Although they may enhance the quality of some commercially useful plants, their effect on plant vigor is generally a negative one, a fact that limits their application in agriculture.

The gibberellins are synthesized in a manner similar to the steroid hormones of vertebrates. They resemble the sex steroids in structure, and their action is fundamentally parallel. They traverse cell membranes readily because of their lipid-solubility and appear to act by turning on specific genes (derepression). They are particularly active in inducing the enzymes for starch degradation in the seed.

A unique hormone, because it exists as a gas rather than a liquid, is *ethylene*. It is usually associated with relatively rapid ripening of fruit. It is produced by the plant itself, usually when the fruit has reached its maximum size. It may also be added externally by produce handlers to bring about ripening, especially if the fruit has been picked when green.

A number of the changes associated with *senescence* (aging and degeneration), such as leaf abscission, are mediated by ethylene. It is regarded by some botanists as the chief messenger of death or the seasonal hiatus in vigor that marks the end of each growing season. A variety of *inhibitors*, which are not all primary hormones, may also play a role in maintaining dormancy and resisting growth. Their action complements the influence of ethylene. One inhibitor regarded as a true hormone is *abscisic acid*.

**EXAMPLE 3**   Abscisic acid not only induces dormancy in seeds, buds, and cambial layers but also acts as a kind of stress hormone. With high concentrations of abscisic acid, the outermost leaf vestiges are converted to protective scales, synthesis of watertight resins increases, and stomata close in preparation for dry conditions. Abscisic acid has also been shown to play a role in leaf abscission, which accounts for its name. All its effects prepare the plant for adverse conditions. Abscisic acid is synthesized by chloroplasts, but small amounts are also synthesized by apical meristems in the shoot and the root.


## 14.3   PHOTOPERIODISM

*Photoperiodism* refers to the ability of plants to respond to relative periods of light and darkness. Specific responses, such as flowering, are elicited by long periods of light in some cases and by short periods of light in others.

In terms of the flowering response, almost all plants can be assigned to three separate groups: *long-day plants* (short-night plants), which require a minimum period of light, e.g., beet and clover; *short-day plants* (long-night plants), which flower only when the daylight is less than a specific duration, e.g., chrysanthemum and ragweed; and *neutral plants*, whose flowering is not dependent on a critical duration of light or darkness, e.g., sunflower and tomato.

**EXAMPLE 4**   Beet, clover, black-eyed Susan, and other long-day plants usually flower during the summer in the northern hemisphere, when the days are long. On the other hand, such short-day plants as the dahlia flower in either spring or fall, when the days are relatively short and the nights correspondingly long. Sunflower and corn are independent of the photoperiod and bloom any time.

It appears that the crucial event in stimulating flowering is not the length of time of uninterrupted light, but rather the span of continuous darkness. The short-day plant is one that must be exposed to a *minimum* period of darkness. Should a short burst of light intrude upon this period of darkness, flowering does not occur in the short-day plant.

On the other hand, the long-day plant has a *maximum* limit on the period of darkness consistent with flowering; that is, the darkness cannot exceed a critical number of hours or flowering will not occur. If the long night is interrupted by a flash of light, flowering will occur because the maximum limit of darkness has not been exceeded.

The total amount of darkness and light may vary greatly among short- and long-day plants. Some short-night plants may have relatively long durations of night which cannot be exceeded, while some long-night plants may have relatively short periods of darkness which *must be* met or exceeded.

H. A. Borthwick and S. B. Hendricks found that red light (660 nm) and far-red light (730 nm) were most effective in mediating photoperiodicity. It is red light that is most effective in interrupting the minimal period of darkness required for flowering in the short-day plant. Red also proves to be the most effective component of the spectrum in inducing flowering during light exposure in long-day plants. On the other hand, far-red light has exactly opposite effects. Further, the effects of exposure to red light can be nullified by subsequent exposure to far red, and to a great extent the reverse also applies.

Borthwick and Hendricks hypothesized that this state of affairs could best be explained by the existence of a single pigment that acts as a light receptor. This pigment, which they named *phytochrome*, must exist in two forms. One form ($P_r$) absorbs red light, and the other ($P_{fr}$) absorbs far red. Evidence more recently obtained suggests that phytochrome is a conjugated protein (see Chap. 3) whose prosthetic group provides the necessary receptor properties. The two forms of the pigment are differentiated only by a difference in the positions of two hydrogen atoms (see Fig. 14.1).

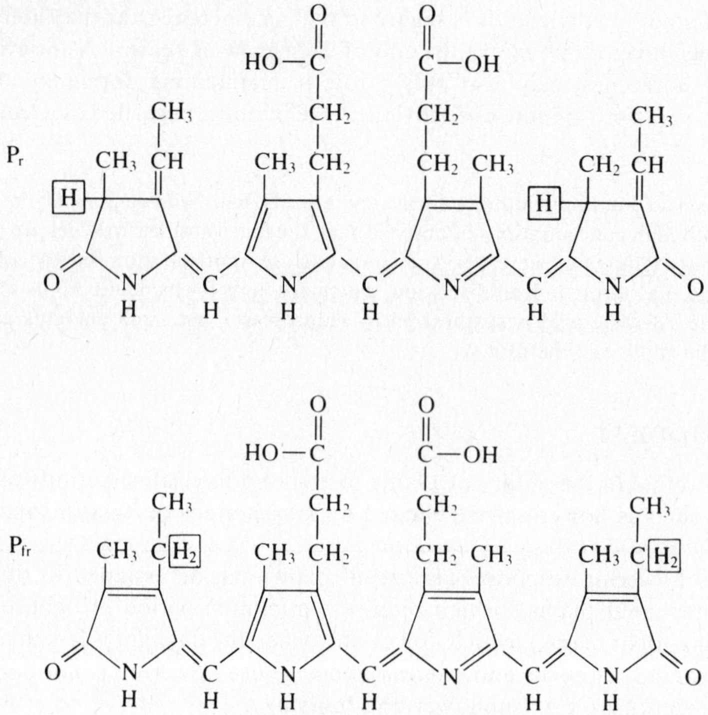

**Fig. 14.1** Phytochrome

The initial synthesis of phytochrome is in the $P_r$ form. When it absorbs red light however, it is converted to the $P_{fr}$ form. $P_{fr}$ will absorb light in the far red to reconstitute the more stable $P_r$ molecule. Because $P_r$ is more stable, $P_{fr}$ will revert to it in the dark and will also tend to break down either spontaneously or as a result of enzyme destruction. Since red light tends to be dominant during the day, most of the phytochrome exists as $P_{fr}$ in the light. At night, $P_{fr}$ levels are low. It would appear, then, that ratios of the two forms of the pigment enable the plant to distinguish night and day.

The determination of the *duration* of night or day periods by the plant is more complex. It probably involves an interaction between phytochrome and an internal clock mechanism that is characteristic of almost all living cells. Biological clocks exist as endogenous (internal) rhythms of fairly intense activity and usually follow a circadian (24-hour) cycle. The phytochrome system functions to mark a dark or light period, but it apparently does not provide a measure of its duration. The two mechanisms may function together to enable the plant to detect the actual times spent in light and darkness.

The rhythmicity of biological clocks seems to arise from alterations of membrane function. Permeability characteristics of the membrane, which are so important in the metabolism of the cell, may be related to either alterations in protein structure or, alternatively, changes in the concentration or characteristics of the phospholipids of the membrane. The phytochromes, in either of their alternative structures, may alter the set point of the clock by changing the characteristics of the cell membrane. Thus, photoperiodicity may ultimately be regulated at the level of the cell membrane.

Photoperiodism is not confined to plants. In groups ranging from insects to mammals, including primates, cycles of light and darkness may initiate migratory movements, alter emotional states, trigger sexual behaviors and courtship patterns, and elicit metamorphosis. Clearly, the hormonal mechanisms involved in these animal responses are quite different from those that mediate flowering in plants, although similarities may exist in the eliciting light waves in the two groups.

## 14.4  PLANT DISEASES

Since all the common plants require water, light, and a variety of nutrients for optimum development and growth, the absence or short supply of any of these factors in the environment challenges the health and survival of the plant. If plants are grown without light, they become extremely long and they fail to form chlorophyll. They are said to be *etiolated*.

Many plants take on a yellowish hue when they fail to form chlorophyll in sufficient amounts. This pathology, known as *chlorosis*, usually arises from short supplies of mineral nutrients in the soil. Iron shortages are a major cause of chlorosis.

Infectious diseases of plants may arise from infestation by viruses, bacteria, or fungi. In addition to the serious economic effects on farmers of these plant diseases, ensuing famine may lead to large-scale death and starvation. Close to a million people died in Ireland during the mid-1840s as a result of the fungal blight that destroyed the potato crop. It was also mainly responsible for the mass emigration of Irish people to the United States during that period.

A particularly bizarre social phenomenon is sometimes elicited by a fungus that attacks such monocot crops as wheat, rye, barley, and malt. This fungus, known as *ergot*, produces mood-altering products, which may be ingested by humans who eat the infected grain or even the alcoholic by-products of the grain. Mass waves of psychosis (ergotism) swept through Europe in the Middle Ages and have occurred even more recently as a result of the contamination of bread and other grain products. However, medically useful derivatives such as ergotrate are extracted from ergot-infected grains. Ergotrate and similar compounds are powerful vasoconstrictors that are used to control bleeding, ameliorate migraine headaches, and stimulate uterine contractions.

Wheat crops worldwide have been destroyed by a fungus known as *rust*. Viruses cause *mosaic disease*, which may destroy many vegetable and fruit crops. Spraying has been reasonably successful in checking the spread of some of these infectious diseases.

Plants may also show developmental abnormalities. These are often particularly evident in plants that are growing in tissue culture. When plants are wounded, they often develop masses of amorphous material with very poor cellular differentiation known as *calluses*. Plant tumors and even plant cancers may arise and spread through the plant as an amorphous invasion of surrounding, well-differentiated tissue. *Galls* are growths on a plant that are induced by parasites and are usually highly organized. *Crown galls* are tumors induced by bacteria. They are usually less well differentiated than other types of galls.

# Solved Problems

**14.1**  A sheathlike covering over the tip of most grass plants, the *coleoptile*, was found by Charles Darwin to be especially sensitive to light. By covering various parts of the coleoptile with dark material, Darwin showed that it was the very tip of the coleoptile that was necessary to elicit the phototropic response. He postulated that in some way the light striking the tip causes some factor to move from the tip to lower portions of the shoot.

A generation or so later the Dane P. Boysen-Jensen cut the coleoptile near its tip and inserted a thin slice of gelatin between tip and stump. If a light were shone upon one side of the shoot, an overgrowth occurred on the opposite side so that the shoot bent toward the light.

In 1926, Went removed coleoptile tips from oat seedlings and placed them base-down on agar blocks for an hour or more (see Fig. 14.2). He then placed these blocks on oat seedlings that had been kept in the dark and had their tips removed, and found that they resumed their growth. Most important, if the blocks were placed off center, the shoots bent toward the opposite side.

What conclusions about phototropism would you suppose were drawn from these experiments?

The best interpretation of these results is that a diffusible material passes down from the coleoptile tip to the cells below. This chemical, or hormone, was called *auxin*. At first, it was felt that auxin moved downward rather than laterally. Light supposedly destroyed auxin on the light-exposed side, so that the concentration of auxin on the dark side was greater. Since auxin produces a lengthening of cells, the higher concentration of auxin on the dark side would produce a twisting toward the light.

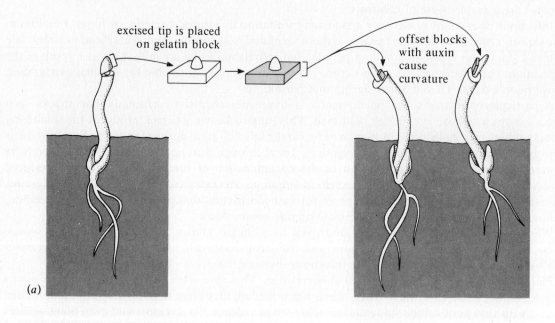

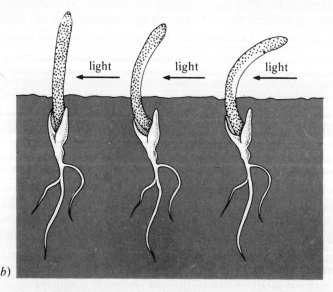

**Fig. 14.2** Went's experiment with oat seedlings

We now realize that auxin may move laterally within the plant. In the case of phototropism, light produces an uneven distribution of auxin rather than its destruction. There is surprisingly little understanding of how auxin moves through the plant, as well as what molecules may possess auxin properties. Alterations in membrane permeability, as well as active transport mechanisms, probably play a role in the distribution of auxin.

One view of the mechanism by which auxin produces cell elongation focuses on the receptors in plant cells that bind to auxin and influence hydrogen-ion flow across the plant cell membrane. Changes in pH which follow from hydrogen-ion migration activate enzymes that alter the stiffness of the cell wall and facilitate its stretching. Some evidence has been produced that auxin may act more directly to increase both enzyme and structural protein concentration.

**14.2**   In nature, auxins cause the ovary to develop into a fruit, and this usually occurs following fertilization and the formation of seeds. How do you suppose auxins have been used to thwart nature in the commercial production of some fruit?

In plants treated with auxin, the ovary may be persuaded to produce fruit without fertilization and the formation of seeds. Seedless tomatoes are one example of auxin-induced fruits.

**14.3**   Suggest some ways inhibitors might be important to plant survival.

Inhibitors are especially important to the survival of higher plants in temperate zones. These plants become quiescent or *dormant* during the harsh winter period or when conditions are not capable of supporting an active plant. Seeds often will not begin to germinate until they have been exposed to cold and then brought to moderate temperature. The breaking of dormancy and the beginning of growth depend on the release from inhibition of vigorous metabolic activity.

In desert zones seeds contain inhibitors that are removed by a heavy soaking. Such soaking can occur only when sufficient rain has fallen to assure a modicum of success in the growing seed. This parallels the situation for nondesert plants in which a prolonged period of cold must precede the warming period in order to break dormancy.

**14.4**   Why do you suppose the presence of one rotten piece of fruit in a container promotes the rapid ripening and decay of all the fruit in that container?

The hormone ethylene not only produces ripening but also, in relatively short order, brings about the softening, browning, and liquefaction of decay as well. Once ripening or decay begins, it results in the rapid formation of large quantities of ethylene. Because gases are highly mobile, ethylene rapidly disseminates throughout the container in a chain reaction of decay and further production of the gas.

Although ethylene is regarded as a major hormone for senescence, it does stimulate growth under some circumstances and may also influence flowering in some semitropical plants.

**14.5**   In many plants the early pattern of growth results in a compact boutonniere arrangement. Under some circumstances this compact, low-lying configuration suddenly produces flowers with extremely long stems. This unusual stem elongation preceding flowering is called *bolting*. What hormone would you suspect is responsible for this phenomenon?

Gibberellins are the responsible agents for bolting as well as for the extreme elongation of stems in general. Although gibberellins are identified with unusual and even pathologic development patterns in plants, they also interact with auxins to regulate cambial production of secondary xylem and phloem.

**14.6**   M. H. Chailakhian of the Soviet Union was one of the first to explore the possible control of flowering by a specific hormone during *photoinduction* (the onset of flowering as a result of specific light exposure regimens). In his experiments, conducted in 1936, he worked with chrysanthemums, which are short-day (long-night) plants. He first removed the leaves from the upper half of the plants but left the leaves on the lower half. He then exposed the lower half with its leaves to short days (long nights) and the upper leafless half, suitably shielded, to long days. The plant produced flowers. When he reversed the procedure, so that the shielded upper half received the appropriate short-day (long-night) procedure while the lower half with its leaves was exposed to a long-day regimen, no flowering occurred. What conclusions do you suppose he drew from these results?

Since it was the part of the plant containing leaves that had to be exposed to the appropriate light cycle, Chailakhian logically concluded that photoinduction was mediated through the production of a flowering hormone by leaves. The hormone was designated a *florigen*. Unfortunately, it has never been isolated or analyzed, despite the strong evidence for its existence.

Confirmatory evidence for the action of florigen comes from grafting experiments with cocklebur, another short-day plant. If two cockleburs are grafted onto one another and then separated by a lighttight partition, treatment of one plant by the appropriate short-day light cycle will cause it to flower, and this will be followed by the flowering of the second plant, which did not receive the appropriate light stimulus. Only a small number of leaves on the short-day plant need be stimulated for this effect to occur. Despite a failure to isolate a florigen, evidence has accumulated that this hypothesized substance may be actively inhibited by noninducing photoperiods.

**14.7**   If the critical photoperiod for a short-day plant is 13 hours of daylight, will it flower when there are 15 hours of daylight?

A short-day plant requires a *minimum* level of *darkness* in order to flower. It follows then that, under natural conditions, the plant may also be said to have a *maximum* level of *daylight*, beyond which it will not flower, since there will not be enough hours of darkness remaining in the 24-hour period. In this example, the critical level of daylight is 13 hours, and any level of daylight beyond this will not allow the plant sufficient hours of darkness. The plant will therefore not flower when there are 15 hours of daylight.

**14.8**   Suppose that a short-day plant has a critical photoperiod of 14 hours of daylight. Determine whether it will flower or not under the following conditions: (1) 15 hours of daylight followed by 9 hours of darkness; (2) 12 hours of daylight followed by 12 hours of darkness; (3) 13 hours of daylight followed by 11 hours of darkness, with a flash of far-red light at hour 18; (4) 12 hours of daylight followed by 12 hours of darkness, with a flash of red light at hour 18 followed by a flash of far-red light; and (5) 10 hours of daylight followed by 14 hours of darkness with a flash of red light at hour 17.

1.   The plant will not flower, because it will not experience the minimum amount of darkness required.
2.   The plant will flower because it will receive the minimum amount of darkness required.
3.   The plant will flower because it will receive the minimum amount of darkness required. The far-red light will accelerate the conversion of $P_{fr}$ to $P_r$ and thus will merely enhance the process that was occurring during the dark period anyway.
4.   The plant will flower because it will receive the minimum amount of darkness required. Although the first flash of light, because it was red, will force the plant to restart its clock in measuring the hours of darkness, the flash of far-red light will reverse this.
5.   The plant will not flower. The flash of red light will in a sense force the plant to begin recounting the hours of darkness, and since the flash came at hour 17, there will be only 7 hours of darkness following, which will be insufficient for the plant (which requires 10 hours).

**14.9**   The conversion of $P_r$ to $P_{fr}$ is generally accepted as the mechanism by which a plant detects whether it is in light or darkness. High levels of $P_{fr}$ tell the plant that it is in light. This mechanism, however, does not explain how the plant measures the hours of elapsed darkness. One hypothesis is that the ratio of $P_r$ to $P_{fr}$ provides a means of measurement. Daylight causes the rapid change of $P_r$ to $P_{fr}$. Since $P_{fr}$, being less stable than $P_r$, slowly reverts to $P_r$ in the dark, the ratio of $P_r$ to $P_{fr}$ would get increasingly larger with an increasing period of darkness. It has also been shown that the plant's measurement of time is independent of temperature. Does this fact tend to support or refute the theory?

The conversion of $P_r$ to $P_{fr}$, being a chemical conversion, should be dependent on temperature. The fact that the plant's measurement of time is not temperature dependent suggests that the conversion ratio is *not* the factor used by the plant to measure elapsed time.

# Supplementary Problems

**14.10** Match the hormone in column **B** with the descriptive phrase in column **A**.

| A | B |
|---|---|
| 1. Induces senescence | (*a*) Ethylene |
| 2. Maintains apical dominance | (*b*) Abscisic acid |
| 3. Influences starch hydrolysis | (*c*) Gibberellins |
| 4. Exists in higher concentrations on dark side of shoot | (*d*) Auxin |
| 5. Causes bolting | |
| 6. Maintains dormancy | |

**14.11** Phytochrome could be used by the plant in the detection of shading patterns.
(*a*) True  (*b*) False

**14.12** Etiolated plants fail to produce sufficient chlorophyll.
(*a*) True  (*b*) False

**14.13** Chlorosis is caused by too low a supply of carbon dioxide.
(*a*) True  (*b*) False

**14.14** The potato famine in Ireland was caused by a virus.
(*a*) True  (*b*) False

**14.15** Galls are produced in animals by toxic growth inhibitors from plants.
(*a*) True  (*b*) False

**14.16** Rusts are iron-deficiency anemias associated with crop grasses.
(*a*) True  (*b*) False

**14.17** Long-day plants must have a period of darkness that cannot exceed a critical duration.
(*a*) True  (*b*) False

**14.18** $P_{fr}$ is the more stable form of phytochrome.
(*a*) True  (*b*) False

**14.19** Both auxins and gibberellins tend to lengthen cells.
(*a*) True  (*b*) False

# Answers

**14.10**  1—*a*; 2—*d*; 3—*c*; 4—*d*; 5—*c*;    **14.13**  (*b*)    **14.17**  (*a*)
         6—*b*                           **14.14**  (*b*)    **14.18**  (*b*)
**14.11**  (*a*)                          **14.15**  (*b*)    **14.19**  (*a*)
**14.12**  (*a*)                          **14.16**  (*b*)

# Chapter 15

# Homeostasis

## 15.1 OVERVIEW

Living cells, as well as larger multicellular organisms, can function adequately only within a relatively narrow range of conditions. If the temperature within a cell should exceed 60 °C, the cell will cease its vital functions. At higher temperatures the lipids and proteins of the cell break down and the cell falls apart. At very low temperatures, freezing and ice crystal formation challenge the functional and even the structural integrity of the cells.

Just as extremes of temperature threaten a living cell, so do swings in pH, ion concentration, sugar levels, etc. It is readily apparent that highly complex molecules arranged in specific ways to carry out a variety of interrelated functions can do so only in supporting environments. The microenvironment of the cell is determined by the cell membrane. Its permeability characteristics control which ions enter, which substances can be extruded, and which interior conditions will result from the selective nature of the membrane in a particular environment. Those cells that are successfully adapted to a particular habitat demonstrate an ability to control the fluctuations in the interior compartment to assure a reasonable degree of constancy. The maintenance of constancy is called *homeostasis*. In the living world constancy is not a static affair but a dynamic phenomenon. Forces tending to strengthen a particular trait or raise the concentration of some material are balanced against forces tending to weaken the trait or reduce the concentration. A dynamic "equilibrium" maintains a balance within all organisms. This balance, or steady state, is the single most important feature unifying all aspects of physiology (function) at the level of cell, organism, and even population.

Homeostasis has been studied most intensively in multicellular animals, particularly vertebrates. However, it is operative at all levels of life. Those processes that maintain homeostasis are known as *homeostatic mechanisms*.

Historically, Hippocrates alluded to the belief that disease, a disturbance in health, is righted (cured) by natural forces. This notion of the Greek physician, who first formulated the physician's oath that bears his name, recognizes a natural tendency to oppose abnormality, a resistance to pathological change. However, it was Claude Bernard (1813–1878) who is recognized as the father of the concept of homeostasis. This French physician developed physiology as an experimental science in a series of elegant laboratory investigations during the mid-nineteenth century. His writings at that time stressed the importance of constancy in the internal environment. He was aware of the fact that every cell in the body is bathed in a fluid that is both nourishing and sustaining. The totality of these tissue fluids, which are intimately associated with the blood and lymph (see Chap. 17), makes up *le milieu intérieur, the internal environment.* Bernard stated the now famous maxim, "*La fixité du milieu intérieur est la condition de la vie libre.*" This may be freely translated as "the constancy of the internal environment is necessary for free life."

The term *homeostasis* was first used by Walter B. Cannon (1871–1945), an American physiologist who was particularly interested in the movements of the digestive tract and their control. In an insightful extended essay called "The Wisdom of the Body," Cannon aptly caught the essence of homeostasis—an evolutionary development of a metabolic wisdom that provides for internal constancy.

Cannon named the fixity described by Bernard and extended our understanding of its dynamic nature. He described the homeostatic mechanisms that are poised against each other to regulate such phenomena as temperature, and he cast them in evolutionary terms.

He was particularly interested in the effects of fear and rage on the mammalian organism. His studies of these emotional reactions led him to appreciate the role of the autonomic nervous system in regulating digestive processes and the effects of the emotions on these processes. Soon he realized that the autonomic nervous system played a fundamental role in mediating all the homeostatic phenomena involved in the maintenance of internal constancy. In an approach that was even more insightful than that of his predecessors, Cannon appreciated that perturbation in the organismic steady state may arise from changes within the organism as well as changes from without. He pointed out that even short periods of muscular activity could generate enough heat to coagulate the proteins of muscle tissue if such heat were not dissipated. Thus, activity within the organism could impose as great a strain on homeostatic mechanisms as changes in the external milieu that impinge, in turn, on the internal milieu.

Cannon also recognized that homeostasis was not the responsibility of a single integrative system like the nervous or endocrine system, but that all the organ systems of the body operate cooperatively to effect internal constancy. Further, the more "advanced" the evolutionary stage of a particular group of organisms, the more subtle and complex the homeostatic apparatus.

In his long-term investigation of the *sympathetic nervous system* (SNS), as well as its associated endocrine structure, the *adrenal medulla*, Cannon discovered that it was specifically involved in preparing animals for emergency (threatening) situations. The *parasympathetic nervous system* (PNS) on the other hand mediated reactions, associated with vegetative functions, that would be carried out under peaceful conditions (feeding, drinking, sleeping, etc.). The two systems, sympathetic and parasympathetic, constitute an automated and relatively rapid subsystem within the nervous system that controls the involuntary responses of the visceral organs. This autonomic nervous system is essentially a motor apparatus whose fine tuning is dependent on conditions within other compartments of the nervous system.

When the organism perceives a threat, it switches to an emergency footing and the sympathetic nervous system is activated. The action of sympathetic nerve stimulation, mediated by the *neurohumoral* (nerve chemical) substance *norepinephrine* (*noradrenaline*) and to a lesser extent *epinephrine* (*adrenalin*), is to prepare the body for the emergency situation and to dampen the activity of the vegetative (parasympathetic) system. Since the animal reacts to danger by either fleeing the threat or, when cornered, fighting against the threat, the emergency response has been dubbed the *fight-or-flight reaction*.

Once the emergency is over, the SNS diminishes its activity and the nervous discharge of the parasympathetic nervous system increases. This produces a calming effect on the body and a demobilization of emergency adjustments. Just as emergency responses inhibit activity of the PNS, so the increased activity of the PNS during vegetative periods is associated with decreased activity of the SNS.

## 15.2  FEEDBACK CONTROL

The maintenance of constancy is possible only in situations where an awareness of or sensitivity to change exists. Deviations must be detected before they can be righted. The actual detection of changes in physiological state is mediated by *receptors*, structures that are sensitive to specific environmental changes. Changes constitute the *stimuli* that cause the receptor to signal the existence of an altered state. The receptor's signal is usually an electrical impulse that activates compensatory mechanisms participating in the homeostatic apparatus.

*Feedback* refers to the monitoring or supervision of a process. Generally this involves the creation of conditions such that any process A will lead to creation of a component B that regulates (feeds back on) the initial process A. Such a relationship can be expressed as a circuit: A⇄B. If B tends to inhibit A, then the control loop (circuit) is said to constitute *negative feedback*. If B tends to increase A, then the control loop constitutes *positive feedback*. Clearly, receptors are integral to the process of feedback, since the detection of a change in the initial process activates the feedback circuit.

Homeostasis is maintained largely through negative feedback mechanisms, since it is only through the formation of opposing forces that changes can be checked and overcome. Positive feedback, on the other hand, encourages continued change and magnifies specific deviations.

**EXAMPLE 1**  The digestive enzyme *trypsin* is produced in the pancreas as an inactive *zymogen* (precursor) called *trypsinogen*. A second enzyme arising from the wall of the intestine cleaves a small portion off the end of trypsinogen to form the active enzyme trypsin. Trypsin, once formed, acts to convert more trypsinogen to trypsin until most of the trypsinogen within the intestine exists as the active trypsin. Such a cascade effect in which an initial change influences still more change is typical of positive feedback.

Negative feedback can be easily understood in physical systems such as those that control the heating of a house. These physical systems employ sensing devices and feedback loops and are known as *servomechanisms*.

**EXAMPLE 2**   A sensing device called a *thermostat* is made part of a circuit involving the turning on or off of a furnace. When the temperature falls below a certain level, the thermostat is set to turn on the furnace by completing the circuit. As the temperature reaches and then exceeds a critical value, the thermostat breaks the circuit to cut off the furnace. Such an arrangement assures a relatively stable temperature range in which the *information* of room temperature, the *input*, induces an *output* of heat from the furnace.

The basic currency of servomechanisms and homeostatic control is information. In homeostasis, perturbations of the internal environment become the information that impinges upon a receptor. From the receptor, a message is sent along a circuit that activates an *effector*. The effector functions to return the internal environment to its original state (see Fig. 15.1).

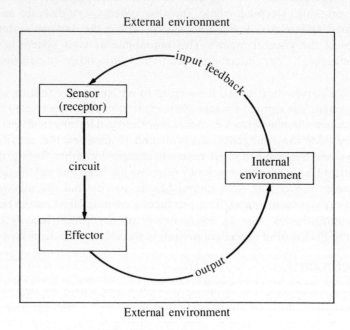

Fig. 15.1 Homeostatic circuit

## 15.3   REGULATION OF TEMPERATURE

*Homeothermy* (endothermy) is the capacity of certain groups of animals to maintain a constant body temperature. It is first observed in the reptiles, although imperfectly expressed. In birds and mammals, however, a remarkably precise apparatus exists for keeping body temperatures constant through a broad range of environmental temperature fluctuations.

The maintenance of temperature in the homeotherms is not a trivial feature, and can be appreciated as a significant advance when homeotherms are contrasted with vertebrates that lack this ability, the *poikilotherms* (ectotherms). Once called "cold-blooded," the poikilotherms actually demonstrate an equilibrium of their internal body temperature with the temperature of the external environment, which may be high or low.

Homeotherms are independent of the temperature fluctuations of the environment, and this has been a major evolutionary advance in aiding their successful conquest of land. A toad or salamander is rendered metabolically immobile during the long stretches of winter in the northern regions of the earth, while most mammals and birds are left free to carry out their activities with internal body temperatures usually ranging from 37 to 40 °C.

An extension of the independence of homeotherms is directly linked to their maintenance of relatively high interior temperatures. Moving about on dry land or in the air, unsupported by the buoyant force of water, exacts a heavy energetic price. The relatively high and constant body temperatures of homeotherms supports an active metabolic rate that produces high energy outputs. These energy outputs are channeled into the arduous tasks of food procurement, mate selection, maintenance of safety, etc.

The major source of animal heat is muscular contraction. When muscles contract, up to 30 percent of the energy released from the degradation of their fuel is transformed into the mechanical energy of contraction, while 70 percent or more is converted into heat. This actually constitutes a highly efficient process, since even a steam engine, which is considered extremely efficient, runs at an efficiency of approximately 25 percent. Furthermore the heat, regarded as an energetic drain in a mechanical engine, is the wherewithal for the homeostatic apparatus in homeotherms.

Homeostatic mechanisms for temperature control exist at many levels. Perhaps the most overlooked device for cooling the body is the behavior pattern involved in fleeing the source of heat. In hot weather most mammals seek shade. "Only mad dogs and Englishmen go out in the noonday sun." In addition, a marked diminution of physical activity occurs. Since muscular contraction is a major source of animal heat, the suppression of physical activity is crucial to prevent further increase in temperature. A third behavioral response is the selection of clothing that in its looseness and color (white tends to reflect light and heat) aids in the dissipation of heat.

The body loses heat at its surface, so heat must be brought to the surface where it can be dissipated. The blood carries a great deal of the body heat. In overheated conditions receptors from skin and some internal structures activate feedback circuits that dilate the blood vessels at the skin's surface, thus bringing an increased volume of blood to the surface. Returning blood is also shunted to more superficial veins for further cooling. Heat is then lost by three physical routes. It may be lost by *radiation*, which involves electromagnetic energy waves traveling from the surface of the body in a medium of air. It may also be lost by *conduction*, which involves the transfer of heat through direct contact with either cooler air or water. Finally, it may be lost by *convection*, which involves currents of air or water moving past the skin to continuously remove heat lost from the interior of the body.

If heat dissipation is not sufficient to maintain temperature constancy, the *sweat* glands secrete copious quantities of salty fluid (sweat). In humans, sweat glands are profusely scattered throughout the skin, and large quantities of sweat may be produced in accordance with cooling needs. As discussed in Chap. 2, the vaporization of sweat results in a massive cooling effect, since 540 cal are absorbed in the evaporation of 1 g of $H_2O$ (liquid). Obviously, where air is saturated with water vapor and no evaporation can occur, cooling is checked. This is why we are particularly uncomfortable on hot, humid days, when sweating is not accompanied by evaporative cooling.

In some animals, like the dog, the tongue and mouth parts are a major evaporative cooling apparatus. By panting, the dog increases the exchange between moist membrane surfaces and the air and thus enhances the cooling effect. Rats and mice, which lack abundant sweat glands on the skin, lick one another to augment the cooling effects of evaporation. Both panting and licking behavior increase with a rise in temperature. This strongly suggests temperature homeostasis is the basis for the development of these responses.

Many of the mechanisms for warming the body are reversals of the cooling processes; however, several are unique to the heating process. Mammals possess an apparatus for increasing the thickness of their covering of fur (hair). It involves *piloerection*, the standing up of the individual hairs as a result of the contraction of piloerector muscles inserted at the base of each hair. Both cold and such emotional reactions as fear will trigger piloerection, which is why you feel the skin at the back of your neck "crawl" when you hear something scary. Cats seem particularly menacing when their fur stands on end and they seem to increase so greatly in size. During piloerection an increase in the thickness of the insulating layer is augmented by the trapping of stationary air within the matting of hair or fur.

If behavioral adjustments, the shunting of blood, and piloerection prove inadequate, a unique response may be invoked—*shivering*. Shivering is a massive, spasmodic, relatively uncoordinated series of muscular contractions that produces a great deal of heat in a short period of time. Like piloerection, it is not simply a reversal of a mechanism operating under high-temperature conditions but has developed as a unique adaptation against extreme cold. Shivering is a reflex controlled at a low level of brain organization (see Chap. 22). At a body temperature below 30 °C the shivering reflex stops. *Cryosurgery* capitalizes on the reduced metabolic rate at this temperature to avoid a number of complications associated with surgery at normal body temperature.

## 15.4  REGULATION OF BLOOD SUGAR

Glucose is the major carbohydrate fuel found in the blood, and for many organs of the body it is the primary fuel. It is carried in the plasma to all parts of the body. In some regions it is removed across the capillary beds and utilized directly as an energy source. In other regions it is taken up and stored as glycogen (see Chap. 5) or converted to such high-energy intermediates as fatty acids.

In adipose tissue glucose provides the raw material for the synthesis of fatty acids (*lipogenesis*) as well as for the activated glycerol needed to convert the labile fatty acids to the more stable neutral fats (*esterification*).

The careful regulation of blood sugar is a singularly important aspect of homeostasis. The handling of glucose is central to the utilization, replenishment, and distribution of all metabolic fuel, and sharp changes in blood sugar level will seriously impair performance and health and even threaten life itself. At low levels of blood sugar, dizziness and associated signs of brain malfunction will occur. This is a result of the brain's almost total reliance on glucose as a fuel. When glucose levels elevate much beyond 80 to 110 mg per 100 ml of blood, regarded as normal, an interference with blood flow through capillary beds occurs. A long-term elevation of blood sugar may result in retinal damage and eventual blindness, kidney damage, and ready susceptibility to infection and even gangrene. Cardiovascular damage is also associated with the failure to maintain steady-state levels of glucose.

A variety of hormones act in concert to bring about the stability of blood sugar, but the most important of these is the peptide *insulin*. Insulin is the one conserving hormone of carbohydrate homeostasis; that is, it reduces the levels of sugar in the blood by promoting the utilization, storage, and metabolic conversion of glucose stores.

Insulin is extremely sensitive to blood sugar levels and is the effector part of a negative feedback circuit that maintains constancy. As soon as blood sugar levels begin to rise, the $\beta$ cells of the islet tissue of the pancreas (see Chap. 21) increase both the synthesis and release of insulin. The effect of an increased insulin titer in the blood is a reduction of glucose concentration. The following reactions mediated by insulin produce this drop:

1. Increased formation of hepatic (liver) glycogen influenced by an insulin-induced rise in liver glycogen synthetase, an enzyme that promotes storage
2. Increased permeability of muscle and fat cells to glucose, with the consequent removal of some glucose from the blood flowing through those tissues
3. Increased oxidation of glucose, thus reducing the supply of free glucose
4. Increased synthesis of proteins from amino acids, which forces heavier reliance on carbohydrates, rather than amino acids, as fuel

A failure to produce insulin, a lack of sufficient insulin, or a refractoriness to insulin's effects causes the disease *diabetes mellitus*. Quite often this disease begins during adulthood (*maturity-onset diabetes*) and may be treated with diet, exercise, and, in more serious cases, oral stimulants of the insulin-producing islet tissue of the pancreas. This milder form of the disease is also called *non-insulin-dependent diabetes mellitus*. More serious is the *juvenile diabetes*, which begins early in life and is almost always treated with insulin injection, hence the name *insulin-dependent diabetes mellitus*. Insulin is either injected or implanted in a continuous drip device under the skin. It cannot be taken orally because of its unfortunate vulnerability to breakdown by the digestive system.

A number of other hormones influence blood sugar levels directly or indirectly. With the possible exception of prolactin, all the hormones investigated have effects on blood sugar opposite those of insulin; that is, they tend to mobilize sugar from storage depots and raise its levels in the blood. One of the most important of these hormones is epinephrine. It is released from the adrenal medulla copiously during fight-or-flight responses. Its target organs are the liver and, to a lesser extent, skeletal muscle. Epinephrine activates an enzyme, phosphorylase, which catalyzes the degradation of glycogen to glucose.

A group of steroid hormones from the adrenal cortex also tend to raise blood sugar levels. Known as *glucocorticoids*, they include *cortisol, cortisone,* and *corticosterone.* Their chief action is an enhancement of *gluconeogenesis,* the conversion of amino acids and other noncarbohydrates to glucose (see Prob. 5.16). Noncarbohydrate sources such as fatty acids are utilized in increased amounts during the period of steroid influence; this tends to spare blood glucose and maintain its levels in the blood. The adrenal steroids are also associated with the fight-or-flight response and the reaction to stress, particularly starvation.

*Thyroxin* from the thyroid gland and the growth hormone *somatotropin* (STH) from the pituitary also tend to act as antagonists to insulin. *Glucagon,* which is produced in the $\alpha$ cells of pancreatic islet tissue, also acts as an insulin antagonist. However, glucagon has no effect on muscle and promotes glycogenolysis only in the liver.

The regulation of blood sugar does not go on independently of the metabolism of other carbohydrates or the pathways of lipid and protein metabolism. As described in Chap. 5, the major metabolic pathways are quite interdependent. The regulation of one component of necessity involves alterations in the components of other pathways.

# Solved Problems

**15.1**   Claude Bernard was an aficionado of the theater. One evening he had dissected out a rabbit liver to analyze for sugar. However, he had to run off to the theater before he could complete his analysis. Upon his return, he resumed his chemical tests and found that the sugar levels were inordinately high. What conclusions do you suppose Bernard drew from this observation, and what homeostatic system did he hypothesize on the basis of these conclusions?

Bernard reasoned that a storage carbohydrate existed within the liver and that this stored material broke down to free glucose in the untreated liver. He called the storage carbohydrate *glycogen,* which literally means "sugar former." After partially characterizing and then isolating this polysaccharide, Bernard soon realized its probable function—it served as a reservoir for blood glucose. Carbohydrates from the digestive tract were taken up by the liver and stored as glycogen during feeding. During periods when blood sugar levels declined, the liver could continuously release small amounts of glucose from its glycogen stores. In this manner glycogen could serve to maintain steady-state levels of blood glucose: liver glycogen increases as intestinal sugars are stored in the liver to prevent increases in blood sugar during feeding, and liver glycogen diminishes as it provides the glucose to maintain blood sugar levels between feedings.

This constituted Bernard's first insights into the concept of a constancy of the internal environment. He later extended the notion of "fixity" to the maintenance of body temperature. Explorations of nervous control of blood vessels during vasodilation (widening of blood vessels) and vasoconstriction (narrowing of blood vessels) were also carried out by Bernard, and helped him to understand the responses involved in temperature adjustments.

**15.2**   Among the changes brought about by increased sympathetic stimulation are an increased frequency and strength of the heartbeat, rise in blood pressure, shunting of blood from the visceral organs of the trunk of the body to the peripheral vessels of arms and legs, decrease in clotting time, and sharp rise in blood sugar. The emotions of fear or rage that arise during emergencies tend to trigger a continuous sympathetic response throughout the period of danger.

All these alterations enhance the capacity of the organism to run from or directly to resist danger. How?

In fight-or-flight situations, the more vegetative processes, such as digestion, can be put in abeyance; the extremities become all-important. This explains the shunting of blood to the arms and legs. Emergencies are also times of increased energy demands by the body and increased production of by-products that need to be excreted. The increases in heart rate and strength and the rise in blood pressure keep the active muscles perfused with blood, which supplies them with the elevated levels of glucose and removes waste products. Finally, faster clotting is of obvious value when injury is likely.

**15.3**   The body consists of many homeostatic systems. What would you say are the most common receptors in these circuits?

The specialized endings of nerves are the most common receptors of the body. A stimulus first acts upon the receptor and sets up an excitatory impulse that travels along the rest of the nerve, on through the entire feedback circuit, and often ends in the stimulation of a muscle. The *threshold* (minimum intensity required for initiating an impulse) for the specific stimulus is extremely low, so that when such a stimulus is present, it will readily set off the feedback circuit.

Osmoreceptors in the hypothalamic area of the brain provide a good example of homeostatic feedback. Under conditions of dehydration, the osmotic concentration of the blood increases. As this solute-laden blood moves through the hypothalamus, it stimulates the receptors there, which have a low threshold to higher solute concentrations. In this particular circuit, the effector stimulated is a region of the hind lobe of the pituitary, which manufactures *antidiuretic hormone* (ADH). The ADH released from the pituitary acts on the kidneys, where it inhibits the formation of urine. Water retained in the body as a result of this action will now oppose the dehydration and push the organism back toward the original steady state.

**15.4**   Many of the responses to cold environmental temperatures involve a reversal of mechanisms employed during adjustment to high temperature. What sorts of reversals might be involved?

Behavioral responses include seeking out warmer areas (sunning on a warm rock) and increasing motor activity. Any increase in activity tends to generate heat. Humans also respond to cold by increasing their amount of insulation. Several layers of clothing are more efficient than one thick layer, since air trapped between the layers itself constitutes a layer of insulation.

Blood is brought to the surface during exposure to high temperature, but it is kept well below the surface and closer to the body's core during exposure to cold. This reduces the tendency to dissipate heat to the environment.

**15.5**   In physical and biological systems, when currents lie next to each other but move in opposite directions, *at any given level* the constituents (oxygen, heat, etc.) of the two currents tend to form an equilibrium. This is readily seen in the gills of fish. Deoxygenated blood in the gill flows in a direction opposite to the flow of oxygenated water passing over the gill. As water passes along the gill, it loses oxygen to the blood in the gill; therefore, a gradient of oxygen is established in the water, with the highest concentration being at the front of the gill. Deoxygenated blood enters the rear of the gill, at the level of low oxygen concentration, and moves forward through the gradient of increasingly higher concentrations of oxygen in the water. An equilibrium between blood and water is established at each level, and by the time the blood reaches the front of the gill, it has an oxygen content close to that of the water entering the gill and is thus reoxygenated. This whole process is called a *countercurrent multiplier system.*

A countercurrent multiplier system is used in the body to reduce heat loss and prevent blood from being carried cold from the extremities to sensitive internal organs. How do you suppose this works?

As seen in Fig. 15.2, an adjacent artery and vein may work together to bring warm blood to a cold extremity (finger, in this case) without lowering appreciably the temperature of the blood returning to the core. Blood entering the extremity via the artery is gradually cooled all along its length by the cooler blood

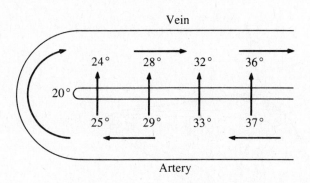

**Fig. 15.2** Countercurrent multiplier system

coming back in the adjacent vein. However, the blood in the vein moving back from the cold tip is warmed by the blood in the artery. Thus, a gradient is set up in which the temperatures within the artery and vein at any specific level are quite close, but a relatively sharp drop is achieved as progress is made toward the tip. It is only at the tip that low temperatures exist for both the artery and vein.

Countercurrent exchanges are essential features at all extremities. They work best where blood flow is rapid and unimpeded. In such degenerative diseases as diabetes in which the flow of blood may be impeded, countercurrent exchanges may not function optimally.

**15.6** What are some of the methods by which hormones oppose the effects of insulin?

Insulin works to increase the utilization, storage, and metabolic conversion of blood glucose. These processes may be countered by increasing the degradation of glycogen to glucose, as occurs through the secretion of epinephrine or glucagon. The permeability of the cells to glucose may be reduced, as occurs under the influence of certain glucocorticoids, and thus the removal of glucose from the blood may be blocked. Certain steroids maintain glucose levels in the blood by stimulating the conversion of noncarbohydrates to glucose (gluconeogenesis) and thus spare the blood glucose from being oxidized. Certain hormones, such as cortisol, inhibit protein synthesis in most organs and thus reduce their reliance on carbohydrates as fuel, since more amino acids are available.

**15.7** The blood that leaves the capillary beds of the digestive system, including stomach, pancreas, and intestine, does not go into increasingly larger veins and on to the heart as it does from most other parts of the body. Instead, this blood moves from the capillary beds of the digestive tract to the capillary beds of the liver through a vein known as the *hepatic portal vein* (see Fig. 15.3). (Any vein connecting two capillary beds is designated a *portal vein.*) What homeostatic function do you suppose this system performs?

The blood leaving the digestive tract is usually loaded with glucose, amino acids, and other fuel metabolites during, and for some time after, a meal. Should this blood move into general circulation, it would dump a tremendous concentration of metabolites into the internal environment and pose a real threat to homeostasis. Instead, the blood is channeled into the functional conduits of the liver, the major site for metabolic transformations, where excess metabolites can be removed and stored. In the case of excess glucose, glycogen is formed under the influence of high insulin levels (induced by rising blood sugar levels during food intake). The blood that leaves the liver and returns to the heart for recirculation is no longer laden with excess sugar. The storage of other food molecules, such as fats, may also occur in the liver. In some cases, excess metabolites may be transformed to other kinds of molecules. Amino acids are particularly likely to undergo conversion to other types of compounds in the liver.

During periods when no food is entering the digestive tract, the liver liberates glucose molecules into the blood to maintain adequate blood glucose levels. Exercise will also stimulate the liver to release stored carbohydrate. These reactions are keyed to the release of glucagon and the glucocorticords of the adrenal cortex.

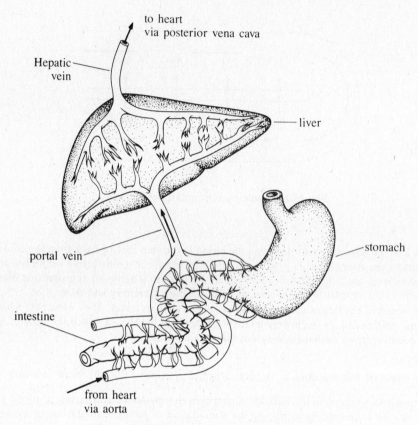

**Fig. 15.3** Hepatic portal system

# Supplementary Problems

**15.8**  Walter B. Cannon originally planned a career in the theater.
(*a*) True   (*b*) False

**15.9**  Negative feedback involves an effector that opposes the original change.
(*a*) True   (*b*) False

**15.10**  Homeostasis is essentially static.
(*a*) True   (*b*) False

**15.11**  Claude Bernard discovered glycogen.
(*a*) True   (*b*) False

**15.12**  The frog is independent of the temperature changes of its environment.
(*a*) True   (*b*) False

**15.13**  The furnace in a home is equivalent to the receptor of a homeostatic feedback circuit.
(*a*) True   (*b*) False

**15.14**  The evaporation of water involves only negligible energy exchanges.
(*a*) True   (*b*) False

**15.15**  Homeostatic mechanisms include behavioral responses.
(*a*) True   (*b*) False

**15.16**  In what way can mammals increase the thickness of their insulating layers during exposure to cold?

**15.17**  Under conditions of starvation will insulin levels be low or high?

**15.18**  Under conditions of starvation will cortisone levels be low or high?

**15.19**  Select the hormone from column **B** that best matches the descriptions in column **A**.

|  A | B |
|---|---|
| 1. An insulin antagonist from the pancreas | (a) Insulin |
| 2. Promotes gluconeogenesis | (b) Cortisone |
| 3. Rises with rising blood sugar | (c) Epinephrine |
| 4. A polypeptide hormone | (d) Glucagon |
| 5. Diabetics lack effective amounts of | |

**15.20**  What two conditions are essential for any countercurrent system?

# Answers

| | | | | | |
|---|---|---|---|---|---|
| **15.8** | (*b*) | **15.13** | (*b*) | **15.18** | high |
| **15.9** | (*a*) | **15.14** | (*b*) | **15.19** | 1—*d*;  2—*b*;  3—*a*;  4—*a*;  5—*a* |
| **15.10** | (*b*) | **15.15** | (*a*) | **15.20** | The two currents must (1) lie close together and |
| **15.11** | (*a*) | **15.16** | piloerection | | (2) flow in opposite directions. |
| **15.12** | (*b*) | **15.17** | low | | |

# Chapter 16

## Animal Nutrition

### 16.1 FOOD PROCUREMENT

All organisms require a steady supply of high-energy (highly ordered) materials, known as *foods*, to provide fuel for their functional needs. Green plants, the common *autotrophs* (self-feeders), synthesize their food with the aid of sunlight from simple inorganic substances like $CO_2$ and $H_2O$ (Chap. 6). The fungi obtain their food by absorption from the immediate environment, as do the nonphotosynthetic bacteria. However, the protozoans (a phylum within the kingdom Protista) and animals (multicellular heterotrophs) generally capture relatively large masses of food material, consisting of whole organisms (*prey*) or parts of organisms. This procurement of large particles of food is known as *bulk capture*. An elaborate apparatus usually exists for the mechanical breakdown of food, since smaller pieces are more easily processed and assimilated. The same ends may be achieved by chemical means, however.

**EXAMPLE 1**  Fungi, such as mushrooms, usually live on dead and rotting material. Such a nutritional habit is known as *saprophytism*. Other fungi live upon or within the tissues of a live host; this lifestyle is called *parasitism*. In either case, soluble materials are readily taken up by the fungi through absorption. Fungi tend to exist as a network of long, thin strands. As such they ramify and may completely permeate surrounding blocks of food material. They then secrete digestive juices to the exterior, which break down the bulk materials. Because they are made up of cells that contain cell walls, fungi cannot engulf particles of food and then create internal food vacuoles as the amoebas do. Those groups of fungi that are parasitic break down live material in much the same way.

Foods contain a variety of chemically defined *nutrients*, which provide materials for energy production as well as for the structural substances needed for cell maintenance and growth. The major nutrients include carbohydrates, proteins, and lipids. Vitamins and minerals are also needed in smaller amounts. Proteins are particularly important as a structural material, especially because of their essential amino acids (see Chap. 3). Carbohydrates and lipids are the major providers of energy, but they fulfill a structural role as well, particularly in assembling membranes.

Many animals live on vegetation alone; they are classified as *herbivores*. Other animals have a diet restricted to animal flesh; they are called *carnivores*. Still other animals, such as humans, eat both plants and animals; they are known as *omnivores*. The more varied the diet, the more likely that nutritional requirements will be met. However, the energetic cost of obtaining particular foods must be considered.

**EXAMPLE 2**  A large carnivore traps a small snake underground. After a considerable amount of frenzied digging, it catches the snake and brings it to the surface. However, the caloric value of the snake is less than the energy expended by the carnivore in catching it. Such a feeding pattern is clearly counterproductive.

Many animals show elaborate devices for capturing and tearing apart their prey. These may consist of extreme running speeds, the ability to mimic the background of the hunting areas, sharp teeth in strong jaws, and massive forelimbs. Other organisms, particularly those parasitic forms that live within the tissues of a host, show only rudimentary procurement devices. Since they are surrounded by simple, soluble food materials, only minimum processing seems to be necessary.

### 16.2 DIGESTION AND ABSORPTION

Our discussion of digestion will be limited to the *one-way digestive tract* of those animals that possess a separate opening (*mouth*) for bringing food into the digestive tract and a second opening (*anus*) for ejecting the food remains. It is termed a "*one-way*" digestive tract because the food moves in only one direction, from the mouth to the anus. In the more primitive groups such as coelenterates

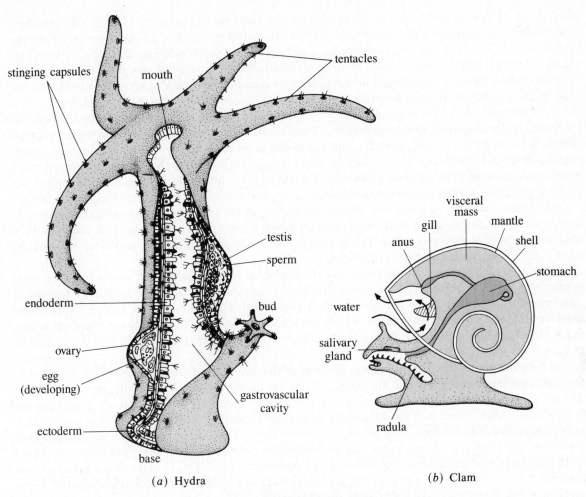

*(a)* Hydra                                        *(b)* Clam

**Fig. 16.1** Digestive tracts

(two-layered multicellular organisms such as hydra, shown in Fig. 16.1*a*) and the three-layered flatworms (such as planaria), a gastrovascular cavity is found that has a single opening to the environment. This is probably a much less efficient arrangement, since materials entering and leaving pass each other, making "assembly-line" specialization impossible.

Food is moved along the digestive tract in a mouth-to-anus direction by waves of contractions beginning in the *esophagus* and passing along each region to the terminus of the large intestine. This sequence of contractions, a reflex in nature, is called *peristalsis*. One-way movement of materials is ensured by a system of special circular muscle valves called *sphincters* at key junctures along the tract. Such sphincters exist between the esophagus and stomach, between stomach and small intestine, and at the anus. When a sphincter contracts, the opening is effectively shut.

Within the digestive tract, specializations exist for

1. Storage of food
2. Mechanical breakdown of ingested chunks of food
3. Chemical degradation of food material

## INGESTION

It comes as no surprise that adaptations for grasping and holding prey are found at the mouth end of the digestive tract rather than at the tail or anus end. These adaptations include a mouth with strong jaws, sharp or grinding teeth, and a posterior *pharynx* that draws the food in the direction of the anus.

Mechanical grinding occurs in the mouth or further back along the digestive tract in a special chamber. In the case of some snails a sawlike device (*radula*) (Fig. 16.1*b*) attached to the pharynx breaks the food down into smaller pieces. In earthworms, birds, and a number of other groups, a heavy muscular chamber (*gizzard*) behind the stomach or the crop grinds the food particles into smaller pieces. Some animals swallow pebbles that, lodged in the gizzard, augment the grinding process. In other cases, the walls of the gizzard are thick and possess projections that accomplish the necessary grinding.

In vertebrates, *teeth* have evolved as the primary tools for holding and chewing. Teeth are modified epidermal and mesodermal structures that are found in all vertebrates except birds and the duck-billed platypus. Each tooth is a hard, complex unit consisting of an inner cavity containing the *pulp* (blood vessels, nerve, and other soft parts); the *dentin*, which is largely collagen that has undergone considerable calcification; and an external layer of *enamel*. The enamel constitutes the biting and grinding surface of the tooth; it is an almost completely calcified layer particularly rich in *apatite*: $Ca_{10}(PO_4)_6(OH)_2$.

In humans the teeth make up two dental arches, one in the lower jaw and one in the upper jaw. Each tooth is set firmly into a shallow depression along a bony ridge extending from each jaw by a network of connective tissues within the proximal *root* of the tooth. The root makes up most of the tooth, with the crown forming the remainder.

Collectively, the teeth serve to grasp and dismember prey, tear bulk food, and grind it to more manageable size. The teeth that accomplish these tasks vary in size, shape, and number. Carnivorous vertebrates usually have sharp, pointed teeth highly adapted for cutting and tearing. Vegetarians (herbivores) tend to have flatter teeth, which are ideally suited for grinding. Since plant cells have much more roughage (cellulose), a considerable amount of chewing must be applied in order to rupture the cell walls.

In humans a total of 32 teeth eventually form in the adult. The first set of teeth, encountered in the child, are fewer in number and are gradually lost during maturation. Each adult jaw contains a set of four *incisors*, which are used for biting. They are shaped like arrowheads and are present in the middle of the arch. These four are flanked on both sides by single *canines*, sharp tearing teeth, which are much longer in carnivorous animals like lions and tigers. Toward the rear of the jaw are a pair of *premolars* and then three larger *molars* on each side. The premolars and molars are flattened, ridged teeth with large surfaces adapted for grinding and pulverizing food. The 16 teeth of the lower jaw are matched by 16 teeth in the upper jaw. During biting and chewing a near perfect articulation of upper and lower sets of teeth is vital for efficiently processing ingested food. Disarticulation may lead to pain, as well as poor processing of food.

Teeth in humans are often subject to extensive decay (termed dental *caries*). This decay may produce pain, infection, abscess, and eventual loss of the teeth. High levels of sugar in the mouth, bacteria, changes in pH, and poor nutrition—all contribute to dental pathology. In caries the calcified crystal of the tooth is solubilized. The addition of fluorine to the drinking water reduces the incidence of caries, presumably by strengthening the crystalline structure of the tooth and reducing the production of harmful acids.

Considerable modification of the digestive tract is found in animals that do not capture and ingest large chunks or particles. Many insects subsist on fluids, such as blood. They generally have special piercing devices as components of their anterior ends.

**EXAMPLE 3**  In the mosquito the mouthparts are organized into a special tube within a tube. The skin of the host is pierced and an anticoagulant is injected to prevent clotting. Blood is then sucked up into the empty tube to provide a meal.

The *filter feeders* obtain tiny food particles by straining large volumes of fluid from the environment through combs, small pores, or other seinelike structures. Filter feeding is not confined to small animals like insects and mollusks; it also occurs in large animals like the whale. Strips of bony *baleen* hanging from the upper jaw act as a strainer for the huge amounts of shellfish and smaller teleosts (bony fish) that are regularly filtered out from the large amounts of plankton ingested.

Chewing food may accomplish the chemical, as well as the physical, breakup of food. This chemical degradation is achieved through the action of *salivary glands*. There are three sets of salivary glands in humans: *parotid* (just below the ear), *submaxillary* (in the region below the upper jaw), and *sublingual* (below the tongue). They each produce saliva, which passes down a short duct into the oral cavity. *Saliva* is a fluid rich in ions and containing some enzymes. Its major function is probably the wetting and lubrication of food particles. However, saliva also contains an *amylase* (whose older name was *ptyalin*). Although the ptyalin begins the breakup of starch that is ingested, food chewed in the mouth reaches the stomach rather quickly, and the stomach acids quickly inhibit the degradation of starch started in the mouth. Saliva also exerts an antibacterial action in the mouth and holds many harmful bacteria in check.

The sight, smell, and taste of food act reflexively on the salivary glands to produce an increased flow of saliva. It is this response that is described by the expression "drooling over food." Even the thought of certain foods, such as a lemon, will bring about increased salivary flow.

In most animals (see Fig. 16.2) the anterior opening of the digestive tract with its teeth and mouthparts is followed by a *pharynx*, a muscular chamber that carries food to a tube called the *esophagus*. The esophagus usually leads to a storage organ, the *crop* or *stomach*. Such storage depots may occur more than once along the digestive tract. Many grazers, such as the horse, also have an enlarged *cecum*, a pouch at the junction of the large and small intestines. Present in the cecum is a dense population of cellulose-digesting bacteria that make the degradation products of the cellulose available to the host.

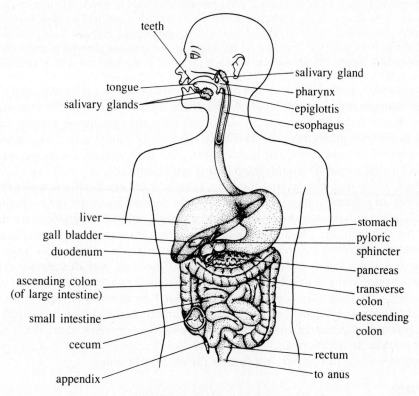

**Fig. 16.2** Human digestive tract

**EXAMPLE 4** A particularly novel adaptation for grazing is found among the *ruminants*, large mammals such as cows and deer that hastily swallow prodigious amounts of grass and weeds and then bring the material back up for further chewing and processing. The material brought back up is called the *cud*, and the rechewing helps to break down the cellulose.

To accommodate the large mass of cellulose that is taken in, the ruminant has a four-chambered stomach (Fig. 16.3). The first two chambers, the *rumen* and the smaller *reticulum*, are actually expanded portions of the

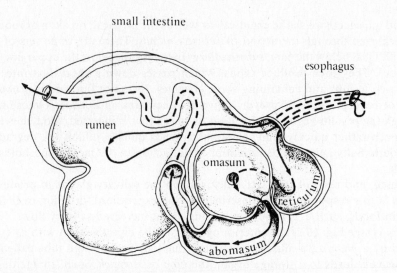

**Fig. 16.3** The ruminant digestive system

esophagus. It is from these two chambers that the cud is regurgitated. A huge population of bacteria and protozoans dwells in the rumen and reticulum, both of which act like fermentation vats in which cellulose, as well as all the major foodstuffs, is slowly processed. It should be noted that a considerable amount of nutrient synthesis is carried out by the bacteria in the rumen and reticulum. The bacteria also supplement the ingested food, since great numbers of them are carried along to the posterior chambers of the stomach and intestine. The fluids and well-chewed cud move into the *omasum*, and, finally, the *abomasum*, which is regarded as the true stomach of ruminants.

Without a considerable storage capacity the organism would have to eat continuously to provide a steady supply of calories. Blind pouches called *diverticula* balloon out from the intestine in many species of insect to augment the storage capacity. During feeding the organism is vulnerable to attack, so that prolonged feeding would increase the likelihood of death or injury. Also, the inability to store the food from a single kill would mean that much of the food from specific forays would be wasted and the eating machine (the animal) would be operating inefficiently.

The next portion of the digestive tract is the long, tubular *intestine*. In all mammals the intestine is further divided into an extremely long *small intestine* and a somewhat shorter *large intestine*. A great deal of chemical digestion and most of the absorption of nutrients into the circulatory system following digestion occur in the small intestine. A variety of adaptations exist for increasing the absorptive surface. A great length of intestine provides a correspondingly high absorptive surface area (the human intestine is more than 6 m long and has almost 300 m$^2$ of surface area). Consistent with this relationship between surface area and absorption is the relatively short length of the *duodenum* (the first portion of the small intestine), where very little absorption occurs (but most digestion occurs). A highly folded surface further increases the absorptive capacity of the intestine. Further, highly vascularized *villi*, fingerlike projections of the mucosa, increase the contact between the nutrients moving along the digestive tract and the boundary surface. In addition, each epithelial cell of the mucosal layer has tiny projections, the *microvilli*, along its surface to increase still further the absorptive lining.

## DIGESTIVE ENZYMES

The mechanical breakdown of food, which occurs primarily in the mouth and stomach (or gizzard), is accompanied or followed by the chemical breakdown of nutrients by catalysts called *digestive enzymes* (see Table 16.1). These enzymes are chiefly involved in hydrolysis reactions:

$$\text{Polysaccharides } [C_6H_{10}O_5]_x + x(H_2O) \rightarrow x(C_6H_{12}O_6)$$

$$\text{Proteins} + H_2O \rightarrow \text{amino acids}$$

$$\text{Lipids} + H_2O \rightarrow \text{fatty acids} + \text{glycerol}$$

**Table 16.1** Digestive enzymes

| Source and Enzyme | Substrate (substance acted upon) | Site of Action |
|---|---|---|
| **Salivary glands** | | |
| Ptyalin (salivary amylase) | Starches | Mouth |
| **Stomach** | | |
| Pepsin | Proteins, pepsinogen | Stomach |
| **Pancreas** | | |
| Pancreatic amylase | Starches | Small intestine |
| Lipase | Fats | |
| Trypsin | Polypeptides, chymotrypsinogen | |
| Chymotrypsin | Polypeptides, oligopeptides | |
| Carboxypeptidase | Polypeptides | |
| Deoxyribonuclease | DNA | |
| Ribonuclease | RNA | |
| **Liver** | | |
| Bile (emulsifies) | Fat globules | |
| **Small intestine** | | |
| Aminopeptidase | Polypeptides | |
| Tripeptidases | Tripeptides | |
| Dipeptidase | Dipeptides | |
| Maltase | Maltose (sugar) | |
| Lactase | Lactose | |
| Sucrase | Sucrose | |
| Enterokinase | Trypsinogen | |
| Phosphatases | Nucleotides | |

The enzymes that act on starch are traditionally called *amylases*, although a more inclusive term for enzymes that act on polysaccharides, oligosaccharides, trisaccharides, etc., is *carbohydrases*. Enzymes that act on proteins are called *proteases*. The hydrolysis of proteins is known as *proteolysis*. In a similar way the hydrolysis of neutral fats (a major type of lipid taken into the digestive tract) is termed *lipolysis*. Digestion does not take place all at once. Instead, there may be many steps and a series of enzymes involved in each of the major degradations.

As shown in Table 16.1, the major groups of digestive enzymes come from the pancreas and small intestine. Mechanical digestion and food storage occur in the mouth and stomach; chemical digestion is relatively minor in those organs. Protein digestion is almost entirely dependent on the proteolytic enzymes manufactured in the pancreas and sent to the duodenum through the *pancreatic duct*. Recall that both trypsin and chymotrypsin are formed as inactive zymogens (trypsinogen and chymotrypsinogen), which are activated by a splitting off of a small length of peptide. *Enterokinase* and trypsin itself play a role in these conversions. *Pepsin*, a proteolytic enzyme found in the stomach, is also secreted as an inactive *pepsinogen* that is converted to the active pepsin by the small amounts of pepsin already present in the stomach (an example of *autocatalysis*).

A highly complex interaction of enzymes is involved in the complete degradation of proteins. Pepsin, trypsin, and chymotrypsin are *endopeptidases*: they hydrolyze peptide bonds in the interior of long polypeptide chains. Individually, the eventual products of each are peptides of moderate length; combined, their effects produce *oligopeptides* (short-chain molecules). The differences among these three enzymes lie in their specificities and the side of the amino acid (carboxyl or amino) that they hydrolyze. Trypsin tends to attack the peptide bond on the carboxyl side of the amino acids lysine and arginine, while chymotrypsin is specific for the bond on the carboxyl side of tyrosine, phenylalanine, or tryptophan. Pepsin, which is not as effective a proteolytic agent as trypsin or chymotrypsin, is specific for the bond on the amino side of tyrosine and phenylalanine. These three different endopeptidases are able to produce shorter peptide fragments from a variety of long-chain polypeptides.

The *exopeptidases* are enzymes that act at the ends of peptide fragments of any length. They split off the last amino acid by breaking a terminal peptide bond. They are essentially of two types: *carboxypeptidases*, synthesized in the pancreas and acting on terminal peptide bonds at the free carboxyl end of the peptide chain, and *aminopeptidases*, synthesized in the small intestine and acting on terminal peptide bonds at the free amino end of the peptide chain.

Both exopeptidases exert their effects within the small intestine. In addition to the amino acids produced by exopeptidases, single amino acids are also produced by the action of various *dipeptidases* that hydrolyze the dipeptides formed within the intestine by the combined action of the endopeptidases. A number of different dipeptidases exist, each with affinities for specific dipeptides.

Although digestion of starch begins with the action of ptyalin in the saliva, the bulk of starch digestion occurs in the small intestine. *Pancreatic amylase* is secreted into the duodenum, where it degrades starch into the disaccharide maltose. This double sugar is then hydrolyzed into single molecules of glucose by the action of the enzyme *maltase*. Similarly, sucrose is converted to the monosaccharides glucose and fructose by the enzyme *sucrase*, and lactose is converted to glucose and galactose by *lactase*.

The primary agent of fat digestion is the enzyme *lipase*. Secreted by the pancreas, it breaks fat molecules into glycerol and fatty acids. It is aided in its work by *bile*, which emulsifies (solubilizes) the fats into smaller globules and thus increases the surface area available to attack by lipase. Bile also helps neutralize the hydrochloric acid that enters the small intestine from the stomach.

Bile is formed in the liver as part of the breakdown of red blood cells, which end their life after about 90–120 days following their formation. It is a fluid containing complex salts, pigments, and some steroid. Although produced in the liver, bile is stored in the gall bladder. During digestion it is forced from the gall bladder through the *common bile duct*, organized from the *hepatic duct* of the liver and the *cystic duct* of the gall bladder, to the duodenum.

Enzymes are released when needed. The coordination of this release in the interest of digestive efficiency is under the control of the autonomic nervous system and a variety of hormones that are usually produced within the digestive tract. The major nerve trunk influencing digestive responses (muscular contractions of digestive organs and the release of enzymes) is the *vagus nerve* of the parasympathetic nervous system. You will remember from the discussion in Chap. 15 that digestive activity is suppressed by sympathetic nervous stimulation. As a rule, nervous stimulation is more prominent at the anterior end of the digestive tract, while hormonal actions are more important at the level of the stomach and are particularly involved in bringing digestive juices into the intestine.

## ASSIMILATION OF NUTRIENTS

The first portion of the small intestine, called the *duodenum* because an early measurement of its length was 12 finger breadths, is the major digestive site. In the much longer *jejunum*, which follows the duodenum, and in the terminal portion (*ileum*), most of the absorption of nutrients occurs. A great deal of absorption of fluids and minerals occurs in the large intestine.

The monosaccharides, end products of carbohydrate digestion, are directly taken up by the circulatory system. An active transport system for each of the common monosaccharides aids their transport across the intestinal mucosa. Interestingly enough, glucose is not moved more quickly than several other monosaccharides, despite its central importance in energy metabolism. Facilitated diffusion and, to a relatively minor extent, simple diffusion also are involved in the transport of simple sugars.

The absorption of the products of lipid digestion is quite complex. Smaller fatty acids may diffuse into capillaries and then be transported to the general circulation. Larger fatty acids may join other lipid materials to form complex lipid droplets known as *chylomicrons*. These chylomicrons accumulate in the lymph vessels of the intestine, which are known as *lacteals*. They may then move into the bloodstream. Many lipids move into mucosal cells as monoglycerides and diglycerides and may undergo changes in their degree of esterification intracellularly. Cholesterol probably reaches the liver as part of a chylomicron or as esterified cholesterol. In the liver, cholesterol is modified for export or stored. The liver also synthesizes cholesterol from smaller building blocks.

Amino acids and oligopeptides are generally transported into capillary beds of the intestine by active transport. In a few cases, passive diffusion may be the means of absorption. Polypeptides and proteins cannot cross the mucosal membrane.

These end products of digestion, which ultimately move into general circulation, are used by near and distant organs both as sources of fuel and for the synthesis of structural materials. We have long known that most nutrients can be converted to others rather readily. The only limitation applies to the interconversion of fatty acids to carbohydrate or protein. All other transformations occur freely.

The liver is the major site for metabolic interconversions (see Chap. 5). Situated between the input apparatus of the digestive system and the general circulation, this highly active and metabolically "talented" organ is ideally placed to respond to the general metabolic needs of the body. In many cases, hormones at distant sites influence the metabolic activity of the liver as an aid to maintaining homeostasis.

**EXAMPLE 5**   Shortly after the ingestion of a meal, the blood sugar begins to rise as carbohydrates are digested and absorbed as monosaccharides. The pancreas secretes insulin in response to this incipient hyperglycemia (high blood sugar). Insulin acts on the liver to increase the enzymatic pathway involved in glycogen formation so that sugar in the blood moving from the digestive tract will be removed and stored as glycogen. Enzymes for lipogenesis are also increased to convert sugar to the more caloric fatty acids. In peripheral tissues, such as muscle and adipose tissue, the permeability to glucose increases and blood sugar is thereby reduced. In the absence of food in the alimentary tract, the situation is reversed (see Chap. 15).

## EGESTION

In humans, the small intestine joins the large intestine several centimeters from the end of the large intestine rather than end to end. The pouch formed at the blind end of the large intestine is known as the *cecum* (see Fig. 16.2). Extending from the end of the cecum and hanging into the abdominal cavity is a narrow tube called the *vermiform appendix*, named for its wormlike, "added on" character.

The first part of the large intestine is called the *ascending colon* because the remaining digested fluid (*chyme*) is moved in an upward or anterior direction. The colon then is directed across the body to form the horizontal *transverse colon*. Another bend produces the *descending colon*, which terminates in the short tubular *rectum*, where material to be passed out of the body (*feces*) is collected. The movement of feces through the terminal opening (*anus*) of the digestive tract is called *egestion*, or *defecation*. Since a great deal of water and mineral material are absorbed during the trip along the colon, the feces that reach the rectum are semisolid unless *diarrhea* (excess defecation of watery stools) exists. More than 60 percent of the fecal mass by weight consists of dead bacteria, which indicates how numerous the bacteria of the large intestine are. Colon bacteria play a role in the absorption of water and minerals, produce certain vitamins, and maintain normal bowel activity. Digestive upsets caused by antibiotics are usually due to the destruction of intestinal bacteria by these antibiotics.

The filling of the rectum with feces triggers a reflex (see Chap. 22) in which the lower portion of the spine sends motor impulses to the colon, producing increased muscular activity. Coupled with the relaxation of the sphincters in the anal region, these contractions exert pressure on the rectal contents and eject the feces. The coordination of these movements is controlled by a defecation center in the *medulla oblongata* of the brain, a neural region lying at the boundary of brain and spinal cord. Defecation is not completely involuntary, except under extreme conditions such as execution by hanging or overwhelming fear. In such cases, it is the massive stimulation of the vagus nerve that causes defecation and, usually, urination as well. Under more usual circumstances one can override the defecation impulse by voluntary contractions of pelvic muscles and the anal sphincter. Continual tampering with regular defecation rhythms may lead to constipation, a chronic inability to defecate regularly or difficulty in carrying out defecation.

## 16.3  THE VERTEBRATE LIVER

The liver of vertebrates is the largest internal organ of the body and probably the most diverse. It is foreshadowed by the *hepatopancreas* of invertebrates, but that organ is not nearly so varied in its metabolic roles. The outpocketing of the gut to form a hollow cecum in amphioxus (a close relative of the vertebrates) is regarded as the first true *homologue* (developmentally related organ) of the liver.

In humans the liver is a large, wedgelike organ divided into two main lobes. The liver, in all vertebrates, is both an *exocrine gland* (it sends secretions out through a localized duct) and an *endocrine gland* (it secretes substances directly into the bloodstream). These glandular functions are carried out in addition to the significant cycles of metabolic interconversions that take place. The liver produces albumin, cholesterol, and fibrinogen (a blood clotting factor) and stores iron and the fat-soluble vitamins A and D. It is the site for the destruction of red blood cells and the conversion of hemoglobin to bile salts and bile pigments. In vertebrate embryos it is also the source of red blood cells.

The liver converts amino acids into small reserves of protein and also into glucose (during gluconeogenesis) under the driving force of glucocorticoids.

The liver also strains the blood of particulate materials that are no longer useful. These include degenerating red blood cells, foreign material in the circulatory system, and unassimilated native bulk material. Large macrophages known as *Kupfer cells* are abundant in the *sinusoids* (capillaries) of the liver, and they engulf particulate material. Other types of macrophages may also be present and function to rid the blood of large molecular aggregates. Within the *parenchymal*, or working, cells of the liver, the absorbed fragments of blood cells are converted to bile. Although most erythrocyte breakdown begins in the spleen and bone marrow, bile formation is associated with the liver.

The liver also degrades many toxins in the body to harmless substances. This is accomplished by *oxidation*, *methylation*, *hydroxylation*, or *conjugation* with organic moieties (parts), such as glucuronic acid. The liver excretes many of these neutralized toxins; it also tends to concentrate them within its *lobules.* This suggests people should avoid eating the liver of any organisms that are capable of producing poisons or that may have accumulated poisons through ingestion.

In its production of enzymes, the liver may also reduce the levels of potentially harmful substances. Serum cholinesterase, which is synthesized in the liver, helps maintain acetylcholine at manageable levels. A variety of hormones are rendered inactive by hepatic systems primarily involved in detoxification. Cholesterol is *esterified* by these systems, as well. Since a diseased liver cannot carry out these functions, people with impaired liver function are very vulnerable to drug intoxication, toxemia, and hormone buildup.

## 16.4  DIET AND HEALTH

It has been pointed out in previous chapters (3, 5, and 15) that we are literally what we eat. This does not only mean that the substance of our body is made up of the basic unit building blocks that come in with our ingested food or that we are dependent on what we eat for the energy to drive our living engine. Far more, it means that our ability to function well; our personality, outlook, and sense of well-being; and even our capacity to learn and develop are related to the food we eat. The inability of many undeveloped countries to increase their agricultural resources or develop an industrial economy may very well be related to food shortages that sap the strength of their citizens.

A number of diseases can be successfully treated by specific dietary alterations. The reduction of salt is now a common medical practice in the treatment of *hypertension* (high blood pressure). *Sprue*, a disease caused by an allergic response to wheat products, is treated by eliminating these products from the diet. Mineral deficiencies, such as lack of calcium during pregnancy, are readily treated with the deficient ingredient, but in many cases hormones must also be given to assure that the mineral will be effectively utilized.

**EXAMPLE 6**  *Osteoporosis* involves loss of bone. It is especially prominent in older Caucasian women. When treated with calcium alone, it often does not respond. However, calcium given along with doses of the hormone progesterone effects a marked improvement in the condition and reduces the danger of breakage associated with bone loss.

In the case of proteins (see Chap. 3), we must take in enough amino acids to offset nitrogen loss, but in addition, we must continually ingest *essential* amino acids, which cannot be produced in the body. Certain fatty acids are likewise essential and must be supplied by outside food sources. *Malnutrition*, the failure to achieve a minimal intake of nutrients for maintaining health, may thus arise from either qualitative or quantitative deficiencies. A further complication is the possible failure to digest, absorb, or otherwise process ingested food.

**EXAMPLE 7**  The majority of people in many parts of Africa, and many Asians as well, are not able to utilize the *lactose* present in the milk of almost all mammals. This disaccharide requires *lactase* for digestive breakdown; most blacks and Asians do not synthesize this enzyme. The undigested lactose may increase osmotic pressure so that fluids collect in the intestinal lumen and produce cramps and diarrhea. Digestive upset may also arise late in digestion from bacterial fermentation of lactose in the large intestine and the consequent production of large amounts of gas. Milk is a near perfect food for most mammalian infants, but its usefulness may be limited in adults, especially those in third world countries. For these people, milk may actually present a hazard. Consequently, many western countries have eliminated powdered milk from the emergency relief food products they send to undeveloped countries.

Fortunately, the fermentation products of milk such as cheese, curds, and yogurt do not have the offending lactose, since the fermenting bacteria break it down. These nourishing products may still be used by those ethnic groups who have lost the allele for synthesis of lactase in the course of their evolution.

## 16.5  SPECIAL CASE OF THE VITAMINS

Vitamins are special organic substances necessary in tiny amounts to sustain life. They usually function as coenzymes (see Chap. 5) in a variety of metabolic reactions. Their absence usually causes a specific deficiency disease.

**EXAMPLE 8**  British sailors who embarked on long ocean voyages in earlier times were given limes to take with them. Thus arose the sobriquet "limeys," for English seamen. Those sailors who regularly ate the limes did not get *scurvy*, a disease that afflicted many sailors from other lands. In scurvy, heavy bleeding of the gums, poor healing of wounds and fractures, and general fatigue are prominent features. Only in the twentieth century have scientists learned that scurvy is caused by a shortage of *vitamin C (abscorbic acid)*, a water-soluble vitamin particularly abundant in citrus fruits and many vegetables.

Approximately 12 vitamins are significant in human nutrition. They are usually identified by letters, but they also have chemical names. Some are water-soluble and others fat-soluble (see Table 16.2). Generally, the fat-soluble vitamins, particularly D and A, produce harmful effects if taken several magnitudes beyond the required dose. The water-soluble vitamins, such as C and the B complex group, are less likely to be dangerous at higher doses because any excess is excreted.

The benefits of taking vitamins to avoid serious vitamin deficiencies are well-documented. The scorbutic effects of vitamin C deficiency, the serious neurological and growth impairments of *pellagra* caused by a deficiency of niacin, and the central nervous system defects of *beriberi* that arise from a lack of thiamine (vitamin $B_1$) are clear, and all are preventable through diets containing the relevant vitamins. However, considerable controversy exists over the advantages to one's health of taking *extra* doses of vitamins. Many health experts contend that vitamins are necessary only in small amounts and that a well-balanced diet supplies these necessary levels.

One controversial question concerns the role of vitamin E. An antisterility function of this vitamin exists in rats, especially males. However, no clear-cut role is evident for humans. Claims that $\alpha$-tocopherol (its chemical name) prevents heart attacks and cancer have not been substantiated. Anecdotal evidence abounds for the miraculous effects of vitamin E as well as of vitamin A's precursor, $\beta$-carotene; both are antioxidants, substances that can mop up free radicals, molecules which can injure cells and also lead to plaque buildup in arteries. However, studies have failed to establish a causal

Table 16.2a Vitamins

| Name, Formula, and Solubility | Important Sources | Functions | Result of Deficiency or Absence (in humans, except as noted) |
|---|---|---|---|
| Lipid-soluble vitamins: **A** ($C_{20}H_{30}O$), anti-xerophthalmic | Plant form (carotene, $C_{40}H_{56}$) in green leaves, carrots, etc.; is changed in liver to animal form ($C_{20}H_{30}O$), present in fish-liver oil (shark); both forms in egg yolk, butter, milk | Maintains integrity of epithelial tissues, especially mucous membranes; needed as part of visual purple in retina of eye | Xerophthalmia (dry cornea, no tear secretion), phrynoderma (toad skin), night blindness, growth retardation, nutritional croup (hoarseness) in birds |
| **D** ($C_{28}H_{44}O$), antirachitic | Fish-liver oils, especially tuna, less in cod; beef fat; also exposure of skin to ultraviolet radiation | Regulates metabolism of calcium and phosphorus; promotes absorption of calcium in intestine; needed for normal growth and mineralization of bones | Rickets in young (bones soft, yielding, often deformed); osteomalacia (soft bones), especially in women of Asia |
| **E**, or tocopherol ($C_{29}H_{50}O_2$), antisterility | Green leaves, wheat-germ oil and other vegetable fats, meat, milk | Antioxidative; maintains integrity of membranes | Sterility in male fowls and rats, degeneration of testes with failure of spermatogenesis, embryonic growth disturbances, suckling paralysis and muscular dystrophy in young animals |
| **K** ($C_{31}H_{46}O_2$), antihemorrhagic | Green leaves, also certain bacteria, such as those of intestinal flora | Essential to production of prothrombin in liver; necessary for blood clotting | Blood fails to clot |
| Water-soluble vitamins: **B** complex Thiamine ($B_1$) ($C_{12}H_{17}ON_4S$), antineuritic | Yeast, germ of cereals (especially wheat, peanuts, other leguminous seeds), roots, egg yolk, liver, lean meat | Needed for carbohydrate metabolism; thiamine pyrophosphate, an essential coenzyme in pyruvate metabolism (stimulates root growth in plants) | On diet high in polished rice, beriberi (nerve inflammation); loss of appetite, with loss of tone and reduced motility in digestive tract; cessation of growth; polyneuritis (nerve inflammation) in birds |
| Riboflavin ($B_2$) ($C_{17}H_{20}O_6N_4$) | Green leaves, milk, eggs, liver, yeast | Essential for growth; forms prosthetic group of FAD enzymes concerned with intermediate metabolism of food and electron-transport system | Cheilosis (inflammation and cracking at corners of mouth), digestive disturbances, "yellow liver" of dogs, curled-toe paralysis of chicks, cataract |
| Nicotinic acid, or niacin ($C_6H_5O_2N$), antipellagric | Green leaves, wheat germ, egg yolk, meat, liver, yeast | Forms active group of nicotinamide adenine dinucleotide, which functions in dehydrogenation reactions | Pellagra in humans and monkeys, swine pellagra in pigs, blacktongue in dogs, perosis in birds |

**Table 16.2b** Vitamins

| Name, Formula, and Solubility | Important Sources | Functions | Result of Deficiency or Absence (in humans, except as noted) |
|---|---|---|---|
| Folic acid ($C_{19}H_{19}O_6N_7$) | Green leaves, liver, soybeans, yeast, egg yolk | Essential for growth and formation of blood cells; coenzyme involved in transfer of single-carbon units in metabolism | Anemia, hemorrhage from kidneys, and sprue (defective intestinal absorption) in humans; nutritional cytopenia (reduction in cellular elements of blood) in monkeys; slow growth and anemia in chicks and rats |
| Pyridoxine ($B_6$) ($C_8H_{12}O_2N$) | Yeast, cereal grains, meat, eggs, milk, liver | Present in tissues as pyridoxal phosphate, which serves as coenzyme in transamination and decarboxylation of amino acids | Anemia in dogs and pigs; dermatitis in rats; paralysis (and death) in pigs, rats, and chicks; growth retardation |
| Pantothenic acid ($C_9H_{17}O_3N$) | Yeast, cane molasses, peanuts, egg yolks, milk, liver | Forms coenzyme A, which catalyzes transfer of various carboxylated groups and functions in carbohydrate and lipid metabolism | Dermatitis in chicks and rats, graying of fur in black rats, "goose-stepping" and nerve degeneration in pigs |
| Biotin (vitamin H) ($C_{10}H_{16}O_3N_2S$) | Yeast, cereal grains, cane molasses, egg yolk, liver, vegetables, fresh fruits | Essential for growth; functions in $CO_2$ fixation and fatty acid oxidation and synthesis | Dermatitis with thickening of skin in rats and chicks, perosis in birds |
| Cyanocobalamin ($B_{12}$) ($C_{63}H_{90}N_{14}O_{14}PCo$) | Liver, fish, meat, milk, egg yolk, oysters, bacteria and fermentations of *Streptomyces*; synthesized only by bacteria | Formation of blood cells, growth; coenzyme involved in transfer of methyl groups and in nucleic acid metabolism | Pernicious anemia, slow growth in young animals; wasting disease in ruminants |
| C, or ascorbic acid ($C_6H_8O_6$) | Citrus fruits, tomatoes, vegetables; also produced by animals (except primates and guinea pigs) | Maintains integrity of capillary walls; involved in formation of "intercellular cement" | Scurvy (bleeding in mucous membranes, under skin, and into joints) in humans and guinea pigs |

*Source:* T.I. Storer, R.L. Usinger, R.C. Stebbins, and J.W. Nybakken, *General Zoology*, 6th ed., McGraw-Hill, New York, 1979.

connection between supplements of these vitamins and protection against cancer and heart disease. The advice to eat a wide variety of fruits and vegetables still holds.

A classic controversy centers on the possible role of massive doses of vitamin C in preventing colds. Since this prescription was argued by a Nobel laureate, Dr. Linus Pauling, many laymen were impressed by the arguments and began regimens of high vitamin C intake, especially when they felt a cold coming on. These early claims, advanced in the 1960s, have not been substantiated in the main, although there is some indication that in certain cases the symptoms of a cold may be assuaged by large doses of vitamin C.

Although vitamins are not prohibitively expensive and malabsorption may indeed lead to vitamin deficiencies in unusual cases, one must still resort to vitamin supplements with a considerable degree of caution. Kidney stones have been associated with megadoses of vitamin C, and the lipid-soluble vitamins (A, D, E, and K) can be harmful in excess. This is particularly true for A and D, where damages to both adults and fetuses have been reported.

## 16.6  OBESITY AND EATING DISORDERS

Animals in the wild usually adjust their food intake to energy expenditures so that marked changes in body weight, particularly the accumulation of fat, do not occur. Among domestic animals or those kept in zoos, weight homeostasis falters and obesity may appear. *Obesity* may be defined as body weight exceeding 25 percent of the desirable standard. This excess weight must be caused by excessive *adiposity* [hyperplasia or hypertrophy of adipose i.e., fat tissue; an adipocyte, or fat cell, is shown in Fig. 16.4]. Extra muscle mass, such as might develop in a weight-lifting program, is not regarded as obesity. You will recall (Chap. 4) that *hyperplasia* is an increase in the number of cells of a tissue while *hypertrophy* is an increase in size owing to the enlargement of individual cells.

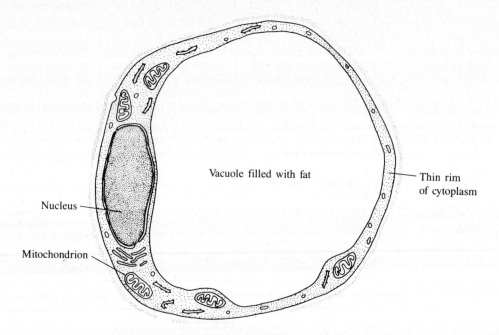

**Fig. 16.4** Adipocyte

At complete rest under no metabolic stress an average human of 70 kg expends approximately 68 cal/h. In a sedentary job such a person might use up about 2300 cal daily. Certainly the average western diet exceeds that number of calories, and those who do not expend energy in regular exercise programs are threatened with weight gains that are detrimental from both a cosmetic and a health point of view. Simple obesity is not by itself regarded as a pathology, but its presence complicates preexisting disease states, especially cardiovascular conditions, diabetes, arthritis, and even susceptibility to cancer.

The ability to convert newly arrived foodstuffs into long-term storage fuel (fat) is regarded as a significant evolutionary advance in birds and especially in mammals. The efficiency of this process is connected to the elegant metabolic versatility of adipose tissue, which was earlier misconceived as a mere storage bin for lipids manufactured elsewhere. However, this capacity to convert and store calories as fat, though of importance to survival under conditions of scarcity, becomes a threat to health in an environment in which tempting foods are freely available and seductively displayed and advertised. In humans a complex of subtle desires and responses determines whether feeding will be initiated, continued, and, finally, terminated. Both appetite and hunger are involved. According to the eminent nutritionist Jean Mayer, *appetite* is a complex of usually pleasant sensations "by which one is aware of desire for and the anticipation of ingestion of palatable food." *Hunger* is a far more intense, unfocused craving for any kind of food following acute deprivation. Satisfaction of both hunger and appetite induces a sensation of surfeit, or fullness, which cuts off further eating. These sensations are called *satiety* and are mediated by satiety centers in the hypothalamus of the brain. Other hypothalamic centers are involved in hunger and appetite.

The normal regulation of food intake appears to be dependent on physiological events arising in the gut (stomach and intestine), brain, and circulatory system. It has also been shown that physical activity is related to the appropriate control of feeding. Long bouts of overexertion may cause an adjustment of caloric intake to a level below that necessary for weight maintenance and thus produce an underweight condition. Too little exercise may be associated with excess food intake and thus may produce obesity. Only within appropriate ranges of physical activity does the "appestat" work effectively. Research in obesity appears to confirm that obese individuals are not merely gluttonous or lacking in self-control but are the victims of altered homeostatic mechanisms which are difficult to defy.

Closely related to obesity in terms of its malfunctioning feeding control system is *anorexia nervosa*. In this serious illness, found largely among adolescent girls in western societies, a deep-seated conviction of threatening obesity causes the normal-weight or underweight victim literally to starve to death. Although it seems a simple matter to intervene and force-feed these people, intervention often occurs too late to reverse the process of starvation. Anorexia appears to arise from psychological problems, but it has been extremely resistant to conventional therapeutic procedures.

*Bulemia* is another disorder of food intake that is associated primarily with young women in affluent countries but is also encountered in slightly older adults as well. It involves periods of herculean engorging with food, followed by forced vomiting. Weight gains are not apparent, nor is starvation a consequence of this anomalous behavior. Damage to the stomach, especially from bleeding associated with violent vomiting episodes, may occur. As a consequence of the bizarre behavior demonstrated, bulemia sufferers may isolate themselves from family and friends and suffer from depression.

## 16.7  DEFICIENCY DISEASES

General starvation in which caloric needs are not met by sufficient intake of carbohydrates and fats is life-threatening but not widespread, except in parts of the developing world. In Africa, *marasmus* is a disease characterized by slow starvation. In many cases sufficient calories are taken in to prevent death through starvation, but specfic nutrients are lacking or dangerously low in amount. Fairly common in sub-Sahara Africa, *kwashiorkor* is a deficiency disease in which protein is inadequate. Unfortunately, foods rich in protein, such as beef and pork, are much more expensive than those high in carbohydrate and even fat. Even if caloric needs are being met, long-term deprivation of protein will result in the breakdown of tissue and loss of vital structures. Not only must sufficient dietary protein be provided to maintain the nitrogen balance, but the protein must contain the essential amino acids needed to synthesize complete proteins.

Most dramatic of the deficiency diseases are those associated with specific vitamin deficiencies. We have already cited scurvy as a deficiency disease associated with deprivation of vitamin C. This *avitaminosis* (syndrome associated with vitamin deprivation) involves an inability to form the intercellular cement that holds cells together, such as the *endothelial* (inner-lining) layer of capillary beds. Connective tissue is not laid down properly, and the response to infection is impaired.

Niacin, part of the B complex, is eventually incorporated into the coenzymes NAD and NADP. Its absence results in the impairment of those metabolic processes that supply the cell with energy. The nervous system is particularly sensitive to such deficiencies, and the niacin-deficiency disease *pellagra* involves impaired neural function, such as diminished intelligence, lack of alertness, and fainting spells. A deficiency of vitamin $B_{12}$ is associated with *pernicious anemia*, a chronic insufficiency of hemoglobin that was fatal before its etiology (cause) was understood. Folic acid is another B vitamin that influences red blood cell formation but is chiefly involved with nucleotide formation.

Among the fat-soluble vitamins, vitamin A deficiency is associated with night blindness. The visual pigment needed for vision in dim light, which requires vitamin A, is not formed in sufficient quantities. Vitamin D deficiency is associated with *rickets*, a defect in bone formation resulting from poor absorption and assimilation of calcium. Vitamin K deficiency may impair normal blood clotting and cause internal bleeding.

# Solved Problems

**16.1** From the following descriptions of teeth patterns, give the likely feeding habits of the animal to which each pattern belongs: (*a*) sharp, thin teeth, curved backwards into the mouth and all approximately the same shape; (*b*) six incisors in lower jaw (none in upper), followed by a gap and then a number of relatively flat premolars and molars in upper and lower jaws; (*c*) incisors, well-developed canines, sharply cusped premolars, and reduced numbers of molars.

(*a*) This *homodont* (having teeth of the same shape) pattern is characteristic of snakes. Because they eat their prey whole, they do not require a varied array of utensils in their mouths to break up the food. Lacking limbs, most cannot effectively grasp their prey by any other means than their mouths, so their teeth are sharp and recurved (bent backwards) to make escape difficult and to aid in working their food down their throats. (*b*) Grazing and browsing herbivores have a *heterodont* pattern; they use their six lower incisors to crop off grass and leaves and make use of the flat premolars and molars as grinding surfaces to break up the cellulose of the vegetation they eat. (*c*) The well-developed canines are generally found in animals (such as canines) that must tear through meat; the sharp premolars and molars serve to cut and shear meat from bone.

**16.2** During emergency situations, the mouth often gets "cottony" (exceptionally dry). Why is this?

As was seen in Chap. 15, during fight-or-flight situations the sympathetic nervous system reduces the activity in vegetative processes such as digestion. The salivary glands, being part of the digestive system, are therefore inhibited by the sympathetic nervous system from producing saliva, and the mouth becomes dry.

**16.3** Most animals cannot digest cellulose, so that grass, an ubiquitous plant, cannot satisfy their nutritional needs. How do some animals utilize grass and, in insects like termites, even wood?

Except for certain species of insects, no animal synthesizes its own cellulase (an enzyme). However, animals that can subsist on grass or wood generally enjoy a symbiotic relationship with bacteria or protozoans that are capable of synthesizing cellulase. Without cellulase, cellulose cannot be degraded. In the case of termites, which do so much damage to wooden structures (their major dietary staple), flagellated protozoa reside within the insect's stomach. Since it is the protozoan that is vital to the digestion of the cellulose of wood, elaborate mechanisms exist for assuring its presence in every termite. Young termites are *coprophagic*—they eat the feces of their parents of other adults of the termite colony. This necessity for reinfecting the young may very well have been the significant factor in the evolutionary tendency toward a social system among termites.

**16.4**  List three advantages to an animal in having a large internal food storage capacity.

1.  The capacity to store food allows extra time for microorganisms to break down cellulose and thereby increases digestive efficiency.
2.  Storage capacity reduces the number of times that an animal must venture out in search of food and thus reduces the number of times it is vulnerable to predation.
3.  Without the ability to store food, animals in some cases would have to leave part of a kill behind, with a resultant waste of some of the energy that was put into making the kill.

**16.5**  *Vomiting* is a forcible ejection of both partially digested food and digestive juices from the upper gastrointestinal tract. It is mediated by the nervous system, primarily the parasympathetic trunk (vagus), and coordinated by a center in the medulla oblongata of the brain. It usually involves initially a strong wave of nausea and excess production of saliva, which floods the mouth. Then the duodenal region of the intestine and the lower (pyloric) stomach contract and force food content into the upper part of the stomach. A profound *inspiration* (breathing in) occurs, the esophagus relaxes, and a series of spasmodic abdominal contractions, along with contractions of the stomach, takes place. These conjoined responses force fluid and partially digested food to move up through the esophagus and out of the mouth. The *glottis*, or opening into the trachea (windpipe) is kept closed throughout this forcible evacuation; this guards against choking. What function do you suppose the vomiting reflex serves?

Vomiting is a response to esophageal, *gastric* (stomachic), or intestinal irritation, including poisoning. Such irritation is best relieved through the elimination of the irritating food material. Vomiting is thus a valuable device to protect the organism. Even if ingested material is benign and the irritation is due to injury of the digestive tract itself, the elimination of food may provide much needed rest for the system. A clear-cut illustration of the value of vomiting is provided by the rat. This hardy creature would be all but invincible except for the fact that it lacks a vomiting reflex. Since it cannot rid itself of irritating poisons that are swallowed, the rat readily falls prey to a variety of poisons other species could eliminate through vomiting.

**16.6**  There does not seem to be any function for the appendix in humans. In humans, even the cecum, which serves as storage space in herbivores and acts as a fermentation vat in horses and a few other species, seems to be of little use. What possible dangers can the appendix pose?

Since the appendix is a dead-end narrow tube, closed at its distal (unattached) end, it may accumulate particulate material that causes infection. An inflamed appendix is usually associated with considerable pain, marked abdominal distension and tenderness, and a high white blood cell count accompanied by fever. Surgery to remove the offending appendix is usually performed on an emergency basis. It has been humorously observed that the only function of the human appendix is, in its pathology, to provide a means of livelihood for young surgeons.

Should the inflamed appendix burst, a serious challenge to life is incurred. The infected material will usually cause a generalized infection, known as *peritonitis*, within the abdominal cavity. The *peritoneum* is a broad but thin membrane that lines and encloses the abdominal cavity. Where it covers the visceral organs, it is called the *visceral peritoneum*. An unchecked inflammation of the peritoneum usually results in death.

**16.7**  List four features of the intestine that increase absorptive capacity.

(1) Increased length, (2) extensive folding (earthworms have a single large fold, the *typhlosole*), (3) projection of fingerlike villi into the intestinal lumen, (4) evagination of microvilli from epithelial cells of the mucosa.

**16.8**   The tadpole has a relatively long, highly coiled intestine, whereas the metamorphosed adult frog has a short intestine. Why do you suppose this difference exists?

   The tadpole is herbivorous; because its food contains large amounts of cellulose, it requires more time (and therefore more intestine) to digest it. The adult frog is carnivorous; lacking great quantities of cellulose in its diet, it does not require a long intestine.

**16.9**   If cleavage of a mixture of long-chain polypeptides yielded shorter chains, many of which had tyrosine or phenylalanine at their amino ends, what was the likely digestive enzyme used?

   Although both pepsin and chymotrypsin cleave polypeptides at tyrosine and phenylalanine, only pepsin does so at the amino side of the amino acid and thus leaves tyrosine or phenylalanine with a free amino group at the end of the hydrolyzed segment.

**16.10**   List some of the functions of the liver.

   1. Maintenance of blood sugar homeostasis
   2. Interconversions of nutrients (e.g., carbohydrates to fats, amino acids to carbohydrates and fats)
   3. Removal of unwanted particulate matter from the blood through the mediation of macrophages
   4. Production of bile
   5. Manufacture of plasma proteins such as fibrinogen and albumin
   6. Manufacture of cholesterol
   7. Manufacture of red blood cells in vertebrate embryos
   8. Deamination of amino acids and excretion of resulting ammonia as urea, uric acid, etc.
   9. Degradation and excretion of toxins
   10. Storage of iron
   11. Storage of vitamins

**16.11**   Individuals suffering from *hepatitis*, an inflammation of the liver, are given diets high in simple sugars. Why?

   The liver has already been described as a major site for the interconversion of the major foodstuffs. Such transformations impose slight metabolic burdens on the liver, which are easily handled in a state of good health. When the liver is in the grip of pathology, burdens are best avoided. A diet high in simple sugars will provide a means for maintaining blood sugar levels at normal levels without exacting a price from the liver. It is particularly important that low blood sugar levels be avoided, because that will stimulate gluconeogenesis, a process that imposes considerable strain on the liver, since amino acids will be stripped of their amino groups and transformed to carbohydrate metabolites.

   Under conditions of sugar shortage the combustion and reformation of lipids are also impeded. You will recall from Chap. 5 that lipids burn in a carbohydrate flame. A shortage of carbohydrate could easily result in a "sputtering" lipid metabolism in which toxic products might form. A production of ketone bodies, for example, would challenge pH stability and the stability of the liver as well. Providing the simplest fuel molecules reduces the work that the liver must carry out.

**16.12**   Why do you suppose excessive doses of vitamins such as A and D pose a greater threat to health than vitamins such as C and the B complex vitamins do?

   Vitamins A and D are fat-soluble; because of this they tend to move out of the bloodstream and accumulate in the body and thus are much more likely over time to reach harmful levels. Vitamin C and the B complex vitamins are water-soluble and, so, are regularly excreted in the urine; they therefore do not build up so readily to harmful levels.

**16.13**   How would you suppose obesity in other mammals differs from most cases of human obesity?

   Animal obesity arises from physical, rather than psychological, sources. It may arise from defects in the regulation of food intake or the control of basal metabolism. It may also be associated with a marked diminution of voluntary motor activity. In some cases a metabolic defect may exist, either genetic or induced, that predisposes the animal to the accumulation of lipid in adipose tissue.

Models of animal obesity have long been studied with the expectation that they would provide insights for dealing with human obesity. Generally, these hopes have not been realized. One type of obesity in mice and rats is caused by the injection of *gold thioglucose*, a complex metallo-organic compound that effectuates the passage of gold into the brain across the blood-brain barrier. Gold especially lodges in the satiety center of the hypothalamus and destroys its inhibition of feeding so that pathological overeating occurs. Such obesity models obviously are not encountered in humans, but they do provide an insight into hypothalamic mechanisms for the control of feeding.

A genetic form of obesity was first observed in a strain of mouse (C57 Bl/6Job) at the Jackson Laboratories in Bar Harbor, Maine, in 1949. These mice attained weights of more than 100 g, which is 250 percent that of the lean littermates. Their livers were loaded with lipid, and they showed a variety of changes in metabolic enzymes and an altered behavior of key endocrine glands. Paradoxically, they demonstrated a marked hyperglycemia, yet they produced high levels of circulating insulin. Though quite different from most human obesities, these obese, hyperglycemic mice demonstrate the possibility of genetic influence in obesity. Genetic obesities of other types have been found in rats, dogs, and Shetland ponies.

Most experts involved in treating human obesity recognize the difficulty of ameliorating what are deep-seated compulsions to take in more calories than are expended. Psychological factors are probably crucial in most cases, and an emphasis on behavior modification through groups like Weight Watchers has grown in recent years. A well-balanced diet must not be sacrificed if good health is to be maintained. The slow and steady alteration of basic eating habits accompanied by increased exercise is probably the only sensible road to normal weight. Pituitary and thyroid abnormalities may underlie obesity in some rare cases, but most cases of adult-onset obesity yield to altered nutritional habits. It has been stressed by nutritional scientists such as Jules Hirsch of Rockefeller University that patterns of food intake established early in life may have particularly crucial effects on the possible development of obesity later in life. The numbers and condition of adipocytes may mediate these effects. A recent finding in mice demonstrating genetic obesity is that their adipose cells produce abnormally low levels of *adipsin*, a protein. It is speculated that adipsin may act like a hormone and influence the brain or the metabolism of other tissues.

**16.14** In Japan, the Japanese demonstrate an extremely low incidence of heart attacks and even hypertension. Descendents of Japanese who move to the United States soon show a tendency toward an increased incidence of heart attack and an increase in blood pressure, even though there is little intermarriage with the larger population in early generations. This tendency develops over a time span of as little as two generations. What conclusions about health can be drawn from these facts?

Since intermarriage is not significant in newly arrived Japanese immigrants to the United States, one can assume a reasonably similar genetic makeup between their offspring and Japanese living in Japan. The conclusion must be that American nutritional influences, social stress, etc., account for the negative changes in the migrant Japanese group. If destructive changes can occur in an altered environment, reason would dictate that altered patterns of nutrition, health care, etc., could similarly improve the health and longevity of populations.

# Supplementary Problems

**16.15** The fang of a poisonous snake is a modified tooth.
(a) True   (b) False

**16.16** The wisdom teeth are incisors.
(a) True   (b) False

**16.17** Humans have a two-way digestive tract.
(*a*) True   (*b*) False

**16.18** The major function of the stomach is the storage of food.
(*a*) True   (*b*) False

**16.19** The chief action of villi is to break down large chunks of particulate material in the *chyme* (semiliquid product of digestion).
(*a*) True   (*b*) False

**16.20** *Segmentation*, a to-and-fro series of contractions in the digestive tract, helps break down food particles.
(*a*) True   (*b*) False

**16.21** Sphincters in the digestive tract act like valves.
(*a*) True   (*b*) False

**16.22** Cholesterol is synthesized in the cecum.
(*a*) True   (*b*) False

**16.23** Match the phrases in column **A** with the regions in column **B**.

| A | B |
|---|---|
| 1. Ptyalin acts here | (*a*) Stomach |
| 2. Absorption of nutrients | (*b*) Duodenum |
| 3. Absorption of fluids and electrolytes | (*c*) Ileum |
| 4. Production of vitamins by bacteria | (*d*) Colon |
| 5. Pancreatic juice acts here | (*e*) Mouth |
| 6. Low pH | |

**16.24** Why do some people with advanced diabetes lose weight?

# Answers

| | | | |
|---|---|---|---|
| **16.15** (*a*) | **16.17** (*b*) | **16.19** (*b*) | **16.21** (*a*) |
| **16.16** (*b*) | **16.18** (*a*) | **16.20** (*a*) | **16.22** (*b*) |

**16.23** 1—(*e*); 2—(*c*); 3—(*d*); 4—(*d*); 5—(*b*); 6—(*a*)

**16.24** They cannot utilize or store sugar adequately, so these calories pass out of the body through the urine. Also, they cannot synthesize and store fat, because insulin is required for the normal activity of lipogenic and esterification pathways.

# Chapter 17

# Circulation and the Blood

In 1628 William Harvey, an English naturalist and natural philosopher, first stated that the blood circulates in a definite path throughout the body rather than merely surging back and forth. This seminal idea, which led to the development of physiology as a rigorous experimental science, could not be completely understood until the capillaries were discovered much later by Marcello Malpighi, an Italian anatomist. Since it is the blood that distributes materials throughout the body, it is readily apparent that circulation underlies all aspects of function within the organism.

In all vertebrates and several groups of invertebrates (e.g., earthworms) the circulatory system is a completely closed system of tubes. Other organisms have an open circulatory system, in which the tubes are contiguous with open regions, or *sinuses*. Such open systems show a lowered transport efficiency and a slower circulatory time.

**EXAMPLE 1** Insects possess an open circulatory system. As a result, their circulation of blood-borne materials, including $O_2$, is relatively slow compared with that of vertebrates. Yet, the insects are the most successful group of terrestrial inhabitants, and their ability to cope with the exigencies of a land habitat is tied to their very high level of metabolic activity. Such activity is possible because the oxygen necessary for combustion of their fuel is provided outside the circulatory system by a series of open tubes called *tracheoles*, or *trachea*. These tracheoles originate from openings on the surface (*spiracles*) and ramify throughout the body of the insect.

## 17.1 COMPARATIVE CARDIOVASCULAR SYSTEMS OF VERTEBRATES

The cardiovascular system of all vertebrates consists of a muscular pump, the *heart*, and a system of tubes that carry blood to and from the heart. Vessels that carry blood away from the heart are called *arteries*; those that carry blood toward the heart are known as *veins*. Smaller arteries are called *arterioles*; smaller veins are designated *venules*. The all-important connectors between arterioles and venules are the *capillary beds*, in which the actual exchanges between blood and tissues occur.

In primitive animals the heart is little more than an expanded section of blood vessel. There may be one or more of such elemental pumping devices, an arrangement that assures a continuous movement of blood. In fish, the earliest of the vertebrates, the heart consists of a single thin-walled receiving compartment, the *atrium*, or *auricle*, which empties into a thicker-walled, more powerful pump chamber, the *ventricle* (see Fig. 17.1). Only one turn through the heart occurs during a complete circulatory cycle. Blood is pumped to the gill capillaries for oxygenation, but because the fish does not have a pump on the other side of the gills, the blood moves slowly and with considerably diminished force through the rest of the body (systemic circulation).

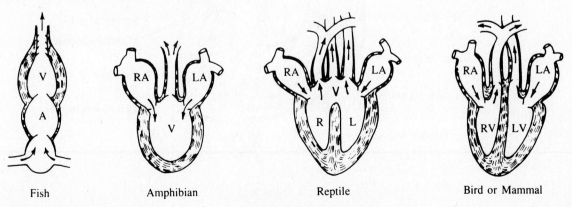

| Fish | Amphibian | Reptile | Bird or Mammal |

**Fig. 17.1** Vertebrate hearts

A considerable improvement occurs in amphibians, which live the early part of their lives in water and, following metamorphosis, the later, adult part on land. Here we encounter two reception chambers, a left and right auricle, but a single ventricle. In the amphibian two circulations are evident, a *pulmonary* and a *systemic* circuit. The lungs in most amphibians are hollow and relatively inefficient; the skin acts as an adjunct oxygenation organ, and some of the blood that moves toward the lungs is actually shunted through the skin.

**EXAMPLE 2**   Deoxygenated venous blood from the systemic circulation enters the right auricle and then passes into the ventricle; highly oxygenated venous blood from the pulmonary circulation enters the left auricle and then passes into the ventricle. Although there is some mixing of the blood from the two auricles (and therefore the two circulations), when the ventricle contracts, most of the blood from the right auricle (systemic blood) passes out the pulmonary artery and proceeds to the lungs; most of the blood from the left auricle (pulmonary blood) passes out the *aorta* and proceeds to all the tissues of the body.

In both birds and mammals a clear separation of the two circulations is apparent and the four-chambered heart can actually be viewed as two hearts. The right side of the heart receives deoxygenated blood from the systemic circulation and pumps it to the lungs through the pulmonary artery. Oxygenated blood returns to the heart through the two pulmonary veins and is pumped out the left ventricle into the aorta, from which it eventually reaches all parts of the body.

## 17.2   THE HUMAN HEART

The human heart is a four-chambered structure located in the chest (see Fig. 17.2). The heavy muscular portion comprises the two ventricles; the two atria appear as flaps lying atop the ventricles.

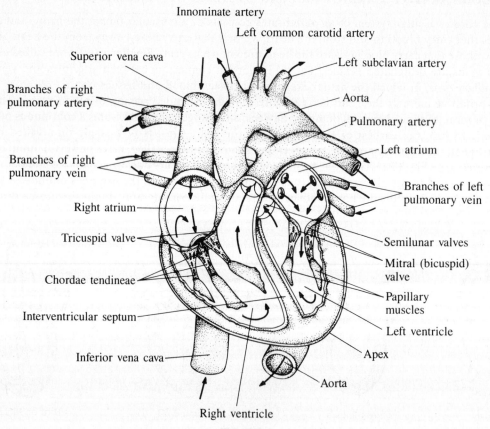

**Fig. 17.2**

Both atria fill at about the same time. During filling, the atria are in a relaxed state. Starting with a node of tissue in the right atrium, the *sinoatrial node* (also called the *sinoauricular*, or *SA*, node), a wave of contraction sweeps over the atria, and blood is sent to the relaxed ventricles. The relaxed state of heart muscle is called *diastole*; the contracted state is known as *systole*. Systole in the atria is well over before the ventricles go into systole. This contraction of the ventricles would tend to send the blood back into the atria if not for the action of *atrioventricular* (AV) *heart valves*, which lie between each atrium and ventricle. These valves consist of flaps of connective tissue that, like a door between two rooms, permit movement in one direction but close when the pressure shifts direction.

The AV valve on the right side consists of three flaps of connective tissue and is known as the *tricuspid valve*. The AV valve on the left consists of two flaps of connective tissue and is called the *bicuspid*, or *mitral*, *valve*. These flaps are connected to "guy ropes," or strings of tendonous material, known as the *chordae tendineae*. The chordae tendineae are, in turn, embedded in thick *papillary muscles*, whose partial contractions prevent the flaps from moving beyond their closed juncture. The flaps are free to move into the ventricular chamber during atrial systole, and thus the blood moves out of the atria and into the ventricles. However, the opposite swing of the flaps (into the atria) is prevented by the taut chordae, so that during ventricle systole the blood causes the flaps to close and does not move back into the atria.

Blood from the right ventricle is forced into the pulmonary arteries and on to the lungs. Blood from the left ventricle exits through the *aortic trunk*, the thick vessel that leads blood to the appropriate arterial subdivisions that will carry it to all parts of the body. *Semilunar valves* in the aorta and pulmonary artery prevent the backflow of blood into the ventricles. The semilunar valves are "half-moon" pockets. When blood tries to reenter the ventricles from the arteries, these pockets are filled with the blood and, becoming distended, come together in the lumen of the artery, and thus block the reentry.

The heart muscle is itself supplied with blood by the left and right *coronary arteries*. These vessels arise from the aorta at a point close to its origin in the left ventricle. They ramify to provide blood throughout the heart. If a branch of the coronary artery is occluded, either through spasm or blockage by a blood clot, damage to the heart muscle supplied by that branch occurs. The damaged muscle is described as an *infarct*. If such damage is extensive, the heart may stop and death ensue. This type of pathology is popularly termed a *heart attack* or *heart seizure*. It is a very common cause of death in middle-aged men in western countries, but women have begun to show an increased incidence of heart attacks in recent years.

The actual output of blood by the heart is called *cardiac output*, and it is equal to the product of the *stroke volume* and the *heart rate*. Stroke volume is the amount of blood pushed out of the heart by a contraction of the ventricles. Under resting conditions cardiac output is about 5 liters/min in humans. Among the factors affecting stroke volume is the amount of blood returning to the heart. This quantity changes constantly as the internal and external environments of the body change. The heart has the capacity to accommodate and pump out (within physical limits) whatever load of blood reaches it from the body. This is one way of stating *Starling's law*. The heart constantly adapts to the rate of blood flow coming to it, so that no damming, or backing up, of blood occurs. The nature of cardiac muscle is such that the more it is stretched, the greater its strength of contraction. Because of this fact, when blood flow to the heart increases and stretches the ventricles, their ability to pump the increased load also increases.

Both the heart as a whole and individual cardiac muscle fibers possess an innate ability to beat on their own. This property is called *automaticity*. Cardiac muscle is unique in that it is capable of conducting an appreciable electrical impulse. Internal signals that arise within the heart spread readily from muscle fiber to muscle fiber. In mammals, the beat begins in the sinoatrial node located in the right atrium. This node of cardiac muscle is probably just an extremely sensitive region in which internal stimuli first lead to an electrical discharge. Since the pace of the heartbeat is controlled by the electrical changes associated with the SA node, it has been called the *pacemaker*.

A wavelike spread of the impulse to all parts of the two atria occurs, and these chambers undergo systole. A slowing down then occurs, as the impulse travels into the *atrioventricular* (AV) *node*, another specialized node. The spread of impulses along the fibers of the AV node imposes a delay of over 0.1 s and thus assures that the atria have finished their systole before ventricular contraction begins. The impulse at the AV node then fans out through a bundle of fibers called the *AV bundle*, or the *bundle of His*. These fibers, called *Purkinje fibers*, leave the AV bundle and carry the impulse rapidly throughout the ventricles.

The entire heart generally beats in a coordinated manner, and blood moves through this pump in an orderly fashion. However, if damage to the heart occurs through oxygen deprivation or mechanical injury, the individual fibers may begin beating in a chaotic, uncoordinated manner. This is called *fibrillation*. If not quickly corrected, as by application of an external electric shock, it will lead to death.

**EXAMPLE 3**   An *electrocardiogram* (EKG) is a tracing of electrical changes in the heart on chart paper affixed to a moving drum or on an oscilloscope. The device for obtaining such a record is the *electrocardiograph*. Changes in the electrical potential of the heart are charted as waves with respect to time. The amplitude of these waves as well as their duration may indicate possible pathologies. Since these electrical changes may be obtained in a noninvasive fashion by merely affixing electrodes to various parts of the body, it is a procedure that is carried out rather easily.

Each beat of the heart is characterized by five separable wave regions on the EKG (see Fig. 17.3). These are designated as P, Q. R, S, and T. The P wave, a moderate "blip" in the cycle, is associated with depolarization (loss of resting potential) of the atria before they actually contract. The sharp changes of the QRS region are associated with the depolarization of the ventricles just before they contract. The repolarization of the ventricles is marked by the modest wave of the T region. Atrial repolarization is obscured by the QRS wave.

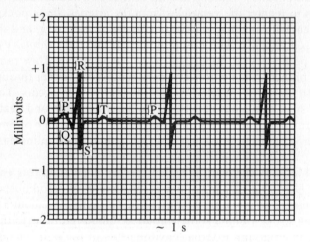

**Fig. 17.3** The electrocardiogram (EKG)

Although demonstrating automaticity, the beat may be modified by influences from the autonomic nervous system. A branch of the sympathetic nervous system, the *cardioaccelerator nerve*, releases norepinephrine in the region of the right atrium where the heartbeat begins and tends to speed up the heart rate. A branch of the vagus nerve, liberating acetylcholine at its ending near the pacemaker, acts to slow the rate. Strong vagal stimulation may actually stop the heart completely for short periods. Overwhelming fright may even stop the heart permanently as the result of a hypervagal response.

## 17.3   ARTERIES, VEINS, AND CAPILLARIES

Arteries are thick-walled vessels that carry blood away from the heart. In all arteries except the pulmonary artery the blood is oxygenated. The innermost layer of the artery is the *intima*, or *endothelium*. This lining consists of an elastic membrane to which a single layer of flattened epithelial cells is attached (see Chap. 4). The endothelium is extremely smooth and offers minimal resistance to the flow of blood. Such smoothness is also essential to avoid the initiation of the clotting process.

The artery's middle layer, or *tunica media*, is the thickest layer. It contains *smooth* (involuntary) muscle fibers, most of which are circular. Yellow elastic fibers are also prominent. The functional contractions of the artery are carried out by this layer.

A very tough, essentially inelastic, third layer makes up the outer region of the artery, the *tunica externa*. It contains mainly white fibrous connective tissue.

Veins are similar to arteries in their tripartite structure; however, their walls are much thinner and will collapse when empty. Although they lack the elasticity of the arteries, they are readily dilated by blood moving through them.

Veins lack the pumping pressure of the heart to keep the blood flowing; instead they rely on a series of one-way valves working in concert with the squeezing pressure exerted by the routine activity of nearby skeletal muscles. The pressures from surrounding muscles cause the blood to move, and the valves ensure that it moves in only one direction, toward the heart.

**EXAMPLE 4**   Movement of blood against gravity is a particular problem in the long veins of the legs. These veins are milked in a regular fashion by the contracting muscles in the leg. If one is forced to stand for long periods, relief can be provided by frequent exercise of the legs, which tends to squeeze the veins and force the blood back to the heart.

Also, negative pressures tend to develop in the great veins leading into the right atrium. Such "suctions" may play a role in bringing blood into that chamber.

The significant exchanges between cells and the circulatory system occur in the *capillary beds*, networks of small tubules that stand between the *afferent* (entering) arteriole and the *efferent* (leaving) venule. Most of the capillaries are composed of a single layer of cells, similar to the endothelial lining of arteries or veins. The cross-sectional area of all the combined capillaries of a given bed is much greater than that of the entering arteriole or the exiting venule. Hence, the blood moves very slowly and laboriously through the capillaries. However, each individual capillary is quite narrow, often less than 0.01 mm in diameter.

**EXAMPLE 5**   In a continuously flowing river the speed at any point is determined by the cross-sectional area at that point. Where the riverbed widens, the rate of flow diminishes. Where the riverbed narrows, the speed increases. Often, the narrows in a river or stream are characterized by swift and dangerous currents. The rate of flow then is inversely proportional to the cross-sectional area of the flowing column. The same relationship holds for the column of blood flowing through the vessels.

Although the total potential cross-sectional area of the capillaries is great, at any one time most of the capillaries are closed off. Only 5 percent of the total volume of blood is found in capillary beds. Less than 20 percent is contained in the arteries, while more than 70 percent is found in the large veins. Veins thus act as volume reservoirs for the circulatory system.

## 17.4  CONTROL OF BLOOD PRESSURE

*Blood pressure* refers to the force exerted on the blood vessels by the blood contained within them. It is expressed as force per unit area of vessel. Since this pressure is actually measured by a device using a column of mercury it is expressed in terms of the height of a column of mercury sustained by the pressure of the blood.

The blood pressure (usually measured as arterial pressure) is produced by two primary events. The first is the force of the heartbeat imposed on the blood leaving the ventricle; the second is the peripheral resistance (back pressure) to that force, imposed by the arteries and, more significantly, the arterioles. It is readily apparent that if there were no pump, there could be no force and, hence, there would be a zero blood pressure. Less obvious perhaps is that if there were no peripheral resistance, there also would be no pressure. The peripheral resistance is produced mainly by the constriction of arterioles in the outer portion of the circulatory system.

**EXAMPLE 6**   When water moves through a garden hose, the pressure within the hose is determined by the head of pressure at the source and the resistance along the hose and at its end. If the nozzle of the hose is constricted, the hydrostatic pressure will rise. Similarly, if the hose itself is sharply bent, the resistance to flow increases sharply and the pressure will go way up.

Blood flow and blood pressure are not uniform, because the left ventricle of the heart, the source of the systemic pump, alternately contracts (systole) and relaxes (diastole). During ventricular systole, arterial blood pressure is at its highest, and during diastole at its lowest. Clinically this is expressed as two pressures, e.g., 120/80. The *120* is an average systolic pressure for a young adult, and the *80* a corresponding diastolic pressure.

If the arteries were rigid, the blood pressure would fall to zero during ventricular diastole and move rapidly to extremely high values in the following systole. Such sharp alternating shifts of blood pressure would probably be injurious to the integrity of the arterial wall and would impose mechanical strains elsewhere in the circulatory system. Fortunately, the elastic arteries stretch during systole, which acts to moderate the rise in pressure. More important, the arteries, which were stretched during ventricular systole, will return to their unstretched state during ventricular diastole. This maintains the pressure and smooth flow within the vessels. A minimal degree of continuous artereolar resistance is necessary to maintain this pressure range stemming from arterial stretching and recoil.

During the passage of blood from the ventricle into the aorta and on along the entire arterial system, a wave of dilation (expansion) sweeps along in a linear fashion followed immediately by a wave of contraction. These alternate contractions and dilations may be felt as the *pulse* if a finger is placed above an artery that lies close to the skin. The pulse is actually a measure of the heartbeat.

The contraction of the arterioles is a significant factor in the regulation of blood pressure. Recall that increasing the cardiac output, through either an increase in stroke volume or an accelerated rate, also augments the pressure. With few exceptions, increased sympathetic stimulation constricts the arterioles at the entrance to most capillary beds and reduces the flow of blood through these beds. This produces an increased resistance to flow and hence results in an augmented blood pressure. The direct application of the neurohumoral substances norepinephrine (noradrenaline) or epinephrine (adrenalin) produces similar results. In *essential hypertension*, a generalized high blood pressure that is not caused by renal disease or other specific pathologies, a continual constriction of the arterioles probably occurs owing to neurohumoral imbalance brought on by physical factors or psychic stress. Quite often, treatment with antihypertensive drugs or diuretics (agents that promote loss of water) proves successful in lowering the pressure and markedly reducing the heightened risk of the rupture of a blood vessel in the brain (stroke; also called *cerebral vascular accident*) or heart attack.

Hypertension is associated with a systolic above 150 or a diastolic above 95 mmHg. An elevated diastolic, particularly if it exceeds 100, is usually regarded as the more serious threat to health. This may be owing to the longer duration of the diastolic in the cardiac cycle. The *pulse pressure* is defined as the difference between the systolic and diastolic pressures. A high pulse pressure may be a high-risk factor for stroke especially in situations in which the arteries have undergone sclerosis.

The constriction or dilation of specific arterioles may also serve to divert blood to or from a particular area. The action of parasympathetic stimulation on the arterioles of the penis is to dilate them, engorging the penis with blood and causing it to become erect. The production of $CO_2$ locally during muscle contraction actively produces a vasodilation in the region and thus ensures that it receives an adequate supply of materials.

The various vascular components in different parts of the body that play a role in maintaining blood pressure are coordinated through a *vasomotor center* in the medulla oblongata. You will recall that Claude Bernard discovered the *vasomotor nerves*, which innervate the muscles of small arteries and arterioles to produce vasoconstriction or vasodilation. A *cardioacceleratory center* and a *cardioinhibitory center* also lie within the medulla and affect blood pressure through their influences on the heart.

## 17.5  CONSTITUENTS OF BLOOD AND LYMPH

*Blood* is a fluid made up of liquid plasma (55 percent), the primary component of which is water, and floating cells (45 percent). The plasma is richly endowed with dissolved proteins, lipids, and carbohydrates. *Lymph* is very similar to plasma but a little less concentrated. The blood, lymph, and fluids bathing the tissues constitute one-fifth of the total body weight; the blood alone comprises one-twelfth of the body weight and typically is just under five liters in volume.

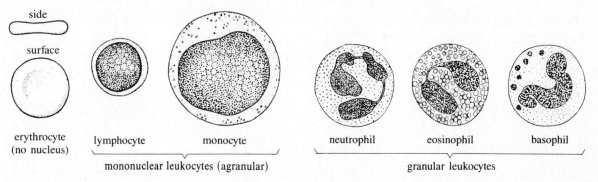

**Fig. 17.4** Cells of blood

The cellular constituents of the blood, usually described as the *formed elements*, consist of three types (see Fig. 17.4):

1.  *Red blood cells*, or *erythrocytes*
2.  *White blood cells*, or *leukocytes*
3.  *Platelets* (*thrombocytes*)

The *viscosity* (resistance to flow) of blood is quite appreciable, 4.5 times greater than that of distilled water. This high viscosity, owing largely to proteins, blood cells, and a variety of macromolecules, contributes to the slow flow of blood through the capillary beds.

The pH of blood is maintained within a narrow range of 7.3 to 7.5 in healthy individuals. At less than 7.3, *acidosis* exists, and life is severely threatened. Starvation and advanced diabetes often produce this condition. At a pH much above 7.5, *alkalosis* ensues. The excessive loss of stomach acid through serious bouts of vomiting may produce this condition, which is also a serious threat to life. The maintenance of a constant pH is dependent on the action of buffer systems present in the plasma and in the red blood cells—the carbonic acid/bicarbonate ion system, the $H_2PO_4^-/HPO_4^{2-}$ system, and a variety of proteins that act as zwitterions (see Chap. 3).

Red blood cells (rbc's) carry oxygen. In mammals, they are relatively small and, when mature, lack a nucleus and other organelles such as mitochondria. On average there are a little less than 5 million rbc's per $mm^3$ in women and about 5.5 million per $mm^3$ in men. The bulk of the rbc is taken up by the conjugated protein *hemoglobin*, a complex molecule containing a globin protein and a *porphyrin* called *heme* (see Fig. 17.5).

**EXAMPLE 7**  Hemoglobin is the chief transport protein involved in carrying oxygen within the rbc. It is a conjugated protein whose prosthetic group (heme) is quite similar in a variety of groups but whose globular apoenzyme (see Chap. 3) has fluctuated considerably in the course of evolution. Recall that the protein is made up of two alpha chains and two beta chains. Each of these chains sequesters a heme molecule that is capable of union with an oxygen molecule. A variety of hemoglobin types are especially evident in mammals. In some cases, specific chains are different for the fetus than for the adult. Many more mutations have been identified for the beta chains of hemoglobin than for the alpha chains.

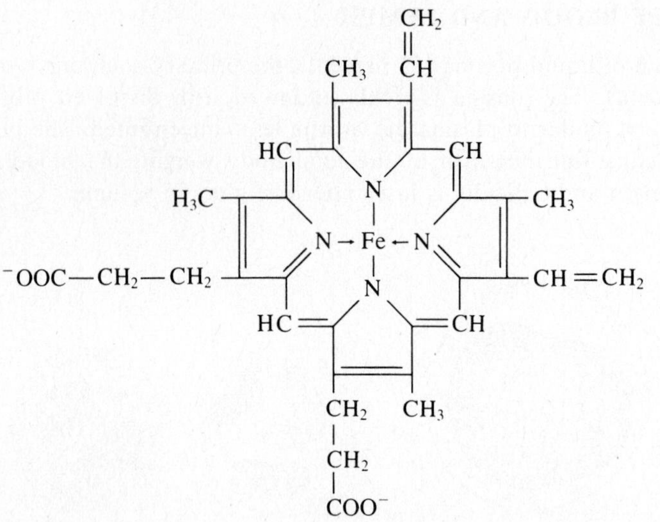

**Fig. 17.5** Heme

A renal hormone, *erythropoietin*, appears to regulate the production of rbc's. In the adult the *bone marrow* (spongy internal tissue) of the long bones is the chief source of new erythrocytes; in the fetus the liver produces them.

White blood cells are primarily involved in guarding against invading organisms that are often associated with disease. They are subdivided in one system of nomenclature into two large subgroups (see Fig. 17.4): the *granulocytes*, which contain a granular cytoplasm, and the *agranulocytes*, which contain a patently smooth cytoplasm. Among the granulocytes are:

- *Neutrophils*, whose cytoplasm stains with neutral dyes and whose nuclei are generally lobed
- *Eosinophils*, leukocytes that stain with acid dyes and are often linked to allergic responses
- *Basophils*, leukocytes with irregular cytoplasmic clumps that stain blue with alkaline stains

The agranulocytes exist as *monocytes* and *lymphocytes*. Monocytes are considerably larger than lymphocytes but make up less than 10 percent of the leukocyte population. Almost three-fourths of the circulating leukocytes consist of neutrophils. About one-fifth consist of lymphocytes. Leucocytes exist in much smaller numbers than erythrocytes, generally varying from 5,000 to 10,000 per $mm^3$ of blood. Much higher leukocyte counts are encountered during the presence of infectious diseases and inflammation and are clinical clues in their diagnosis.

*Platelets* are rounded bodies that are actually fragments of larger cells produced in the red bone marrow. They are also called *thrombocytes*. They function chiefly as initiating factors in the clotting of blood, but they are also known to play a role in immunological reactions. They are capable of limited phagocytic behavior as well.

The *clotting*, or *coagulation*, of blood involves the formation of a solid mesh of fibrils from soluble factors contained in the plasma. Clotting is probably the most important of the several mechanisms involved in stopping *hemorrhage* (bleeding). The constriction of vessels and the release of neurohumoral agents are also involved in stanching (dampening) the bleeding.

The formation of a clot is a complex procedure involving at least 30 different substances. Not all the steps are known with complete certainty, but the basic process is not difficult to understand.

The clot proper consists of a network of fibrin and blood cells. The fibrin is formed from the polymerization of fibrinogen monomers. The conversion of fibrinogen, in turn, is caused by a small protein called *thrombin*. Thrombin is normally absent from the bloodstream except during clot formation. It is produced from a precursor called *prothrombin*. The conversion of prothrombin to thrombin is one of the more complex and highly critical phases of the clotting process. It requires, in addition to a variety of ions and cofactors, the presence of *thromboplastin*, which is derived from platelets and injured tissue. The clotting process is initiated by trauma (injury) to blood-vessel walls, which produces an activated thromboplastin.

Specific clotting factors are associated with particular kinds of hemophilia, diseases in which clotting does not readily occur. The absence of factor IX is a cause of hemophilia B. Factor IX is also known at *PTC*, or *Christmas factor*. The much more common hemophilia A is associated with the absence of *antihemophilic globulin (AHG)*, or factor VIII.

As part of the homeostatic system relating to clotting, the blood contains a clot-dissolving enzyme called *plasmin*, or *fibrinolysin*. It circulates as a precursor *plasminogen*. Menstrual blood is actually blood that was clotted but then unclotted by an activated plasmin. This explains why the menstrual flow continues for several days without clot formation.

*Lymph* is a yellowish fluid with an osmotic potential considerably less than that of plasma but otherwise quite similar to it. It is carried in lymph capillary vessels arising within the tissues of the body and, from there, in the lymphatic vessels, which resemble a system of veins. The lymph returns to the general blood circulation through the large thoracic duct that empties into the subclavian vein. Lymph fluid is moved largely through contraction of muscles, which exert a massaging effect. Valves in the lymphatic vessels promote a one-way flow. Knots of glandular tissue known as *lymph nodes* are present in the lymphatic system. They are associated with the immune defense system of the body.

## 17.6  OSMOREGULATION

The maintenance of a relatively high osmotic pressure within the circulatory system is a prerequisite for physiological stability. In *shock*, which constitutes a serious threat to life because of a marked fall in blood pressure, plasma leaves the blood vessels, and a sluggish servicing of all tissues follows from the reduced volume and increased concentration of blood. A loss of plasma protein may be a causative factor.

Blood that passes into the capillary bed from the arteriole is under a hydrostatic pressure of more than 100 mmHg. This tends to force fluid and dissolved materials out of the capillary bed into the surrounding tissue. However, by the time the blood reaches the venule end of the capillary, the pressure within the vessel has dropped considerably, and the *osmotic* pressure of the blood is now greater than the *hydrostatic* pressure within the vessel. As a result, the fluid and its various molecular constituents are brought back into the vessel, and no appreciable net change in blood volume or concentration occurs. This balance may be altered by any condition that affects either the blood pressure or the osmotic pressure of blood. A shortage of proteins within the blood, largely albumin, would inhibit the reabsorption of fluid at the venule end of a capillary bed. The extensive failure to reabsorb fluid will produce a generalized swelling known as *edema*. In both cardiac insufficiency and kidney disease, edema is a common symptom. It should be noted that fluid tending to pool within tissues may also be brought back into the bloodstream by the vessels of the lymphatic system.

# Solved Problems

**17.1**  Although all vertebrates have closed circulatory systems, the circulation of blood in the earlier vertebrates is not as efficient as that in humans. Explain.

The oldest vertebrates are the fish, and they have a correspondingly primitive circulatory system. Like higher vertebrates, fish have a pumping mechanism to carry deoxygenated blood to a source of oxygenation (gills in the case of fish), but unlike higher vertebrates they have no mechanism for pumping freshly oxygenated blood to the rest of the body; that is, there is no heart distal to the gills. Although, in amphibians, blood pumped to the lungs is returned to the heart for pumping to the rest of the body, amphibians have only one ventricle, so that oxygenated and deoxygenated blood are to some extent mixed. This problem is solved in humans and other mammals by the presence of separate ventricles for the two types of blood.

**17.2**    If the average adult heart at rest beats 72 times per minute, what is the average stroke volume?

In the text we learned that cardiac output is approximately 5 liters/min. Since

$$\text{Cardiac output} = \text{stroke volume} \times \text{heart rate}$$
$$\text{Stroke volume} = 5 \text{ liters/m} \div 72 \text{ beats per minute}$$
$$= 5000 \text{ ml/m} \div 72 \text{ beats per minute}$$
$$= 69 \text{ ml per beat}$$

**17.3**    The human fetal heart contains two structures not normally found in the adult heart. One, the *ductus arteriosus*, is a short tube that connects the pulmonary artery to the aorta. The second, the *foramen ovale*, is an opening between the right and left atria. Since both structures disappear around the time of birth, what do you suppose their purpose is?

In the fetus oxygen is supplied by the mother; the lungs are clearly not functional. The ductus arteriosus and foramen ovale serve to shunt blood into the systemic system and away from the pulmonary circulatory system, which, in the absence of functioning lungs, the fetus does not need.

**17.4**    In the preceding problem, the foramen ovale was shown as a mechanism for shunting blood from the right atrium to the left atrium and thus bypassing the pulmonary circulatory system. Why do you suppose blood doesn't flow from the left to the right atrium instead?

Two factors prevent left-to-right flow in the atria. The first is a thin piece of tissue across the foramen, which serves as a weak valve to stop the flow into the right atrium. The second is that the pressure in the right atrium is greater than that in the left, so that blood is driven toward the left atrium. This difference in pressure exists because the left ventricle is pumping blood into a functioning system but the right ventricle is pumping against the uninflated lungs and alveoli (tiny sacs that fill with air in the functioning lungs); thus a tremendous back pressure is created and transferred to the right ventricle and, in turn, to the right atrium.

**17.5**    When an impulse is initiated in the SA node, it is spread by the muscle fibers of the atrial wall, radiating in a contractile wave throughout the atria. After a delay in the AV node, the impulse enters the bundle of His, from which it rapidly spreads throughout the ventricles along the Purkinje fibers. The movement of the impulse along the Purkinje fibers is roughly six times as fast as along normal cardiac muscle. How would you explain these three differences in impulse movement (in the atria, AV node, and ventricles) in terms of heart function?

Because of the relatively slow spread of the impulse from the SA node along the muscle fibers of the atrial wall, neither atrium contracts all at once, but rather in a wave spreading toward the atrioventricular septum and valves. This contraction in either of the atria is consequently not so powerful as if the atrium contracted all at once; however, maximum contractile strength is unnecessary, since the atria need push the blood only into the ventricle, usually assisted by gravity. The delay at the AV node is necessary to ensure that the atria have had time to contract completely, and thus empty themselves of their blood, before the ventricles contract. Because of the speed of conduction in the Purkinje fibers, the contractile impulse reaches almost all parts of each ventricle at the same time; therefore, the ventricle contracts essentially as a unit. This provides a great deal more thrust than the wavelike contraction of the atria. Because the ventricles are pushing blood either to the lungs or throughout the entire body, this greater force is necessary.

**17.6** Damage to the heart valves usually leads to abnormal heart sounds called *murmurs*. Either of two types of damage typically occurs: (1) several of the cusps may partially adhere to each other. This restricts their ability to open and causes blood to be ejected through their narrowed openings in high-pressure jets; this narrowing process is called *stenosis*. (2) The edges of the cusps may become scarred. This restricts their ability to close tightly, so that blood is able to leak back through them; this leakage is called *regurgitation*. Because the highest pressures are generated in the left ventricle, the aortic and mitral valves are most subject to damage and, therefore, to murmurs. For each of these two valves, which murmur (owing to stenosis or regurgitation) would you expect to hear during ventricular diastole? During systole?

During ventricular diastole, the ventricle fills with blood and produces a negative pressure. The mitral valve opens to allow the blood to flow from the atrium into the ventricle, and the aortic valve prevents backflow from the aorta. Consequently, mitral stenosis (which would restrict the flow from the atrium) or aortic regurgitation (the backflow of blood) would cause murmurs. During systole, the mitral valve blocks the blood flow into the atrium, and the aortic valve opens to permit the flow. Consequently, during systole, mitral regurgitation or aortic stenosis would cause murmurs.

**17.7** If you press a finger down on a prominent vein, say, on the back of your hand and then slide the finger distally to a new pressure point closer to the fingers, would you expect the section of vein you just moved along to refill with blood? Suppose you had moved the finger proximally toward the upper arm?

In the first case, blood would have to flow backward in the vein to refill the section you emptied; however, the valves in the vein prevent backflow, and the vein should remain empty or refill only slowly. In the second case, the emptied section of vein would be quickly refilled by blood traveling toward the heart.

**17.8** Bernoulli's principle states that the pressure exerted by a stream of air or by fluid within a container is inversely proportional to the velocity of the stream. It is this principle that explains why an airplane can fly. Because the wing is curved on top, as the wing passes through the atmosphere, the air flowing over the top moves a greater distance than that flowing under the wing, so the air on top flows faster. The resulting lower pressure (Bernoulli's principle) on the top of the wing relative to the bottom creates lift. Bernoulli's principle also operates in certain vein pathologies, such as varicose veins. From what you know of vein structure and blood flow, explain this relationship.

Varicose veins are commonly found in the legs of people who must stand for long periods of time. The blood in the veins must rely on passive means of movement, such as that provided by the surrounding muscle. During long periods of standing, when the blood must move against gravity, movement is very slow, and pressure against the walls of the veins correspondingly high, as Bernoulli's principle would predict. Because of the distention of the walls, the venous valves are no longer long enough to come together to close, so that backward leakage of blood occurs, further slowing the blood flow and stretching the walls. The pressure and stretching eventually damage the valves and walls of the vein and exacerbate the problem. Elastic support hose may help counter the venous distention; in serious cases, surgical removal of the vein may be necessary.

**17.9** A *sphygmomanometer* is used to measure blood pressure. This apparatus consists of an inflatable cuff that has a tube capped by an inflating pressure bulb and a second tube connected to a mercury *manometer* (apparatus for measuring pressure). The cuff is wrapped around a subject's upper arm and then inflated to a pressure above the usual upper limit for human systolic pressure. A stethoscope is placed over the antecubital artery, just below the level of the cuff. The pressure is slowly released. At first there is no sound; then, at a particular pressure a pounding is heard in synchrony with the heartbeat. At a lower limit of pressure the pounding disappears. The pressure at which the sound is first heard is the systolic pressure; the pressure at which the sound disappears is the diastolic pressure. Why do you think these two points are considered measures of the systolic and diastolic pressures?

The pressure in the cuff at its upper extreme is sufficient to close completely the antecubital artery. When the cuff pressure falls just below the pressure exerted by the heart on the artery (systolic pressure), the heart will be able to overcome the cuff's constriction and force a small jet of blood through the artery. This first spurt is believed to be the pounding sound heard through the stethoscope. So long as there is enough cuff pressure to cause even a slight constriction of the artery, the resulting turbulence as the blood passes this point will create the pounding sound. As the cuff pressure is released, a point is reached at which the pressure in the artery at rest (ventricular diastole) is enough to prevent even slight constriction; there will thus be no turbulence in the artery, and the pounding sound will disappear. This pressure, which is just enough to maintain the natural roundness of the artery, is considered the diastolic pressure.

**17.10** One of the prime functions of blood is to carry oxygen to the tissues. However, water or even plasma takes up comparatively little oxygen. How is this problem solved by living organisms?

Special pigments exist that readily combine with oxygen and then release it at suitable tissue sites. *Hemoglobin* (Hb) is the best known of these pigments. Whole blood in humans is almost 70 times more efficient than plasma in carrying oxygen, which suggests that the combining power of hemoglobin for $O_2$ is very high. In mammals and other vertebrates hemoglobin is contained within the erythrocytes. Among many annelid worms and mollusks it is a constituent of the plasma.

Other pigments are found in the animal kingdom. *Chlorocruorin*, a green pigment found in the plasma of some annelid worms, utilizes iron in its prosthetic group. *Hemoerythrin* also contains iron but is associated with the corpuscles in the blood of annelids. *Hemocyanin* is a blue, copper-containing pigment widely distributed in the plasma of many mollusks.

When combined with oxygen, hemoglobin is known as *oxyhemoglobin*. It is a brighter red than the almost purplish hemoglobin. The oxygen attachment to respiratory pigments is looser in the vertebrates than in invertebrates. This has led to speculation that a major function of invertebrate pigments is to store oxygen over long periods of time to guard against highly anaerobic environments.

**17.11** Both an embolism and a thrombus are abnormal clots formed within the circulatory system. A *thrombus* arises in a particular area and stays there. It does its damage by occluding an artery and the capillary beds branching from that artery. If a thrombus arises in the coronary artery and blocks one of its major branches, a type of heart attack called a *coronary thrombosis* ensues.

An *embolism* is a clot that forms in one region and then travels through the bloodstream to block a vessel at some distant site. Quite often enforced inactivity following surgery or serious illness promotes clot formation, especially in the veins of the legs. These clots may break off and lead to an embolism. Why is there no danger that such clots will lodge in the brain and cause cerebral embolism? What sort of embolism *is* posed by such clots?

All venous blood must first return to the heart before it can reenter the systemic system. Any clot formed in the leg would first be carried to the right side of the heart. Since all such blood entering the heart must pass through the lungs, any clot carried by venous return (from the legs, or any other part of the body for that matter) would be filtered out by the pulmonary arteries or capillaries and could not continue on to lodge in the brain. Although a clot from the legs could not lodge in the brain, it could easily lodge in the lungs, causing a *pulmonary embolism*, a potentially fatal event.

**17.12** When whole blood is subjected to increasing partial pressures of oxygen, it combines with that oxygen to form oxyhemoglobin. The combination of oxygen with hemoglobin may be expressed as *volumes percent* (vol %; milliliters of oxygen per 100 milliliters of blood) or as percent saturation of hemoglobin with oxygen. In Fig. 17.6 the union of oxygen and hemoglobin is shown for three different physiological conditions (see Prob. 17.13). It can be seen that as the partial pressure of oxygen rises, the amount of oxygen in the blood (expressed in volumes percent) rises sharply over a narrow range. The overall pattern is a sigmoid curve. This rapid

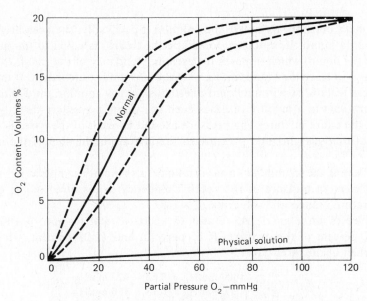

**Fig. 17.6** Oxyhemoglobin dissociation curve

rise is because of the formation of oxyhemoglobin. The straight line near the bottom of the graph shows the much more limited uptake of oxygen by a solution when hemoglobin is *not* present. Given that the uptake of oxygen shown in the curves is due almost entirely to the formation of oxyhemoglobin, at what volume percent is hemoglobin essentially 100 percent saturated?

Figure 17.6 shows that the rapid formation of oxyhemoglobin begins to slow (the curves begin to flatten) at a partial pressure of oxygen of roughly 60 mmHg. The curves do rise slightly beyond this point, but clearly they become very nearly flat at about 20 vol %. This then is roughly the point at which hemoglobin is 100 percent saturated with oxygen. By inspection, hemoglobin is seen to be roughly 50 percent saturated at 10 vol %.

**17.13** The two dotted curves on either side of the solid curve in Fig. 17.6 show the formation of oxyhemoglobin under two different levels of carbon dioxide—one curve shows oxygen uptake under conditions of high carbon dioxide (as found in the veins) and the other, under conditions of low carbon dioxide (as found in the arteries). Differentiate the two curves. [Hint: It is as important that hemoglobin be able to give up oxygen (for example, in oxygen-starved, carbon dioxide–saturated muscle) as it is that it be able to combine with it.]

It is of obvious advantage for hemoglobin to have a high affinity for oxygen when in the lungs and arteries, since at these sites it must pick up and transport oxygen. However, when saturated hemoglobin reaches oxygen-deprived areas of the body (areas that have correspondingly high levels of carbon dioxide), it is of advantage that the hemoglobin have a lower affinity for oxygen so that it will more readily release it to the deprived tissues. Another way of saying that hemoglobin has a lower affinity for oxygen is to state that at a given partial pressure of oxygen it is less saturated. Looking at Fig. 17.6, we can see that, for example, at 20 mmHg, the curve on the right shows only 4 vol %, whereas the left curve shows 10 vol %. The curve on the right thus shows hemoglobin when it has a reduced affinity for oxygen (as would be seen in muscles and veins); this curve therefore, is seen during the conditions of high carbon dioxide found in the muscle and veins. The curve on the left indicates a higher affinity for oxygen and thus represents hemoglobin in the lungs and arteries. The shift of the curve to the right under conditions of high carbon dioxide (and low pH) is called the *Bohr effect*.

**17.14** Two laws, governing the uptake of gases by liquids, enable us to understand the transport of gases by the blood. *Henry's law* deals with the equilibrium between gases in solution and those in the atmosphere immediately above the solution. Specifically, the pressure, or tension, of a gas dissolved in a liquid (its tendency to escape the liquid) is equal to the partial pressure of the gas above the liquid, which opposes the escaping tendency of the gas. If the pressure above is reduced, gas will leave the liquid until a new equilibrium is established. If the pressure above is increased, gas will dissolve in the liquid to establish a new equilibrium. *Dalton's law of partial pressures* asserts that in a mixture of gases each gas exerts a pressure directly proportional to its amount in the entire mixture. The pressure exerted by each of the gases is called the *partial pressure* ($P$) of that gas; the total pressure of the mixture is the sum of the individual partial pressures.

Factors affecting the solubility of a gas in a liquid also include temperature and the individual *solubility coefficient* (a measure of the specific tendency to dissolve) of the gas. Generally, a higher temperature reduces the solubility of a gas.

If a mixture of three gases, *A*, *B*, and *C*, exerts a total tension of 1000 mmHg and *A* constitutes 62 percent of the mixture, *B* 15 percent, and *C* 23 percent, what are the partial pressures of each of the three gases?

$$P_A = 1000 \text{ mmHg} \times 0.62 = 620 \text{ mmHg}$$
$$P_B = 1000 \text{ mmHg} \times 0.15 = 150 \text{ mmHg}$$
$$P_C = 1000 \text{ mmHg} \times 0.23 = 230 \text{ mmHg}$$

**17.15** Construct a table showing the various cellular constituents of blood.

| Erythrocytes | Leukocytes | | Thrombocytes |
|---|---|---|---|
| | Granulocytes | Agranulocytes | |
| | Neutrophils | Monocytes | |
| | Eosinophils | Lymphocytes | |
| | Basophils | | |

**17.16** Summarize the major chemical reactions involved in the formation of a blood clot.

$$\text{Prothrombin} \xrightarrow[\text{Ions, cofactors}]{\text{Thromboplastin}} \text{thrombin}$$

$$\text{Fibrinogen} \xrightarrow{\text{Thrombin}} \text{fibrin}$$

# Supplementary Problems

**17.17** The major constituent of plasma is  (*a*) protein.  (*b*) NaCl.  (*c*) water.  (*d*) cholesterol.  (*e*) none of these.

**17.18** The site of the thickest musculature in the heart is the  (*a*) right atrium.  (*b*) left ventricle.  (*c*) aorta.  (*d*) right ventricle.  (*e*) none of these.

**17.19** A secondary site in amphibians for the oxygenation of blood (in addition to the lungs) is the  (*a*) skin.  (*b*) brain.  (*c*) liver.  (*d*) kidneys.  (*e*) spleen.

**17.20** Venous blood coming from the head area in humans returns to the heart through which major vessel?  (*a*) aorta,  (*b*) superior vena cava,  (*c*) hepatic portal vein,  (*d*) carotid artery,  (*e*) subclavian vein

**17.21**  The ventricles spend a longer time in diastole than in systole.
(*a*) True   (*b*) False

**17.22**  Vagal stimulation of the heart would tend to slow the rate.
(*a*) True   (*b*) False

**17.23**  Unlike capillaries, arteries and veins lack an endothelium.
(*a*) True   (*b*) False

**17.24**  In *atherosclerosis* (arteriosclerosis) the deposition of lipids such as cholesterol, followed by precipitation of $Ca^{2+}$ salts, hastens the hardening of a vessel.
(*a*) True   (*b*) False

**17.25**  Valves in the cardiovascular system are found only in the heart.
(*a*) True   (*b*) False

**17.26**  Cardiac output would increase through either an increase in heart rate or a rise in stroke volume.
(*a*) True   (*b*) False

**17.27**  The lymphatic system can best be described as an auxiliary system of veins.
(*a*) True   (*b*) False

**17.28**  The most significant part of the cardiovascular system is the capillary bed.
(*a*) True   (*b*) False

**17.29**  Obesity is often associated with hypertension. Loss of weight may ameliorate the hypertension.
(*a*) True   (*b*) False

**17.30**  Smaller animals generally have a lower pulse rate than larger animals.
(*a*) True   (*b*) False

# Answers

| | | | | | |
|---|---|---|---|---|---|
| **17.17** | (*c*) | **17.22** | (*a*) | **17.27** | (*a*) |
| **17.18** | (*b*) | **17.23** | (*b*) | **17.28** | (*a*) |
| **17.19** | (*a*) | **17.24** | (*a*) | **17.29** | (*a*) |
| **17.20** | (*b*) | **17.25** | (*b*) | **17.30** | (*b*) |
| **17.21** | (*a*) | **17.26** | (*a*) | | |

# Chapter 18

# Immunology

In the broadest sense *immunology* refers to all the defense mechanisms that the body can mobilize to fight the threat of foreign invasion. The skin and its accompanying structures are formidable obstructions to the growth of and penetration by viruses and bacteria. Sweat and other secretions tend to maintain a low pH at the epidermal surface, which discourages the propagation of many *pathogens* (disease-causing organisms). *Natural flora*, consisting of indigenous bacteria that keep each other's populations in check, are present at the surface of the skin and within many of the crevices and ducts of the body; these bacteria not only keep each other's populations in check but also act as barriers to the growth of foreign microorganisms.

Where openings do occur, the internal structures contiguous with the openings are lined with a layer of protective mucus, which not only lubricates but also may entrap invaders, which are then excreted. Secretions along such tubes as the tear ducts actually destroy bacterial invaders.

Should the fortress of the skin be penetrated, a variety of internal responses may occur that are part of the repertoire of the body's defense system. Inflammation, accompanied by ingestion of foreign invaders by phagocytes, is a major nonspecific response. Viruses, in particular, elicit the formation of *interferons*, small proteins produced in a cell attacked by a virus. Interferons act to protect neighboring cells against viral invasion by attaching to them to block the formation of viral protein.

Interferons have also been implicated as possible anticancer agents, but the results of research have thus far not substantiated the efficacy of interferons in inhibiting cancer growth.

## 18.1 HUMAN IMMUNE SYSTEMS

The *immune system* consists of defenses that are highly specific in their action. The anatomical arena of this defense system includes the lymph vessels and spongy lymph nodes, the white cells of the blood and bone marrow, and the thymus gland. The immune response is mediated almost entirely by two kinds of lymphocytes: the *B lymphocyte* and the *T lymphocyte*. Both types of cell are derived from *lymphocytic cells* in the bone marrow; they then differentiate (T lymphocytes in the thymus and B lymphocytes in bone marrow) and finally take up residence in the lymphoid tissues of the body. During the immune response to foreign agents, B lymphocytes are chiefly involved in the formation of globular proteins called *antibodies*; this process constitutes the *humoral response*. In the second type of immune response, the *cell-mediated response*, T lymphocytes initiate the attack on foreign bodies by various cell types. In both types of response, the invading entity is recognized by its *antigens*. Each toxin or organism has unique chemical compounds not found in other entities; these are antigens. They usually consist of proteins, large polysaccharides, or large lipoproteins and are frequently found on the surfaces of unicellular organisms. A *specific* antibody exists in the body for almost every antigen.

### THE HUMORAL RESPONSE

Antibodies are globular proteins coded for by specific genes. They are also known as *immunoglobulins* because of their globular characteristics and their association with the immune response. They are made up of four polypeptide chains (see Fig. 18.1). Two of these chains are rather long and are known as *heavy chains*; they are usually identical. The two remaining chains, known as *light chains*, are short, and they too are identical in structure. Each of the four chains, which are bound to each other by S—S bonding, possesses a *constant* region characteristic of the class of antibody and a *variable* region, which provides the specificity necessary to match the more than 1 million kinds of antigens encountered during a person's lifetime.

Each antibody has two binding sites for antigen—the two clefts formed by the association of heavy and light chains. The opening to each of the clefts is at the $NH_2$ (head) end, which contains the variable

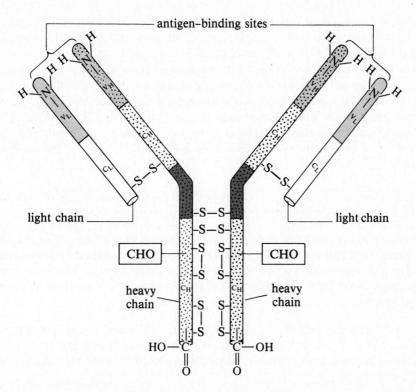

**Fig. 18.1** Structure of an antibody

sequences. These sequences produce unique conformations in the antibodies, so that each antibody has a cleft that fits a specific antigen. It is now apparent that the entire antigen need not participate in the attachment process; rather, a small portion of the antigen, the *antigenic determinant*, fits into the cleft of the antibody. The cleft, in turn, is not due merely to the primary sequence of amino acids at the variable sites, but is dependent on complex folding patterns of the light and heavy chains.

One antigen, especially if it is complex, is capable of reacting with more than one antibody, but this is not the usual situation. Also, the antigens reacting with antibodies initially to produce the proliferation of lymphocytes must be large molecules. Later, the antigenic determinant alone may serve to continue the immune response.

In humans five different antibody classes have been found. The most common group of immunoglobulins are the *gamma globulins* (IgG). *IgM* is the first group to be found when infection strikes, but this group has a high turnover rate and does not persist at high levels. Tears, saliva, and even milk contain *IgA*, a third group, whose function is to maintain a low level of bacterial growth in structures handling these secretions. *IgD* tends to associate with the surface of B lymphocytes, but its function is not yet known. *IgE* tends to promote the release of histamine by mast cells when it binds to its antigen. It has also been associated with the body's fight against parasites as well as with allergic reactions.

Antibodies are the primary weapon of the humoral response. They may attack an antigenic organism or molecule directly, or they may activate related systems, which will attack the invader.

One mode of direct attack is *agglutination*, which involves the clumping of antigens into an antigen-antibody complex. Since each antibody has two binding sites, it can hold two antigenic organisms (such as bacteria); most antigenic agents, in turn, have multiple antigenic sites, so that the agents are also bound to more than one antibody. The result is a network of interconnected antibodies and antigens. These complexes reduce the mobility of the invasive agents and make them more susceptible to *phagocytosis*, a process in which leukocytes engulf and ingest the invaders. The clumps may also become insoluble and precipitate out, again restricting mobility. Certain antibodies are able to lyse cells they attack, and others can bind the toxic sites of antigens and effectively neutralize them.

When antibodies complex with their specific antigens, they also activate the *complement system*, a group of roughly 12 enzyme precursors found in plasma and other body fluids. The activated enzymes of this system eat into the cell membranes of invading organisms and eventually cause them to rupture. The complement also makes cell surfaces more susceptible to phagocytosis and exerts a *chemotaxic* effect that draws neutrophils and macrophages to the invading organism. Viruses may be made nonvirulent through attacks on their molecular structures by the complement, and the complement reduces the mobility of organisms by inducing their agglutination. Finally, the complement elicits a local *inflammatory response*.

**EXAMPLE 1** The inflammatory response is a nonspecific response to the injury of cells and the possible penetration of the mechanical defenses of the body by invading organisms. It is mediated largely by the release of histamine from both injured epithelial cells and mast cells from interwoven connective tissue. The inflammatory response involves an increase of blood flow to the damaged area, largely through the vasodilation of small arterioles induced by the histamine. Capillary walls are distended and readily penetrated by granulocytes (particularly neutrophils; see Fig. 17.4), which gather at the injured site. Local swelling and an increase in temperature may be accompanied by systemic fever and a marked increase in granulocytes. Fever is generally attributed to the release of proteins known as *pyrogens* by bacteria.

Certain antibodies, particularly of the IgE type, attach to cell membranes, including the membranes of basophils in the blood. When an antigen reacts with such an antibody, the attached cell swells and ruptures and releases histamines, chemotaxic substances, lysosomal enzymes, and a factor that prolongs smooth muscle contractions.

## THE CELL-MEDIATED RESPONSE

In the cell-mediated response, whole cells—sensitized T lymphocytes—rather than chemicals (immunoglobulins) attack eukaryotic cells, whether they be the cells of invaders such as helminth worms or cells of the host body that are somehow different from what the lymphocytes recognize as "self—cancer cells, cells that have been invaded by viruses (and thus carry viral antigens on their surface), and transplanted organs and tissues. The reaction between the T cell and the invader is strongest when the target cell has two antigens the T cell can recognize. Since virus-infected cells have both the "self" antigen and the viral antigen for which the T cell is sensitized, they are particularly vulnerable to T-lymphocyte action.

**EXAMPLE 2** A series of antigens intimately associated with the *glycocalyx* (carbohydrate portion of the cell membrane) plays an important role in the recognition of cells as self as well as in the interaction of infected body cells with T lymphocytes in the cell-mediated immune response. A second group of these antigens is present only on the cells that make up the immune defense; this group promotes the cooperativity and intimate association that characterize the immune system. These antigens are glycoproteins and constitute the *major histocompatibility complex* (MHC). They are coded for by approximately two dozen genes, and since each of these genes exists in many allelic forms, the number of different genetic combinations producing correspondingly unique antigenic complexes is almost limitless. The MHC is extremely useful in predicting the possibility of successful organ grafts and also plays a significant role in forensic (legal) medicine as an accurate means of identifying individuals for criminal or paternity cases, from relatively small amounts of tissue.

The destruction of the invading organism is effected in several ways during the cell-mediated response. Specialized T cells called *cytotoxic T cells* attack the invaders directly. Like antibodies, the T cell is sensitized to the specific antigen it is attacking. When it finds this antigen, it releases cytotoxic and digestive enzymes, which lyse the alien cell. These T cells also release factors that attract macrophages, increase their activity, and keep them in the infected area, where the macrophages may engulf the invaders.

## 18.2　THE IMMUNE RESPONSE

At one time two theories existed to explain the tremendous variety of antibodies (more than 1 million) that could be elicited by antigens circulating within the body. The *instructional theory* explained such diversity by hypothesizing an ability on the part of antigens to fashion antibodies from white cells by intervening in the actual process of antibody formation. This view assumed that the antigen was a kind of template that could direct the activity of the immune system. Current theory, however, holds that the variability is the result of an unusual instance of DNA rearrangements. Although normally DNA consists of fairly stable, linear arrangements of triplets, the DNA coding for the variable segments of the immunoglobulins appears to be able to rearrange itself, making possible a million different immunoglobulin sequences and surface proteins and consequently a million different lymphocytes. Thus, it is not each antigen tailoring its own specific antibody that produces the great variability in lymphocytes, but rather a genetic rearrangement within the lymphocyte's own genome, which yields so many possible combinations that all antigens are covered.

When an antigen invades the body, the specific antibody for that antigen is mobilized to incapacitate it. This rapid proliferation of the antibody is explained by the *clonal selection theory* (see Fig. 18.2), first proposed by Macfarlane-Burnett, an Australian immunologist. According to this theory, a monocyte that has grown into a large macrophage ingests and digests the invading organism, but incorporates its antigens into its own plasma membrane. It then seeks out and, by secreting *interleukin 1* (IL-1), activates *helper T cells* (lymphocytes) carrying molecules on their surface that specifically match the antigen carried by the macrophage.

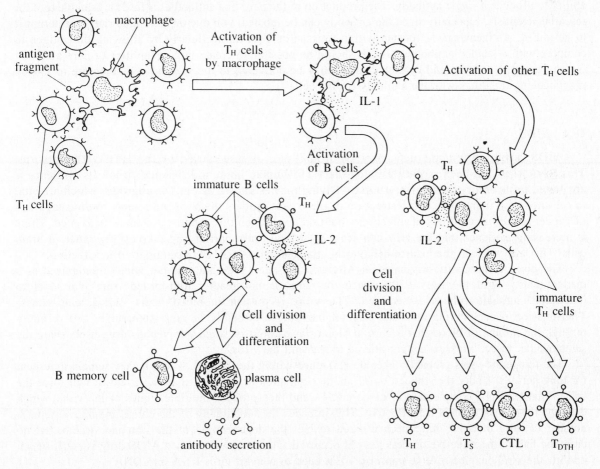

**Fig. 18.2** Clonal selection

These activated helper T cells have two roles. In one role, they seek immature B lymphocytes that contain surface antibodies identical with their own (and therefore specific for the antigen causing the response) and secrete *interleukin 2* (IL-2). This induces the "naive" lymphocytes to begin dividing and form both *plasma cells* and *memory cells*. Both these cell types divide repeatedly and flood the body with their progeny. The shorter-lived plasma cells secrete the antibodies in great quantity. The memory cells produce antibodies, but are quiescent during the initial antigen assault; they remain in the circulation after the antigen has been eradicated, ready to begin secreting antibody should the antigen reappear.

In its second role, the activated helper T cell looks for immature T cells bearing its surface recognition molecules and activates them with interleukin 2. These T cells then differentiate into memory cells, cytotoxic T cells, more helper T cells, and *suppressor T cells*. We have already considered the roles of the first three of these. The suppressor T cells act to stop the immune reaction when the invader has been subdued and leave only the memory cells, which will reactivate should the antigen reappear.

## 18.3 HYBRIDOMAS

In 1984 two European immunologists (Köhler and Milstein) received the Nobel prize in Physiology or Medicine for developing the *monoclonal antibody technique*. Essentially this involves the selection of a line of lymphocytes that produce one kind of antibody and then the fusion of these lymphocytes with cancer cells. The result is a *hybridoma*. Normal cells in culture usually die after only a few divisions, whereas cancer cells are virtually immortal. Hybridomas combine this longevity with the lymphocyte's ability to produce a single antibody. The production of the pure line antibodies affords a valuable tool for research scientists, especially since the antibody can be labeled with dye or other materials that permit it to be traced. Antigens can be located within a tissue or even intracellularly by allowing the antigen to complex with a labeled antibody. Also, it should be possible to create large quantities of pure antibody for any targeted antigen, and cytotoxic drugs may be delivered to the specific cells they are to kill by piggybacking the drugs with the antibody for the cells.

## 18.4 AIDS

AIDS (acquired immune deficiency syndrome) is a disease caused by the HTLV-III retrovirus. This RNA virus has been recently labeled the *HIV* (human immunodeficiency virus). Its virulence is attributed to its ability to attack a group of T lymphocytes known as the T4 subgroup, which are vital to cell-mediated immunity. The destruction of these cells renders the body vulnerable to unusual kinds of cancer as well as a variety of infections that take over when the immune system is weakened. There is increasing evidence that the HIV can cross the blood-brain barrier and take up residence in brain cells. This may explain the bizarre behavioral changes that sometimes accompany the infection.

The disease apparently originated in Africa in the late 1950s but was not widely recognized as a distinct entity until the late 1970s. Since the virus is not readily transmitted from individual to individual, intimate contact is required. The virus is present in blood, tears, saliva, and semen. Transmission occurs during anal and vaginal intercourse, possibly in oral intercourse, and certainly through the mixing of blood that occurs in blood transfusions or when intravenous drug users share the same needle. Insect bites are not considered to transmit the virus.

The temperate virus probably takes up residence within the T4 cell, where it may remain quiescent for long periods. Under the stimulation of another infection or similar triggering event, it will force the cell to make a DNA complement of its core RNA and then produce multiple copies of the virus, which destroys the cell or renders it ineffective. This impedes the functioning of the entire immune system. A number of drug therapies have been utilized to halt the degeneration of the immune system, but no dramatic halt to the infective process has yet occurred. The most widely used AIDS drug is AZT, which inhibits the viral enzyme reverse transcriptase needed to convert virus RNA into DNA.

## 18.5  HUMAN BLOOD GROUPS

There are a number of antigens usually present on the red blood cell that determine blood type. Around 1900 Karl Landsteiner discovered the first major class of these polysaccharide antigens, those making up the ABO blood type. They are actually only two in number, an A antigen and a B antigen (see Table 18.1). People with A antigens on their red blood cells contain B antibodies in their plasma. They are *type A*. Those with B antigens contain A antibodies in their plasma. They are classified as *type B*. Persons with both A and B antigens on the red blood cell are *type AB*, and they contain no ABO antibodies in their plasma. Type O people have no antigens on their red blood cells but carry both antibodies in their serum. Clearly, a problem arises when antigen and antibody of the same kind are present together. In the course of human evolution such contiguities were selected against, so that antigen A is never present with antibody A and antigen B is never present with antibody B.

**Table 18.1**  ABO blood types

| Genotype | Phenotype (blood type) | Antigen | Antibody |
|---|---|---|---|
| $I^A I^A$ or $I^A i$ | A | A | $\beta$ |
| $I^B I^B$ or $I^B i$ | B | B | $\alpha$ |
| $I^A I^B$ | AB | A and B | none |
| $ii$ | O | none | $\alpha$ and $\beta$ |

Problems arise during blood transfusions when red blood cells with antigen A are put into people with antibody A in their plasma (type B or type O) or when red blood cells with antigen B are put into people with antibody B in their plasma (type A or type O). Under such conditions the red cells clump and precipitate out of circulation as antigen-antibody complexes. It should be noted that it is primarily the red cells of the donor and the plasma of the recipient in a transfusion that occasion concern because the plasma of the donor is diluted so greatly by the recipient's plasma that the donor antibodies are not a significant threat to the red cells of the recipient. As Table 18.1 indicates, type O blood is a universal donor because its red cells have no antigens, while type AB is a universal recipient because its plasma has no antibodies to react with transfused red cells.

In transfusions, cross-matching tests are almost always carried out with small samples of blood taken from the prospective donor and recipient to assure that no untoward precipitation will occur.

Other antigens on the red cell may also play a role in the agglutination (clumping) of blood. An antigenic factor called *Rh* is either present (Rh$^+$) or absent (Rh$^-$). When present, it will elicit antibodies to Rh in Rh$^-$ recipients. Rh looms large as a complication of pregnancy when the father is Rh$^+$ and the mother Rh$^-$. The child of such a union will usually be Rh$^+$, since Rh is inherited as a dominant allele. When a little mixing of blood of mother and fetus occurs in the placenta across breaks in their separate circulatory systems, the mother produces anti-Rh antibodies. It is only at the second pregnancy that a sufficient amount of antibody is produced and transported across the placental barrier to cause serious damage to the Rh$^+$ child through a clumping of its red cells. Such a condition is called *erythroblastosis fetalis*.

Other antigenic factors include the MN system. These various antigenic systems provide a basis for the identification of individuals, particularly in legal situations. The MHC however has tended to supplant blood typing for this purpose.

# Solved Problems

**18.1**   Given the phagocytic nature of granulocytes, what do you suppose the granules are?

The granules are lysosomes (see Chap. 4). When the granulocyte engulfs a bacterium or other foreign particle, a vacuole is formed. Lysosomes, with their array of hydrolytic enzymes, fuse with these vacuoles and help digest the "offending" body. The proteolytic enzymes are probably most effective in destroying these particles.

**18.2**   Immature B- and T-cell lymphocytes are identical in appearance; however, after activation by helper T cells, the B cells develop huge quantities of rough endoplasmic reticulum (RER), whereas the activated T cells develop relatively little. Why do you suppose this is?

The large amount of RER is a sign of intense protein synthesis, which would be expected in the B cell, since it must produce copious amounts of immunoglobulin. The activated T cells do not produce large amounts of protein and therefore do not require extra RER.

**18.3**   Compare B lymphocytes and T lymphocytes.

B lymphocytes and T lymphocytes arise from blood-forming stem cells that are the progenitors of all blood cells (leukocytes and erythrocytes) and are found in the bone marrow of the adult. *Stem cells* refer to those relatively undifferentiated cells that give rise to specialized cell types.

The B lymphocyte develops and matures in the bone marrow. When sensitized to an appropriate antigen, it undergoes enlargement and rapid cell division to produce two kinds of cells that are no longer naive. One of these types, produced in great numbers, is the plasma cell, which secretes great numbers of its specific antibody into the bloodstream. This is known as the *humoral immune response*. The second type is the long-lived memory cell, which provides for long-term immunity. Should the particular antigen that fits the antibody of the memory cells appear in the future, a great proliferation of both plasma and memory cells will occur.

The T cell matures in the thymus under the influence of a hormone complex secreted by that gland. Two kinds of receptors, similar in function to the antibodies of the B lymphocytes, have been found on the surface of the T lymphocyte. Antigens complementary to these receptors bind to them. When activated, the T lymphocyte, like the B lymphocyte, divides to form memory cells and active T lymphocytes. The active T lymphocytes are of at least three types, and all play a role in the cell-mediated immune response in which reactions occur between the various kinds of T lymphocytes and whole cells.

**18.4**   The currently accepted explanation for the bewildering diversity of antibodies found in a species is that sections of DNA in the region of the chromosome coding for these antibodies are capable of essentially unlimited rearrangement. It is thus possible that in each lymphocyte the chromosomal region coding for an antibody is unique to that lymphocyte. What are two other possible genetic explanations (not counting the instructional theory) for the vast diversity of antibodies?

(1) Each lymphocyte may contain genes for *all* the antibodies produced by the species but have all but one repressed. (2) The chromosomal region coding for antibodies in a species may be highly subject to base-substitution mutations.

**18.5**   Immunity can be artificially induced by injecting a person with an antigen that somehow no longer possesses the ability to cause the disease it would produce in its normal state. This weakened or dead organism still possesses the antigen that evokes the immune response and thus prepares the person for future assaults. This immunity is called *active immunity* because the subject actively produces the antibodies and sensitized lymphocytes that counter the antigen. (Alternatively, antibodies or sensitized lymphocytes may be injected directly into the system, imparting a *passive immunity* to the person.)

Usually vaccinations of antigen are given in multiple doses, separated by several weeks or even months. Why do you suppose the doses are spread out over such a long period?

During the evocation of the *primary immune response*, there exist only a very few B- and T-lymphocyte clones able to respond to the specific antigen introduced. These clones then begin to divide and create more sensitized lymphocytes and memory cells. It is the memory cells, which maintain their vigil after the initial invasion of antigen, that are responsible for the *secondary immune response*, should the antigen be reintroduced to the system. During this secondary response there are many clones ready to be sensitized, rather than just a few as in the primary response; consequently, the response is much stronger, producing much greater quantities of antibody and sensitized lymphocytes, and it lasts longer. When vaccinations are spread over several months, they are able to take advantage of this secondary response. A given amount of antigen thus produces far greater quantities of antibody and sensitized lymphocytes than it would if given all at once.

**18.6**   How do antibodies contend with bacteria and viruses?

Antibodies coat these cells extensively; this process acts as an attractant for macrophages that avidly ingest the coated cells. The exact nature of the interaction between the leukocytic phagocyte and the antibody on the cell surface is not known.

Antibodies attached to a virus or a bacterium may interfere with its normal invasive patterns either through mechanical (steric) blocking of necessary attachment points or through chemical interaction with sites on the invading particle. A virus may not be able to attach to a host cell and hence may fail to inject its nucleic acid within the cell; its program of invasion and capture of the genetic machinery of the host is thus interrupted. The bacteria or viruses also may be agglutinated in antigen-antibody complexes.

In company with the dozen or so kinds of complements that are present in the plasma, the antibody may participate in the direct lysis (splitting) of the cell. The complement can act as an amylase (starch-digesting enzyme) in combination with an attached antigen-antibody complex. The creation of a channel in the bacterial cell is sufficient to produce osmotic distension and eventually a bursting of the entire bacterium.

Finally, the antibodies may participate in inflammation reactions.

**18.7**   Infected wounds produce a thick, yellowish fluid called *pus*. What do you suppose the primary constituent of pus is?

When the skin is torn, phagocytes are attracted to the site to engulf the bacteria that enter the wound. As these phagocytes absorb the bacterial toxins, they die in large numbers, producing the pus.

**18.8**   What is the significance of cross-linking among the receptors of both naive B lymphocytes and naive T lymphocytes?

Lymphocyte receptors are of two types, yet are quite similar in their responses. The naive, or virgin, B-lymphocyte receptors are antibodies which are probably identical with the free antibodies which the plasma cells descending from that naive cell will be producing in prodigious amounts. The receptors of the naive T lymphocyte are not antibodies as such, but are proteins of a similar nature. Both types of receptors, on both B and T lymphocytes, must join to two antigens. In the case of the B receptors the two antigens are alike; in the case of the T receptors, which are of two different types, the two antigens are unlike.

When an appreciable number of these antigen molecules are bound to the receptors, the receptors become cross-linked, since each antigen may have more than one antigenic determinant that fits a receptor cleft. The linking of any pair of receptors, then, is dependent on an antigen's bonding to each of these receptors at the same time. Cross-linking seems to be the inciting stimulus for the proliferation of both B and T lymphocytes. Since cross-linking can occur only with a formidable presence of antigen, it is a signal for cell multiplication that will not be turned on casually.

Isolated antigenic determinants known as *haptens* cannot by themselves initiate such proliferation, because they can bind to only one antibody cleft and will not bring about cross-linking. However, these haptens can stimulate the continuation of an immune reaction that has been initiated by intact antigen. In plasma cells the cross-linking phenomenon promotes the production of free antibodies, which are released into the general circulation or into the lymph system depending on the location of the cell.

**18.9**   Experiments with larval frogs have shown that if an organ is removed at a certain stage of development and transplanted into another animal, the organ will be rejected if reintroduced later into the original animal. If only half the organ is transplanted and then reintroduced, it will not be rejected. What do these results suggest?

The fact that the organ was rejected suggests that during its absence from the developing donor, a process of "self" identification took place in the donor, during which all cells in the body were identified as "friendly"; the transplanted organ, absent during this labeling period, was thus seen as "foreign." The alternative explanation—that the organ was somehow made foreign during its transplanted time in the recipient animal—is discounted by the fact that when part of the organ was kept in the donor, the reintroduced half was not seen as foreign and, so, had not changed.

Although the actual mechanism by which the lymphocytes learn to recognize their own organism is unknown, some evidence suggests that the suppressor T cells are involved.

Occasionally, the body loses its sense of recognition of self, and the immune system directs its antibody fire against native cells. This is what is meant by an *autoimmune disease*. In *rheumatoid arthritis* a bacterial infection may trigger the continuous formation of antibodies against lymphoid tissue at the joints. *Lupus erythematosus* is a generalized connective tissue autoimmune disease.

**18.10**   There are two separate MHC (major histocompatibility complex) systems. Class I molecules of a specific configuration are present on the surface of all cells. They are a distinctive kind of identification representing a variety of antigens coded for by a very flexible genetic control system. All the cells of any one organism possess a similar array of these surface proteins. They are the principal codes for "self." Only those cells which become cancerous develop a different array of antigens which mark these cells as outlaws; they are then subject to attack from such segments of the immune system as the T lymphocytes. Class II antigens of the MHC are located only on the cells of the immune system. Why do you suppose this dichotomy exists?

The class II antigens apparently serve as components of an intricate communications network that permits close interaction among B lymphocytes, the three kinds of T lymphocytes, mast cells, and the various kinds of macrophages. In a system in which there exist both stimulation, as in the activation of B lymphocytes by T helper cells, and modulation, as in the damping of the immune response by T suppressor cells, an intimate, rapid, and separate system is necessary to assure an appropriate measured response within the immune system.

**18.11**   One mechanism by which the HIV retrovirus may induce AIDS is the destruction of the self-recognition site on helper T cells. Why should this be so devastating?

Apparently, the helper T cells become unable to recognize other B and T lymphocytes and, so, are unable to activate them. The humoral and cell-mediated immune responses are thus diminished, and the body is left susceptible to infection.

**18.12**   Suppose a woman with type O blood is suing a man whose blood type is AB for paternity. If the child is type O as well, can the man have fathered the child?

The answer is no. Three alleles control the blood type antigen expressed on the surface of the red blood cell (erythrocyte). Antibodies in the plasma are regulated in accordance with erythrocyte antigens. $I^A$ and $I^B$ are alleles that produce A antigens and B antigens on the red blood cell, respectively. The allele $i$ is recessive and produces no antigens. As seen in Table 18.1, type A may result from the genotype $I^A I^A$ or $I^A i$. Type B is produced by $I^B I^B$ or $I^B i$. $I^A I^B$ produces the codominant type AB, while $ii$ results in type O. Type O is a universal donor, because it does not provide any antigen to be agglutinated, while type AB is a universal recipient because it has no antibodies to react with incoming antigens.

The disputed father is type AB and so can produce only gametes with $I^A$ or $I^B$ alleles. These alleles, combining with the mother's $i$ alleles, can produce only type A or type B children; both gametes must be $i$ to produce the type O phenotype seen in the child.

**18.13** Charlie Chaplin was ruled to be the father of Joan Barry's type B baby. Barry was type A, and Chaplin type O. Was this ruling correct? What was Barry's genotype? The baby's?

Since Barry was type A, she could not have provided the $I^B$ gamete necessary to produce a type B baby. Therefore, the father must have supplied this gamete, an impossibility for Chaplin, since, being type O, he could produce only i gametes. Therefore, Chaplin could not have been the father.

Barry, being type A, could have a genotype of $I^A I^A$ or $I^A i$; however, if she were of the former genotype, she could have given the type B baby only an $I^A$ gamete, and since the father supplied the $I^B$ gamete, the baby would have been type AB. Therefore, Barry must have supplied an i allele; her genotype must have been $I^A i$ and the baby's $I^B i$.

## Supplementary Problems

**18.14** Why are $I^A$ and $I^B$ alleles illustrative of codominance?

**18.15** The reaction of antigen with antibody is most similar to the union of enzyme and  (a) vitamin.  (b) ribosome.  (c) substrate.  (d) metal.  (e) hormone.

**18.16** The four chains of an antibody are held together by S—S bonds.
(a) True  (b) False

**18.17** The rearrangements of exons accounts for a significant proportion of heavy- and light-chain diversity.
(a) True  (b) False

**18.18** Cell-mediated immunity involves only the B lymphocytes.
(a) True  (b) False

**18.19** Allergic reactions involve overresponses of the immune systems, especially the release of histamine from mast cells.
(a) True  (b) False

**18.20** Active immunity involves stimulating the body to produce large amounts of antibody.
(a) True  (b) False

**18.21** Passive immunity involves the injection of suppressor T cells.
(a) True  (b) False

**18.22** Hybridomas produce pure antibodies in great abundance.
(a) True  (b) False

## Answers

**18.14**  Because they are both expressed phenotypically when present together

| | | | |
|---|---|---|---|
| **18.15**  (c) | **18.17**  (a) | **18.19**  (a) | **18.21**  (b) |
| **18.16**  (a) | **18.18**  (b) | **18.20**  (a) | **18.22**  (a) |

# Chapter 19

# Gas Exchange

We have already studied the process of *respiration*, whereby organisms oxidize metabolites in the presence of oxygen to capture the energy contained in the metabolites' bonds. Respiration produces not only energy but also the by-products carbon dioxide and water. (Similarly, the processes of glycolysis and fermentation break down metabolites for energy and produce by-products.) Clearly, a means of *gas exchange* is necessary in order to provide the cells with oxygen and to remove carbon dioxide. Rather confusingly, this process is also referred to as *respiration*. To distinguish between the two forms of respiration, we will call the oxidative process *internal*, or *cellular*, *respiration* and the gas exchange process *external*, or *organismic*, *respiration*. It is on this second type that we will concentrate in this chapter.

## 19.1 MECHANISMS OF EXTERNAL RESPIRATION

In unicellular and small multicellular organisms, gas exchange is accomplished quite readily across the cell membranes. Since the gases are dissolved in fluid, it is imperative that moist membranes be available for the movement of the gases into and out of cells and organisms. *Desiccation* (drying out) inhibits the movement of gases in addition to having other deleterious effects.

Single-celled organisms are completely dependent on diffusion for the movement and exchange of gases involved in internal respiration. Oxygen permeates the cell membrane to enter the diffusion arena. The gases that arise from internal respiration diffuse through the cytoplasm and exit across the plasma membrane. The opposing passages of $O_2$ and $CO_2$ across the membrane are influenced by the partial pressure of the gases in the external environment. Thus, a low partial pressure of $O_2$ in the atmosphere or fluid surrounding an organism diminishes the rate of diffusion of $O_2$ across the cell membrane into the cell.

For single cells, and multicellular organisms that are flat, diffusion is an adequate method for fulfilling the organism's need for gas uptake and liberation. Since the distance from the external environment is the limiting factor in determining the efficiency of gas exchange, *thickness*, and therefore *volume*, are of crucial concern to an organism, whereas area is not.

The ability of a single-celled organism to exchange gases becomes increasingly strained as its size increases. This is because the cell surface (where gas exchange occurs) increases only as the square of the cell's radius, whereas the cell volume (and therefore the cell's total metabolic activity) increases as the cube of the radius. Diffusion time to and from the surface also increases with volume. Thus an upper limit to cell size is established by the physical constraints simple diffusion places on gas exchange.

Similarly, as multicellular organisms become increasingly complex, cells buried in the interior are farther removed from the cell layer where gas exchange with the environment occurs and so become increasingly limited in their ability to obtain and eliminate gases through simple diffusion. Beyond a certain complexity, strategies must be introduced to supplement diffusion.

Almost all invertebrates are metabolically less active than homeothermic vertebrates such as birds and mammals. This reduces the need for extremely rapid gas exchange but does not eliminate the requirement for mechanisms to supplement simple diffusion. Various types of circulatory systems have evolved to fulfill this need. Diffusion still provides the route for the initial passage of oxygen into the organism, but once this vital gas has penetrated the surface, it is carried by bulk flow to all parts of the organism by the blood. In most cases, special respiratory pigments such as hemoglobin or hemocyanin (see Chap. 17) aid the blood in carrying the oxygen. The blood also carries $CO_2$ and $H_2O$ from respiring tissues to the surface, where diffusion of these products to the external environment occurs. Such a gas-disseminating circulatory systems exist in most animals from the level of annelids (segmented worms such as earthworm) on.

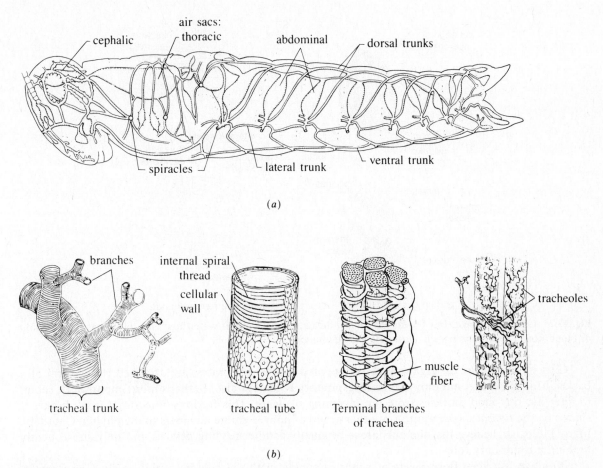

**Fig. 19.1** (*a*) Respiratory system and (*b*) trachea of insects. [*From Storer et al.; (*a*) after Albrecht, (*b*) after Snodgrass.*]

A second adaptation for bringing air more rapidly into the interior is the *tracheal system* (see Fig. 19.1), characteristic of many insects and arachnids (spiders). This is a system of tubes that ramifies throughout the body of the organism and carries air to individual cells. It is analogous to the stomata and air spaces of leaves in green plants. The larger tubes are known as *tracheae* and arise from apertures along the surface of the body called *spiracles*. The spiracles, like the stomata of the leaf, may open or shut in accordance with the action of valves. The large tracheae are maintained as open tubes by supporting rings of *chitin*, a tough nitrogen-containing polysaccharide also found in the cell walls of fungi. This tracheal system is actually a substitute for blood-borne gas distribution to and from a discrete respiratory organ.

In most aquatic animals the respiratory organ consists of a series of protruding flaps known as *gills*. These gills, richly endowed with blood vessels engaging in the vital exchange of gases, may be relatively simple extensions of an epithelial surface, as in the sea worm *Nereis*, or they may be elaborate repeating units covered by complex protective devices as in the bony fish (teleosts). In most bony fish, water enters the mouth cavity and is forced out through five sets of cartilaginous gill arches on each side of the head, usually just posterior to the eye. Gills (see Fig. 19.2) consist of feathery *filaments* extending from supporting rods that arise from each gill arch. The filaments, in turn, increase their surface area by the formation of *lamellae*, which are lateral projections arising from the gill filaments. In many invertebrates the gills are present as naked flaps extending from the side of the head, or they may be covered by an exoskeleton as in crustaceans. In some forms the gills occur within the pharynx (throat). In all bony fish the gills are covered by a protective shield called the *operculum*.

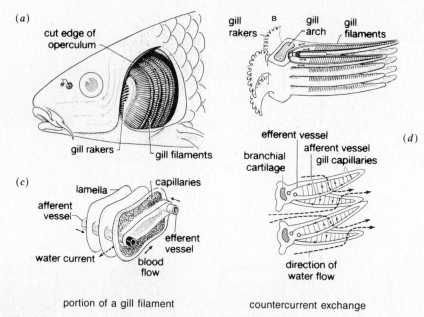

**Fig. 19.2** Gills in a teleost. In (*d*), dashed lines indicate the direction of water flow and solid lines the direction of blood flow. [*From Storer et al., partly after Goldschmidt.*]

Gills in *amphioxus*, an organism closely related to the vertebrates, are arranged in the form of a basket. This gill basket not only acts as a respiratory organ but also, perhaps more importantly, serves to filter out tiny food particles from the incoming seawater. This feeding function of a gill basket is shared by the *tunicates* (sea squirts), another group of nonvertebrate members of the phylum Chordata. Many biologists believe that the gills arose originally as filter-feeding devices and only subsequently took on a respiratory role.

Primitive fish keep fresh layers of water in contact with the gills by rapidly swimming forward. This turnover of water in contact with the gills is necessary because oxygen is only poorly dissolved in water. Some fish may actually drown if held in one place.

Most bony fish demonstrate complicated breathing movements in which water is actively drawn past the gill apparatus. These movements involve an elaborate system of valves, changes in the size of the oral cavity, and the opening and closing of the operculum (see Fig. 19.3).

Both gills and lungs are essentially tissue outpockets or extensions that increase the respiratory surface. Many fish *gulp* (forcefully swallow) air and absorb oxygen in gaseous form directly across the

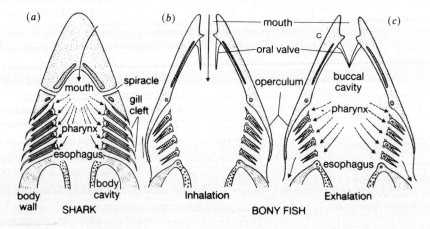

**Fig. 19.3** Breathing movements in fishes. (*From Storer et al., after Boas.*)

highly vascularized tissue of the oral cavity. It is quite possible that a fishlike ancestor of today's amphibians, reptiles, birds, and mammals developed a special sac in the pharyngeal (throat) region that could hold on to this precious air. Such a structure may, in time, have developed into the ramified pulmonary apparatus of advanced vertebrates.

The *lungfish* is a primitive remnant of a group of fish that evolved lungs while still largely dependent on their gills for respiratory exchange. Lungfish live in stagnant waters in warm climates or in streams that tend to dry up for short periods. Their lungs, which develop as ventral evaginations of the gut, serve as a supplementary mechanism for obtaining oxygen during difficult periods. Additionally, the lungfish is capable of *estivation*, a process similar to hibernation but occurring in response to extremely dry conditions. The lungfish builds a muddy cocoon around itself and sleeps deeply through the period of drought.

In many bony fish a lung coexists with gills, but the lung does not possess a respiratory function. Instead it provides a hydrostatic mechanism for maintaining the fish at various levels in the water. In this role it is designated a *swim bladder*. To move up in the water the fish brings (via the bloodstream) air into the swim bladder; to descend, the fish expels air from the swim bladder.

The lungs of the vertebrate represent one of the most vital evolutionary adaptations for life on dry land. By internalizing the moist respiratory surface participating in gas exchange with the surrounding (*ambient*) air, lung breathers could modify their skins for other purposes. The transition from gill breathing to lung breathing is best observed in the *amphibians*, that class of vertebrates whose very name (*amphi* meaning "both" and *bios* meaning "life") suggests a twofold lifestyle. Most amphibians start out life as gill-breathing, fishlike *larvae*. They then undergo a radical series of changes, known collectively as *metamorphosis*, in which a lung-breathing adult is produced. The change from the larval tadpole to the adult frog provides us with a *phylogenetic* (relating to the course of evolution) insight contained within the individual development of a single specimen.

## 19.2  MAMMALIAN RESPIRATION

The breathing apparatus in mammals is exemplified by the human respiratory system (see Fig. 19.4). Mammals are highly efficient colonizers of the terrestrial domain, and their respiratory system

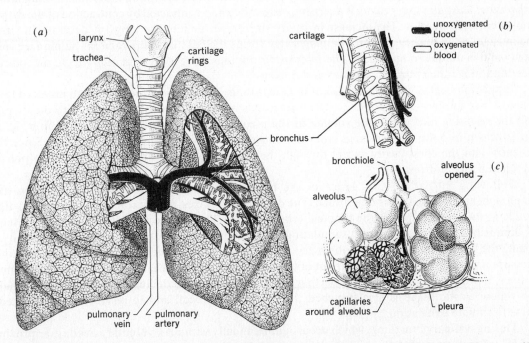

**Fig. 19.4**  Human respiratory system. Vessels depicted in black carry unoxygenated blood. (*From Storer et al.*)

reflects this success (however, the breathing apparatus of birds is even more efficient; see Prob. 19.10). Air may enter the respiratory tree through either the nose or the mouth, although warming and filtration of inspired air occur best through the longer nasal passages. The nasal passages open into the nasal pharynx and are separated from the mouth by a partition known as the *palate*. It is called the *hard palate* in the anterior region just behind the teeth and is called the *soft palate* as it gradually becomes less rigid in the more posterior region.

Air taken in (*inhaled*), whether through the mouth or nose, passes through the pharynx and then into the *trachea*, a rigid tube resting in front of the esophagus. The opening of the trachea (the *glottis*) is guarded by the flaplike *epiglottis*. The *larynx*, or voice box, lies at the origin of the trachea. The trachea bifurcates into a left and a right *bronchus*. Each of these bronchi subdivides into many *bronchioles*, which constitute the small branches of the respiratory tree. *Bronchitis* is an inflammation of these ramified tubules.

Each bronchiole terminates in a blind cluster of tiny air sacs called *alveoli*, which are heavily supplied with capillaries. It is across the alveolar membrane that the main events of external respiration occur. The capillaries surrounding the alveoli bring oxygen-depleted, carbon dioxide-laden blood from the pulmonary arteries. The alveoli, on the other hand, upon inhalation contain a high partial pressure of oxygen ($pO_2$) and a low partial pressure of carbon dioxide ($pCO_2$). The resulting pressure gradients lead to rapid diffusion of these two gases across the capillary and alveolar walls—oxygen enters the capillaries and is carried to the tissues; carbon dioxide enters the lungs and is expelled upon exhalation.

**EXAMPLE 1** Inspired air contains 20.96 percent oxygen with an expected tension of about 160 mmHg. Expired air contains 15.8 percent oxygen with a relatively appreciable tension of 120 mmHg. Evidently the lungs remove roughly 25 percent of the oxygen present in inspired air. Carbon dioxide is present at approximately 0.03–0.05 percent in inspired air, but it increases to 4 percent in expired air. The $CO_2$ tension in expired air is thus 100 times greater than in inspired air.

The respiratory surface afforded by the alveolar system of mammals is much greater than the surface arising from the essentially smooth structures of amphibian lungs. In reptiles, in which the skin is no longer employed as a significant respiratory surface, a porous and spongy lung similar to the mammalian counterpart occurs.

The mammalian lung plays an essentially passive role in the breathing movements—*inhalation* and *exhalation*. The pulmonary cavity is alternately expanded and contracted by contraction of the diaphragm and intercostal muscles. The increase in cavity size draws air into the lungs. A decrease in size when the muscles relax pushes air out. The *diaphragm* forms the base of the pyramidal pulmonary (pleural) cavity and is unique to mammals. The *intercostals* are muscles between the ribs which, on contraction, raise and move the entire rib cage up and outward.

Inhalation begins with stimulation of the diaphragm by the *phrenic nerve*; this stimulation is largely due to a rise in blood levels of $CO_2$. The intercostals are stimulated concurrently with the diaphragm, and the resulting increase in the volume of the pleural cavity reduces the pressure within the cavity. Air is drawn into the passive lungs from the outside as a result of the suction created. Since there is no entry into the chest cavity except through the trachea, air must enter the lungs when pressure in the pleural cavity drops.

Following inhalation, $CO_2$ levels in the blood are reduced, and phrenic stimulation of both diaphragm and intercostals is diminished. The diaphragm relaxes, and its consequent increased diameter results in its doming up into the thoracic (chest) cavity. The intercostals also relax as a consequence of diminished excitation of their stimulating nerve trunks and the simple response to gravity. As a result, the total volume of the pulmonary cavity is considerably diminished and the intrathoracic pressure rises sharply. This forces air out of the lungs, resulting in exhalation. Yellow elastic connective tissue is a prominent part of the supporting tissue of the lung. As the lung stretches during inhalation, the yellow elastic tissue is also stretched. Its tendency to recoil and to resume its unstretched state helps to initiate exhalation.

During ordinary breathing, which occurs automatically without a voluntary overlay, approximately 500 mL of air passes in and out of the lungs of adults with each breath. This is known as *tidal air*. An additional amount of air, which can be inspired voluntarily beyond the tidal air, is known as the *inspiratory reserve*. It approximates 3 liters. Following a normal exhalation, approximately 1 liter of air can be forcibly expelled—the *expiratory reserve*. These combined capacities, measured by the maximum expiration following a maximal inspiration, is called the *vital capacity* and generally is between 4.5 and 5 liters. Even after a maximum expiration, more than 1 liter of *residual air* remains in the lungs. This air plays a key role in maintaining an exchange of gases between blood and alveolar air between breaths.

## 19.3  REGULATION OF BREATHING

Although some voluntary control over breathing can be exerted, through signals from the cerebral cortex, breathing is ultimately an involuntary response to neural and chemical stimulation.

Bilateral *inspiratory centers* and *expiratory centers*, located in the medulla oblongata and pons, communicate with respiratory muscles and sensors via various cranial nerves, including the phrenic and vagus nerves. During rest, it is primarily the inspiratory centers that control breathing, with expiration being largely a passive response. These centers send via the phrenic nerve a stream of excitatory impulses to the diaphragm and chest muscles, causing them to contract, resulting in an inhalation. After about 2 s, the center stops firing, and the muscles relax, resulting in an exhalation.

During times of increased activity, when breathing is faster and deeper, the expiratory centers become operational, (1) sending a periodic stream of inhibitory impulses to the inspiratory centers, thus stopping inhalation, and (2) sending stimulatory impulses to muscles in the chest and abdomen that cause exhalation greater than that which results from simple relaxation of the diaphragm and other inspiratory muscles. The expiratory centers are abetted by stretch receptors, located mainly in the bronchi and bronchioles. When the lungs expand beyond a certain limit, these receptors fire impulses through the vagus nerve to the expiratory centers, which in turn inhibit the inspiratory centers. This is called the *Hering–Breuer reflex*. The *pneumotaxic center*, located in the pons, also is able to effect inhibition of the inspiratory centers.

Although neural impulses actually effect breathing, the chemical balance in the blood and cerebrospinal fluid determines the rate of neuronal firing. Sensors located in the carotid artery and aortic arch detect changes in pH and in levels of $CO_2$ and $O_2$; sensors in the ventricle of the medulla are sensitive to pH changes in the cerebrospinal fluid. Increases in $CO_2$ or drops in pH or $O_2$ levels cause these sensors to fire proportionately more frequently, stimulating the respiratory centers to increase the depth and frequency of breathing. Deeper and more frequent breaths cause more $CO_2$ to be blown off, with a resulting rise in pH, and more $O_2$ taken in (see next section). These changes cause the sensors to slow their firing, and breathing consequently slows. Interestingly, $CO_2$ levels have a much greater effect on breathing rates than $O_2$ levels do. Actually, it may be the increase in $H^+$ ion concentration (decrease in pH) resulting from rises in $CO_2$ levels that triggers the breathing changes.

## 19.4  EXCHANGE OF $O_2$ AND $CO_2$ IN THE BLOOD

The $CO_2$ that enters the bloodstream from metabolically active tissues has several fates. Approximately 10 percent is dissolved in the plasma ($CO_2$ is roughly 30 times more soluble in water than $O_2$ is). The remainder enters the red blood cells. Some of this loosely associates with the amino groups of the amino acids in the hemoglobin, forming *carbaminohemoglobin* (HbNHCOOH). The remaining red blood cell $CO_2$ combines with water to form carbonic acid:

$$CO_2 + H_2O \rightleftharpoons \underset{\text{carbonic acid}}{H_2CO_3}$$

The carbonic acid quickly dissociates into the bicarbonate ion:

$$H_2CO_3 \rightleftharpoons H^+ + HCO_3^-$$

Although these reactions occur in the plasma too, in the red blood cell the formation of carbonic acid is greatly accelerated by the enzyme *carbonic anhydrase*. As we will see, it is very important that this reaction, facilitated by carbonic anhydrase, moves rapidly in either direction, depending on $CO_2$ levels. The $HCO_3^-$ diffuses into the plasma; to maintain ionic balance an equal number of $Cl^-$ ions move into the red blood cell (*chloride shift*).

When the blood reaches the lungs, it encounters low levels of $CO_2$ in the alveoli. As a result, the carbaminohemoglobin dissociates back into $CO_2$ and hemoglobin, and the resulting $CO_2$ diffuses into the lungs, from which it is exhaled. As $CO_2$ levels in the blood fall, $H_2CO_3$ tends to revert back to $CO_2$ and $H_2O$. This $CO_2$ also diffuses into the lungs and is exhaled. The diminishing levels of $H_2CO_3$ cause $H^+$ and $HCO_3^-$ to recombine, forming more $H_2CO_3$. At the end of this chain, the falling levels of $HCO_3^-$ in the red blood cell cause bicarbonate in the plasma to diffuse back into the red blood cell and the reactions shown above proceed in the reverse (i.e., left to right) direction. The net result of these reversals is that in the lungs most of the $CO_2$ that was bound as carbaminohemoglobin or as the bicarbonate that diffused into the plasma is converted back to free $CO_2$ and diffuses into the lungs, where it is exhaled.

In Chap. 17 we studied the Bohr effect, the decrease in affinity of hemoglobin for $O_2$ in the presence of high levels of $CO_2$ (or low pH). This effect facilitates the release of $O_2$ to tissues that have been metabolically active and consequently have little $O_2$ and high levels of $CO_2$.

In the oxygen-rich lungs, where uptake, rather than release, of $O_2$ is required, hemoglobin demonstrates another valuable property: *cooperativity*. When one $O_2$ molecule forms its noncovalent bond with one of the four heme groups, it alters the shape of the heme in such a way that the $O_2$ affinity of the other hemes increases because of transmitted conformational changes. Each additional binding of $O_2$ further increases the $O_2$ affinity of the remaining free hemes. This cooperativity permits rapid uptake of $O_2$ in the lungs.

# Solved Problems

**19.1** Differentiate between internal (cellular) respiration and external (organismic) respiration.

Internal respiration is primarily a *chemical* process, in which molecules are broken down and their energy captured; external respiration is primarily a *physical* process of gas exchange, in which the raw materials and by-products of internal respiration are processed.

**19.2** What is the advantage of a respiratory pattern in which oxygen is utilized, as opposed to one involving merely glycolysis and fermentation?

Although this has been discussed in Chap. 4, it is appropriate to review it here. The utilization of oxygen as a terminal electron acceptor represents a maximal tapping of the energy stored in the organic fuel degraded by the cell. In glycolysis or fermentation only some of the energy of the fuel molecules is liberated to produce ATP. More than 19 times the energy yield of glycolysis is generated by the oxidative processes associated with the Krebs cycle and the terminal election-transport chain.

Only with the advent of photosynthesis and the creation of an atmosphere rich in molecular oxygen was an aerobic metabolism possible. Such a metabolism is indispensable for the highly active terrestrial forms and has become a major feature of almost all eukaryotic organisms.

**19.3** At one time nurses would remove all plants from hospital patients' rooms at night. What justification do you suppose was given for such a practice?

All plants and animals, and even most single-celled organisms, take in oxygen for the oxidation of metabolites and expend $CO_2$. Superimposed on this basic aerobic respiratory pattern is the process of photosynthesis, which uses up $CO_2$ and produces molecular oxygen. When photosynthesis occurs, its intensity usually far exceeds the intensity of the respiratory process in the plant, so that a net gain of $O_2$

and a net loss of $CO_2$ occur. However, in the absence of light, as in a darkened hospital room, very little or no photosynthesis is going on. Therefore the plants, which are now only respiring, compete with the patient for the $O_2$ of the atmosphere.

It is not likely that so little oxygen would be present as to result in any danger to a patient sharing aerobic resources with a few plants. However, the practice of removing plants at night still occurs in some hospitals.

**19.4**    Why is the circulatory system so much more important for external respiration in the earthworm than in the insect?

The earthworm breathes through its skin, which is kept moist at all times by mucous secretions. Once oxygen has penetrated the skin by the relatively slow process of diffusion, it must be carried to all parts of the earthworm's interior. In the same way, $CO_2$ and $H_2O$ must be rapidly transported by the blood from the tissues where they are produced to the skin, where they may be discharged to the environment. Although most of the earthworm is close to its surface and the surface of the earthworm is relatively great, the absence of an adjunct distribution system would not permit a sufficient rate of gas exchange to sustain metabolic needs. Therefore, the earthworm requires a circulatory system to effect the necessary exchanges.

The insect, on the other hand, has a tracheal system. This ramified system of tiny, relatively rigid tubes that terminate in tiny pockets adjacent to the internal tissues enables the atmosphere to "reach into" every nook and cranny of the insect's body and obviates the need for further internal transport. Since the air must diffuse into the termini of each tubule, this system would lose its effectiveness were insects to achieve a very large size. However, the tracheal system has permitted insects and such arthropods as centipedes and spiders to conquer a terrestrial environment and spread successfully within that environment.

**19.5**    Describe the system depicted in Fig. 19.2 by which fish actively push water over their gills.

In the first step, the mouth cavity expands, drawing water in, much as expansion of the pulmonary cavity in mammals draws in air. The operculum simultaneously closes to prevent water from being drawn in backward over the gills. Next, the mouth cavity contracts, and the oral valve closes, preventing water from passing back out the mouth. Instead, the operculum opens, and water is forced out over the gills.

**19.6**    What is the significance of Henry's law for gas transport?

As was noted in Chap. 17, Henry's law states that if the temperature is kept constant, the amount of gas dissolved in liquid is proportional to the partial pressure of the gas in the space above the liquid. As that partial pressure increases, more gas is dissolved; as it decreases, more gas comes out of solution. The actual amount of gas dissolved depends too on the *solubility coefficient* of that gas, a measure of how much gas is dissolved in comparison with some standard at the same partial pressure. At any point the escaping tendency of a gas in a liquid is balanced by a tension in the atmosphere that forces the gas into solution. Generally, an increase in temperature decreases the amount of gas a liquid can hold. This explains the bubbles that form as a glass of cold water is allowed to warm up. It can no longer hold the same amounts of gas, and the various gases coming out of solution produce the bubbles. Clearly, this tendency of gases in solution to reach equilibrium with their counterparts in the atmosphere above is crucial to gas exchange in the lungs. Carbon dioxide–laden blood coming from the body passes into an air space (the lungs) where the partial pressure of carbon dioxide is very low, causing the carbon dioxide to leave the blood in response to the reduced pressure. Similarly, the high partial pressure of oxygen in the lungs relative to the blood causes oxygen to enter solution.

Dalton's law of partial pressure complements Henry's law. It maintains that in a mixture of different gases, each exerts a pressure directly proportional to the percentage of that gas in the mixture. The pressure of each gas is known as the *partial pressure*; the total pressure of the mixture is equal to the sum of all the partial pressures. As an example, assume that the pressure of air under base conditions is 760 mmHg. If the air contains 21 percent $O_2$, then the partial pressure of the oxygen gas in that mixture is $0.21 \times 760$ mmHg $\cong 160$ mmHg. The partial pressure of molecular nitrogen ($N_2$) is roughly 593 mmHg, so its proportion in air is $(593/760)(100) \cong 78$ percent.

**19.7**    In the gills of fish the deoxygenated blood coming from the tissues flows through the capillary beds of the lamellae in the opposite direction to that of the water moving across the gills (see Fig. 19.2). What is the significance of this countercurrent arrangement, and what would be the result if the blood and water flowed in the same direction?

   This arrangement, already discussed in Chap. 15, ensures a maximum efficiency of exchange. In the gills it is the transfer of oxygen that we are concerned with. The blood that first reaches the gill has the lowest levels of $O_2$. It encounters water that has come close to completing its trip across the gills and so has lost a good deal of its oxygen. But enough oxygen is present to exceed that of the blood, so oxygen moves from the water to the blood. As the blood moves farther along the capillary bed, it gradually becomes more oxygenated but it encounters water which has increasing levels of $O_2$, since that water is moving in an $O_2$-rich-to-$O_2$-poor direction. Hence, oxygen keeps moving from water to blood. At the exit end for the blood it now has a maximum level of $O_2$, roughly equivalent to that of the water entering the gills. Thus oxygen moves from water to blood at every level of the countercurrent setup. If blood and water moved in the same direction across the gill, the maximum $O_2$ uptake would occur at the start and an equilibrium level would shortly be established that would be roughly half that found in the countercurrent arrangement.

**19.8**    In the great majority of living fishes the nostrils (nasal cavities) are relatively superficial cul-de-sac structures used for smell. In the lobe-finned fish the nasal cavity first extended into the mouth through the creation of *internal nares*. Air could then come into the mouth and throat from the nose. The lobe-finned fish with their fleshy lateral fins are probably ancestral to amphibians and others involved in the further adaptive radiation of the vertebrates. What adaptive benefits might the internal nares have imparted?

   For the lobe-finned fish no functional advantage need be associated with this new passage, but it did open up the possibility of nose breathing in descendent forms. This freed the mouth for subsidiary functions and lowered the vulnerability associated with keeping the mouth continually open. The refinement of a palate in almost all mammals and birds and some reptiles afforded the opportunity of eating while continuing to breathe. Additionally, the nasal passages have become efficient in warming, filtering, and moistening the air that moves directly from their labyrinthine length to the entrance of the trachea.

**19.9**    What physical adaptation would you say evolved as a result of the relative positions of the trachea and the esophagus?

   Because the esophagus lies behind the trachea, swallowed food must cross over the glottis. If the epiglottis did not exist to cover the glottis whenever food was swallowed, a great danger of aspirating food into the trachea and choking would exist.

**19.10**   The metabolic demands of flight have necessitated that birds evolve a highly efficient respiratory system. In birds the respiratory apparatus consists of not only trachea, bronchi, and lungs, but also an additional feature—*air sacs*. These supplementary pockets are found both anteriorly and posteriorly and even ramify into hollow spaces within the bone. Should a bird's trachea be blocked, air could actually get into the respiratory system through a severed bone. The hollowed bones of the bird also provide a lighter-weight skeleton, thus making flight easier.
   Lacking a diaphragm, birds inhale purely by contracting their abdominal and costal (i.e., chest) muscles. In flight, the beating wings and the thoracic flight muscles exert a bellows effect. Air moves from the bronchi directly to the posterior air sacs, without moving into the lungs themselves. At the same time, air that was present in the lungs moves into the anterior air sacs. On exhalation, the air in the posterior air sacs passes into the lung, and the air in the anterior air sacs moves out of the body. The alveoli are not closed sacs at the termini of the bronchioles. Instead, they are tiny ductules (*parabronchi*) across which the maximum exchange of gases occurs. Why is this system more efficient than that found in mammals?

In the mammalian lung, air enters a blind sac. Because the lung cannot collapse completely, some old air always remains in the lung following exhalation. In the avian lung, which has a separate entry and exit point, however, air flow through the lung is unidirectional and continuous. This permits a near complete turnover of air in the lungs with each breath. Thus air does not stagnate in the lungs. The difference between the two systems is analogous to the difference between the bidirectional digestive tract of the hydra and the unidirectional tracts of vertebrates. An additional advantage is that the flow of blood over the parabronchi is in the opposite direction to the flow of air, thus establishing a highly efficient countercurrent system.

**19.11** Skin divers occasionally drown because of a phenomenon known as *shallow water blackout.* After repeated dives below the surface, between which the victims hyperventilate by inhaling and exhaling deeply, they suddenly lose consciousness. What would you guess is the physiological basis for this condition?

   During hyperventilation, most of the carbon dioxide in the lungs is blown off. Consequently, after prolonged hyperventilation the carbon dioxide levels in the blood fall dramatically. Although the divers may not be taking in enough oxygen, they may fail to realize this, since carbon dioxide is the primary stimulus for breathing. They therefore may black out from lack of oxygen to the brain, even though they feel no respiratory distress.

**19.12** A prominent feature of a certain group of mammals that share a particular habitat is that they all have a large volume of blood—sometimes double that found in mammals not living in this particular habitat. They also have correspondingly larger blood vessels, and their red cell count is unusually high. Further, they possess an increased concentration of myoglobin, a compound with a high affinity for oxygen, found in muscle and similar to hemoglobin. Under certain circumstances the heart rate of these animals can be slowed and blood shunted away from organs other than the heart and brain. Finally, mammals found in this habitat show an increased reliance on anaerobic metabolic pathways. What sorts of animals do you suppose exhibit these traits and why?

   All these adaptations are responses to conditions of reduced oxygen levels, and they are found in most marine mammals. Since many of these animals remain beneath the surface for long periods of time, they require such abilities. The increased volume of blood and the high red cell count provide the mammals with a larger storehouse of oxygen to use during a dive, and the presence of increased myoglobin helps to ensure that more oxygen gets stored there before each dive. Further, the shunting of blood away from organs, except the heart and brain (which do not tolerate oxygen deprivation well), reduces oxygen consumption, ensuring that the muscles will have an adequate supply and enabling the animal to stay down longer. The increased reliance on anaerobic pathways is yet another response to reduced oxygen tensions.
   These adaptations have proved so efficient that many diving mammals actually exhale before they dive, rendering themselves less buoyant and also less susceptible to decompression sickness (bends) when they come up.

**19.13** Oxygen tensions in the alveoli, the pulmonary artery, and the pulmonary vein are all different. Given the dynamics of gas exchange in the lungs, match each of the following partial pressures with these three sites: 40 mmHg; 160 mmHg; 100 mmHg.

   The pulmonary artery carries deoxygenated blood to the lungs. We would therefore expect it to have the lowest oxygen tension of the three given. Blood entering the capillaries around the alveoli from the pulmonary artery does, in fact, have a $pO_2$ of around 40 mmHg. Not surprisingly, since it is the source of the $O_2$ used by the body, inspired air in the alveoli has the highest of the three partial pressures, 160 mmHg. By the time the blood has passed through the capillary beds around the alveoli and into the pulmonary vein, its $pO_2$ has been raised from 40 to 100 mmHg, which is also the *average* $pO_2$ of the air in the alveoli.

**19.14** In Chap. 17, in studying the Bohr effect, we noted that a plot of oxygen saturation of hemoglobin versus partial pressure of oxygen yields a sigmoid curve. What characteristic of oxygen uptake by hemoglobin accounts for the sigmoid shape of the curve?

Ths sigmoid curve reflects the cooperativity seen in oxygen uptake by hemoglobin. If cooperativity did not exist, the second heme group in a hemoglobin molecule would be no more likely to take up an oxygen molecule than the first and oxygen uptake would tend to be directly proportional to $pO_2$. This would yield a straight line. However, because of cooperativity the binding of the first oxygen molecule greatly enhances hemoglobin's affinity for the second molecule, and so forth. Thus, the curve rises sharply after the threshold level at which the first molecule is bound.

# Supplementary Problems

**19.15** The oxygen revolution is associated with which process that arose during the evolution of living organisms?
(a) photosynthesis   (b) glycolysis   (c) Krebs cycle   (d) gluconeogenesis   (e) hydrolysis

**19.16** In tracheal systems, gas exchange actually occurs in the spiracles.
(a) True   (b) False

**19.17** A shift of the oxyhemoglobin dissociation curve to the right means that at a given $pO_2$ less oxygen is associated with Hb.
(a) True   (b) False

**19.18** The lamellae are stacks of thin transverse membranes lying along the gill filaments that increase the total gas exchange surface.
(a) True   (b) False

**19.19** In many crustaceans the gills are protected and covered by the exoskeleton.
(a) True   (b) False

**19.20** $O_2$ is more soluble in water (or plasma) than $CO_2$.
(a) True   (b) False

**19.21** The vocal cords are found in the larynx.
(a) True   (b) False

**19.22** Cigarette smoking has been clearly linked to lung cancer and may also be responsible for cardiovascular pathology.
(a) True   (b) False

**19.23** The lungs themselves play the active role in bringing a body of air into their interior.
(a) True   (b) False

**19.24** The vagus nerve stimulates inspiration by causing the diaphragm to contract.
(a) True   (b) False

**19.25** In the human fetus the oxyhemoglobin dissociation curve shifts to the left; this suggests that the fetal hemoglobin has a greater avidity for oxygen and may thus obtain the gas more readily from the maternal blood.
(a) True   (b) False

# Answers

| | | | | | | | |
|---|---|---|---|---|---|---|---|
| **19.15** | (*a*) | **19.18** | (*a*) | **19.21** | (*a*) | **19.24** | (*b*) |
| **19.16** | (*b*) | **19.19** | (*a*) | **19.22** | (*a*) | **19.25** | (*a*) |
| **19.17** | (*a*) | **19.20** | (*b*) | **19.23** | (*b*) | | |

# Chapter 20

## Excretion

Excretion is the process by which an organism rids itself of metabolic wastes. In humans such wastes may accumulate as *urine*, *sweat*, or *tears*. The elimination of feces (*egestion*, *defecation*) is not an excretory process, since fecal material was in the intestine but never within the body proper. (Since humans and other mammals are structured as a tube within a tube, the cavity of the digestive tract is really *outside*, although enclosed by, the body. The body proper runs from the inner lining of the digestive tract to the surface of the skin.) Only materials arising from metabolic activities within the body's cellular structure qualify as excretory products.

The major excretory product is $CO_2$, which arises from the degradation of organic fuel molecules. Some $CO_2$ is utilized for synthetic reactions, but most of it is channeled via the blood to the external environment. Water is also a product of the oxidation of foods, but the wide variety of uses of water make it difficult to consider it a waste product. In the kangaroo rat (and in organisms like it), which inhabits desert areas of the United States, the water arising from the oxidation of seeds and grains is the animal's only source of fluid and hence must be regarded as a necessity rather than a waste material.

Another significant excretory product is the nitrogen that is removed from the amino acids derived from protein. When protein is utilized as a fuel, its amino acids must first undergo deamination or transamination. Eventually, the nitrogen removed is excreted as ammonia, urea, or uric acid. During the normal turnover of protein, a certain amount of nitrogen must be excreted each day. The final form of the nitrogen end product excreted is determined by the availability of water. If water is freely available, the simplest mode of nitrogen release is $NH_3$ formation. However, since this compound is quite toxic, the danger of its buildup where water cannot wash it away has imposed an evolutionary constraint upon some organisms. One way of neutralizing the $NH_3$ is to combine it with $CO_2$ to form *urea*:

$$2NH_3 + CO_2 \rightarrow O{=}C\underset{NH_2}{\overset{NH_2}{\big\langle}} + H_2O$$

This reaction usually occurs in the liver. Humans are *ureotelic* (the nitrogen excretory end product is urea); urea is produced in our livers and brought to the kidneys for excretion in the urine.

*Uricotelic* organisms produce a much more complex product known as *uric acid*:

Uric acid is also nontoxic, but because it is relatively insoluble in water, far less fluid is needed to dispose of it. However, there is a higher energetic cost in its formation.

Our discussion of excretion will be confined to animals for several reasons. The metabolic activity of plants is much lower than that of animals, so the burden of waste disposal is not so crucial in plants. (You will recall that homeostatic mechanisms were also reviewed in terms of the more active animal kingdom.) Also, the demands of photosynthesis cast $CO_2$ in a very different light among green plants, where it serves as a major metabolite for sugar formation. Too, the plants tend to use nitrogen groups in a variety of ways without getting rid of them. This cycling and recycling of nitrogen, ammonia, nitrites, and nitrates means that green plants turn over their nitrogen to such an extent that they do not need to eliminate it.

It should be obvious, at this point, that excretion is intimately tied to fluid and electrolyte homeostasis. Many of the structures associated with waste disposal play a key role in water balance as well. In some cases, these structures were first identified as excretory organelles or organs and later found to function primarily in fluid and electrolyte homeostasis. The level of many substances within the body is the result of a balance between production and elimination. As we will see, excretory organs such as the kidney not only process materials for elimination, but actively regulate the levels of many substances in the blood.

## 20.1  EXCRETION IN INVERTEBRATES

Among single-celled Protista, and even the sponges, excretion is usually achieved by a *contractile vacuole*, which is a fluid-filled organelle that periodically contracts to force fluid, salts, and dissolved waste materials out of the cell. In some cases the vacuole is surrounded by a specific set of vesicles or sacs (often intimately associated with the mitochondria), which channel fluid into the vacuole. This arrangement is related to the ability of the vacuole to maintain a dilute fluid content and to expel this fluid from the cell through a surface pore. However, some protists lack these special organelles and expel their waste materials directly across the permeable cell membrane.

In the flatworms a clearly defined excretory system exists (see Fig. 20.1). It consists of two or more highly branched longitudinal tubules. Some side branches of the tubules terminate in a hollow bulb into which long cilia project. These cilia, beating constantly, assume the appearance of a flame and give the system its name—*flame cell system*. These cilia create continuous currents within the flame cell system, and the currents carry fluid and waste materials out of the body through the *excretory pores*, which are at the termini of other side branches of the longitudinal tubules.

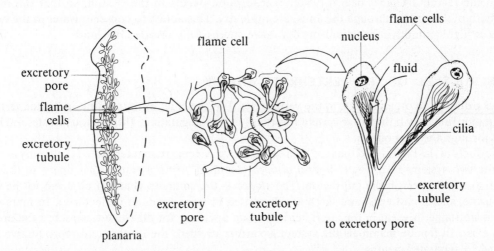

**Fig. 20.1**  Flame cell system of planaria (complementary structure on right omitted).

In the earthworm, which has a closed circulatory system, excretion is carried out by a close association of the circulatory system with a unique set of *nephridia* (see Fig. 20.2). Each segment of the earthworm's body contains a pair of nephridia. Body fluids enter a nephridium through the membrane of the bulblike *nephrostome*, which is the ciliated opening into the nephridium. The nephridium then gives rise to a coiled tubule, which is intimately associated with a blood capillary. This association of excretory structure and blood vessel, which allows reabsorption of material, is a functional forerunner of the vertebrate kidney. The nephridium terminates in a large bladder that opens to the exterior by means of a *nephridiopore*.

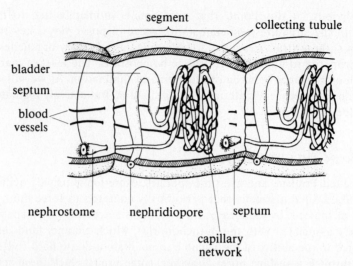

**Fig. 20.2**  Earthworm segment showing a nephridium.

Insects may have evolved from ancestors who possessed nephridial systems, but these nephridia have not persisted. Instead insects have evolved a new set of excretory structures—the *Malpighian tubules.* These are outpocketings of the gut at the junction of the midgut and hindgut. These tubular sacs are washed by the blood of the insect and absorb fluid and waste materials at their closed ends. Uric acid is formed within the tubules, and fluid and salts are reabsorbed by the open blood system surrounding the tubules. The urine formed in the tubule moves into the hindgut and out of the body through the rectum. A great deal of water reabsorption occurs in the rectum, so that the feces and urine, both of which exit through the anus, are quite dry. This ability to conserve water in the excretory process is highly adaptive for organisms that have successfully invaded dry land.

## 20.2  STRUCTURE OF THE VERTEBRATE KIDNEY

The kidney is an organ unique to the vertebrates. It is the chief excretory unit in higher vertebrates, but its major function in the lower vertebrates (fish) is osmoregulation. Nitrogenous wastes are handled largely through the gills of fish.

The ducts of the kidney and those of the reproductive system are so interrelated that many authorities study the two systems as a single *urogenital* (*uro* meaning "urine") *system* that opens to the outside through external pores in the tail region. The *cloaca* is the common chamber adjacent to the tail into which urine, sperm and eggs, and feces are deposited in their egress to the exterior. In mammals the cloaca is no longer present in the adult, and separate openings for digestive wastes and the urogenital system exist. In female mammals the system is further divided; the urinary and reproductive systems each have a separate opening.

In all vertebrates the kidney begins its development at an anterior site of the kidney-forming (*nephrogenic*) mesoderm called the *pronephros.* A series of tubules usually forms in association with specific segments of the body. The pronephros is soon superseded in the embryo by the *mesonephros.* It starts off as a series of tubules posterior to the pronephric region but soon is modified by a lengthening and convolution of the tubules and a loss of segmentation. The relatively long mesonephros is the functional kidney of adult fishes and amphibians and of reptile, bird, and mammal embryos. It is nonsegmented and forms a duct called the *ureter,* which conducts urine formed in the kidney to the cloaca or *urinary bladder.* This latter saclike structure stores the urine. Urine leaves the urinary bladder and reaches the outside through the *urethra.*

The two human kidneys are located laterally at the back of the abdominal cavity. On gross scale the kidney is divided into three regions—an outer *cortex,* a middle *medulla,* and an inner cavity, the *pelvis* (see Fig. 20.3). Urine formed in the outer two layers collects in the renal pelvis and is transported to the urinary bladder by the ureter (one from each kidney). From the bladder the urine is transported to the exterior by the urethra.

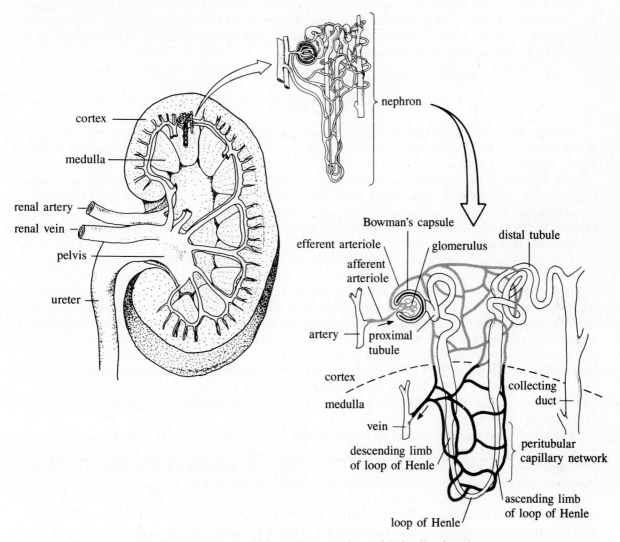

**Fig. 20.3**   The human kidney. Cutaway view, with details of nephron.

The actual production of the urine occurs in the many *nephrons*, the functional units of the kidney (Fig. 20.3). Each nephron consists of three parts. (1) The *glomerulus* is a tight ball of capillaries that filter blood through their walls into the second component of the nephron. Because of the high hydrostatic pressure in the glomerulus, almost all components of the blood, except proteins and the formed elements, are squeezed out of the capillaries. (2) The filtrate passes into the second element of the nephron, the *convoluted tubule*, beginning at the *Bowman's capsule*, a sac that surrounds the glomerulus and receives the glomerular filtrate through its cell walls. The remainder of the tubule consists of a *proximal* section, a middle *loop of Henle*, and a *distal* section. It is from the convoluted tubule that most of the ions, molecules, and water of the filtrate are reabsorbed back into the bloodstream. (3) The forming urine next passes into the *collecting tubule*, from which additional water may be reabsorbed before the tubule empties the urine into the pelvis of the kidney. The glomerulus, Bowman's capsule, and proximal and distal convoluted tubules are in the cortex; the loop of Henle and the collecting tubules are largely in the medulla.

Blood flows directly to the kidney from the aorta along the *renal artery*. This artery divides into many arterioles in the kidney, each one entering a Bowman's capsule and subdividing into the numerous capillaries of the glomerulus. These capillaries unite on their distal side to form a small arteriole, which then leaves the capsule. The arteriole again subdivides into numerous *peritubular capillaries* that surround the convoluted tubule and are responsible for reabsorption from the filtrate. The capillaries subsequently join to form a venule, which joins with the many venules from the other nephrons to form the *renal vein*, which in turn empties into the posterior vena cava.

## 20.3   FUNCTIONING OF THE VERTEBRATE KIDNEY

Strong evidence indicates that the earliest vertebrates evolved in fresh water. Rudimental features of the advanced kidney probably existed but were not integrated into a functioning nephron until higher vertebrates evolved.

The function of the nephrons of the kidney is clearly seen in freshwater fish. In these organisms large amounts of fluid must be excreted, while salts must be conserved. Fluid and dissolved solutes are forced out across the vessel membranes of the glomerulus as blood enters this tuft of coiled capillaries. The filtrate then enters the Bowman's capsule. The glomerulus with its surrounding capsule is called the *Malpighian corpuscle*.

The narrow neck immediately beyond the capsule is highly ciliated. This arrangement increases the amount of water that can move along the nephron. Immediately following the neck is the *proximal segment* of the tubule. It is in this moderately expanded region that monosaccharides and small proteins are reabsorbed from the nephron directly into the body. Divalent ions tend to be reabsorbed through channels found more posteriorly in this same proximal segment. In the relatively short *intermediate segment*, the next portion of the nephron, the beating action of numerous cilia moves the filtrate rapidly along. A *distal segment* follows the intermediate segment. It is here that channels can be found for the reabsorption of $Na^+$, $K^+$, and $Cl^-$; this reabsorption is an active transport process involving ATPases to catalyze energy release. The ends of the many nephrons come together to form ducts that gather the urine remaining in the nephron tubules and carry it to the urinary bladder for regular expulsion into the surrounding water.

These basic features may be recognized, although in modified form, in later, more sophisticated renal organs. In freshwater fish, no special devices for water reabsorption occur. The nitrogenous end product in these organisms is ammonia, which passes out across the gills as well as in the urine. Since freshwater fish produce copious urine, ammonia levels do not reach toxic concentration.

In marine fish the filtrate flows more slowly through the nephron, permitting a greater opportunity to reabsorb fresh water from the filtrate. Marine fish drink salt water to overcome the loss of water to their hypertonic environment. Divalent cations in the body are pumped *into* the nephron at the posterior part of the proximal segment; consequently, seawater intake results in a net gain of fresh water. An aglomerular kidney is also found in some marine fish—a situation which reduces considerably the amount of fluid which passes into the urine.

The habitat of most amphibians is generally rather moist, and their kidneys are quite similar to those of freshwater fish. Reptiles, however, are often found in *xerotic* (dry) environments, and their kidneys reflect the stresses of such a habitat. In some marine reptiles, extrarenal devices have evolved to deal with the excretion of excess salts. These salt-excreting glands, similar to those of marine birds, are usually found adjacent to the nose. Land reptiles form uric acid as their main nitrogenous end product, thus conserving water. Other reptiles, especially those living in marine environments, excrete urea. Crocodiles excrete ammonia.

Both mammals and birds possess kidneys that are remarkably efficient, especially in terms of water conservation. The basic mechanism for producing a highly concentrated urine involves the creation of a markedly hypertonic environment around the nephron, so that water leaves the tubules by osmosis. Differences in the selective permeability of various regions of the nephron tubules also play a role.

## 20.4 HUMAN KIDNEY FUNCTION

The remarkable ability of the human kidney to produce a highly concentrated urine is believed to depend on the formation of a very steep concentration gradient in the renal medulla, through which a hypoosmotic urine must pass both in the descending arm of the loop of Henle and in the collecting tubule. A great deal of reabsorption from the crude filtrate is thought to occur in the proximal convoluted tubule. Water, amino acids, glucose, and ions all leave the tubule and are absorbed into the peritubular capillaries.

In the *countercurrent multiplier theory* (see Fig. 20.4), the loop of Henle, because it turns back on itself, creates a countercurrent multiplier system. As the filtrate descends the loop, large amounts of water pass by osmosis into the increasingly concentrated tissue fluid around the tubule and then into the peritubular capillaries. At the same time, some sodium and chlorine ions diffuse into the tubule. The ascending arm of the loop, which is impermeable to water, actively expels Na$^+$ ions (and Cl$^-$ ions follow passively). These ions enter the fluid of the medulla, thus maintaining its concentration gradient,

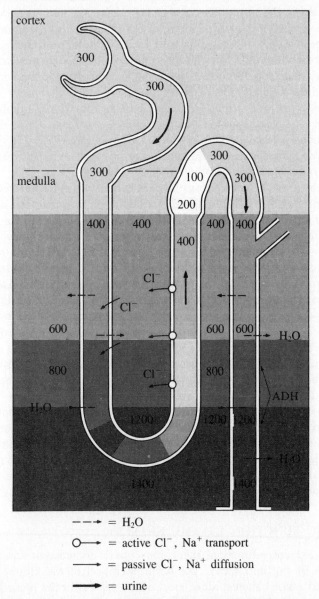

- - - → = H$_2$O
O—→ = active Cl$^-$, Na$^+$ transport
——→ = passive Cl$^-$, Na$^+$ diffusion
▬▬▶ = urine

**Fig. 20.4** Countercurrent multiplier model. (*Modified from Storer et al.*)

and are free to reenter the descending arm of the loop to be recycled. Thus, $Na^+$ and $Cl^-$ travel in a circuit, from ascending arm to descending arm around the loop and back to ascending arm. As the filtrate descends the loop, it becomes increasingly concentrated; however, by the time it reaches the distal convoluted tubule, so many ions have been actively transported out of the tubule that the osmolality has fallen to a level even lower than when it entered the proximal tubule. (Remember, the ascending arm is impermeable to water, so water cannot follow the ions and thus keep the osmolality high.) As the filtrate (by now it can be considered urine) passes down the collecting tubule, it makes another journey through the concentration gradient of the medulla. Depending on the permeability of the collecting tubule, further osmosis of water out of the nephron and into the peritubular capillaries may occur, leaving a highly concentrated urine behind.

In the mid-1970s evidence accumulated that the ascending arm of Henle's loop does not contain the channels for ion transport necessary to achieve active extrusion of $Na^+$ and $Cl^-$ ions from that portion of the tubule. Considerable doubt also exists as to the permeability of the walls of the descending loop to $Na^+$ or $Cl^-$. Thus, the countercurrent multiplier hypothesis has been recast.

The *two-solute model* is the result. This postulates that renal efficiency, in terms of producing a concentrated urine, derives from the participation of both NaCl *and urea* in urine production. It also involves *differential* permeability of the nephron, especially the loop of Henle, to salt and water; but the passage of urea across the tubular wall and its buildup immediately around the tubule are crucial. Compared with the original model, the two-solute model postulates a simpler countercurrent action between ascending and descending arms of Henle's loop.

The two-solute model maintains that the filtrate passing into Henle's loop is essentially isotonic with blood plasma. As it moves down the descending arm, it loses water to the hypertonic exterior. This external hypertonicity is largely due to high levels of urea. The walls of the descending tubule are highly permeable to water but virtually impermeable to salt and urea. This results in the production of a relatively concentrated filtrate within the tubule.

At the bottom of the loop the tubule loses its permeability to water but becomes quite permeable to NaCl, which concentrates in the filtrate there. As the filtrate moves up the ascending arm, salt can simply diffuse out to the hypotonic tissues. Near the top of the ascending arm, however, salt must be pumped from the filtrate by active transport. The considerable egress of salt lowers the osmotic pressure within the tubule and so encourages additional water loss further along the nephron. However, the urea originally deposited within the capsular fluid is now quite concentrated in the tubule because of the aforementioned water loss. Water leaves the filtrate in the distal tubule, and the concentrated filtrate from many nephrons is finally brought to the collecting ducts. The collecting ducts are, at least at their lower portions, quite permeable to urea. Since urea levels in the filtrate are much higher than in the surrounding tissue, urea passes out of the ducts and contributes to the high osmotic pressure of the surrounding tissue—a high osmotic pressure which resulted in water loss in the descending arm of Henle's loop and which continues to produce emigration of water from the urine in its continuing passage along the collecting tubule. It is clear that both urea and salt contribute to the extraction of water from the filtrate that moves from the capsule to the ureter.

The two-solute theory does not abandon all the features of the countercurrent multiplier hypothesis. The bottom of the loop does contain the highest levels of salt, and these salt concentrations eventually create a hypertonicity in the surrounding tissue that results in water being extracted from the filtrate, especially from the collecting ducts. The loss of water from the filtrate produces the high concentrations of urea within the tubule which eventually result in urea building up high osmotic levels in the tissue surrounding the descending arm.

## 20.5  HOMEOSTATIC FUNCTION OF THE KIDNEY

In addition to its excretory function, the kidney has a homeostatic function; it is a locus for homeostatic regulation of salt and water. The basic operation of the kidney lends itself to such a maintenance of salt and water balance. Since most of the solutes in the plasma pass into the capsule of the nephron in the initial filtration process of urine formation, the blood need reabsorb from the rest of the tubule only those materials necessary to maintain appropriate homeostatic levels for each constituent. Further, by regulating the amount reabsorbed, a balance in the blood may be effected at the renal level. Glucose is generally completely reabsorbed in the initial part of the proximal tubule. However, in a diabetic individual, whose blood levels of glucose may be very high, a considerable portion of the glucose present in the initial filtrate may remain in the urine and pass out of the body. This pathological phenomenon, known as *glucosuria*, is often the first sign of diabetes. The sugar in the urine increases its osmotic pressure, which results in the drawing of more water into the nephron and formation of copious urine. Thirst increases in response to this loss of water.

*Antidiuretic hormone (ADH)* is the primary agent for maintaining water balance. It is elaborated by a relatively small region of the *hypothalamus*, a highly versatile part of the brain lying just above, and helping to regulate, the pituitary gland. ADH increases the permeability of the walls of the collecting ducts to water. Since the collecting ducts are surrounded by a hypertonic medium, water leaves the ducts and is conserved by the blood vessels lying beside the ducts. This retention of water under the influence of ADH results in the production of a concentrated urine. On the other hand, a diminution of ADH prevents reabsorption of water by the blood vessels in the kidney, and a more dilute and copious urine forms.

Just as ADH acts on water permeability, so *aldosterone*, a steroid of the adrenal cortex, regulates the active transport of sodium ions. The site of aldosterone action is the ascending arm of the loop of Henle. In the presence of high levels of the hormone, $Na^+$ is reabsorbed from the filtrate within the tubule into the surrounding vessels—a salt-conserving process. Very low levels of aldosterone are associated with great losses of sodium ion in the urine and a rise in urine volume as a result of its increased osmotic pressure.

The amount of aldosterone produced in healthy individuals is generally influenced by dietary salt intake so that a balance exists between sodium intake and sodium outflow.

## 20.6 ACCESSORY EXCRETORY STRUCTURES

Specialized epithelial structures known as *sweat glands (sudoriferous glands)* are located in the skin. In humans, these sweat glands are present everywhere but are particularly concentrated in the palms and soles of the feet. Since the sweat glands are activated by the sympathetic branch of the autonomic nervous system (the same system that initiates "fight or flight" responses in emergency situations), excessive fear or similar nervous response produces an abundance of sweat, especially on the palms.

Sweat is a source of abundant water loss, but it may contain salts and even urea. The salts are chiefly made up of NaCl, with traces of $K^+$ and $Mg^{2+}$ ions as well. Under pathological conditions, sweat may contain bile pigments, protein, and even sugars. On occasion, blood has been found in sweat.

Tears, which are produced by the *lacrimal glands* lying in the superior outer corner of the orbit of the eye, also constitute a minute excretory product. They drain into the lacrimal canals at the inner corner of the eye socket and from there through a sac into the nose. During crying spells, a copious fluid discharge from the nose indicates that an excess of tears is being produced. Tears, which are modest in volume under ordinary circumstances, are essentially dilute NaCl solutions with occasional exotic materials also present. The AIDS virus has been found in tears at low levels. A major function of the tears, of course, is lubrication and moistening of the surface of the eye.

# Solved Problems

**20.1**   What are the advantages and disadvantages of eliminating nitrogenous wastes as ammonia?

When proteins are utilized as fuel molecules, or are otherwise being broken down, their amino acids are *deaminated*, or stripped of their amino groups. This produces ammonia and an organic product that may enter the glycolytic pathway or be used for synthetic processes. It would be simplest for the organism to deal with ammonia as the major nitrogenous excretory product, since it is the initial by-product and would thus require little additional energy to be excreted as is.

However, a disadvantage of ammonia is its relatively high toxicity. This can be a serious problem where the water available is not sufficient to dilute the ammonia. For this reason, other nitrogenous end products have evolved. The formation of these alternative products requires both energy and organic materials.

**20.2**   What are the advantages of urea and uric acid as end products of nitrogen excretion?

Since evolution is an opportunistic process, the final character of metabolic patterns cannot be predicted on the basis of neat, logical expectations. In mammals, a complex series of reactions, largely endergonic in nature, evolved in which two molecules of $NH_3$ combine with one molecule of $CO_2$ to produce a water-soluble substance, urea $[(NH_3)_2CO]$. Urea is not toxic at moderate levels, and in that respect, in animals that do not live in a freshwater environment, it is a significant improvement over $NH_3$. It should be noted that the formation of urea involves both major excretory products, ammonia and carbon dioxide. Since urea is soluble in water, it requires some of that precious solvent in its passage out of the organism.

In insects, which may be judged the most successful conquerors of the terrestrial environment, nitrogen is excreted almost exclusively as uric acid. Since uric acid is insoluble in water, it requires virtually no aqueous carrier and is an ideal end product under circumstances of water shortage. In most insects uric acid is excreted as dry crystals.

In all land reptiles and birds, the final nitrogenous excretory product is also uric acid. For these egg-laying vertebrates, uric acid is particularly effective as an excretory product, since it can pass out of the embryo and accumulate as a solid, nontoxic product within the allantois. In this way, it does not interfere with the development of the embryo.

**20.3**   Marine elasmobranchs (e.g., sharks) have evolved unusually high levels of urea in their blood-streams. Why do you suppose this has occurred?

Among marine elasmobranchs, water is constantly drawn from the organism by the hypertonic seawater in which they live. Thus, despite their being surrounded by water, they are threatened by dehydration through osmotic stress.

To surmount this osmotic challenge the cartilaginous fish have evolved the ability to build up as much as 2 percent urea in their bloodstream. They are essentially using the nitrogenous excretory product as osmotic ballast, bringing the osmotic pressure of their body fluids up to that of the surroundings so that they no longer lose water to their surroundings. This is a particularly ingenious linking of excretory function with water balance.

**20.4**   Why do many marine fish drink seawater, whereas freshwater fish do not drink at all?

Marine fish, living in a hypertonic environment, suffer from a shortage of available water. They drink seawater and extract the water from the salt solution for their metabolic needs. However, they are left with excess salts, which must then be excreted. This is largely accomplished through glands in the gills that excrete NaCl. A variety of ATPases participate in the reaction, which is highly endergonic. In some marine fish special rectal glands excrete excess salts into the terminus of the digestive tract, where they pass into the external environment.

Since freshwater fish are surrounded by a hypotonic environment, their problem is quite different. They must get rid of the fresh water that is constantly moving into their bodies. To drink water would be counterproductive, and so they never drink. Instead, the little salt that is present in the water around them is drawn in across the gills. This process also requires energy and involves the participation of ATP-splitting enzymes in glands within the gill.

**20.5**   *Osmoconformers* are organisms whose internal osmotic characteristics are the same as those of the environment. Further, when changes in osmotic pressure occur in their environment, these organisms soon adjust internally by assuming the altered tonicity. Such a passive osmotic character is particularly prominent in marine invertebrates. *Osmoregulators* are organisms that can maintain a constant internal osmotic environment independent of the external environment. This independence is gained, of course, only with considerable energy expenditure. What are some of the mechanisms by which osmoregulators control their internal environment?

The major systems employed in osmoregulation are those that also participate in excretion. Although the contractile vacuoles of freshwater protozoans are almost exclusively involved in water balance, the nephridial systems of the annelid worms handle both wastes and water. This is also the case for the Malpighian tubules of insects. These outgrowths of the digestive tract accumulate both wastes and water, which are then brought to the hindgut, where water is reabsorbed into the circulatory system.

In the vertebrates, osmotic regulation is intimately linked to excretion within the kidney tubule. In birds and reptiles the possibility of producing a highly hypertonic urine provides a significant mechanism for retaining water in the face of a desiccating environment. In many fish, the gills play a role in maintaining appropriate levels of salt in the body, and this aids the kidney in its hydrostatic function. Since the gills of freshwater fish excrete ammonia, the gills can be viewed as performing both excretory and osmotic functions. Urea, a major excretory product, has already been discussed in terms of providing osmotic ballast in marine elasmobranchs, which are continuously threatened with water loss. Urea is also utilized in creating a hypertonic urine in the convoluted tubule of the nephron and the collecting tubule.

**20.6**    What is the advantage of an aglomerular kidney?

In some marine fish the glomerulus is considerably reduced or even absent. The major functional arm of the kidney becomes the tubule along which reabsorption and some secretion occur. Since the glomerulus is the major site for water loss from the blood, an adaptive advantage exists in reducing the glomerular function in marine vertebrates facing a problem of dehydration. The aglomerular kidney, along with other modifications of the tubular nephron, occurs in only a few species of marine fish, but it does indicate that structure is relatively plastic and may be modified to effect success in survival.

**20.7**    Create a flow chart showing the movement of blood from the aorta, through the various blood vessels associated with the kidney, to the posterior vena cava.

Aorta → renal artery → arteriole → capillaries of glomerulus → arteriole from Bowman's capsule → peritubular capillaries → venule → renal vein → posterior vena cava

**20.8**    Summarize the major points of the countercurrent multiplier theory and the two-solute theory.
   Both postulate that

1.   a countercurrent concentrates salt and raises osmolality of tissue fluid in the renal medulla;
2.   the bottom of the loop of Henle contains the highest level of salt;
3.   the surrounding, hypertonic tissue extracts $H_2O$ from the filtrate moving through the loop of Henle and the collecting ducts.

|  | Countercurrent Model | Two-Solute Model |
|---|---|---|
| Descending loop | 1. Permeable to $Na^+$, $Cl^-$, $H_2O$<br>2. Filtrate hypoosmotic (to start)<br>3. Egress of $H_2O$ from filtrate by osmosis<br>4. $Na^+$, $Cl^-$ diffuse in<br>5. Filtrate becomes increasingly concentrated | 1. Permeable to $H_2O$; relatively impermeable to $Na^+$, $Cl^-$, urea<br>2. Filtrate isotonic (to start)<br>3. Egress of $H_2O$ from filtrate by osmosis<br>4. Filtrate becomes increasingly concentrated |

| | Countercurrent Model | Two-Solute Model |
|---|---|---|
| Bottom of loop | Highest concentration of salt in filtrate is reached | 1. Permeable to $Na^+$, $Cl^-$; impermeable to $H_2O$<br>2. Influx of $Na^+$, $Cl^-$ into loop<br>3. Highest concentration of salt in filtrate is reached |
| Ascending loop | 1. Impermeable to $H_2O$<br>2. Active transport of $Cl^-$ out; passive diffusion of $Na^+$ out<br>3. Progressive decrease in osmolality within tubule<br>4. Urea is concentrated | 1. Impermeable to $H_2O$<br>2. $Na^+$ and $Cl^-$ diffuse out, except at top of loop, where active transport occurs<br>3. Progressive decrease in osmolality (osmotic pressure drops) within tubule<br>4. Urea is concentrated |
| Collecting duct | Further concentration of urea before excretion | 1. Further concentration of urea before excretion<br>2. Permeable to urea<br>3. Some urea diffuses out<br>4. Urea-induced hypertonicity of tissue draws $H_2O$ out of loop of Henle |

**20.9** Production of ADH is influenced by many physical and chemical factors. What would you guess is the effect of alcohol on production of this hormone? Given ADH's homeostatic function, how do you suppose high blood pressure affects ADH production?

Increased ADH production increases the permeability of the collecting tubule of the nephron, allowing more water to be reabsorbed into the body and retained. Beer drinking greatly increases urinary output; this reflects the fact that alcohol depresses ADH production, so that less water is reabsorbed through the collecting tubule walls and therefore passes out of the body as urine. (Agents such as alcohol that inhibit reabsorption of water from the tubules are called *diuretics*.) High blood pressure has a similar effect. Since elevated pressure is often associated with increased fluid in the body, this pressure can be lowered through excretion of fluid. Thus ADH levels diminish when blood pressure rises. The corresponding decrease in permeability of the collecting tubule walls causes more water to be excreted, reducing fluid levels in the body.

**20.10** In one type of adrenal carcinoma known as *pheochromocytoma*, extremely high levels of aldosterone are induced by the tumor cells. This in turn has a very dangerous effect on the body. What would you guess this effect is? Explain.

Since aldosterone makes the ascending arm of the loop of Henle permeable to sodium ions, large amounts of it cause a correspondingly large migration of sodium ions into the tissue around the loop, leaving a very dilute urine to enter the collecting tubule. Because of the resulting low osmotic pressure of the urine, large amounts of water leave the tubule. Thus, fluid retention in the body is greatly augmented by high levels of aldosterone, with a concomitant jump in blood pressure.

# Supplementary Problems

**20.11** Excretion in most animals is closely tied to (*a*) water regulation. (*b*) salt balance. (*c*) neural integration patterns. (*d*) both (*a*) and (*b*). (*e*) both (*a*) and (*c*).

**20.12** In protozoans lacking contractile vacuoles, nitrogenous wastes are excreted by *(a)* direct diffusion across the plasma membrane. *(b)* Malpighian tubules. *(c)* breakdown in the nucleus. *(d)* phagocytosis. *(e)* all of these.

**20.13** Which of the following organs can be regarded as excretory in function? *(a)* kidney. *(b)* skin. *(c)* lungs. *(d)* lacrimal glands. *(e)* all of these.

**20.14** An excretory function of the liver derives from its excretion of breakdown products of hemoglobin. These are *(a)* RNA fragments. *(b)* bile pigments. *(c)* ozone particles. *(d)* all of these. *(e)* none of these.

**20.15** The filtration process occurring in the glomerulus is dependent on the high pressure of the blood in that capillary cluster.
*(a)* True *(b)* False

**20.16** The filtrate produced in the Bowman's capsule is virtually isosmotic with the circulating blood plasma.
*(a)* True *(b)* False

**20.17** Both marine and terrestrial environments require water conservation mechanisms. In all cases, this involves the utilization of urea to create an internal osmotic pressure that is relatively high.
*(a)* True *(b)* False

**20.18** Filtration and selective reabsorption are the only processes occurring along the nephron.
*(a)* True *(b)* False

**20.19** Proteins do not usually pass into the nephron across the glomerular network.
*(a)* True *(b)* False

**20.20** In diabetic persons not all the glucose of the initial filtrate is completely reabsorbed. This causes sugar to be present in the urine.
*(a)* True *(b)* False

**20.21** The nitrogenous excretory product of freshwater fish is likely to be uric acid.
*(a)* True *(b)* False

**20.22** Long loops of Henle are generally associated with organisms that can produce a highly concentrated urine.
*(a)* True *(b)* False

**20.23** The loop of Henle is found as the intermediate segment in mammals and *(a)* freshwater fish. *(b)* marine fish. *(c)* birds. *(d)* amphibians. *(e)* none of these.

# Answers

| | | | | | | | |
|---|---|---|---|---|---|---|---|
| **20.11** | *(d)* | **20.13** | *(e)* | **20.15** | *(a)* | **20.17** | *(b)* |
| **20.12** | *(a)* | **20.14** | *(b)* | **20.16** | *(a)* | | |

**20.18** *(b)* (They are the major events, but some secretion from blood vessels may also occur along the nephron.)

| | | | | | | | |
|---|---|---|---|---|---|---|---|
| **20.19** | *(a)* | **20.21** | *(b)* | **20.22** | *(a)* | **20.23** | *(c)* |
| **20.20** | *(a)* | | | | | | |

# Chapter 21

# Hormones and Chemical Control

Two types of integration exist within animal organisms. In one, the *nervous system*, a communication network is formed that affords rapid and specific connections within the body. The nerve impulse is the means by which messages travel from one region to another. The organizing and directing influences of the nervous system reside in the brain and spinal cord of the central nervous system.

The nervous system's type of integration, with its "hard-wiring," in which impulses travel along well-defined tracts, is in sharp contrast to the second type of integration, the *endocrine system*. Endocrine glands are localized secretory structures that form and release chemicals called *hormones* directly into the general circulation. These hormones are then broadcast to all parts of the body and may produce far-reaching effects on a variety of structures. Since the hormones produced are not channeled into ducts to serve a specific site, the endocrine glands are also called *ductless glands*.

Hormones, derived from a Greek root meaning "to excite," may actually inhibit certain processes while stimulating others. Hormones may exert their effects on responsive structures by (1) altering gene function, (2) directly affecting metabolic pathways, or (3) controlling the development of specific organs or their secretory products. More detailed descriptions of hormone action will be given in discussion of individual endocrine glands. Organs that respond to a hormone are called the *target organs* of the hormone.

In some cases, a combination of neural and hormonal mechanisms functions in intimate associations to achieve integration. Such mechanisms make up a *neurohumoral* system, in which nerve tracts exert their effects through the secretion of particular hormones known as *neurohumors*. Neurohumoral mechanisms are particularly prominent in the functional axis that joins the brain and the pituitary gland.

## 21.1 EARLY ENDOCRINE SYSTEMS

Endocrine structures and integrating hormones exist in invertebrates, but they have not been as extensively explored as the vertebrate hormones. Although the identification of specific invertebrate hormones and an understanding of their mode of action are not complete, we will study endocrine function in these organisms because it provides an insight into the evolution of hormone action in all animal organisms. Best known are the endocrine systems associated with growth and metamorphosis in insect groups.

**EXAMPLE 1** Even in insects in which the immature metamorphic forms are not radically different from the adult's, hormones play a role in the successive stages leading to full maturity. The earliest (often wormlike) forms are known as *larvae*. Since invertebrates such as the insects possess a restraining exoskeleton made of hard chitin, each growth stage is accompanied by a shedding of this skeleton. This shedding process is called *molt*. Each immature stage produced on the ontogenic (pertaining to the development of the individual) journey toward adulthood is known as a *nymphal stage*. The molt following the final larval stage produces a *pupa*; during the pupal stage, extensive morphogenetic rearrangements occur that yield a fully developed adult at the next molt.

Each successive molt is induced by a hormone called *ecdysone*, secreted by the *prothoracic glands* located in the first body segment behind the head (see Fig. 21.1). Ecdysone production is in turn stimulated by *brain hormone*, a peptide produced in neurosecretory cells of the forebrain.

Although ecdysone induces each molt, it is the blood levels of *juvenile hormone*, produced by a pair of glands (the *corpora allata*) in the hindbrain, that determine what the stage following the molt will be. When the titer of juvenile hormone is high, another larval stage follows. When juvenile hormone blood levels drop, the pupal stage ensues. When juvenile hormone completely disappears, the next molt produces an adult.

**EXAMPLE 2** The discovery of *insulin*, a hormone that regulates carbohydrate metabolism in mammals and birds, in such invertebrates as starfish led to the realization that insulin was a primitive *feeding hormone*. In the invertebrates

molting hormone (α-ecdysone) of silkworm

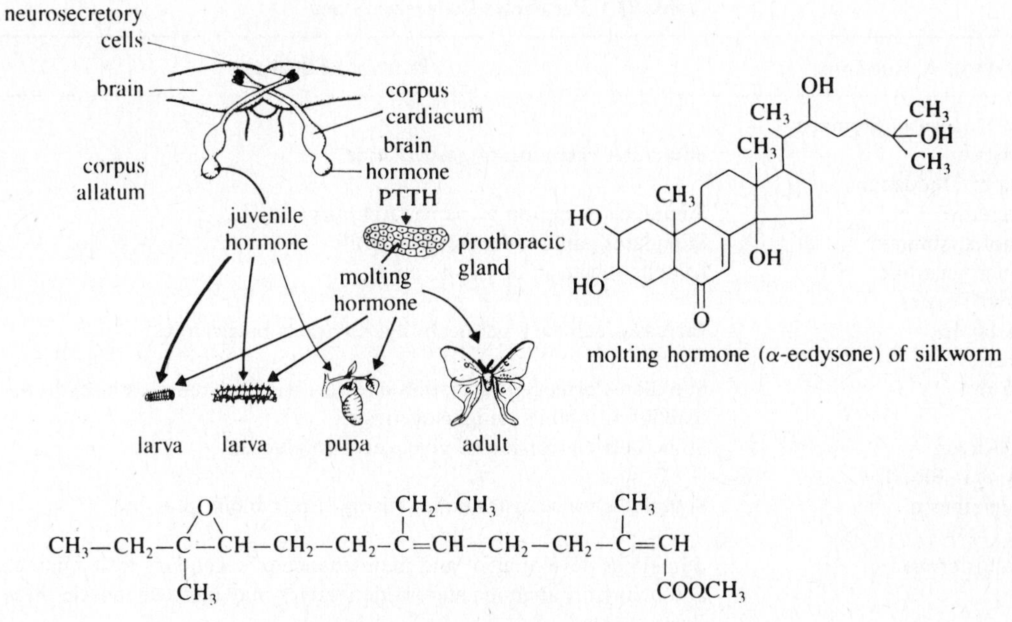

juvenile hormone of silkworm

**Fig. 21.1**   Hormonal influence on insect metamorphosis.

the hormone promotes food gathering. In the higher vertebrates it is still functioning as a feeding hormone but in a much more complex way. It promotes the storage and utilization of carbohydrate so that sugar levels in the blood do not rise too high during and following a meal.

In a variety of invertebrates, hormones control sexual cycles and often are directly involved in the shedding of eggs. All the arthropods demonstrate rather extensive endocrine systems, which play a role in water balance, migration of pigments involved in protective coloration, and growth.

Hormones are usually regarded as chemical secretions which are produced in one part of an organism by a specialized endocrine structure and which have profound metabolic effects on target structures at distant sites in the body. Further, hormones are effective at low concentrations and may have radically different effects on different target tissues or at different concentrations. It is therefore rather intriguing that the same hormone may also be present in different *organisms* and have a different function in each. Protozoans and even bacteria produce substances that appear identical with the hormones of multicellular forms. An insulin produced by *E. coli* tests exactly like horse or human insulin in standard glucose oxidation measurements. Pituitary hormones have been isolated from protozoans. Since the typical endocrine function could not be occurring in these unicellular forms, one must conclude that chemicals that arose at an early stage in life's evolution and may be useful in limited kinds of cellular communication later became part of the endocrine complex typical of advanced vertebrates.

## 21.2   VERTEBRATE ENDOCRINE SYSTEMS

The endocrine glands of vertebrates, particularly mammals, have been studied extensively. Their role in the maintenance of homeostasis has been elucidated in considerable detail. Some of the endocrine glands are exclusively endocrine in function. They may produce one or several hormones. Other endocrine glands, such as the pancreas, demonstrate significant endocrine and nonendocrine functions. Still other organs, such as the kidney and liver, are chiefly involved in nonendocrine functions, but they demonstrate minor hormonal production.

**Table 21.1 Vertebrate Endocrine Gland**

| Source & Hormone | Principal Effect(s) |
|---|---|
| **Pyloric Mucosa of Stomach** | |
| Gastrin | Stimulates secretion of gastric juice |
| **Mucosa of Duodenum** | |
| Secretin | Stimulates secretion of pancreatic juice |
| Cholecystokinin | Stimulates gallbladder to release bile |
| Enterogastrone | Inhibits secretion of gastric juice |
| **Damaged Tissues** | |
| Histamine | Increases capillary permeability; contracts bronchioles |
| **Pancreas** | |
| Insulin | Stimulates glycogen formation and storage; stimulates carbohydrate oxidation; inhibits neoglucogenesis |
| Glucagon | Stimulates conversion of glycogen into glucose |
| **Kidney plus Blood** | |
| Angiotensin | Stimulates vasoconstriction, causing rise in blood pressure |
| **Testes** | |
| Testosterone | Stimulates development and maintenance of secondary male characteristics and behavior; anabolic steroid derivatives may increase muscle mass (hypertrophy) |
| **Ovaries** | |
| Estrogen | Stimulates development and maintenance of secondary female characteristics and behavior |
| Progesterone | Stimulates secondary female characteristics and behavior; maintains pregnancy |
| **Thyroid** | |
| Thyroxine; triiodothyronine | Stimulate oxidative metabolism |
| Calcitonin | Prevents excessive rise in blood calcium |
| **Parathyroids** | |
| Parathormone | Regulates calcium-phosphate metabolism |
| **Thymus** | |
| Thymosin | Stimulates effective immunological responses in lymphoid tissues |
| **Adrenal medulla** | |
| Epinephrine | Stimulates the group of reactions commonly termed "fight or flight" |
| Norepinephrine | Similar in action to epinephrine, but causes more vasoconstriction and is less effective in stimulating conversion of glycogen into glucose |
| **Adrenal Cortex** | |
| Glucocorticoids (corticosterone, cortisone, hydrocortisone, etc.) | Inhibit incorporation of amino acids into protein in muscles and so lead to atrophy; stimulate formation (from noncarbohydrates) and storage of glycogen; help maintain normal blood sugar level |
| Mineralocorticoids (aldosterone, deoxycorticosterone, etc.) | Regulate sodium-potassium metabolism |
| Cortical sex hormones (adrenosterone, etc.) | Stimulate secondary sexual characteristics, particularly those of the male |
| **Hypothalamus** | |
| Releasing factors | Regulate hormone secretion by anterior pituitary |
| Oxytocin | Stimulates kidneys to reabsorb $H_2O$ and smooth muscles, including blood vessels, to contract |
| ADH | |
| **Anterior Pituitary** | |
| Growth hormone | Stimulates growth |
| Thyrotropic hormone | Stimulates the thyroid to produce and release $T_3$ and $T_4$ |
| Adrenocorticotropic hormone (ACTH) | Stimulates the adrenal cortex to produce and release glucocorticoids |

**Table 21.1  (continued)**

| Source & Hormone | Principal Effect(s) |
|---|---|
| **Anterior Pituitary  (continued)** | |
| Follicle-stimulating hormone (FSH) | Stimulates growth of ovarian follicles and of seminiferous tubules |
| Luteinizing hormone (LH) | Stimulates conversion of ovarian follicles into corpora lutea and secretion of sex hormones by ovaries and testes |
| Prolactin | Stimulates mammary glands to secrete milk |
| Melanocyte-stimulating hormone | Controls skin pigmentation in lower vertebrates by increasing melanin synthesis and dispersal, so that skin darkens |
| **Posterior Pituitary[1]** | |
| Oxytocin | Stimulates uterine muscles to contract and mammary glands to release milk |
| ADH | Stimulates increased water reabsorption by kidneys and constriction of blood vessels (and other smooth muscle) |
| **Pineal** | |
| Melatonin | Controls pigment distribution and circadian rythms. |

[1] Apparently stores hormones produced by the hypothalamus.

The major endocrine glands of vertebrates are listed in Table 21.1. There are approximately 15 of these glands; their secretions are of four basic types: (1) proteins,* (2) less complex peptides* of various length down to single amino acids, (3) catecholamines, and (4) steroids. Figure 21.2 shows the location of the major endocrine glands in the human body.

## INTESTINE

The first hormone discovered was a substance named *secretin*, which is released from the mucosa of the duodenum when it is stimulated by the acidic digest coming into the first part of the intestine from the stomach. W. M. Bayliss and E. H. Starling of London's University College first showed (in 1901) that the release of secretin in the circulatory system elicits the flow of digestive enzymes from the pancreas. Some years later a second digestive hormone, *cholecystokinin*, was discovered. It, too, is released from the duodenal mucosa and produces a flow of bile from the gallbladder.

## STOMACH

The stomach itself was found to be an endocrine organ in 1905. Partially digested materials, particularly meat, stimulate the mucosa of the lower portion of the stomach to secrete a *hormone* named *gastrin*. This gastrin travels in the bloodstream until it reaches the stomach once again, where it stimulates gastric glands to release gastric juice.

## PANCREAS

The *pancreas* is a dual gland which demonstrates both ducted (exocrine) properties and ductless functions. Its exocrine product constitutes a major digestive juice, which reaches the duodenum through the pancreatic duct. It was not until the end of the nineteenth century that the role of the pancreas as an endocrine organ clearly apart from its digestive function was established. Isolation of the hormone which exerts so powerful an effect upon carbohydrate metabolism was delayed because of the destructive effects of pancreatic trypsin upon the yet-to-be identified insulin. It was not until 1922 that the hormone was isolated from the pancreas of dogs in the medical college of the University of Toronto. The ingenious investigators Fred Banting and Charles Best first tied off the pancreatic ducts of many dogs and waited for the exocrine tissue to degenerate. They then removed the remaining pancreases and, keeping the temperature below freezing, extracted insulin from the tissue mash.

*Proteins are more than just long linear polypeptides. A single protein may contain more than one polypeptide chain and have a tertiary and a quaternary structure. And, of course, proteins are functionally more complex than polypeptides.

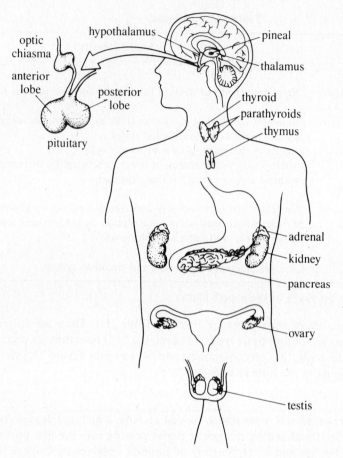

**Fig. 21.2**  Major endocrine glands in humans.

Scattered through the pancreas is the *islet tissue* (*islets of Langerhans*), which contains the hormone-secreting cells. The *alpha cells* secrete *glucagon*, which tends to raise blood sugar levels (Chap. 15); the *beta cells* produce *insulin*, which is probably the sole hormone responsible for reducing blood sugar levels. A small amount of *somatostatin* is also produced within the islet tissue, in the *delta cells*. This hormone has been associated with a dampening of growth and also serves as a modulator of nerve impulse transmission. Recently it has been shown to be involved in the regulation of insulin and glucagon production. Insulin and glucagon are actually small proteins, while somatostatin is a simpler peptide.

## ADRENALS

The *adrenal glands*, one of which lies above or, in some vertebrates, beside each kidney, are relatively small structures consisting of two distinct portions, each arising from a different germ layer. The entire gland is surrounded by a protective capsule.

The *adrenal cortex*, the outer layer, arises from mesoderm and, in turn, consists of three distinct layers. Immediately beneath the capsule is the *zona glomerulosa*, which is a major site for secretion of such *mineralocorticoids* as *aldosterone*, the chief sodium-controlling steroid in humans (see Sec. 20.5). The primary action of aldosterone is to promote the uptake of sodium from the filtrate in the distal renal tubule back into the blood. Aldosterone production is increased when blood potassium levels rise and, more significantly, when the blood pressure in arterioles leading to the glomeruli of the kidneys goes down, as when a blockage occurs.

**EXAMPLE 3**  Each kidney, in combination with the blood, constitutes an endocrine system that raises blood pressure in response to occlusion of the renal artery. When such a blockage occurs, the kidney cortex releases a

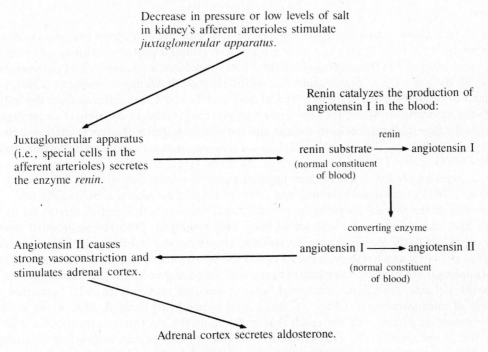

**Fig. 21.3**  Renin-angiotensin system.

chemical, *renin*, which reacts with a blood protein to form *angiotensin*, a vasoconstrictor (see Fig. 21.3). The combination of constricted flow and compensatory increased output by the heart raises blood pressure, thus tending to overcome the blockage of the renal artery. Angiotensin also causes the adrenal cortex to secrete additional aldosterone, thus further raising blood pressure.

*Glucocorticoids*, such as *cortisol* (*hydrocortisone*) and *corticosterone*, are secreted by the middle cortical layer, the *zona fasciculata*. These steroids raise blood glucose levels by promoting the breakdown of proteins to amino acids and their conversion in the liver to glucose. A marked breakdown of lymphatic tissue is also associated with elevated glucocorticoid levels, and this in turn is associated with a diminished immunological response. Stress is the initiating stimulus for release of the glucocorticoids. Nerve impulses caused by the stressing stimulus induce the hypothalamus, an endocrine gland in the brain, to secrete *corticotropin releasing factor* (*CRF*). This travels to the anterior pituitary, an endocrine gland closely associated with the hypothalamus, where it stimulates secretion of *adrenocorticotropic hormone* (*ACTH*), also called *corticotropin* or *adrenocorticotropin*. ACTH induces the secretion of the glucocorticoids, which in turn have a negative feedback effect on both the hypothalamus and anterior pituitary secretion of CRF and ACTH.

The innermost zone of the adrenal cortex, the *zona reticularis*, secretes some glucocorticoids and significant amounts of masculinizing androgens. A small amount of female sex steroids is also produced. Occasionally tumors of the cortex produce an excess of androgens, leading to marked masculinization in females.

The *medulla* of the adrenal gland arises from ectoderm and may be regarded as part of the autonomic nervous system. Direct stimulation by the autonomic nervous system causes it to secrete (in a 4:1 ratio) the catecholamine hormones *epinephrine* (*adrenaline*) and *norepinephrine* (*noradrenaline*), which are significant in mediating emergency ("fight-or-flight") responses. In fact, their effects on the organs of the body are almost identical with the effects of direct sympathetic stimulation.

Because the pituitary and adrenal glands often work together in response to a variety of physiological stresses, the concept of a *pituitary-adrenal axis* has developed. Hans Selye, a Canadian physiologist, is credited with the theory of a *general adaptation syndrome* in response to stress, which is largely rooted in a common response by pituitary, adrenal cortex, and adrenal medulla in all mammals.

**THYROID**

The *thyroid gland* arises from the embryonic gill slits. Its hormones are *thyroxine* ($T_4$) and *triiodothyronine* ($T_3$). Both $T_4$ and $T_3$ are complexed with globular proteins within the *follicles* (swollen sacs) of the thyroid and in this state are identified as *thyroglobulin*, a viscous colloid representing stored hormone. In the presence of *thyrotropin* (also called *thyroid-stimulating hormone*, or *TSH*), produced in the anterior pituitary, thyroglobulin is hydrolyzed, and $T_4$ and $T_3$ are released from the follicles and pass into the bloodstream, where they are carried by plasma proteins to a variety of target organs. TSH also causes the thyroid follicle cells to enlarge and increase in number, leading to increased production of the thyroid hormones. The release of TSH by the pituitary is in turn induced by *thyrotropin releasing hormone* (*TRH*) secreted by the hypothalamus. As blood levels of thyroid hormones rise in response to TSH, a negative feedback mechanism begins to inhibit production of both TSH in the anterior pituitary and TRH in the hypothalamus, and levels of thyroid hormones subsequently fall.

One effect of the thyroid hormones is to influence the rate at which carbohydrates are oxidized in the body and, consequently, the amount of body heat produced. (The resting level of oxidation is referred to as the *basal metabolic rate*, or *BMR*.) Oxygen uptake is increased through an enhanced synthesis of glycolytic and Krebs cycle enzymes and also by an uncoupling of oxidative phosphorylation. This uncoupling is achieved by a swelling of the mitochondria induced by thyroid hormone; the swelling disrupts the delicate subcellular machinery linking electron transport to ATP formation (see the discussion of chemiosmosis in Chap. 5). Since ATP is not being formed, little or no inhibition of oxidative processes occurs, with the result that traffic along the electron-transport chain increases.

Thyroid hormones are also crucial in the development and sexual maturation of all vertebrate species. They are particularly important for maturation of brain and other central nervous system structures. In vertebrates such as amphibians, the thyroid not only regulates growth and sexual maturation but also controls metamorphosis.

As far back as 1876 investigators realized that a second type of cell exists in the thyroid gland, in addition to the follicular cells forming the epithelial layer. Little was known of the function of these cells until 1962, when they were discovered to be the source of *calcitonin*, a polypeptide that lowers blood levels of calcium. They have since been designated *C cells* because of their hormone product.

**PARATHYROIDS**

As the name implies, the *parathyroid glands* are embedded in the thyroid, a pair on each side. The *chief cells* of the parathyroid secrete a polypeptide hormone, *parathormone*, a primary regulator of calcium and phosphate levels in the blood. Parathormone acts in two ways:

1. It induces such diverse tissues as kidney tubules, bone, and intestine to release calcium into the bloodstream. In the case of bone, both calcium and phosphate are released as the mineralized material is broken down.
2. Although it promotes reabsorption of calcium into the efferent arteriole of the kidney tubule, it facilitates excretion of phosphate into the urine.

Thus, parathormone tends to raise plasma levels of calcium and lower levels of phosphate.

Calcium plays a ubiquitous regulatory role in terms of neuromuscular function, permeability of cell and intracellular membranes, control of both hormonal and enzyme reactions, and clotting of blood.

**THYMUS**

The endocrine function of the *thymus* was unrecognized for a long time. It is now known that the hormone *thymosin*, released by the thymus, is responsible for inducing functional maturity in developing lymphocytes.

**GONADS**

The endocrine function of the *gonads* is closely associated with their sexual role. The *androgenic* (male-producing) steroids of the *testes* are responsible for the secondary sexual characteristics of males. The major androgen is *testosterone*.

The female sex steroids produced by the ovary are *estrogens* (chiefly *estradiol*) and *progesterone.* The estrogens are responsible for initiating and maintaining the female secondary sex characteristics and for initiating the menstrual or estrous cycle. As the name implies, progesterone prepares the reproductive tract for pregnancy. It is also the major constituent of the birth control pill, effecting suppression of ovulation. The fluctuations in the levels of ovarian sex steroids are responsible for the periodic changes of the female reproductive tract during the menstrual and estrous cycles. As we saw in Chap. 12, these changes in ovarian hormone levels are in turn mediated by the hormones of the hypothalamus and anterior pituitary.

## PITUITARY

The *pituitary gland* consists of a posterior and an anterior lobe. The *posterior pituitary* is formed as a downgrowth of the hypothalamus in the brain and is primarily neuronal. It does not produce any hormones itself but receives, stores, and eventually releases two peptide hormones synthesized in the hypothalamus—*oxytocin*, which causes contraction of the uterine muscles at birth and the forceful release of milk in nursing mammals, *antidiuretic hormone* (*ADH*), which causes the body to reabsorb water from the urine.

The *anterior lobe* arises from pharyngeal epithelium as an invagination called *Rathke's pouch.* It is the anterior pituitary that contains the hormone-producing structure, the *adenohypophysis.* This structure produces both primary and *tropic* hormones (hormones that induce other endocrine glands to release hormones).

The primary hormones of the adenohypophysis are *growth hormone* (*GH*), *prolactin, melanocyte-stimulating hormone* (*MSH*), and a variety of *endorphins* and *enkephalins.* Growth hormone regulates growth; however, excess GH in early life produces *gigantism.* In mature years it leads to abnormal bone formation, since bone growth in length is limited. Such malformed overgrowths constitute the condition known as *acromegaly.* Prolactin has widespread effects in a variety of vertebrates, particularly in terms of fluid and electrolyte balance. In mammals it is chiefly involved in milk production. MSH functions in the regulation of pigment cells in the skin of some vertebrates, but its role in mammals is not completely clear. Both endorphins and enkephalins, derived from the splitting of a large protein precursor molecule in the pituitary, appear to function as pain inhibitors.

The anterior pituitary is often referred to as the master gland because, as we have seen, in addition to various primary hormones, it produces the *tropic* hormones, which control secretion in many of the other endocrine glands. Among these tropic hormones are TSH, ACTH, follicle-stimulating hormone (FSH), and luteinizing hormone (LH).

## HYPOTHALAMUS

The *hypothalamus* is a part of the brain located just above the pituitary. It is here that many of the sensory stimuli of the nervous system are converted into hormonal responses. The hypothalamus accomplishes this in two ways, reflecting its dual relation with the pituitary.

The hypothalamus is connected to the posterior lobe by a stalk through which nerves run. It is believed that oxytocin and ADH are produced in the hypothalamus and travel down the nerves to the posterior lobe for storage. They are released from their storage area by nerve impulses from the hypothalamus.

Although the hypothalamus is not structurally a part of the anterior lobe, it is connected by an unusual capillary system. Capillaries leaving the hypothalamus reform into several veins; however, unlike most veins, they do not empty into an increasingly larger venous system. Instead, they enter the anterior pituitary and break into a second capillary bed before reforming and emptying into the normal venous system. We have studied several of the releasing hormones produced by the hypothalamus (see Table 21.2). Nerve impulses traveling to or originating in the hypothalamus cause secretion of these releasing hormones, which then enter the gland's capillary system and travel to the capillary bed of the anterior pituitary, where they stimulate release of their particular tropic hormone.

**Table 21.2 Hypothalamic Releasing Hormones***

| Name of Hormone | Abbreviation |
|---|---|
| Corticotropin releasing hormone | CRH |
| Thyrotropin releasing hormone | TRH |
| Luteinizing hormone releasing hormone | LH-RH |
| Follicle-stimulating hormone releasing hormone | FSH-RH |
| Growth hormone releasing hormone | GH-RH |
| Growth hormone release-inhibiting hormone | GH-RIH |
| Prolactin release-inhibiting hormone | PRIH |
| Prolactin releasing hormone | PRH |
| Melanocyte-stimulating hormone release-inhibiting hormone | MRIH |
| Melanocyte-stimulating hormone releasing hormone | MRH |

*Also called *factors*, so in other books, the abbreviations may have an F instead of an H at the end.

## PINEAL GLAND

The *pineal gland*, lying within the skull, was long associated with the folklore of a mysterious "third eye." Beginning with the 1950s a clear-cut relation was established between pineal function and daylight-night ratios. In a number of lower vertebrates the pigment distribution in pigment cells is altered by a pineal hormone called *melatonin*. This hormone is responsible for lightening the skin in a light environment and darkening the skin in dim light or darkness. In many vertebrates, including the mammals, melatonin may play a role in the maintenance of circadian rhythms, the cyclic alterations in function that regularly occur over roughly a 24-h period.

## DAMAGED TISSUE

Any damaged tissue acts as an endocrine gland when it releases *histamines*. These hormones relax the muscles of the blood vessels; this increases the vessels' permeability, allowing the elements of the immune system to reach the injured site.

## PROSTAGLANDINS

Another group of chemicals, the *prostaglandins*, produced by most cells of the body, also demonstrate characteristics of hormones. Although they are modified fatty acids that usually have only local effects and are rapidly destroyed, they fulfill a variety of regulatory functions and may exert their effects over long distances, thus at least resembling hormones. Three broad classes of prostaglandins have been identified since they were first isolated from the male reproductive tract—*PGA*, *PGE*, and *PGF*. The PGA derivatives are known to reduce blood pressure and may act directly on vascular smooth muscle. PGE compounds influence acid secretion in the stomach and have been implicated in fever reactions. The anti-inflammatory roles of substances such as aspirin and Indocin (indomethacin) derive from the fact that they inhibit formation of PGE. Various components of PGF are involved in responses of the reproductive tract and have been used clinically to induce labor.

## 21.3  MODE OF ACTION OF HORMONES

Hormones exert their effects on target tissues, directly or indirectly, by an alteration of the metabolic activity of specific cells or by interacting with the genome to turn genes on or off or to modulate their activity. In order to carry out these physiological assignments the hormone must either penetrate the cell or set into motion a chemical train of events from an attached position on the membrane. Some hormones may pass directly across boundary and internal membranes of the cell, while others pass along preexisting channels or create new channels upon attachment to the cell.

Many hormones attach to specific receptors on the cell membranes of target cells and invoke the aid of a so-called *second messenger*, that is, an "accomplice" in the cell cytoplasm. Calcium ions are known to serve as second messengers. *Cyclic AMP (cAMP)* is another such molecule (see Fig. 21.4).

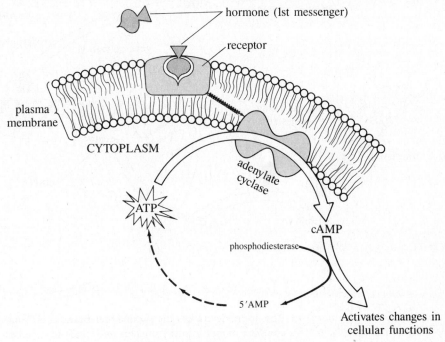

**Fig. 21.4**   Second messenger model.

When a hormone (the first messenger) attaches to a receptor, it causes the enzyme *adenylate cyclase* to convert ATP to *cAMP*. This cAMP then activates or deactivates certain enzyme systems specific for the particular cell, thus realizing the function of the attached hormone. *Phosphodiesterase* subsequently breaks cAMP down to simple AMP, thus ending the hormonal action. The AMP is then recycled into ATP. The second messenger often activates enzymes that are part of a system whose final step produces the actual end action of the hormone. Generally, this mechanism for hormone action is faster than the mechanisms involving modulation of the genome. Modulation of gene action may involve enhancement of transcription or of translation. Steroids, such as the glucocorticoids, usually tie up with a receptor protein in the cytoplasm, and this complex moves into the nucleus, where the effect upon the genetic machinery occurs (see Fig. 21.5). Differential binding of these hormone complexes to the chromatin of different types of cells may account for the specificity of hormone action. In the case of some hormones, such as *thyroxine*, binding may be to a protein receptor within the nucleus itself rather than to a cytoplasmic receptor.

**EXAMPLE 4**   Thyroid hormones are believed to move directly into the nucleus, where they interact with a nuclear receptor, which then produces changes in the activity of particular genes in sensitive tissues. There is some evidence that a multiplicity of thyroid nuclear receptors exists, including one for nerve tissue. Since the nervous system is relatively refractory to the metabolic effects (increase in oxygen consumption) of thyroid hormone, the existence of these receptors in brain and spinal cord was unexpected. Some of the thyroid receptors appear to be related to the products of oncogenes, and evidence exists for a similarity in structure between steroid and thyroid hormone receptors.

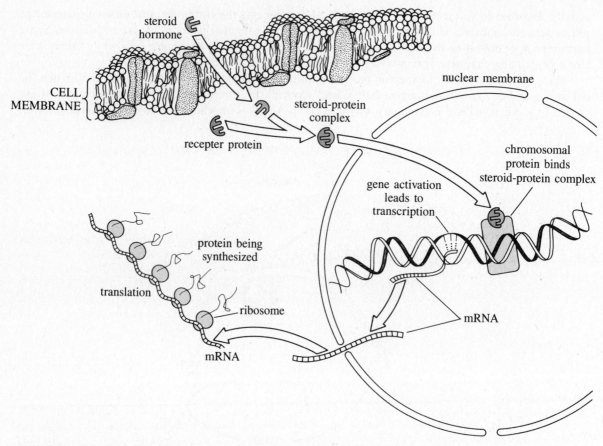

**Fig. 21.5**   Gene activation by a hormone. The hormone crosses the plasma membrane and forms a complex with a receptor protein in the cytoplasm. This complex moves into the nucleus and binds to a chromosomal protein, thereby inducing a specific gene to transcribe its code into mRNA. The mRNA exits the nucleus and is translated into protein by tRNA in the cytoplasm.

# Solved Problems

**21.1**   How do hormones differ from enzymes?

Hormones and enzymes are both effective in small amounts and are not consumed in the many metabolic processes they influence or initiate. But the resemblance soon ends. Enzymes are almost always protein (although some evidence for enzymatic properties of an RNA species has been reported), while hormones may be proteins, shorter peptides, single amino acids and derivatives, or steroids. Enzymes may be synthesized within a cell to function there or may be secreted to the exterior through active transport processes and even pass along a duct to a specific locale. Hormones are released directly into the bloodstream and are distributed throughout the body, where they may exert their effects upon a number of different target tissues.

Enzymes are generally highly specific, catalyzing a single reaction or type of reaction. Hormones have a range of effects, some of which may be quite profound. Hormones may demonstrate different but significant actions on different tissues in the same organism or show very different effects in different organisms. This is not usually the case for enzymes.

Hormones play a key role in the maintenance of their levels through negative feedback involving tropic hormones and the releasing factors of the hypothalamus; this is not paralleled by enzymes.

**21.2**    What would you predict would happen if the corpus allatum of an insect were removed during its first or second immature stage? (Assume it has four immature stages.)

When this operation is performed, the next molt produces a pupa, followed by a midget adult at the succeeding molt.

**21.3**    What other causes of diabetes might there be besides lack of production of insulin by beta cells of the pancreatic islets?

The failure to produce sufficient insulin is a major, but not an exclusive, cause of diabetes. In some cases insulin is produced in sufficient quantity, but the body is refractory to the insulin. A failure to maintain sufficient numbers of insulin receptors on the membranes of target tissues or a defect in their viability may be a responsible factor. In obesity, insulin refractoriness is often encountered.

In some cases the body may produce an abnormal insulinase which destroys insulin at a rapid rate. The usual insulin-degrading pathways have not been implicated in clinical diabetes. The production of antibodies to insulin may tie up the hormone; a diabetes of such origin would be an autoimmune disease.

Experimental diabetes has been produced in animals by injection of such drugs as *alloxan* or *streptozocin*. These substances destroy islet tissue. Some have speculated that viral destruction of islet tissue may be responsible for *juvenile diabetes* (insulin-dependent diabetes), a particularly severe form of the disease compared with *maturity-onset diabetes* (non-insulin-dependent diabetes). The virus may produce an autoimmune reaction which then causes the destruction.

**21.4**    Why do you suppose a diabetic person does not benefit from orally administered insulin?

Insulin is a small protein of approximately 5000 daltons. It is derived from a larger precursor (zymogen) that is enzymatically cleaved to produce a pair of polypeptide chains held together by S—S bonds. If insulin is given orally, the proteolytic digestive enzymes degrade it. For this reason, insulin must be injected. A technique has been developed by which insulin is automatically injected into diabetic patients over a period of days. Insulin preparations also exist that can be injected once in a 24-h period, with slow absorption occurring during that time. These clinical initiatives simplify somewhat the maintenance burden for diabetic persons.

**21.5**    Neurohumoral nerve fibers in the brain seem to function as part of the nervous system and yet their effects are finally achieved through the elaboration of humoral substances that act like hormones despite the short distances they travel to achieve their effects. Could an argument be made that the adrenal medulla is actually part of the neural apparatus?

In the case of the adrenal medulla we are acutely confronted with the difficulty of distinguishing between endocrine and neural function. In the first place, the adrenal medulla arises from ectoderm. This is the germ layer from which the entire nervous system derives. Secondly, the adrenal medulla is mobilized along with the sympathetic nervous system during periods of stress or challenge to the organism. The nerves associated with the sympathetic apparatus achieve their effects by producing norepinephrine at their endings in the immediate vicinity of target organs. The adrenal medulla can be likened to a large sympathetic nerve ending that liberates the catecholamines norepinephrine and epinephrine into the circulatory system, where they eventually extend and prolong the general fight-or-flight response.

**21.6**    Figure 21.6 shows the amino acid tyrosine and the hormone thyroxine. Tyrosine is a precursor of thyroxine. What steps would you say intervene between precursor and end product?

First, tyrosine is doubly iodinated to form *diiodotyrosine*. (Iodine ions are taken up from the blood by thyroid tissue and converted to elemental iodine, which then attaches to tyrosine.) Condensation and modification of two diiodotyrosine molecules produces thyroxine ($T_4$). Triiodothyronine ($T_3$) probably results from the loss of a single iodine within the thyroxine structure. Tyrosine is also the starting point for the synthesis of the catecholamines norepinephrine and epinephrine.

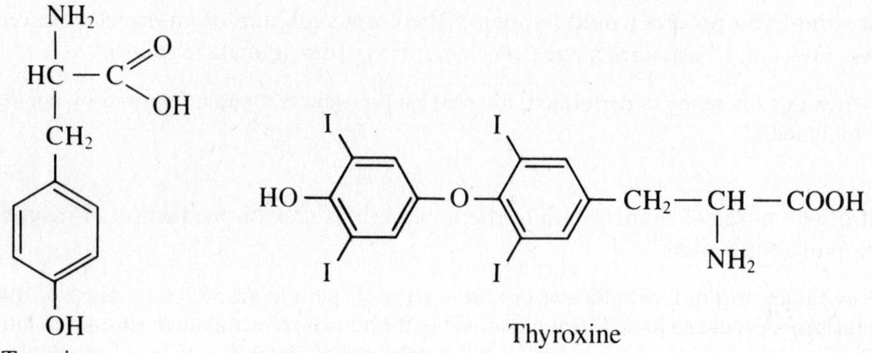

**Fig. 21.6** Tyrosine and thyroxine.

**21.7**   *Myxedema* is a condition caused by low activity of the thyroid gland (*hypothyroidism*) during the adult years. It is associated with low BMR, a diminution of mental acuity, sluggishness, weight gain, and a characteristic swelling of superficial tissue layers. The name is derived from the accumulation of mucus (*myxo* means "mucus") in the edematous fluid. Far more serious is hyposecretion of thyroid hormone in the early years of life. This produces *cretinism*, a condition of marked inhibition of growth, poor sexual development, and seriously diminished intelligence.

Given the symptons demonstrated by hypothyroidism, what would you guess are the symptoms of *hyper*thyroidism?

Not surprisingly, people with hyperactive thyroids (*Graves' disease*) show many symptoms opposite those in hypothyroidism. Their BMR is elevated, and they show poor tolerance to heat. They are extremely nervous and hyperactive and usually experience great weight loss. In addition, they may suffer from *exophthalmos*, a protrusion of the eyeballs from the sockets.

**21.8**   In countries where dietary intake of iodine is low, *goiters*, enlargements of the thyroid, are common. What would you say is the chain of events leading to formation of the goiter?

In the absence of iodine neither thyroxine nor triiodothyronine can be produced. The low levels of these thyroid hormones in the blood cause the anterior pituitary to produce large amounts of TSH, and without the negative feedback effects of $T_4$ and $T_3$, the pituitary continues to produce TSH without interruption. Since two effects of TSH are an increase in the size and number of thyroid follicle cells, so that more thyroid hormone is produced, excessive amounts of TSH cause hypertrophy and hyperplasia of the cells and lead to to the enlarged thyroid.

A number of compounds (*goitrogens*) inhibit thyroid hormone production and lead to formation of goiters. *Thiourea* is a goitrogen that has been used extensively to produce experimental goiter in test animals.

**21.9**   Thyroid hormone plays a significant regulatory role in all vertebrates but is not essential to life. Why then do most mammals die when the thyroid gland is extirpated? (Hint: Consider everything that is removed with the thyroid gland.)

When the thyroid gland is removed, the parathyroids are also lost. Because of the many important roles of calcium in the body, removal of the parathyroids, which are essential in maintaining calcium homeostasis, quickly leads to death.

**21.10**   List the hormones released by the posterior and anterior pituitaries.

Posterior pituitary: oxytocin and antidiuretic hormone. Anterior pituitary: growth hormone, prolactin, melanocyte-stimulating hormone, endorphins, enkephalins, thyroid-stimulating hormone, adrenocorticotropic hormone, follicle-stimulating hormone, and luteinizing hormone.

**21.11** Synthesized opiates such as morphine and codeine provide their analgesic effects by attaching to special receptors in the brain. Scientists for a long time wondered how there could be natural receptors for such synthesized opiates when the opiates were not part of mammalian evolutionary history. How does the existence of endorphins and enkephalins help to explain this mystery?

   Endorphins and enkephalins are naturally occurring pain killers. It is for these "opiates" that the receptors in the brain evolved. The synthesized opiates are structural mimics of these natural pain killers and thus are able to attach to the special receptors in the brain.

**21.12** Regulation of the menstrual and estrous cycles provides an excellent example of the integration of hormonal action seen in mammals. *Briefly* discuss the hormonal interactions involved (see Chap. 12).

   FSH releasing hormone from the hypothalamus causes the anterior pituitary to secrete FSH, which in turn causes the development of the ovarian follicle and subsequent production of the estrogens. The increasing levels of estrogen eventually cause the hypothalamus to slow production of FSH releasing hormone and to increase levels of LH releasing hormone. This stimulates the anterior pituitary to release LH, which in turn promotes ovulation and development of the corpus luteum. The corpus luteum begins secretion of progesterone. Progesterone inhibits production of FSH releasing hormone, so that no new follicles are prepared; however, in the absence of fertilization, progesterone begins to inhibit LH releasing hormone. This causes the corpus luteum to degenerate and reduces progesterone levels. With reduced progesterone, production of FSH releasing hormone in the hypothalamus is no longer inhibited, so that FSH can be released from the pituitary and initiate growth of a new follicle.

**21.13** The hypothalamus serves as a transducer, essentially converting electric impulses into hormonal responses. Name two other endocrine glands that release hormones in response to direct stimulation by nerves.

   The adrenal medulla and posterior pituitary.

**21.14** The circulatory system between the hypothalamus and anterior lobe of the pituitary is unusual. In what way? Discuss the usefulness of such an arrangement and name two other places we have seen it.

   The arrangement of blood vessels between the hypothalamus and anterior lobe (the *median eminence*) is unusual in that after the arterioles in the hypothalamus have formed capillaries, the veins on the distal side do not empty into larger veins but, instead, form a second capillary bed in the anterior lobe. Such a dual capillary system allows substances (releasing hormones, in the case of the endocrine system) to be taken up into the bloodstream at a specific site (the hypothalamus) and dropped off at a second specific region (the anterior lobe). The hepatic portal system serves such a function, as do the glomerulus and peritubular capillaries of the kidney, although in reverse order (the substance first leaves the blood system and later reenters it).

**21.15** At least 13 different hormones use cyclic AMP as the second messenger in performing their functions. Since these hormones all have different functions, how do you suppose it is possible for them to use the same intermediary to achieve these different ends?

   A first level of specificity is achieved by the fact that specific hormones can bind only to highly specific receptors. Thus, all cells but those carrying the receptors are ruled out for binding and subsequent hormonal action. Even specific parts of a selected cell are determined by the position of the receptor. The second level of specificity is imposed by the fact that although cAMP may be able to trigger a number of enzymes, it will encounter only specific enzyme systems in differentiated cell types.

**21.16** The hormones acting in concert are the major long-term integrating force of homeostasis. We have studied three primary ways in which hormones interact to control their own levels. Discuss these three and suggest other ways in which hormone levels might be controlled.

First, a number of hormones exert *antagonistic* actions on one another so that the body may move from one directional swing to another by increasing release of one hormone that opposes its antagonist. When high amounts of insulin build up and blood sugar goes down, a release of epinephrine, glucagon, and several other *hyperglycemic* hormones quickly drives blood sugar levels back to their balance point. Antagonistic hormone clusters do not of themselves maintain homeostatic balance but provide a means for it through negative feedback.

The second control mechanism of the endocrine system involves the *tropic hormones*, produced by the anterior pituitary gland, which stimulate activity in a number of other endocrine glands. There are as many as eight tropic hormones that engage in a "pas de deux" with another hormone, in which the level of the tropic hormone tends to rise when the level of the primary hormone falls or when the specific endocrine gland must be called into increased activity. High levels of the primary hormone tend to suppress further release of the tropic hormone. This kind of negative feedback ensures a homeostasis of hormone levels.

A third mechanism for control of hormone production comes from the hypothalamus. As many as 10 factors have been isolated that stimulate the pituitary to produce its tropic hormones. Each of these factors is known as a *releasing hormone*, e.g., follicle-stimulating hormone releasing hormone (FSH-RH). These hormones are secreted by the hypothalamus into an area of the hypothalamus called the *median eminence*. Capillaries in this area pick up the releasing factors and transport them a short distance to the anterior pituitary, where they stimulate release of their respective tropic hormones. Levels of circulating primary hormone may exert negative feedback on the releasing hormones as well as on the tropic hormones issuing directly from the pituitary. The releasing hormones from the hypothalamus are generally oligopeptides, fewer than 10 amino acids in length.

Other factors may determine how much active hormone reaches a target tissue. In some cases the hormone must be modified in order to exert an effect. Insulin, for instance, is first produced as a single long chain which must be modified and broken into two chains which then attach to one another. Interference with this process of modification could block or diminish insulin function. Transport of the hormone out of the cell, usually as coated granules, may constitute a factor for blockage or for enhancement of hormone levels. Plasma proteins which might be involved in carrying the hormone in the blood might also be a target for modulation of hormone availability. The condition of the membrane of the target cell would be important for hormones that attach to receptors on the cell surface. Protein receptors in the cytoplasm and nucleus could also influence hormone action. Nutritional status, particularly in terms of nitrogen balance, could influence synthesis. Degradation of the hormone, especially by the liver, would also influence the final titer of hormone available for activity.

# Supplementary Problems

**21.17** The hormones of the adrenal medulla are modified  (*a*) fatty acids.  (*b*) amino acids.  (*c*) monosaccharides.  (*d*) nucleotides.  (*e*) steroids.

**21.18** Match the terms in column **A** with the endocrine glands listed in column **B**.

| A | B |
|---|---|
| 1. Hyposecretion causes cretinism | (*a*) Thyroid |
| 2. Dissociates oxidative phosphorylation | (*b*) Pituitary |
| 3. Uses iodine in its hormone | (*c*) Adrenal cortex |
| 4. Stimulated by ACTH | (*d*) Pancreatic islet tissue |
| 5. Produces glucagon | |
| 6. Contains a protein that lowers blood sugar | |
| 7. Stores hypothalamic hormones | |
| 8. Hypersecretion causes giantism in the young | |

**21.19** The median eminence is part of a portal system by which hypothalamus and pituitary achieve a humoral connection.
(*a*) True  (*b*) False

**21.20** Insulin must first attach to receptors on a fat cell before it exerts its effects.
(*a*) True  (*b*) False

**21.21** Insulin, as a natural hormone, is resistant to the effects of digestive enzymes.
(*a*) True  (*b*) False

**21.22** In Graves' disease, the victim tends to be obese.
(*a*) True  (*b*) False

**21.23** Parathyroid hormone is necessary to maintain life.
(*a*) True  (*b*) False

**21.24** Both epinephrine and the glucocorticoids tend to reduce blood sugar levels.
(*a*) True  (*b*) False

**21.25** The adrenal medulla secretes approximately 80 percent epinephrine and 20 percent norepinephrine.
(*a*) True  (*b*) False

**21.26** Estradiol is secreted by the Graafian follicle of the ovary.
(*a*) True  (*b*) False

**21.27** The placenta may be considered an endocrine structure.
(*a*) True  (*b*) False

**21.28** The incorporation of iodine into table salt might be expected to reduce the incidence of simple goiter.
(*a*) True  (*b*) False

**21.29** Tyrosine is significant in hormone formation in the thyroid and the adrenal medulla.
(*a*) True  (*b*) False

# Answers

**21.17** (*b*)
**21.18** 1—(*a*); 2—(*a*); 3—(*a*); 4—(*c*); 5—(*d*); 6—(*d*); 7—(*b*); 8—(*b*)
**21.19** (*a*)
**21.20** (*a*)
**21.21** (*b*)
**21.22** (*b*)
**21.23** (*a*)
**21.24** (*b*)
**21.25** (*a*)
**21.26** (*a*)
**21.27** (*a*) Chorionic gonadotropins are secreted.
**21.28** (*a*)
**21.29** (*a*)

# Chapter 22

# The Nervous System

## 22.1 OVERVIEW

The nervous system, present in all vertebrates and most invertebrates, is concerned with that universal property of life called *irritability*, the capacity of cells and whole organisms to respond in a characteristic fashion to changes in the environment called *stimuli*. These stimuli may arise from internal as well as external alterations. The specific reaction elicited by a stimulus is termed a *response*. Generally, the response yields an adjustment that promotes the well-being of the entity. Stimulus-response reactions are usually rapid and afford a continuous mechanism for maintaining internal constancy in the face of environmental change.

In most nervous systems interconnecting fibers form a communications network that allows continuous monitoring of internal and external conditions. Signals, in the form of a flow of electric current called *nerve impulses*, are carried from one part of the nervous system to another. The generation of an impulse is called *excitation* and is the result of a localized flow of ions.

Stimuli produce impulses in nerve cells (neurons) whose endings are exquisitely sensitive to these stimuli. Such impulses, which tend to move toward the central axis of the nervous system, are called *sensory*, or *afferent*, impulses. Those impulses that move from the central axis to excite responses by glands or muscles are termed motor or efferent impulses. The significant difference between primitive and more advanced forms in terms of their neural (nervous system) capacity is that the latter organisms are capable of more complex and highly subtle interactions connecting afferent impulses bringing information into the *central nervous system* (CNS) (brain and spinal cord) with the efferent impulses producing appropriate responses. The central *neurons* (nerve cells) residing within the brain and spinal cord are called *connector neurons*, or *interneurons*. Their complexity differentiates the relatively sterotyped responses of more primitive forms from the highly diverse neural reactions of the more recently evolved vertebrates.

## 22.2 PHYLOGENETIC DEVELOPMENT OF THE NERVOUS SYSTEM

In early multicellular organisms, such as sponges, there is little in the way of a coordinating system. In the *Cnidaria*, a phylum of which hydra is a well-known example, a tissue level of organization is apparent. Slender sensory cells, the afferent apparatus of a neural coordinating system, are especially numerous in the *ectoderm* (outer layer). They communicate with a *nerve net*, which is the significant feature of the coordinating system. The nerve net is a relatively diffuse structure and is somewhat limited in the range of responses it can produce—usually stimulation of epitheliomuscular cells or specialized defense structures. Fine muscular movements are not yet possible at the nerve net level of organization. However, in some cnidarians an additional nerve ring exists, which permits a greater complexity of neuromuscular responses. At the nerve net level of organization, impulses usually travel in either direction along the neuron.

The comb jellies (phylum Ctenophora) also have a nerve net, but in this phylum evidence of localized differentiation exists. Among such hints of higher levels of specialization are an *oral ring* around the mouth and a series of eight neural strands immediately below the combs.

In the flatworms (phylum, Platyhelminthes) an organ level of organization is readily apparent. At the anterior end are two lobes of concentrated nerve tissue that make up the brain (see Fig. 22.1). Extending posteriorly are two ropes of neurons that make up the nerve cords. These structures provide a centralized nervous system that can process information coming from sensory cells at the surface and permit much more complex behaviors than those of more primitive organisms.

In mollusks (snails, clams) further complexity is achieved by increased *cephalization* (concentration of nerve cells at the head end) and a greater number of knots of nerve cell bodies known as *ganglia*,

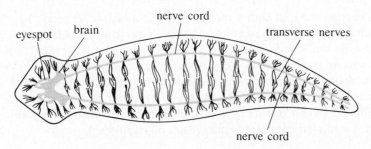

**Fig. 22.1** Nervous system of Platyhelminthes.

which are scattered through the nervous system. An array of diverse sensory structures also characterizes this phylum. The octopus is even capable of complex learning patterns.

The earthworm (phylum Annelida) also demonstrates a complex nervous system. A variety of sensory cells are concentrated at the head end, but sensory organs are not prominent. Two large ganglia make up the brain. A double, but fused, nerve cord runs along the body on the ventral surface. In each segment of the earthworm a ganglion buds off the nerve cord to coordinate sensory and motor impulses.

A remarkable degree of cephalization is found in arthropods, particularly the insects. Most prominent of the sense organs in the anterior region are simple or, in some groups, compound eyes. A "ladder" type of nervous system lies along the ventral surface—a double nerve cord punctuated by a few or many ganglia lying along its length. Coordination of the delicate movements found in the appendages is largely a function of the ganglia of each segment—affording a high degree of decentralization of motor function.

Although the echinoderms are relatively advanced by many criteria, their nervous systems are not as complex as those of most arthropods. Certainly, a clearly defined central nervous system is not present in most representative forms.

## 22.3  THE NEURON AS THE FUNCTIONAL UNIT OF NERVOUS ACTIVITY

The functional unit of the nervous system in both invertebrates and vertebrates is the neuron (see Fig. 22.2). This highly specialized cell, which contains the typical repertoire of organelles found in most eukaryotic cells, is highly adapted for communication because of its wirelike projections. The *dendrites* are projections, often branched like a tree, that carry impulses *toward* the central cell body. The *cell body* is a thicker region of the neuron containing the nucleus and most of the cytoplasm. The *axon* is a projection, generally very long, that carries impulses *away from* the cell body. Usually a neuron has only a single axon. Many axons and even dendrites may combine to form a single *nerve*.

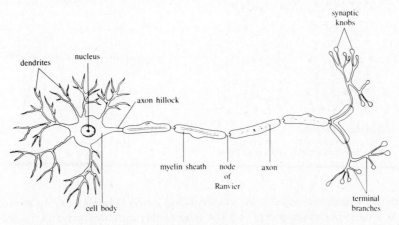

**Fig. 22.2** A model of a typical motor neuron.

The neurons are supported, in both a mechanical and a metabolic sense, by *glial cells*. These cells are far more numerous than the neurons and vary considerably in both their architecture and their specific function. In the brain and spinal cord glial cells are called *neuroglia*. In the outlying neurons of the *peripheral nervous system*, which carry impulses to and from the central nervous system, the supporting tissue consists of *Schwann cells*. These Schwann cells tend to grow round and round the axon so that it is wrapped in a many-layered insulating cover called the *myelin sheath*. This fatty, membranous sheath, which is a feature of most peripheral nerve fibers in vertebrates, affords a rapid and highly efficient "insulated wire" for transmission of impulses.

## 22.4   THE NEURAL IMPULSE

Early theories likened the passage of a nerve impulse to the flow of electric current in a wire. The nerve, according to this theory, plays a relatively passive role in the propagation of the impulse. However, studies of the giant axons found in the squid (*Loligo*) produced results inconsistent with this interpretation. The early theories predicted a falling off of voltage as the impulse travels along the axon, but instead the strength of the impulse, once generated, is unchanged from start to finish. Further, according to the early theories, the axon should not be an adequate conductor of electricity, because its lipid (myelin) composition is more conducive to insulation than to conduction of electric charge.

When a neuron is not conducting an impulse, it is said to be in the *resting state* (see Fig. 22.3). In this condition a *resting potential*, that is, a difference in charge, exists between the inside and the outside of the membrane. A higher concentration of sodium ions exists outside the membrane, while there is a higher concentration of potassium ions inside. In addition, a number of negatively charged proteins

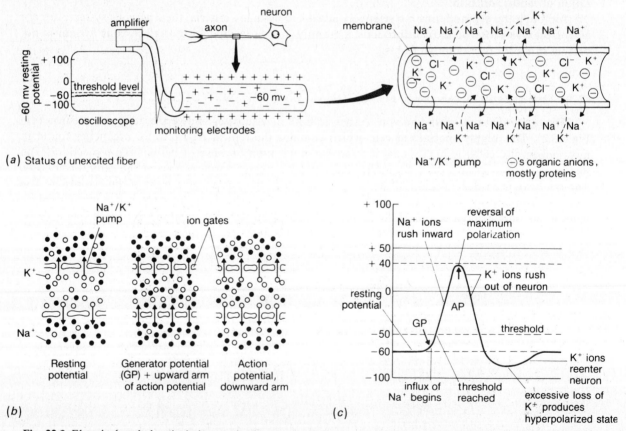

**Fig. 22.3** Electrical and chemical changes in the neuron during a nerve impulse. (*a*) An unexcited fiber expends energy operating a sodium-potassium pump. (*b*) The state of the unexcited nerve (*left*), the stimulated nerve (*center*), and the nerve during repolarization (*right*). (*c*) The nerve impulse as seen on an oscilloscope.

reside on the inside. These concentration gradients are maintained by two factors: the impermeability of the resting membrane to $Na^+$ and the action of a $Na^+/K^+$ pump that, driven by ATP, transfers $Na^+$ to the outside and pumps $K^+$ in. Because of these gradients the inside of the neuron is negative relative to the outside; a potential difference of roughly $-60$ millivolts (mV) exists across the membrane. The natural tendency to correct this energetically unstable imbalance is the driving force behind the nerve impulse.

When a neuron is stimulated, the point of stimulation suddenly becomes permeable to sodium ions, which rush in, *depolarizing* the membrane (that is, eliminating the potential difference), as the incoming positive ions balance the negative internal charge. Enough $Na^+$ rushes in to actually make the inside of the membrane positive for a few milliseconds.

This shift of charge constitutes the neural impulse, or *action potential*. Although it occurs at only one place on the neuron, it triggers a depolarization of the adjacent area, thus initiating a new action potential. This process continues as a wave of depolarization down the length of the axon. The impulse is thus not actually transported anywhere but like a wave of water is *recreated* at each point.

At any point on the neuron when the action potential reaches a maximum (about $+40$ mV) of the interior relative to the exterior, the membrane suddenly again becomes impermeable to $Na^+$. At the same time, $K^+$ is pumped out, until it essentially balances the number of sodium ions that rushed in and the membrane is *repolarized*. This efflux of positive ions restores the resting potential of $-60$ mV (albeit with potassium ions rather than sodium ions); in fact, it does such a good job of reestablishing the resting potential that there is a brief overshoot of negativity. After the resting potential is established, the $Na^+/K^+$ pumps restore the original sodium and potassium gradients existing before the initiation of the action potential. Until the membrane reaches its resting potential again, it is incapable of developing a new action potential; while this is the case, the membrane is said to be in its *refractory period*. This prevents back stimulation by a neighboring membrane region.

In addition to the $Na^+/K^+$ pumps, there appear to be ion channels in the membrane that are specific for either $Na^+$ or $K^+$. These channels are guarded by voltage-sensitive proteins that, by changing shape in response to specific voltages, act as gates to open and close the channels. The potassium channels appear to have only one gate; however, the sodium channels probably have a gate at either end, an *activation gate* and an *inactivation gate*. In the resting state the activation gate is closed, and the inactivation gate is open. When the neuron is stimulated, the $Na^+$ activation gate opens and $Na^+$ is free to follow its gradient into the neuron. When the depolarizing wave peaks at $+40$ mV, the inactivation gate closes, and the membrane again becomes impermeable to $Na^+$; also, the $K^+$ gate opens, allowing potassium ions out to reestablish the resting potential. When the resting potential is reached, all gates assume their original configurations and the $Na^+/K^+$ pumps get busy reestablishing the original ion gradients.

Each neuron has a threshold of stimulation below which it will not fire. Above this threshold all stimuli, regardless of strength, evoke a depolarization (action potential) of common intensity. This phenomenon is known as the *all-or-none principle*. Instead of evoking stronger action potentials, stimuli of increasing strength elicit multiple firings, with the frequency increasing with the strength of stimulation.

We noted earlier that many vertebrate neurons are wrapped in a myelin sheath. Since this lipid covering does not permit an exchange of ions between the extracellular fluid and the interior of the neuron, a continuous wave of depolarization is not possible as in the unmyelinated fibers and another means has evolved to propagate the nerve impulse. Actually, the myelin sheath is regularly interrupted along the length of the neuron by short stretches of unmyelinated membrane. These *nodes of Ranvier* are capable of initiating an action potential just as unmyelinated neurons do. The action potential produces an electric current that instantaneously moves through both the extra- and the intracellular fluid to the next node, where the current initiates another depolarization, and so forth. This skipping process (*saltatory conduction*) is much faster than transmission in unmyelinated fibers because the impulse is largely transmitted as an electric current rather than by the much slower chemical processes involved in continuously recreating an action potential. Even during the chemical segments of transmission (in the nodes), the process is faster than in unmyelinated membranes because the nodes are 500 times more permeable to ions.

## 22.5  THE SYNAPSE

The point at which an axon and a dendrite associate is called a *synapse*. In invertebrates possessing a nerve net, such as hydra, the axons and dendrites touch each other at such junctions, so that passage of a nerve impulse across the synapse is an electric event; nerve impulses can pass indiscriminately in either direction along the nerve. As a result, the nerve net fires as a unit when stimulated. Discrete neural pathways, with concomitant specialized function, are few and simple in such animals. In vertebrates, because of the unique characteristics of the synapse, nerve impulses generally move in only one direction and at synaptic junctions may tend to follow one pathway over another.

In the typical mammalian synapse (see Fig. 22.4), a definite gap (the *synaptic cleft*) exists at the synaptic junction, usually about 18–20 nanometers (nm) wide. Movement of the nerve impulse across this gap is primarily a *chemical* event mediated by neurotransmitters such as *acetylcholine (ACh)*, *γ-aminobutyric acid (GABA)*, *norepinephrine*, and *serotonin*.

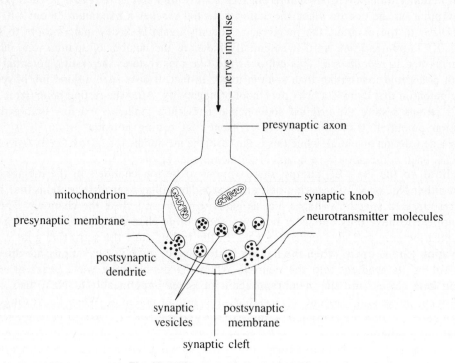

**Fig. 22.4** The synapse in mammals.

The axon ends in many small *synaptic knobs*, containing a number of sacs, *synaptic vesicles*, which are filled with neurotransmitter. When an action potential reaches the knob, voltage-sensitive calcium gates in the knob's *presynaptic membrane* open, permitting an influx of $Ca^{2+}$. This causes an enzyme, *calmodulin*, to effect the attachment of microtubules to the synaptic vesicles, which are then drawn to and fuse with the presynaptic membrane. The vesicles then rupture and spill their neurotransmitters into the synaptic cleft. Diffusing rapidly across the cleft, the neurotransmitters bind to special receptors in the *postsynaptic membrane* of the apposing dendrite or cell body. When a threshold number of receptors have been bound, sodium or potassium gates open in the postsynaptic membrane and positive ions flood in, lowering the internal negativity of the neuron. This change in electric potential is called an *excitatory postsynaptic potential (epsp)*. By itself it is not sufficient to exceed the activation threshold of the receiving neuron. However, if a series of impulses stimulate a single knob in a small enough span of time, their combined effect may be sufficient to exceed the threshold of the receiving neuron, thus triggering a new action potential and continuing the impulse. This is known as *temporal summation*.

The receiving neuron's threshold may also be achieved by the combined epsp's of many axons stimulating the dendrite or cell body simultaneously. This is referred to as *spatial summation.*

Some neurons release neurotransmitters that inhibit, rather than activate, the apposing neuron. These neurotransmitters (GABA is believed to be one) cause *chloride ion gates* in the postsynaptic membrane to open, permitting $Cl^-$ to pour in and hyperpolarize the neuron. This increase in negative polarization is called an *inhibitory postsynaptic potential (ipsp).* Such a potential is subtractive relative to the epsp's, which means that in order for the neuron to fire, a compensatingly larger number of epsp's must be generated. With possibly thousands of *inhibitory neurons* and *excitatory neurons* impinging on a single dendrite or cell body, the permutations of control available to the organism are greatly enhanced.

**EXAMPLE 1**   Simplistically, suppose that motor neuron E enables a boy to pick up objects. In order to fire, it requires the epsp's created by neurons A and B. However, suppose the boy is considering stealing some candy. An inhibitory neuron C (from his conscience) might generate an ipsp, thus reducing the spatially summated epsp's caused by neurons A and B to a level below the threshold required to fire motor neuron E. Unless an additional epsp is generated (say, by a neuron D from his sweet tooth), motor neuron E will not fire (and he will not pick up the candy).

To prevent continued random firing of a receiving neuron, the synaptic neurotransmitter must be neutralized. This is accomplished by enzymes that catabolize the neurotransmitter, thus preventing further binding to receptors on the postsynaptic membrane.

**EXAMPLE 2**   *Acetylcholinesterase* splits acetylcholine into acetic acid and choline, thus preventing further stimulation of the postsynaptic membrane. The choline is then taken up by the presynaptic membrane and reassembled into ACh by the enzyme *choline acetyltransferase.* The synaptic vesicles that ruptured at the presynaptic membrane detach, merge in limited numbers, and are then filled with ACh.

Because neurons also communicate with cells other than other neurons, synapses do not exist only between neurons. A relatively simple synapse exists, for example, between the end of a motor nerve (the *motor end plate*) and the membranous surface of the muscle fiber, or *sarcolemma.* This kind of connection, at the *neuromuscular junction,* is called an *excitatory synapse.* The neurotransmitter in this case is ACh, and unlike the situation at neuron-neuron synapses, a single impulse releases enough ACh to fire the muscle.

## 22.6   REFLEX ARC

The neuron, or nerve cell, is considered the *anatomical* unit of the nervous system. The reflex arc is regarded as the *functional* unit of this system. A *reflex* is an unvarying and automatic response to a specific stimulus.

**EXAMPLE 3**   In the *knee-jerk reflex* (see Fig. 22.5), if the tendon of the quadriceps muscle lying below the kneecap (patella) is given a sharp blow, the leg will extend abruptly. This is brought about by the creation of an impulse in an afferent neuron that is carried to the spinal cord. A synapse with a motor neuron in the cord results in the impulse's moving along the axon of that motor neuron to the extensor muscle of the leg. Contraction of the extensor muscle produces the knee jerk.

Most of the reflexes that maintain posture are two-neuron reflexes. Those that are involved in withdrawal of a body part from a painful or noxious situation usually require the interpolation of an interneuron in the spinal cord. They are three-neuron reflexes.

The *simple reflex arc* is named for the physical path actually taken by the impulse as it moves from the receptor of the afferent neuron to the effector of the motor neuron. The reflexes are mechanisms for maintaining appropriate posture, regulating blood pressure, and orienting the body to environmental

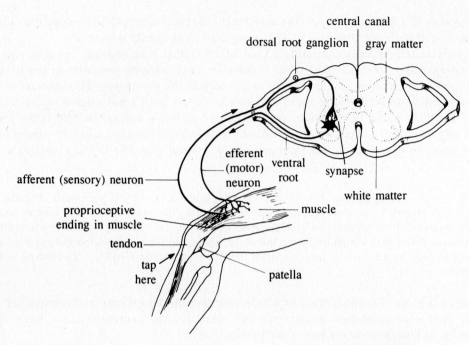

**Fig. 22.5** The knee-jerk reflex. Arrows show direction of nerve impulse.

conditions that threaten the organism. The loss of certain *extensor* (increasing the angle at a joint) or *flexor* (decreasing the angle at a joint) reflexes may be used clinically to assess damage to the central nervous system.

## 22.7 RECEPTORS AND EFFECTORS

### RECEPTORS

*Receptors* are the modified dendritic endings of sensory nerves that are constructed to respond to specific kinds of stimuli. In many cases the receptor is little more than the free ending of a sensory nerve, embedded in epidermis or other surrounding tissue. In other instances the receptor may be an elaborate sense organ, such as the *retina* of the eye or the *organ of Corti* in the ear, which is the receptor for sound waves.

The specificity of receptors is based on two properties. First, each receptor has a low *threshold* for its particular stimulus. The threshold is the minimal strength of stimulus necessary to set off an impulse. Second, the receptor sends only one kind of message to the central nervous system no matter what the nature of the stimulus. This is known as *Muller's doctrine of specific nerve energies.*

**EXAMPLE 4** The retina of the eye is highly sensitive to light, since its threshold to that stimulus is extremely low. Any stimulus applied to the retina, whether a sharp blow or a red-hot poker, is translated by the sensory apparatus of the brain as a burst of light. This is why we "see stars" when punched in the eye.

Receptors have been classified into three major groups, depending on their anatomical position. Those receptors in the skin, which therefore receive direct stimulation from the environment, are called *exteroceptors*. Exteroceptors range from the relatively undifferentiated bare nerve endings, which transmit pain, to the elaborate special receptors of the eye, ear, nose, and tongue. At the surface, particularly the fingertips, separate oval structures exist to perceive light touch as contrasted with heavier pressure.

*Proprioceptors*, the second group of receptors, are located in muscles, tendons, and the regions surrounding joints. These receptors, when stimulated, set up impulses that result in restoration of stretched structures to their original state. They play a significant role in the maintenance of posture and also provide us with a sense of the position of the body and its parts in space.

*Interoceptors*, the third type of receptor, are essentially free nerve endings that terminate in the surface of blood vessels and a variety of internal organs. Many of the reflexes that control homeostatic responses of lungs, heart, etc., originate in these receptors.

## EFFECTORS

*Effectors* are the end structures in a reflex arc or more complex nervous response that bring about the actual specific response. They are usually muscles whose contraction produces the response, or they may be glands that secrete particular substances as a result of stimulation. Strictly speaking, the effector is not really a neural structure.

**EXAMPLE 5**   A single nerve impulse traverses the motor end plate and produces a contraction of all the fibers involved in the *motor unit*. All the muscle fibers of the motor unit that contract simultaneously constitute the unit of contraction. The major neurohumor of the motor end plate is acetylcholine.

## 22.8   SPECIAL SENSE ORGANS

An organism is constantly bombarded by various forms and levels of energy emitted by its environment. The special sense organs are transducers that convert the various energy forms into nerve impulses. Unlike other neurons, which fire on an all-or-none basis, sensory organs generally produce *generator potentials*, which can increase in intensity with varying levels of stimulation, thus permitting perception of different volumes of sound, brightness of light, etc. The organism would be unaware of its environment without these special receptors.

## LIGHT

Visual perception can range from the simple ability to distinguish light from dark (as in planarians) to production of extremely sharp images, as in birds of prey. In higher vertebrates, light (part of the electromagnetic spectrum) enters the *eyeball* through the clear *cornea*, where it undergoes initial focusing. It then passes through a hole called the *pupil*, which is surrounded by the iris, a diaphragm that controls the amount of light entering. The light is then focused by the *lens* onto the *retina*, which contains the actual light receptors (the *rods* and *cones*) of the organ. When stimulated, the rods and cones send impulses to two other nerve cells in the retina—the *bipolar cells*, which then synapse with *optic neurons*. The optic neurons from throughout the retina come together to form the *optic nerve*. In vertebrates that have them, the cones permit color perception, while the rods are primarily involved in night vision.

## SOUND

Hearing is one of several ways of picking up vibrations. Many receptors have evolved to detect vibrations, ranging from simple sensory hairs to the *ears* of higher vertebrates. In humans, vibrations enter the large, fleshy *pinna* and travel down the *auditory canal*, striking the *tympanic membrane* (*eardrum*), which begins to vibrate. The tympanic membrane touches the first of three contiguous bones (the *malleus, incus,* and *stapes,* respectively), which transfer the vibrations to the *cochlea*. The cochlea is essentially a fluid-filled tube that is bent back on itself so that the two open ends lie one on top of the other. (This double tube is actually then rolled up from the bent end, giving the cochlea the appearance of a snail shell.) The open ends are covered by membranes and are called the *oval window* and the *round window*. The stapes touches the oval window and transfers vibrations to the fluid (*perilymph*) of the cochlea. These vibrations are picked up by the *organ of Corti* within the cochlea. This structure consists of hair cells that rest on a *basilar membrane*, with the sensory hairs embedded in a gelatinous *tectorial membrane* above. The dendrites of sensory neurons terminate on the hair cells. Vibration of the fluid in the cochlea causes the basilar membrane to move, thereby bending the hairs and creating a generator potential. This triggers impulses in the sensory neurons; the impulses then travel along the *cochlear nerve* to the brain.

### GRAVITY AND MOVEMENT

Body position, body movement, and balance are assisted by the *vestibular apparatus*—the *saccule*, *utricle*, and *semicircular canals*—of the inner ear. The first two of these tell the body what position it is in relative to gravity. They each have a sensory area called a *macula* in their walls. Each macula consists of a layer of hair cells, with their hairs embedded in a gelatinous layer above. Resting on top of this gelatinous mat is a pile of calcium carbonate granules called *otoconia*. The weight of this "rock pile" causes the gelatinous mat to shift as the orientation of the head changes. This shift stimulates the hair cells to change the rate of impulses in the closely apposed vestibular nerve (the nerves around the hair cells fire continuously, even at rest). Since the hair cells are oriented at many different angles, each position of the head produces a distinctive pattern of firing.

In the semicircular canals, *ampullae* detect movement of fluid in the canals caused by movement of the head. Each ampulla consists of a ring of hair cells buried in a gelatinous, thimble-shaped *cupula*, which bends under pressure from the fluid, causing impulses to be generated in the vestibular nerve. The three semicircular canals are arranged at right angles to each other, so that all three spatial dimensions are represented.

### TASTE AND SMELL

Taste and smell rely on *chemoreceptors*. Taste is sensed by hair cells in the *taste buds* of the tongue. These cells lie close to and stimulate sensory nerves, which actually carry the impulses to the brain. Although there are four basic taste types—salty, sweet, bitter, and sour—each located in a separate region of the tongue, various foods may stimulate several of these simultaneously, and at different intensities, creating a blend of flavors.

The organs of smell are located in the *olfactory epithelium* of the nose. Unlike the taste mechanism, the sensory organ of smell is a neuron, the dendrites of which are buried in the epithelium. It is not certain how smells are detected. One widely considered theory is that the shape and size of a molecule, rather than its chemical reactivity, determine the smell. According to this theory, there are seven primary receptor types, corresponding to seven primary odors. As with taste, these primary odors may blend to produce unique odors.

### TOUCH AND PRESSURE

*Mechanoreceptors* detect distance, as in the *lateral lines* of fish, and touch and pressure. In humans, touch is perceived through *Meissner's corpuscles* and *free nerve endings*, both of which are located near the skin's surface. Pain is also produced through stimulation of the free nerve endings. Hairs constitute another type of touch receptor. Because their chief constituent, keratin, is essentially crystalline, hairs discharge small amounts of electricity when deformed, thus stimulating nerves near the hair roots. Such crystals are said to be *piezoelectric*. *Pacinian corpuscles*, which are located deeper in the skin, detect pressure.

### HEAT AND COLD

Temperature receptors (*thermoreceptors*) are not well understood. In humans the primary receptors are believed to be free nerve endings, *Ruffini corpuscles* (heat) and *Krause end bulbs* (cold). Thermoreceptors measure not just absolute temperatures but also changes in temperature. Thus, a change from hot to warm may be perceived as coolness.

## 22.9  BRAIN AND SPINAL CORD

The brain is encased in a bony *skull* and is further protected by three *meninges*. The innermost of these, the *pia mater*, covers the brain and is richly supplied with blood vessels. It brings nutrition and oxygen to the underlying brain and protects it. The cerebrospinal fluid (CSF) lies between the pia mater and the middle layer, the *arachnoid membrane*, providing both cushioning and a supply of ions for the brain and spinal column. The outermost layer, the *dura mater*, is tough and fibrous and provides mechanical support.

The vertebrate brain (see Fig. 22.6) is divided into three basic regions: the hindbrain, the midbrain, and the forebrain.

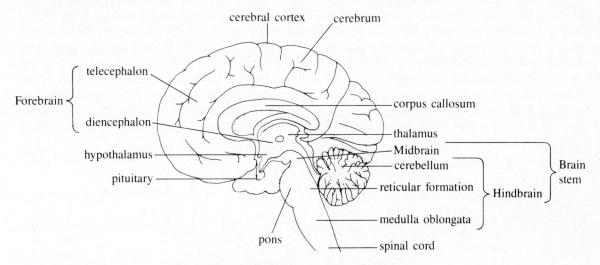

**Fig. 22.6** The human brain (midsagittal section).

The *hindbrain* is chiefly concerned with involuntary, mechanical processes. It consists of three primary structures. (1) The *medulla oblongata*, which lies on top of the spinal cord, contains many of the centers that control such involuntary processes as breathing, blood pressure, and heart rate. Many of these functions are regulated by *nuclei* (masses of neural cell bodies that innervate a common part of the body). All communication between brain and spinal cord passes through the medulla. (2) The *pons* sits on top of the medulla and contains the *tracts* (longitudinal bundles of myelinated fibers) running between brain and spinal cord and also receives tracts from the *cranial nerves* (12 pairs of sensory, motor, and mixed nerves that travel between the brain and various organs without passing through the spinal cord). The pons also coordinates forebrain functions with those of the cerebellum. (3) The *cerebellum* lies behind the medulla and controls balance and muscle coordination.

The *midbrain* lies between the hindbrain and the forebrain and connects the two. It processes the visual and auditory information from the eyes and ears before it is sent to the forebrain, and it is involved in behavioral patterns in the lower vertebrates.

The *forebrain* is most advanced in humans. We have already studied one of its major structures—the *hypothalamus*. In addition to its hormonal responsibilities, the hypothalamus controls such crucial parameters as heart rate, blood pressure, and body temperature, as well as many fundamental drives such as hunger, thirst, sex, and anger.

Resting above the hypothalamus is the *thalamus*. It is one of a series of cerebral nuclei known collectively as the *basal ganglia*. It provides connections between many parts of the brain and between the sensory system and the cerebrum. It may also be the seat of primitive moods and feelings.

Lying in both the thalamus and midbrain is the *reticular formation*, a ramified region of nuclei in close association with major ascending sensory tracts. This structure modulates impulses moving along these tracts and funnels them to higher cortical areas in such a way as to control the level of arousal, or wakefulness. Our ability to keep alert and respond to external stimuli is largely dependent on the reticular formation. Although sleep may be influenced by centers in the brainstem such as the medulla and pons, the reticular formation plays a permissive role in extending states of sleep.

The thalamus, hypothalamus, and part of the cerebral cortex form the *limbic system*, a set of nuclei apparently involved in a kind of coarse adjustment for *affect* (emotional state). Depression and euphoria may originate here. Among the nuclei making up the system are the *hippocampus* (which may be essential to short-term memory), the *amygdala* (associated with anger), and the *cingulate gyrus* of the cortex. The close apposition of the olfactory nerve and the amygdala suggests that olfaction has an effect on mood and emotions. The limbic system may also interact with the reticular formation to select individual items toward which an organism might direct its attention.

The largest and most complex part of the forebrain in humans is the *cerebrum*. It is divided into two hemispheres and has an outer *cortex* composed of *gray matter*, comprising a thin, dense mat of roughly 15 billion nerve cell bodies and dendrites, and a thicker inner layer of white myelinated nerve fibers. Each hemisphere is divided into four major lobes (see Fig. 22.7). At the back are the *occipital lobes*, which receive and analyze visual information. At the lower sides of the brain are the *temporal lobes*, which are primarily concerned with hearing. At the front of the brain are the *frontal lobes*, which regulate fine motor control, including movement involved in speech, and serve as a clearinghouse for incoming sensory stimulation. Above the temporal lobes and behind the frontals are the *parietal lobes*, which receive the stimuli from the sensory organs of the skin and also provide an awareness of body position.

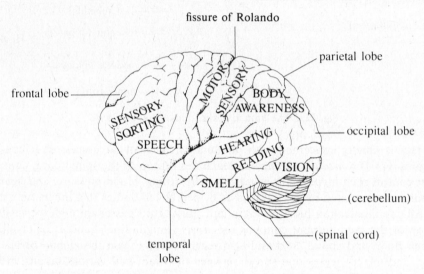

**Fig. 22.7** The four lobes of the cerebrum and their functions.

The frontal and parietal lobes are divided by a ridge (the *fissure of Rolando*) running from the top of the head down the sides. In relatively narrow strips on respective sides of this fissure are the motor area of the frontal and the sensory area of the parietal lobe. Many precise areas in each strip have been mapped, corresponding to specific regions of the body. For example, it is known which specific areas of the right parietal strip receive sensory information from the left ankle, knee, toes, etc., and which segments of the left frontal strip control movement of the right hip, leg, foot, etc. You will notice that, as with most functions of the brain, a given side of the body is in communication with and controlled by the opposite side of the brain.

Although the two hemispheres of the cerebrum are functionally similar, each half specializes to some extent. For example, speech and analytical thinking tend to be centered in the left hemisphere, whereas spatial perception is more of a right hemisphere function. Some learned characteristics, such as handedness, may be centered in either side, and vary with individual experience. The two hemispheres communicate through the *corpus callosum*, an intermediate region that permits information received by one side of the cerebrum to be communicated to the other side.

Those centers of the brain lying in the stem tend to be permanently circuited from birth with only limited capacity to form new connections (synapses). However, the neurons lying within the cortex, especially those at the surface of the *neocortex* (most recently evolved region) may be extremely labile in terms of their establishment of new synapses. A variety of sensory data coming into the brain are not simply channeled into motor efflux. Instead this data may reverberate within central circuits and be stored in the association areas of the parietal lobes. In the process, new synapses may be formed. The storage may not always involve a single region. Thus, the auditory and speech area of the temporal lobe may interact with the visual association area of the occipital lobe to produce a single memory trace. Memory may be of the *short-term* variety, quickly associated and soon discarded, or there may be *long-term memory*, in which permanent retention occurs. Learning most certainly involves long-term memory.

Learning or memory may also involve extraneural *engrams* (specific physical traces) such as the creation of a nucleic acid. Several decades ago the idea surfaced that learning in simple invertebrates involves the formation of a specific RNA associated with a particular learning experience. It was even suggested that ingestion of that RNA could result in the transfer of learning from experienced to naive animals, an idea that filled many professors with a great sense of dread. This concept has not been confirmed.

*Learning* is the modification of the neural apparatus by acquired experience. While it may be expressed in terms of the creation of new synapses, other changes such as a modification of existing synapses or the creation of new molecules may also be part of the modifying mechanism.

The *spinal cord* is a thick dorsal neural tract extending from the brainstem to the lower back. It is almost completely enclosed by the vertebrae of the spinal column, just as the brain is enclosed by the bones of the skull.

Most of the spinal cord consists of ascending tracts conducting sensory impulses toward the brain and descending tracts carrying impulses downward to motor neurons within the spinal cord. At every level the spinal cord also serves as a reflex center, possibly involving a variable number of interneurons.

In the brain, the *gray matter* (neuron cell bodies) is located externally and in the basal ganglia within the internal white matter. In the spinal cord the gray matter is contained within an internal H-shaped region within which lies the central canal (see Fig. 22.5). Within the central canal the *cerebrospinal fluid* is in continuous contact with the cavities lying within the brain, the *ventricles*. The fourth brain ventricle, which overlies the medulla oblongata, is continuous with the central canal of the spinal cord.

Nerve cells in the dorsal extensions of the gray matter of the spinal cord process sensory impulses coming in from the spinal nerves of the peripheral nervous system. The cell bodies of these sensory nerves actually lie within swellings outside the spinal cord known as the *dorsal root ganglia*. Nerve cells lying in the *ventral horn* (extension) of the gray matter process motor impulses going out to effectors. Before its juncture with the spinal cord, each peripheral nerve bifurcates into a dorsal root, which contains the dorsal root ganglion, and a ventral root (see Fig. 22.5). The nerve fiber in the periphery thus carries both sensory and motor fibers.

## 22.10   AUTONOMIC NERVOUS SYSTEM

The *autonomic nervous system* (*ANS*) comprises two subsystems (see Fig. 22.8). The *sympathetic nervous system* (*SNS*) prepares the body for emergency situations, the responses associated with "fight or flight." It produces rapid and total mobilization to thwart danger. The heartbeat is strengthened and accelerated, blood pressure is increased, the blood sugar level goes up, and blood moves from vessels within the trunk to those of the arms and legs to support fighting or running away. On the other hand, the *parasympathetic nervous system* (*PNS*) maintains those conserving functions that restore the organism during periods of tranquility. Clearly, these functions (eating, sexual activity, urination, etc.) must be turned on discretely rather than together.

In both systems the effector impulse begins at a motor neuron within the CNS and is then usually relayed to a second motor neuron. The synapse for these two neurons occurs in a ganglion. For the SNS these ganglia form a chain lying along the spinal cord; a few of the ganglia are in nerve plexuses lying just outside the line of the chain. The preganglionic fibers tend to be short while the postganglionic fibers reaching to the visceral organs are usually quite long. The entire chain is activated as a unit by

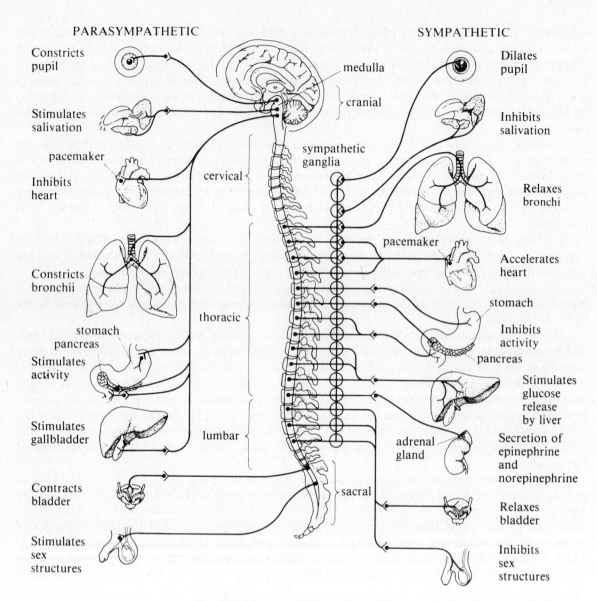

PARASYMPATHETIC                                                SYMPATHETIC

Constricts pupil

Stimulates salivation

pacemaker

Inhibits heart

Constricts bronchii

stomach pancreas

Stimulates activity

Stimulates gallbladder

Contracts bladder

Stimulates sex structures

medulla

cranial

cervical

sympathetic ganglia

thoracic

lumbar

pacemaker

adrenal gland

sacral

Dilates pupil

Inhibits salivation

Relaxes bronchi

Accelerates heart

stomach

Inhibits activity

pancreas

Stimulates glucose release by liver

Secretion of epinephrine and norepinephrine

Relaxes bladder

Inhibits sex structures

**Fig. 22.8** The autonomic nervous system.

a sensing of danger, and signals radiate out to all organs. As is apparent from Fig. 22.8, the SNS emanates from the cervical, thoracic, and lumbar regions while the PNS extends its nerve fibers from the cranial and sacral ends of the CNS.

The ganglia for the PNS actually lie within the target organ itself, a sharp contrast to the centralized chain of ganglia found in the SNS. This decentralized anatomy for the PNS supports the discreteness of its response. The transmitter substance involved is acetylcholine (ACh). Otto Loewi first demonstrated the chemical nature of PNS stimulation, and Henry Dale showed that the substance is ACh. The norepinephrine which mediates the response of the SNS is produced not only by the individual endings of the postganglionic fibers (which generally are activated synchronously) but also by the adrenal gland, which drips norepinephrine and epinephrine into the general circulation. It is this massive infusion of catecholamines that contributes to the mobilization produced by the SNS. It should be noted that for both systems the neurotransmitter at the intermediate synapse in the ganglion (preganglionic-postganglionic) is ACh.

# Solved Problems

**22.1** What is the difference between a neuron and a nerve?

A neuron is a single nerve cell consisting of a cell body, branches called *dendrites* (which usually bring impulses toward the cell body), and a single long *axon* (which in most cases leads impulses away from the cell body). The neuron, although highly specialized to serve its function of conduction, possesses the usual array of organelles found in less specialized eukaryotic cells.

Nerves are fibrous bundles made up of many axons or a mixture of axons and dendrites. These long extensions look like white strings. It is along their length that impulses travel to and fro from outlying receptors to thé central nervous system and out again to effectors. Outside the CNS they are called *nerves*, whereas within the CNS these bundles of axons and dendrites are known as *tracts*.

**22.2** The vertebrate nervous system is the culmination of many evolutionary trends seen in invertebrates, plus some qualitative differences. Discuss.

The complex synapses of vertebrates permit restriction of nerve impulses to one direction and selective firing of specific neural circuits. Such synapses, especially in the complex networks seen in vertebrates, have allowed much more differentiation of function than is seen in the nerve nets of more primitive invertebrates. Differentiation has also permitted development of increasingly complex and numerous sense organs.

The major trends in neural evolution are related to centralization of the nervous system. Increasingly, peripheral nerves became coordinated with a few nerves that ran the length of the body (the beginnings of a central nervous system). Increasing numbers of interneurons provided greater variability of response. Cephalization further concentrated control within a central area, leading to formation of a brain, and within this central region differentiation produced specific functional areas. In vertebrates, the brain is far more dominant than in invertebrates.

The vertebrate spinal cord is quite different from the cords of those invertebrates possessing them. In vertebrates, it is a single, hollow, dorsal tube compared with the double, solid, ventral cord of invertebrates such as annelids and arthropods. The vertebrate cord is also not organized into such a ladderlike pattern of ganglia and connecting tracts.

**22.3** What mechanism causes the action potential to be an all-or-none process?

The size of the action potential is determined by the amount of sodium that pours into the depolarizing neuron. The amount of sodium entering, in turn, is determined by the voltage-sensitive gates. When the gates are open, a set level of permeability exists and a constant amount of sodium ions diffuses in. Above a threshold, any stimulus to the neuron, regardless of strength, has the same result—opening the gates. These gates always allow roughly the same amount of $Na^+$ in each time and then close at a constant potential, thus preventing influxes of $Na^+$ beyond that potential.

**22.4** Why is it assumed that there are two gates at the $Na^+$ channels in the membrane of the neuron?

The sodium channel is closed to passage of $Na^+$ at two distinct stages of polarization: during the resting potential ($-60\,mV$) and at the peak of the repolarization wave ($+40\,mV$). This bimodal pattern suggests two different protein gates, sensitive to two different voltages.

**22.5** Construct a table showing the positions of the $K^+$ gate, the $Na^+$ activation gate ($Na^+$ AG), and the $Na^+$ inactivation gate ($Na^+$ IG) under the following circumstances: in the resting state, on stimulation of the neuron, and just after the peak of the depolarization wave. Let "O" represent an open gate, and "C" a closed gate.

|  | $K^+$ | $Na^+$ AG | $Na^+$ IG |
| --- | --- | --- | --- |
| In resting state | C | C | O |
| On stimulation | C | O | O |
| After peak | O | O | C |

**22.6**   Two differences between an electric current and a neural impulse were specified in the text. What are these, and what qualities of the nerve explain these differences?

Unlike an electric current, the action potential does not lose strength as it progresses along the neuron. This is because the action potential is recreated at each point along the neuron; thus, it is energetically just as new at the end of the neuron as at the beginning.

The myelin sheath that characterizes many of the neurons in vertebrates, being lipid, is an insulator and should not conduct an electric impulse. However, unlike an electric current, the neural impulse is conducted down the insulated axon; the nodes of Ranvier allow the action potential to be recreated along the nerve fiber, and electric currents created by an action potential at one node can then jump to the next node to initiate the next action potential.

**22.7**   Besides speed, what other advantage would you predict saltatory conduction offers over nerve propagation in unmyelinated fibers?

Because far fewer $Na^+/K^+$ pumps are required in myelinated fibers, far less energy (as much as 5000 times less) is used in the form of ATP.

**22.8**   Experiments have shown that action potentials in mammalian neurons can travel in either direction within the neuron. Why, then, do nerve impulses tend to travel only in one direction in mammals?

At the synapse, neurotransmitter is produced and released on only one side of the synaptic cleft; therefore, the mechanism for continuing a nerve impulse exists for only one direction. An impulse may travel in either direction *within* a given neuron, but only those impulses traveling from cell body and dendrite to axon can be passed on to the next neuron. Thus the synapse serves as a valve restricting passage of the nerve impulse to one direction.

**22.9**   Distinguish between temporal and spatial summation.

In temporal summation the postsynaptic neuron is fired by a single synaptic knob. Although a single firing of the presynaptic knob does not release enough transmitter to fire the apposing neuron, if the knob is fired repeatedly in a short enough time span, the released neurotransmitter may reach sufficient levels to induce the apposing neuron to fire.

In spatial summation, the neurotransmitter required to fire the apposing neuron is accumulated not by one knob releasing the neurotransmitter many times over a short period of time, but by the one-time effort of many knobs simultaneously releasing their small amount of neurotransmitter.

**22.10**   Insecticides such as *organophosphates* work by inhibiting enzymes such as acetylcholinesterase. Why do you suppose they are effective?

Enzymes such as acetylcholinesterase break down the neurotransmitters that generate excitatory postsynaptic potentials. In the absence of such enzymes the unquenched neurotransmitter continues to stimulate the postsynaptic neuron or muscle to fire, and to fire in a disorganized fashion. Therefore, insects ingesting organophosphates die from a loss of muscle control, because their neurotransmitters are not catabolized.

**22.11**   The special sense organs are highly modified sensory receptors that initiate such very different sensations as sight, hearing, taste, smell, and balance, yet they transmit essentially identical nerve impulses. How can such different sensations be reconciled with such similar impulses?

The essentially identical impulses go to different regions of the brain for translation. It is thus the final destination of the impulses, not their intrinsic nature, that provides the diversity of sensation.

**22.12**   List the primary functions of the cerebrum.

Complex associations, language and speech, spatial perception, analytical thought, sensation, memory, and fine motor control.

**22.13**  Given what you know about the human brain, which part would you say has changed the least? Which the most?

Although the cerebellum has greatly enlarged, the hindbrain as a whole has not evolved much from its early form. Presumably this is because it controls many of the visceral functions, such as breathing and heartbeat, which are as essential in advanced vertebrates as they were in early ones and thus have not lent themselves to evolutionary experimentation.

The cerebrum, on the other hand, began as a paired swelling on the forebrain concerned mainly with smell. From this primitive "smell brain" the cerebrum has evolved into the complex organ seen in humans.

**22.14**  Why do you suppose the U.S. Air Force is interested in the reticular formation?

The military interest in the reticular formation stems from the need of their personnel to maintain wakefulness during their tours of duty. This is particularly crucial in pilots of supersonic aircraft, for whom complete alertness is often a matter of life and death.

There is considerable evidence that serotonin is the biogenic amine that mediates states of calm and may even be involved in lowering the levels of consciousness. Endorphins and enkephalins may also indirectly diminish arousal. These substances act like endogenous opiates, i.e., morphine and other opium products.

**22.15**  Activity within a tissue is accompanied by changes in electric activity. If electric contacts are placed on the skull and the measured charge is compared with that at a reference point along the body, the electric charges in the brain may be charted. An *electroencephalograph* measures such patterns and records them as an *electroencephalogram* (EEG). Presumably the different patterns of brain waves recorded reflect changes in function, but a specific association with particular activities is not clear. A rather slow and regular wave pattern is encountered under quiet conditions in neurologically intact people. These waves are known as *alpha waves*. When thought occurs or a problem is being handled, the wave pattern changes to rapid, irregular blips of low magnitude—the *beta waves*. During sleep a very slow and synchronous wave pattern is discerned. These are the *delta waves*. How might the EEG be used clinically?

EEGs can be used both to characterize the nature of an abnormality and to localize it in the brain. For example, petit mal, grand mal, and focal epilepsy all demonstrate distinctive brain wave patterns. A few psychopathologies also can be diagnosed by their characteristic EEG patterns. Brain tumors generally affect patterns, either by blocking electric activity or, more commonly, by causing excessive excitation in the area around the tumor. By comparing the EEGs of electrodes placed on different areas of the skull, neurologists can pinpoint areas of abnormal activity and thus localize the tumor.

**22.16**  The sympathetic and parasympathetic systems generally serve the same organs and generate the same action potentials yet have very different roles in the body. Describe two ways in which these seemingly similar systems manage to have dissimilar effects.

The sympathetic system basically prepares the body to fight or flee. It consequently must activate and inhibit many systems simultaneously. This requirement for *mass discharge* is reflected in the interrelation of ganglia in the ganglionic chain. Although the parasympathetic system directly opposes the action of the sympathetic system to some extent, it more commonly influences only one or a few organs at a time, rather than a whole constellation. This discrete versus mass activation is reflected in the less interconnected nature of its nerves. Thus anatomical differences enable the two systems to respond differently.

Physiological differences also contribute to the divergent functions of these two related systems. Although they have similar action potentials, their respective neurons release very different neurotransmitters at a given target organ. The sympathetic system releases norepinephrine or epinephrine, whereas the parasympathetic system releases acetylcholine. The differential effects of these neurotransmitters enable the two systems to influence the target organ in very different, often opposite, ways.

# Supplementary Problems

**22.17** The nerve net of hydra lacks   (*a*) neurons.   (*b*) dendrites.   (*c*) connections.   (*d*) direction in impulse flow.

**22.18** Annelids have a central nerve cord that is   (*a*) dorsal.   (*b*) ventral.   (*c*) hollow.   (*d*) impermeable to K$^+$.   (*e*) none of these.

**22.19** The cleft between pre- and postsynaptic membranes is much wider in what kind of synapse?   (*a*) electric synapse.   (*b*) chemical synapse.

**22.20** Impulses travel much more rapidly along myelinated nerve.
(*a*) True   (*b*) False

**22.21** The resting potential is maintained largely by the sodium pump.
(*a*) True   (*b*) False

**22.22** All glial tissue consists of Schwann cells.
(*a*) True   (*b*) False

**22.23** In both the generation of a nerve impulse along the axon and transmission across a synapse, depolarization occurs.
(*a*) True   (*b*) False

**22.24** The action potential (spike) obeys the all-or-none law.
(*a*) True   (*b*) False

**22.25** Thick axons conduct impulses much more slowly than thin ones.
(*a*) True   (*b*) False

**22.26** Saltatory conduction is associated with those nerve fibers that have nodes of Ranvier.
(*a*) True   (*b*) False

**22.27** The myelin sheath of vertebrate nerves is a particularly good conductor of electric current.
(*a*) True   (*b*) False

**22.28** Match the items in column **A** with the descriptions in column **B**.

|   | A | | B |
|---|---|---|---|
| 1. | A neuropeptide which reduces pain | (*a*) | Acetylcholinesterase |
| 2. | Similar to opium in its action | (*b*) | Synaptic vesicles |
| 3. | Helps to recharge synaptic membrane | (*c*) | Norepinephrine |
| 4. | Source of neurotransmitter | (*d*) | Endorphin |
| 5. | Released from sympathetic nerves | | |
| 6. | Fuse with presynaptic membrane | | |
| 7. | Produced in adrenal gland | | |

**22.29** *Curare* is an arrow poison that binds to ACh receptors and prevents this neurotransmitter from exerting its usual physiological action. The likely effect of curare is   (*a*) diminished heart rate.   (*b*) an increase in nerve conduction.   (*c*) paralysis.   (*d*) enhanced learning.

**22.30** The hypothalamus is part of the   (*a*) hindbrain.   (*b*) midbrain.   (*c*) forebrain.   (*d*) spinal cord.

# Answers

| | | | | | |
|---|---|---|---|---|---|
| **22.17** | (*d*) | **22.22** | (*b*) | **22.27** | (*b*) |
| **22.18** | (*b*) | **22.23** | (*a*) | **22.28** | 1—(*d*); 2—(*d*); 3—(*a*); 4—(*b*); 5—(*c*); 6—(*b*); 7—(*c*) |
| **22.19** | (*b*) | **22.24** | (*a*) | **22.29** | (*c*) |
| **22.20** | (*a*) | **22.25** | (*b*) | **22.30** | (*c*) |
| **22.21** | (*a*) | **22.26** | (*a*) | | |

# Chapter 23

# The Musculoskeletal System: Support and Movement

The bony skeleton is the chief structural system of vertebrates. In association with the skin, which is the largest organ of the body, it provides the basic form and shape of the organism. In addition to the mechanical function of *support*, the skeletal system also serves as the main means of *protection* for vulnerable organs within the body. In conjunction with the muscular system, the skeletal system allows *body movement*. Individual bones or bone formations serve as levers that are moved by appropriately placed muscles. Contractions of specific muscles bring about movement of limbs or body parts.

## 23.1 INVERTEBRATE SUPPORT SYSTEMS

### HYDROSTATIC SKELETONS

Support systems in early invertebrates depended on the relative incompressibility of water and related fluids. Such *hydrostatic systems* still extant consist of a fluid-filled cavity surrounded by muscle.

**EXAMPLE 1**  The internal cavities of hydra and similar tubular cnidarians, the flatworms, and even the annelid worms exemplify the use of a gastrovascular cavity (in hydras and planarias) or the more elaborate *coelom* as a hydrostatic skeleton. Squeezing of this fluid tube by judiciously placed muscle can alter the shape of an organism or bring about movement. Segmentation of the body affords even finer local responses, such as extension of locomotor processes. The specialization and refinement of movement may also be promoted by softening of the cuticle, since a rigid cuticle would prevent the to-and-fro movements required for slithering through the soil or swimming.

### EXOSKELETONS

The *exoskeleton* of the arthopods is a marked advance in structural support mechanisms. Arthropods, which include more than a million species of insects, are considered the most effective colonizers of the terrestrial environment. Only the mammals, among the vertebrates, rival the insects, spiders, and centipedes in their adaptability in moving onto dry land. The external skeleton of these jointed-legged animals plays a key role in this ready adaptation. In almost all cases the exoskeleton not only provides structural support and protection for underlying soft parts but also prevents drying out by means of a waxy outer layer found in terrestrial forms. Muscles are attached to different parts of the skeleton in an arrangement that permits movement around flexible joints. In the case of the exoskeleton, the muscles are attached to the *inside* of the skeleton instead of the outside as in vertebrates.

## 23.2 THE VERTEBRATE ENDOSKELETON

Perhaps the most efficient support system exists in the *endoskeleton* (internal skeleton) of the vertebrates, an internal latticework of *bone* and *cartilage* (see Chap. 4) which frames, shapes, and protects the body and provides a set of levers which maximize the potential for intricate and rapid movement. The bony skeleton, which tends to persist in preserved form long after the soft parts of the body have decayed, provides significant clues to the evolutionary history of the vertebrate line. Anatomical comparison of superficially divergent vertebrate forms reveals the basic similarity, a bone-for-bone relatedness, of all vertebrates—perhaps among the strongest arguments that can be marshaled for a common descent.

Rigidity of the internal skeleton of vertebrates is combined with a flexibility of the bony framework achieved by movement at their connections, or *joints*. The 206 individual bones of the adult human skeleton (see Fig. 23.1) are found in either the *axial skeleton*, consisting of the skull, vertebral column

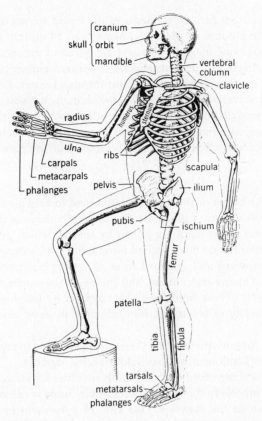

**Fig. 23.1** The human skeleton. (*From Storer et al.*)

(spine), and rib cage, or the *appendicular skeleton*, made up of the arms and legs and the *girdles* (pelvic and pectoral) to which these appendages are attached. The girdles are so named because they encircle the region in which they are located.

## SKELETAL DEVELOPMENT

The entire skeleton arises from mesoderm. Alongside the neural tube, which is ectodermal in origin, a series of regular mesodermal blocks begins to form early in development. These blocks, known as *somites*, are a reflection of the segmentation pattern found in the body plan of most animal phyla from the annelid worms on. A tough but flexible rod of mesodermal tissue, the *notochord*, precedes the formation of the neural tube and induces its formation. Portions of the notochord and the somites combine to form the spinal column, the bony covering of the spinal cord.

Migration of mesodermal cells from the somites and other mesodermal ridges gives rise to most of the axial skeleton. Separate groups of mesodermal cells give rise to most of the protective bones of the skull (*neurocranium*) and the smaller processes and ear bones (*viscerocranium*). The appendicular skeleton is derived from *limb buds*, which appear in humans at the end of the first month of development. These limb buds, consisting of a central core of unspecialized mesodermal cells almost completely covered by a thin layer of ectoderm, presage the vertebrate limbs that will soon form. At first cartilage is laid down to demark the bones of the appendages, but this is largely replaced by true bone near the end of limb development.

## THE SKELETAL SYSTEM AND HOMEOSTASIS

The skeletal system serves a homeostatic function in at least three ways. First, maintenance of a constant state often involves behavioral responses to environmental change. These responses are dependent on the musculoskeletal system, which produces movement and, more specifically, locomotion.

Second, erythrocytes and other formed elements of the blood are produced in the soft red marrow making up the core of the long bones of the body. More than 50 million red cells are produced each minute by the reservoirs of the bone marrow. The *myeloid* tissue of the red bone marrow also produces granular leukocytes and platelets. A constancy of these formed elements is maintained through the action of hormones such as *erythropoietin* on the blood-forming tissues. Erythropoietin is produced in the blood (from a precursor secreted by the kidneys) when oxygen levels in the tissues fall.

Third, the skeletal system dynamically stores such minerals as calcium and phosphorus and, to a lesser extent, sodium, magnesium, and manganese. The reservoirs of such minerals in the bones and teeth are part of a pool that responds to regulatory hormones of the parathyroid and adrenal cortex.

## MOVEMENT

As might be expected, the bones of the axial skeleton tend to provide maximum protection for the brain, spinal cord, and the soft organs of the chest. Consequently they are not designed to permit much movement. The separate bones of the skull are joined by *sutures*, which permit virtually no movement around their tight fit. The vertebrae do have limited movement, which permits the back to be bent, and the rib cage can be moved upward and outward from its resting position by the *intercostal* muscles, which lie between the ribs. On the other hand, the bones of the appendicular skeleton have much greater flexibility at their joints. These bones are held together by tough connective tissue bands, the *ligaments*, which permit a variety of levered movements and, in some cases, allow complete rotation at the joint.

The muscles responsible for movement of the skeleton are attached to the bones by stringy connective tissue strands called *tendons*. Tendons may also hold muscles together.

Movement produced by a contracting muscle usually involves a stationary base and a moving base. The attachment point of the muscle to the relatively stationary bone is called the *origin*. The attachment to the moving bone is known as the *insertion*. One kind of movement involves a simple hinge joint. Contraction of the inner muscle mass tends to reduce the angle at the joint, a movement known as *flexion*. Contraction of the muscles lying along the outer border of the hinge system increases the angle at the joint, a movement called *extension*. More complex movements occur at joints such as the hip, shoulder, and knee in humans.

All muscles produce their effects by contraction. Opposite movements at a joint are due to the existence of *antagonistic* muscle pairs, which oppose one another in their contraction.

**EXAMPLE 2**   The *biceps* is a muscle whose insertion is on the *radius*, a bone of the forearm, and whose origin is on the scapula (shoulder blade). Upon contraction it causes the forearm to flex (come closer to the upper arm). The *triceps*, lying opposite the biceps along the outer surface of the upper arm, originates on the *humerus* (bone of the upper arm) and inserts on an extension of the *ulna* (bone of the lower arm) that lies just beyond the joint. Upon contraction it produces extension of the lower arm. The biceps and triceps are an antagonistic pair of muscles which regulate the position of the forearm by actions which are opposite in nature but both produced by contraction.

## 23.3   ANATOMY AND PHYSIOLOGY OF VERTEBRATE MUSCLE

Movement, both internal and external, is a distinguishing feature of all animals. It is a necessary characteristic of organisms that must obtain their food ready-made rather than manufacture it from simple raw materials. In amoebas movement is achieved by a streaming flow of internal fluid cytoplasm along an extended external tube of stiff cytoplasm (see Fig. 23.2). The extensions, which are formed and re-formed, are called *pseudopodia*. This *amoeboid movement* enables the amoeba to move relatively slowly along a base by a process that is essentially a directed oozing. This kind of movement is also found among the leukocytes that engulf bacteria in our own bodies, since extended pseudopodia can be used to surround external objects. The oarlike beating of cilia and flagella (see Chap. 4) represents a second major mechanism for movement in animal cells. However, the contraction of muscle is probably the most significant means of movement in most invertebrates and the basis for all locomotion

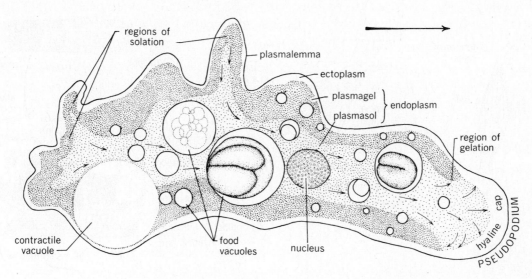

regions of
solation

plasmalemma

ectoplasm

plasmagel

plasmasol

} endoplasm

region of
gelation

contractile
vacuole

food
vacuoles

nucleus

hyaline cap

PSEUDOPODIUM

**Fig. 23.2** Amoeboid movement. (*From Storer et al., after Mast.*)

and most internal movements in vertebrates. A fairly clear idea of the major features of vertebrate skeletal muscle has emerged in the last several decades.

## SKELETAL MUSCLE

The muscle that moves the skeleton is known as *skeletal muscle*; it is also called *voluntary* or *striated muscle*. As these names imply, this muscle is attached to bone, is under voluntary control, and contains a pattern of cross-sectional bands when viewed under the microscope.

The basic unit of the muscle is the muscle cell, or *muscle fiber*. However, as Fig. 23.3 illustrates, skeletal muscle has several levels of organization both above and below this cellular level. Muscle fibers are grouped into bundles called *fasciculi* (see Fig. 23.3b). These bundles, together with connective tissue, nerves, and blood vessels, form the muscle, which is surrounded by a tough layer of fascia beneath the skin.

The muscle fiber itself is a long, slender cell with multiple nuclei and a cell membrane called the *sarcolemma*. Within each fiber are many rodlike bundles called *myofibrils* (see Fig. 23.3c). Each of these consists of a repeating pattern of structures that are seen in the myofibril as cross-sectional bands of different shading (see Fig. 23.3e). These bands are in register among the various myofibrils of a single fiber, causing the striation seen in skeletal muscle fibers.

A single repetition of this myofibril banding pattern is called a *sarcomere*. The sarcomere is delineated on either end by a platelike structure, the *Z line*, that cross-sections the myofibril. Extending longitudinally on either side of the Z line are strands of a thin protein, *actin* (see Fig. 23.3g). Toward the center of the sarcomere these actin strands interdigitate with another longitudinally aligned, thicker protein, *myosin*. (As we will see later, it is the interaction of these two proteins that results in muscle contraction.) In cross section each myosin filament is surrounded by six actin filaments (see Fig. 23.3f).

Each sarcomere is surrounded by a *sarcoplasmic reticulum* (*SR*), a dense network of ducts, analogous to the endoplasmic reticulum (see Fig. 23.3d). The SR contains high concentrations of $Ca^{2+}$ ions, which are released during muscle contraction. Invaginations of the sarcolemma roughly along the Z lines form deeply penetrating tunnels called *transverse* or *T tubules* (Fig. 23.3d). These are involved in conducting action potentials throughout the fiber.

According to the sliding filament theory, muscle contraction occurs through the repeated attachment of myosin myofilaments to actin filaments, followed by a pulling of the actin filaments from both ends of the sarcomere in toward the center, with a consequent shortening of the fiber. The two filament types do not contract, but rather simply slide past each other.

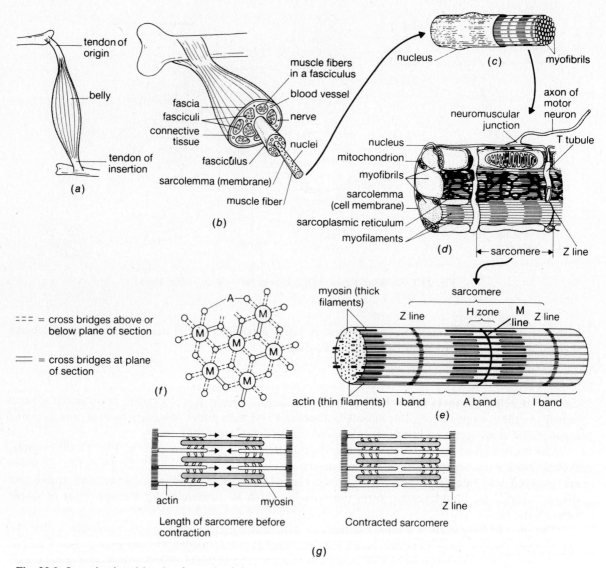

**Fig. 23.3** Organizational levels of muscle: (*a*) the entire muscle; (*b*) cross section of muscle; (*c*) arrangement of myofibrils in a muscle fiber; (*d*) detail of a muscle fiber; (*e*) detail of a myofibril; (*f*) cross section of a myofibril, showing spatial arrangement of actin around myosin; (*g*) resting (*left*) and contracted (*right*) sarcomeres.

The myosin myofilament is an aggregate of myosin proteins, each of which consists of a long tail woven into the aggregate, a pivot (or hinge) region, and a complex globular head, which sticks out from the myofilament (see Fig. 23.4). It is the pivoting of the head, when attached to an actin filament, that draws the actin toward the center of the sarcomere. The movement is mediated by ATP: The myosin head, which is also an enzyme, converts ATP to ADP and uses the released energy to pivot and bind an actin filament. The ADP then is released, and the head pivots back to its original position, pulling the attached actin longitudinally toward the center. With the ADP released, a new ATP can bind the myosin head, causing it to release the actin and begin the process anew. Numerous rapid repetitions of this sequence cause the actin to be ratcheted inward. At the end of the contraction, the actin is released and slides passively back to its original position.

In skeletal muscle, the contraction is initiated by a motor nerve impulse. As with neurons, the resting sarcolemma is polarized. When an action potential reaches the neuromuscular junction, transmitter is released from the axon, reducing the level of polarization across the sarcolemma. If the reduction

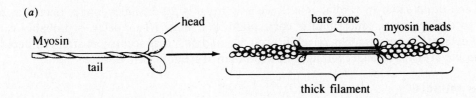

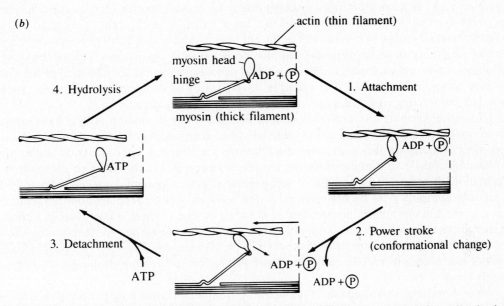

**Fig. 23.4** (*a*) A diagram of a single myosin molecule (*left*) and the assemblage of many myosin molecules into a thick filament (*right*). The four major steps in the contraction of muscle.

hits the threshold level, an action potential is triggered in the fiber. The T tubules permit transmission of the action potential deep into the fiber, so that all the myofibrils are stimulated at essentially the same time and can contract as a unit.

In the *resting* muscle, binding of myosin and actin is inhibited by a protein, *tropomyosin*, which blocks the actin's myosin-binding site. This protein is held in place by another protein, *troponin*. The depolarization of the sarcolemma and T tubules stimulates the release of $Ca^{2+}$ from each sarcoplasmic reticulum into its section of myofibril. The $Ca^{2+}$ ions released from the SR effect a change in the shape of troponin; it shifts, pulling away the tropomyosin and thus freeing the actin for binding with myosin. At the end of the contraction a $Ca^{2+}$ pump draws the $Ca^{2+}$ back into the SR, and the tropomyosin-troponin complex again blocks the binding sites on the actin filament, inhibiting further contraction.

## CARDIAC MUSCLE

Cardiac muscle is primarily composed of striated, mononucleate cells. The cells tend to branch and form networks with other cells. The junction of two such cells is characterized by an *intercalated disk*, a tightly fitting adaptation of the two membranes that offers relatively minimal resistance to the passage of action potentials from one cell to the next. The branching of the cells and the ease of passage of action potentials between them resemble a nervous system and enable depolarization to sweep across the heart and induce almost simultaneous contraction of the cells of a chamber. As we saw in our study of the heart, this unified contraction is necessary, particularly in the ventricles, to generate sufficient pressure for pumping blood throughout the body.

Although the heartbeat is modulated by the autonomic nervous system, it is initiated by excitatory cells in the sinoatrial node of the right atrium. The heart is thus capable of beating independently of external nervous stimulation.

The action potential in cardiac muscle cells is roughly 20–50 times longer than that of skeletal muscle, so the period of contraction is correspondingly longer. This delay is probably caused by a transient decrease in membrane permeability to $Ca^{2+}$ and $Na^+$ that occurs just after full depolarization. Heart muscle also has a longer refractory period than skeletal muscle does.

## SMOOTH MUSCLE

Smooth muscle fibers are mononucleate and much smaller than skeletal muscle fibers. Although they may occur as loosely arranged, independently innervated strands (e.g., in the ciliary muscles and iris of the eye and the piloerector muscles that cause hair to stand upright), they more commonly exist in tightly packed groups in organs such as the gut, uterus, and ureter. Each fiber possesses *tight junctions* with its neighbors, in which apposing cell membranes fuse or almost fuse. This permits a spreading of action potentials across groups of fibers and thus permits nonneural coordination of the muscle group, creating waves of contraction. For example, in *peristalsis* of the gastrointestinal tract such a wave is seen as a contracting ring that moves along, pushing material ahead of it.

The same elements and processes of contraction operate in smooth muscle as in striated muscle, but in smooth muscle there appears to be much more actin and less myosin. Although the longitudinal alignment of filaments does exist, often the filaments are arranged, instead, at odd angles to each other.

Smooth muscle is stimulated by the autonomic nervous system and also responds to hormonal stimulation; however, it is capable of self-generated action potentials as well. Possibly because of oscillating pumping rates by $Na^+$ pumps, a *slow wave rhythm* of depolarization undulates constantly across some smooth muscle membranes. If the crest of one of these waves reaches a threshold level, an action potential results and is spread to the rest of the muscle group. Smooth muscle also shows action potential plateaus similar to those seen in cardiac muscle. These various depolarization characteristics cause smooth muscle contraction and relaxation to be more prolonged than those of other muscle types.

Smooth muscle is also capable of maintaining contractions over long periods. It has been suggested that the sliding filaments causing the contraction tend to stick to each other, thus holding the muscle in contraction without using extra amounts of ATP.

Another unique characteristic of smooth muscle is its tendency to contract when stretched. Stretching apparently decreases the membrane potential, and this in combination with the normal slow wave rhythm generates an action potential. This feature is important for a muscle often found in hollow organs that must expel their contents when full.

# Solved Problems

**23.1** As seen in Fig. 23.3*e*, a sarcomere is divided into lines, bands, and zones that correspond either to myofibril structures or to areas of overlap between these structures. From what you know of the structure and action of the sarcomere, explain what each of these lines, bands, and zones consists of. (The *M line* consists of an enlargment in the center of the myosin myofilament.)

The Z line is the structure anchoring the actin filaments. The *I (isotropic) band* is formed by unoverlapped actin extending to either side of the Z line (and thus lying in two sarcomeres). The *A (anisotropic) band* is caused by myosin. The darker margins of this band are the regions of overlap between actin and myosin (which, being thicker than the unoverlapped regions, are darker). The lighter center, the *H zone*, of the A band is the section of myosin not overlapped by actin.

**23.2** During contraction, what changes would you expect to see in the various lines, bands, and zones of the sarcomere, on the basis of the sliding filament hypothesis?

During contraction, the actin filaments are drawn inward along the myosin myofilament. This causes the Z lines to be drawn closer together. Also, overlapping of actin and myosin increases; therefore, the I band, which is unoverlapped actin, gets smaller. The A band corresponds to the complete myosin myofilament, and since myosin does not change length during muscle contraction (nor does actin), the A band does not change. However, the H zone, which is unoverlapped myosin, gets smaller, and the darker margins of the A band get larger. The M line remains the same.

**23.3**   Each skeletal muscle fiber, or rather each *motor unit* consisting of one or a number of muscle fibers, is innervated by a single axon. A stimulus sufficient to excite depolarizes the membranes (*sarcolemmas*) of each muscle fiber making up the motor unit and produces an all-or-none response in each fiber, yet skeletal muscle is capable of graded responses of varying strengths and movement. How is this possible?

An entire muscle consists of many motor units. Stronger stimuli involve a greater number of motor units. Thus, a graded response is in part dependent on the number of motor units responding to a particular stimulus.

The situation is somewhat complicated by the fact that axon stimulation usually occurs as a series of action potentials moving along the nerve rather than as a single impulse. These volleys of impulses may be summated for any single motor unit and thereby produce a greater response than would be the case for a single impulse. (If the frequency of impulses is sufficiently high, a sustained contraction of maximal magnitude occurs that is called *tetanus*.) Thus, summation effects of a single unit also contribute to graded responses.

The existence in muscle of *fast* (quickly responding) and *slow* (more sluggishly responding) fibers also modifies the total contractile response to a stimulus.

**23.4**   The innervation and internal electrical characteristics of muscle fibers are closely related to function. How is this relation reflected in each of the three muscle types?

Skeletal muscles are charged with responding to changes in the external environment. It is not surprising then that they are stimulated by voluntary motor neurons. They are generally individually innervated and contract independently of their neighbors. This gives the organism diverse potential for highly coordinated fine motor responses tailored to each new situation. The ability of the muscle mass to provide a graded response is another valuable adaptation to external stimulation. Penetration of the fibers by T tubules ensures that the action potential reaches all myofibrils essentially at the same time so that the whole fiber fires as a unit.

The average human heart doggedly beats over 100,000 times a day, locked into the single repetitious task of pumping blood. It reflects this singularity of purpose by initiating its own action potentials (in the sinoatrial node), subject only to modulation by the autonomic nervous system. The presence of intercalated disks in cardiac muscle permits transfer of action potentials to other cells. Thus, unlike skeletal muscle, the fibers act in conjunction with one another, rather than independently. This feature and the presence of special conductive and excitatory cells ensure that the chambers of the heart, especially the ventricles, contract more or less as a unit. The relatively slower conductive qualities of the A-V fibers ensure that there is a delay between atrial and ventricular contraction. The longer refractory period of heart muscle helps to ensure against arrhythmias by keeping recently contracted areas of the heart unresponsive to action potentials radiating from areas contracting later in the beat cycle.

Although the tasks of smooth muscle are diverse, they all involve involuntary activities; therefore, smooth muscle is innervated by the autonomic nervous system. Since smooth muscle responds to changes in the internal environment, it is also stimulated by hormones. Like cardiac muscle, smooth muscle fibers possess tight junctions between cells, enabling action potentials to travel across the muscle mass. This permits the coordinated wavelike contractions, such as peristalsis, necessary in many smooth muscle organs. Also like cardiac muscle, smooth muscle can generate its own action potentials. Since many of the functions of smooth muscle involve localized, continuous undulation, these *myogenic* action potentials free the nervous system of the chore of constantly triggering the contractions. The plateaus seen in smooth muscle action potentials create the prolonged, slow contractions necessary in visceral responses. Many hollow smooth muscle organs must squeeze fluids from their chambers. The tendency of smooth muscle to contract when stretched, owing to a decrease in membrane potential during stretching, serves this function well.

**23.5** How might amoeboid movement have a common basis with muscle contraction?

In amoeboid movement, which involves the flowing of an inner core of fluid endoplasm into the stiff "casing" of ectoplasm in a newly forming pseudopodium, the propulsive force may be a push from behind or a pull from the front. At one time this push or pull was ascribed to secondary folding of cellular proteins. It is now felt that actin specifically is involved in the formation of the gel-like cylinder of ectoplasm that creates the tube into which endoplasm flows. This flow probably takes place when the actin making up the tube interacts with myosin to bring about contraction. In the push hypothesis, the contraction occurs posteriorly to create a force from behind, like the squeezing of the back of a toothpaste tube. In the pull hypothesis, contraction from the front (leading edge) of the pseudopodium pulls the endoplasm forward. In either case, a sliding of actin and myosin filaments may be involved in bringing about the flow.

# Supplementary Problems

**23.6** The hydrostatic skeleton of invertebrates is dependent on which property of water? (*a*) transparency. (*b*) incompressibility. (*c*) high specific heat. (*d*) all of these. (*e*) none of these.

**23.7** A major disadvantage of an exoskeleton is its inhibition of continued (*a*) specialization. (*b*) protection. (*c*) growth. (*d*) glycolysis. (*e*) all of these.

**23.8** In bony vertebrates, the laying down of bone is often preceded by the presence of (*a*) chitin. (*b*) starch. (*c*) cartilage. (*d*) platelets. (*e*) glycogen.

**23.9** The triceps flexes the forearm.
(*a*) True (*b*) False

**23.10** Involuntary muscle has no myosin or actin, since it lacks striations.
(*a*) True (*b*) False

**23.11** Intercalated disks make up the striations in skeletal muscle.
(*a*) True (*b*) False

**23.12** Actin is actually a globular protein that assembles as a long, "beady" structure consisting of two helical chains.
(*a*) True (*b*) False

**23.13** Cross bridges pull the actin chain as the myosin heads return to their unenergized conformation.
(*a*) True (*b*) False

**23.14** Troponin attaches to actin, blocking the attachment of myosin.
(*a*) True (*b*) False

**23.15** ATP is involved in the sliding filament mechanism as well as in the maintenance of the $Ca^{2+}$ pump.
(*a*) True (*b*) False

# Answers

| | | | | |
|---|---|---|---|---|
| **23.6** (*b*) | **23.8** (*c*) | **23.10** (*b*) | **23.12** (*a*) | **23.14** (*b*) |
| **23.7** (*c*) | **23.9** (*b*) | **23.11** (*b*) | **23.13** (*a*) | **23.15** (*a*) |

# Chapter 24

# Animal Behavior

## 24.1  WHAT IS BEHAVIOR?

*Behavior* is a set of activities that orient an animal to its external environment. While behavior is most clearly apparent in terms of an observable series of movements, it may also include internal responses of an adaptive nature.

**EXAMPLE 1**  An organism exposed to excessive heat may move into the shade. At the same time, the organism may also alter the distribution of blood so as to permit loss of heat through radiation from the surface to the environment.

Behavioral patterns usually center on acquiring food, seeking a mate, caring for young, guarding against danger, and other tasks vital to the life of the individual. Integral to behavior are the nervous, muscular, skeletal, and endocrine systems.

*Taxes* (sing.: *taxis*) are simple kinds of behavior, elicited from even single-celled protozoans, that orient an organism to conditions in the environment.

**EXAMPLE 2**  A single-celled euglena tends to swim toward light. This involves an automatic response of a relatively simple nature mediated by the light-sensitive eyespot of the organism and is characteristic of a taxis.

Closely related to taxes are *kineses* (pl.): changes in the rate of movement brought on by environmental stimuli.

**EXAMPLE 3**  Among some insects there is an increased rate of activity in dry areas, a decrease in moist areas. Because of the greater movement and consequent increased migratory activity in dry areas, the insects gradually accumulate in moist environments, where their tendency to move away is considerably reduced.

The study of behavior is a relatively new division of biology and tends to be more descriptive and less assuredly analytical than other areas. One danger of analyzing the patterns of activity of other animals is the tendency investigators might have to ascribe familiar actions to human motives, desires, and goals. This is particularly crucial in the area of intention, where we really have no capability of determining what an animal actually wants when it goes through a set of activities. The intensity of the inner push, whatever its nature, is a *drive*. *Ethology*, which is the comparative study of behavior from an evolutionary perspective, often deals with the drives relating to feeding, sex, care of young, etc. These drives are apparently motivations that spring from a disturbance in the internal equilibrium of an animal. They are modified by factors either internal or present in the environment and drives have sometimes been labeled *instincts*.

**EXAMPLE 4**  An animal which has not eaten for a long period shows shifts in internal equilibria which are expressed in the behavior of food foraging. Such challenges to homeostasis as a lowering of blood sugar presumably are the internal motivational elements that trigger the hunger drive. Complex neural and hormonal pathways are involved; the hypothalamus seems to be the essential mediator of behavioral drives. If the animal is ill or in a weakened condition, the drive may be less intense than would be the case for a completely healthy organism. Previous experience might also modulate the intensity of the drive. In the case of higher primates such as humans, a raging snowstorm might very well deter us completely from leaving the house to purchase food even though a need might exist.

## 24.2  COMPONENTS OF BEHAVIOR

Internal capacities for particular behavioral patterns vary from species to species. Birds can develop a characteristic song, and lions can stalk and capture fast-moving prey. Clearly, there is an innate genetic basis for behavior. However, behavior also changes as a result of experience; interaction with the environment may modify and even elicit specific patterns of behavior. An old conflict about the relative roles of nature (internal genetic programs) versus nurture (environmental influences) has essentially been resolved by the recognition that behavior is the result of both innate features of the organism and the interactions with the environment that accumulate during its lifetime.

A major contribution to an understanding of behavior came from the laboratory of Konrad Lorenz, who with his pupils developed the discipline of ethology from 1926 to 1939. Their discovery of the underlying innate components of behavior permitted a comparison of behavioral responses among groups of organisms. They also placed behavior in the arena of evolutionary phenomena to be studied as part of a pattern of increased adaptation within species.

### FIXED ACTION PATTERNS AND RELEASER

Lorenz and Nicolaas Tinbergen showed that certain patterns of behavior occur in an unvarying manner and are the innate components that make up the genetically determined behavioral repertoire of the organism. A *fixed action pattern* (*FAP*) is a constant response to an eliciting stimulus called a *sign stimulus*. A sign stimulus coming from other members of the same species is called a *releaser*. If the releaser is a constant physical feature of a member of the population, it may serve to produce stereotyped responses that smooth social interactions within a group and produce a measure of constancy in exchanges among group members.

**EXAMPLE 5**  Among several species of birds the young nestlings have distinctive markings within their mouths. When they open their mouths, these markings serve as releasers to produce feeding movements by the parents. The stuffing of food into the young bird's mouth is a fixed action pattern innate in the parents that ensures that the young are cared for and the population maintained. The *proximal* (immediate) cause of the parental feeding behavior is the releaser, but one might argue that an *ultimate* (final) cause is the necessity to care for the young so that the species may continue. Thus, FAPs may be viewed as mechanisms developed over time to accomplish significant adaptations associated with survival.

Releasers are conventionally restricted to sign stimuli that serve a communication function among members of the same species. The action of releasers affords a particularly clear illustration of the existence of genetic factors in determining behavior.

**EXAMPLE 6**  Protective behavior is elicited from a hen by sounds of distress from her chick. The hen's response has the *appearance* of being mediated by an understanding of the situation, followed by a measured response to it. However, if the chick is in dire straits but the sound is masked by a glass or plastic cover, the hen makes no response even though she can clearly see her struggling chick. The restriction of the response to the stimulation of sound alone suggests an innate neural circuit that is responsible for the behavior. Yet even in the relative constancy of the releaser-FAP relation some modification may occur. The condition of the hen, the loudness of the sound, and the presence of other dangers in the environment may alter the degree of response.

**EXAMPLE 7**  In the male stickleback a deep red pigmentation is produced along the ventral surface in the spring. This causes other male sticklebacks to react aggressively when one male invades the neighborhood of another. An exact representation of a male fish without red pigmentation elicits no reaction, but any red object produces a violent reaction. Clearly the color alone serves as the elicitor of aggressive action. The adaptive purpose of the response appears to be the guarding of territory against male invaders from the same species. A limitation in the usefulness of these innate automatic responses can be seen in the fact that this specific FAP was first noted by Tinbergen when a red mail truck passed by a window of his laboratory in Holland and produced aggressive movements in a male stickleback confined to a tank in the window.

## MODIFICATION THROUGH EXPERIENCE

Behavior may be rooted in internal neural circuits that provide the animal with a range of possible responses to specific environmental cues, but these behaviors may be modified considerably by experience. At times the animal may be confronted with competing situations and must choose or assign priorities. A hungry animal engages in food-seeking behavior, but if danger looms, the animal usually turns off the foraging activity and flees. A hierarchy of urgencies clearly exists, and danger in most groups exerts a greater effect in marshaling activities than the drive for creature comfort. Chapter 15 covered the role of the sympathetic nervous system in managing priorities.

*Imprinting* refers to an amalgam of innate and learned behavior in which a particular experience or association made during a critical period permanently affects future behavior. First studied by Lorenz in terms of parental identification by young chicks, it has since been shown to apply to a variety of types of behavior. Lorenz used the term *imprinting* because the phenomenon resembles a permanent impression made upon the brain by the specific association.

Lorenz found that when he separated groups of goslings from their parents and exposed them to a human (Lorenz) who imitated the parental call, the goslings would follow the human as if the person were the natural parent. This phenomenon is called *parental imprinting* and only occurs during the first two days after hatching. The short period of time during which the organism is sensitive to the imprinting stimulus is called the *critical period*. It often occurs early in life, but some types of imprinting may also occur in maturity. Parental imprinting also occurs under natural conditions in other species; in mammals olfactory stimuli tend to play a greater role than the visual stimuli usually associated with parental imprinting in birds.

Sexual imprinting occurs during early associations—a recognition of cospecific individuals for reproductive purposes. Lorenz found that his goslings not only followed him as they matured, but that they attempted to mate with him when they achieved maturity. Among some birds, imprinting is also involved in the learning of the characteristic song by the male. An innate capacity for mastering a particular song is triggered by exposure to that song or to even a few notes of the song. Since some birds learn their song even when raised in isolation, imprinting is not a universal mechanism for developing the song.

*Habituation* involves modification of behavior through a diminution of response to repeated stimuli. A loss of receptivity to repetitious stimuli can be useful in preventing a drain of energy and attention for trivial purposes. Rodents respond to alarm calls by others in their group, but if these calls are continued and no danger is confirmed, further calls may be ignored.

*Conditioning* involves the pairing of an irrelevant stimulus with a natural primary stimulus that elicits an automatic response. In time, the animal becomes conditioned to the secondary (associated) stimulus and responds to it as if it were the natural stimulus. This association of FAPs or similar innate responses with new kinds of sign stimuli broadens the ability of an organism to react appropriately to environmental change, since the conditioning process removes dependence on one kind of release symbol for action.

More complex behavior patterns, such as the ability of a rat to learn to run a maze, are due to trial-and-error learning. This modification of behavior is also called *operant conditioning*, a process demonstrated and studied extensively by B. F. Skinner, a Harvard psychologist. The strategy of operant conditioning in the laboratory involves a continuous monitoring of activity, with rewards given for correct bits of behavior and witheld for inappropriate behaviors. Such monitoring eventually produces a complete bahavior pattern, such as solving a maze or pressing appropriate sequences of bars to achieve a reward. When a trainer is present, operant conditioning can rapidly produce a sequence of fruitful steps from random movements by the introduction of appropriate awards. Under natural conditions, the achievement of a particular goal is the reward that directs random activities into a behavioral pattern. Trial-and-error repetitions, step by step, lead to a final achievement. Once a successful pattern has been learned, it may be passed on by imitation.

An extreme case of behavioral modification involves the application of insight or reasoning to a novel situation. If an animal can direct its behavior to solve a problem for which it has no previous experience, then reasoning is involved. Reasoning in humans appears to involve a recasting of an external situation in the imagination and a manipulation of concepts to produce a solution that can be applied to the situation. However, such insight or reason may also be found in other primates.

**EXAMPLE 8**   A chimpanzee is placed in a cage in which a choice piece of fruit hangs from the ceiling. The chimp cannot reach the fruit, but the keeper has placed some boxes of different sizes in the cage. After a short period of head scratching the chimp moves the largest box to a point just below the fruit and then piles the smaller boxes atop this base until it can climb up and reach the fruit. This type of insight is clearly a step beyond operant conditioning.

## 24.3  CYCLIC BEHAVIOR PATTERNS

Many behaviors exhibit a rhythm in their occurrence. Part of the day is spent in sleep and the other part is characterized by great activity. Migrations occur at some seasons but not at others. Many animals demonstrate a daily periodicity in activity patterns that are approximately 24 h long; hence, they are called *circadian rhythms* (*circa*, "about"; *dia*, "day"). In humans the natural cycle is closer to 25 h but is adjusted to the 24-h rhythm in the physical world.

Internal clocklike mechanisms have been postulated for this widely occurring phenomenon. The existence of such an internal timer is clearly indicated by the cyclic behavior that is observed in organisms from single-celled protozoans and plants to the most recently evolved primates. In single-celled organisms the daily rhythms appear to originate in signals from the nucleus. These rhythms encompass cell division, enzyme synthesis, metabolic cycles, and movement. Once established the periodic activity continues for fairly long periods of time even when the nucleus is removed from the cytoplasm. However, most cells do not continue life too much longer following extirpation of the nucleus, so these findings are limited in their application.

In animals an endogenous timer appears to exist in the brain. Birds are clearly governed by cycles of light and dark in the environment, which tends to implicate the pineal and its light-sensitive production of melatonin. Much speculation also centers on the mechanism whereby a central clock could communicate with other parts of the body to synchronize the totality of responses involved in rhythmic behaviors. Evidence for both neural mechanisms and hormonal effects has been discovered. Presumably, the tidal rhythms (13 h) and lunar rhythms influence the clocks of seashore creatures, involving mechanisms similar to those that operate in circadian rhythms.

A number of studies have been carried out in which human subjects have been isolated from the usual cues that orient an organism to daily periodicities. Such individuals, usually kept in deep caves, are deprived of light and temperature cycles as well as the company of others who manifest daily activity cycles. Even under those conditions of prolonged separation from the "natural" environment, these subjects continue to display a circadian rhythm. It tends toward a period of 25 rather than 24 h, but when brought back to their usual environment and cued by daylight-night shifts, the subjects reestablish a 24 h pattern.

## 24.4  SEXUAL BEHAVIOR

Behavioral patterns associated with the reproductive process are particularly intriguing. The communication involved in readying the sexual apparatus is often complex and subtle and may be associated with elaborate courtship rituals. In most organisms sexual activity is restricted to a relatively short season. This means that mating behavior must be turned on at some point and then turned off at another time. Sexual behaviors involve periodic changes in such internal characteristics as endocrine activity, central nervous status, and the tone of the reproductive tract. But they are also modified by such environmental factors as the amount of light available, amount of food present for building up of fat reserves, presence of mates, etc. Actual mating is generally preceded by an elaborate and ritualized *courtship*, which is often the arena of intense natural selection. In courtship, the strongest and most fit males may lay claim to receptive females, while other males are banished to reproductive oblivion. Similarly, *sexual selection*, in which the female may exercise some choice in tendering herself to and affixing to herself a desirable male, provides a mechanism for the perpetuation of the fit (as judged through female selection).

The totality of the sexual response, extending even to care of the young, exacts a significant drain on the resources and energy of the organism. It also represents a commitment not so much to the continued welfare of the individual but rather to the long-term survival of the group—in the sexual performance of the individual is the seed of "self-sacrifice" for the greater need of the many. Too great a price for such an unselfish commitment could be maladaptive, which explains the constraints upon the time devoted to sex among most organisms. Groups such as humans, which demonstrate a continuous and constant involvement in sexual activity or interest, are believed to derive a secondary advantage from sex—maintenance of communal (family) structure.

Sexual behavior is made up of many diverse components, both innate and experiential. Much of the activity probably consists of FAPs elicited by a variety of releasers or external signs. Some aspects of courtship and mating approximate the aggressive activity associated with the repulsion of one male by another invading a staked-out territory. At a critical time, releasers tend to channel the aggressive actions into sexual movements when a member of the opposite sex is involved. The stickleback is a specific example of such a deflection of aggressive behavior into courtship and mating actions. Aggression, however, is never completely obscured by courtship and mating behavior. The male drives the female stickleback away from the nest after she has laid her eggs and he has delivered his packet of sperm (see Example 11). Among some insects and several types of spiders, the female kills and may eat the male immediately following mating.

Among many animal groups sexual associations are relatively casual and may occur between any pair. Such a sexual lifestyle is designated *promiscuous*. In many birds, on the other hand, sexual pairing is a lifelong association and demonstrates great fidelity. Where a single male services a number of females, creating a harem, the term *polygyny* applies. A single female's mating with a number of males is termed *polyandry*.

## 24.5  SOCIAL ORGANIZATION

The degree of interaction among members of a species may vary from the virtually solitary lifestyle of some crickets to the elaborate societies of ants and humans. Through these levels of social structure some measure of communication must exist to bring about behavior that modulates the particular social pattern. This social communication may consist of calls, facial expressions, postural states, or locomotor patterns. In some cases *cooperative activity* is elicited; in other cases *agonistic* (competitive) interactions arise. Some behavior is *cohesive*—bringing members of a group closer together in a physical sense. Other behaviors are *dispersive*—tending to scatter the members of a group. The flocking of birds preparatory to flight or the gathering of the individual members of a herd of cattle into a tight circle under conditions of danger or cold temperatures demonstrates cohesive behavior. The scattering of male birds or wolves as they stake out and defend a territory is dispersive behavior.

### COMPETITION

Many kinds of behavior involve a struggle between individuals for resources, such as food, mates, or building materials. Since members of a species have similar needs, they are more likely to engage in *competition* with one another to secure scarce resources. Only rarely does the struggle involve actual combat. Usually the contest involves highly stylized displays of strength in which the *agonists* (competitors) threaten one another with a variety of aggressive thrusts such as the baring of teeth, loud growling, or fake charges toward the opponent. Such fierce displays, which are merely symbolic, are known as *rituals*. The development of stereotyped rituals avoids actual bloodletting and could serve as an object lesson for human societies. Ritualized *submission* behavior soon ensues from one of the combatants, who acknowledges defeat.

**EXAMPLE 9**  Wolves engage in combat by increasing the thickness of their fur coat through piloerection, confronting their opponent, growling and baring their fangs, and making short charges. The wolf who has had enough lowers its tail, smooths back its fur, and even exposes its vulnerable throat. These rites of opposition avoid direct combat and enable both members to maintain their fitness and carry their genetic potential to further reproductive opportunities.

Another competitive interaction involves the establishment of a *dominance hierarchy*, or *pecking order*. This is essentially an arrangement of social status in which ranking arises from random aggressions. In chickens, where it was first studied, a top-ranked hen (alpha) could subdue all others in the flock and control access to food and mates merely by symbolic pecks. A second-ranked hen (beta) could exert similar control over others below her. This ranking continued down to the lowest member, a perpetual victim who had no hens over which to exert control. Dominance hierarchies are useful to all members of the group, since they facilitate interactions without having to go through competitive struggles each time.

*Territoriality* is another type of competitive interaction involving the acquiring of a fixed neighborhood by an individual for a more or less extended period of time. The territory becomes an arena for carrying out most of the signficant life functions and is vigorously defended. Territoriality is a significant mechanism for ensuring that those that are most fit gain access to scarce resources, since the less fit are unlikely to acquire and defend a territory.

## ALTRUISM

*Altruism* is a characteristic form of behavior among social animals in which one or several organisms sacrifice their own interests for the welfare of the group. In human societies it is not altogether rare. The soldiers who during battle fling themselves on live grenades to save others are familiar figures in popular literary and cinematic lore. However, human decisions to lay down their lives for others involve a conscious course of action that could hardly be operative in other species. Such self-sacrificing activity in other animals must be made up wholly or in part of innate behavioral patterns. In several species of field rodents, for example, the entire pack is alerted to danger by the alarm cries of scouts who linger in the path of danger while warning others to flee. Such altruistic behavior as the care of the queen's brood by workers who have no direct genetic stake in the welfare of the young bees also springs from inherited pathways, since these altruistic roles are assumed without any evidence of learning or acculturation.

In sacrificing itself the altruistic animal also often sacrifices its chance to pass its own altruistic genes on to the next generation. How, then, can genes for altruism be continued in a population? W. D. Hamilton developed the concept of *kin selection* to account for the maintenance of genes leading to altruism. His theory began with the recognition that the evolutionary success of a specific gene is dependent not just on the success of individuals who possess that gene, but on whether the frequency of that gene increases in the next generation. To make the distinction clearer, our focus should be on the gene rather than the individual, certainly a novel perspective in scanning for evolutionary change.

When considering the genes that mandate or lead to altruistic behavior, one must view not merely the reproductive destiny of the altruist, but the breeding success of related individuals who carry a similar array of genes. If the activity of an altruist (in sacrificing reproductive privileges for the welfare of the group) increases the fitness of closely related individuals who then spread the altruistic genes into the next generation, altruism will flourish. Since the core of this explanation focuses on a collective evolution of closely related individuals in a group, the name *kin selection* was used to describe this series of events. In terms of the survival and success of a particular gene or cluster of genes, the degree of relationship that exists in the kin group is crucial, since closely related individuals tend to share more of their genes.

Research on the tendency of individuals to indulge in altruism has shown that in many species altruistic behavior is usually extended to close relatives with whom the altruist has common genes. This enhances the overall reproductive success of kin groups or clans. This consideration of the fitness of kin as a group, rather than focusing on individuals, has been called *inclusive fitness*. It is in the raising of inclusive fitness that continuation of altruism is explained—the altruist may be stymied, but kin carrying similar genes, including those for altruism, are better able to survive and spread their genes.

## ANIMAL SOCIETIES

Among many species, intimate exchanges between members and adaptive cooperative action are facilitated through the formation of permanent social structures called *societies.* These societies clearly require highly sophisticated mechanisms for communication, a capacity for joining individual needs and behaviors into an integrated superstructure, and the means for ensuring continuation of the society from one generation to the next.

Complex societies are most common among insects and vertebrates, especially those vertebrates with highly evolved brains. In the case of insects, social stability is based on a rigid delegation of functions that are carried out in an unvarying fashion. A rigid caste system is present, and the members of each caste demonstrate behaviors that are largely determined by innate instinctual components. Little variation in the nature of task performance can be detected. Integration is usually achieved through chemical interactions among the members of the group. Among honeybees, as many as 100,000 individuals may make up the community living in a hive. Almost all the members are female. The few males, known as *drones,* have no function other than to mate with the single *queen bee* at the time of her emergence from the egg. Apart from the queen and the few drones, the community is made up of many thousands of *worker* females. This caste may be subdivided into *nurse* workers, who feed the larvae emerging from the thousands of eggs deposited by the queen; *housekeepers,* who tend and maintain the wax cells and guard against predators; and *food gatherers,* who leave the hive and forage for food to be used to maintain the community. Although the worker bee lives less than two months, it goes through each of these "occupational" phases.

In vertebrate societies individuals tend to be more independent and less rigidly programmed than their insect counterparts. Dominance hierarchies rather than caste system characterize vertebrate societies. The alpha and beta members of such a hierarchy not only enjoy the privileges of primary access to food and mates but also share the responsibility of protecting both themselves and those lower down in the hierarchy from external dangers.

Among monkeys and apes, social structure tends to be both complex and capable of modification through experience. Variation in basic social structure is particularly marked in human societies, although the mating bond tends to influence family configurations of most societies. Values and the inculcation of social norms among humans involve the development of a *culture*—ideas and symbols of an abstract nature that are passed from one generation to the next and are often represented in icons, books, works of art, and so forth.

## ROLE OF COMMUNICATION

Societies are maintained through continuous interactions among their members. These interactions involve *communication,* the passing of information from one individual to another. Such information may involve aggressive responses, sexual receptivity, the existence of rare resources in the vicinity, or more subtle feelings or desires. Communication may utilize visual displays, sound, smell, or even taste. Many of the behavioral responses of insects and even those of vertebrates involve the elaboration of chemicals that elicit the particular responses.

**EXAMPLE 10**　Part of the foraging efficiency of honeybees is the ability of worker scouts to come back to the hive and indicate through a "dance" both the direction and the distance of food sources to those who are ready to seek food. This *waggle-taggle dance* by returning bees was shown by Karl von Frisch to supply surprisingly detailed information about the site of such food. The nature of the food could be communicated by the regurgitation of nectar by the returnee directly upon the outgoing group.

Constant communication provides a monitoring of the state of the social structure and permits adjustments to maintain stability. In the case of the honeybee, awareness of an increase in temperature within the hive by groups of workers leads to a wild beating of wings within the hive, which tends to lower the temperature. Such a primitive air-conditioning mechanism can be effective only if temperature states are widely and rapidly communicated.

## 24.6  MODES OF COMMUNICATION

Hormones play a role in communication by influencing the formation or modification of structures utilized as signaling devices.

**EXAMPLE 11**  In the spring the male stickleback secretes hormones that produce the bright red coloration of his undersurface. When the male, under the continuing influence of these same hormones, seeks out shallow water in which to build a nest, he becomes a ready target for females, who are attracted by the red coloration. In turn, the egg-swollen body of the female attracts the attention of the male, and a courtship dance (swim) ensues that culminates in the release of eggs into the nest by the female. As the female leaves the nest, the male deposits sperm and stands by to assume the care of the soon-to-be-hatched young. Hormones, in this instance, generate all the visual cues that control courtship and mating behaviors.

Hormones also play a role in communication by enhancing drives that produce behavior possessing a significant communicating component. They may also induce the formation of structures that are releasers of drives and behaviors. This is readily apparent in the case of sexual development. With hormonal changes, structures may appear that act as releasers for courtship, induce mating movements, and help to compose programs for the care of the young. Reciprocal sexual actions depend on a constant exchange of information between the participants. Among humans, and perhaps the apes, secondary sexual characteristics that broadcast the readiness and availability of potential mates for sexual activity are particularly apparent in general body curvature, the prominence of the breasts, and the pitch of the voice. These features are influenced, of course, by the sex steroids of the gonads.

*Pheromones*, extremely effective chemical modes of communication, have been called *social hormones*. Just as hormones circulate within an individual to exert an integrative and regulatory effect on internal functions, pheromones pass among individuals of a group to produce integrative behaviors.

One of the first pheromones discovered was the sexual attractant elaborated by some female moths, which causes males to fly to the female. As is true for all pheromones, the sexual attractant is a relatively small organic molecule carried in the air and is extremely effective even in minute amounts. Among gypsy moths even a few molecules released by the female can attract males up to a mile away. The effectiveness of the pheromone is dependent on the olfactory sensitivity of the target individual.

Pheromones have been shown to function as markers of territorial boundaries, signals for danger, and regulators of caste development in social insects. A particularly interesting case of pheromone function is the "funeral" emanation arising from dead or dying ants. This chemical mobilizes surrounding workers to eject the moribund individual from the colony, a useful behavior to maintain group health and cleanliness. When this funeral pheromone was isolated and painted on a perfectly healthy worker ant, her colleagues carried her off and deposited her outside the anthill. She returned but was promptly ejected again. After several unsuccessful attempts to regain entry she finally gave up further attempts at reentry. This is a reflection of the highly programmed nature of behavior in insect societies, which leaves little room for novelty or reason.

## 24.7  BIOLOGICAL DETERMINISM AND BEHAVIOR

*Biological determinism* in the context of this chapter's concerns, is a theory that ascribes major characteristics of an organism and its patterns of behavior to genetic influence. In the social aggregations of insects and primates, the theory proposes that the genes impose a significant constraint on behavior. Although experience and culture are accepted as modifying influences, basic social interactions are viewed as the result of an evolutionary selection of genetic programs.

E. O. Wilson, a highly regarded expert on insect societies, proposed (*Sociobiology*, 1975) that behavior can best be understood if examined from an evolutionary perspective. Even those aspects of human society that have traditionally been studied by sociologists and political scientists, such as war, family structure, and norms of citizenship, can be more completely understood as the outcome of a differential survival of particular genes, according to Wilson. Considerable controversy rages over this extension of the biological (evolutionary) perspective to the sociological. While flexibility may exist in individual expression of a particular inherited pattern, many critics of biological determinism maintain that the plasticity (variability) of our social institutions contradicts an interpretation of selection of genetic programs. Criticism is also directed against the implication that only limited improvement of human potentials and talents is possible if our development is channeled and constrained by genetic programs selected for past conditions.

# Solved Problems

**24.1**   *Anthropomorphism* is the tendency to define reality in terms of human concepts and desires. It involves, for example, the assignment of human feelings, thoughts, and modes of action to other species when they may not apply. How might anthropomorphism detract from a scientific study?

An objective appraisal of behavior per se may be distorted by reading into an action some human motivation. The gathering of data must occur in an open and unbiased manner before evaluation. Otherwise, the mechanism for producing particular behaviors may go unrecognized. Another difficulty lies in the possible development of a judgmental (moral) position while studying animal behavior. This is not only unfair but also detrimental to an honest evaluation of the behavior being studied. Also, anthropomorphism may limit the rigor of a search for explanations for behaviors more appropriate to nonhuman models. However, science's stand against anthropomorphizing does not mean—as some people mistakenly believe— that possible feelings or motivations underlying animal behavior have nothing in common with those of human behavior. The stance makes no statement whatsoever; it is simply an attempt to remain objective and let the experiment bring forth its results.

**24.2**   In 1973 the Nobel prize for physiology or medicine was shared by Konrad Lorenz, Nicolaas Tinbergen, and Karl von Frisch. These three investigators are credited with founding the field of ethology and by receiving the Nobel prize for their work, immeasurably enhanced the validity of the evolutionary approach to behavior.

Previously, most students of behavior were trained as psychologists rather than as biologists. Their investigative tools were the white rat and conditioning devices such as the maze (a box containing a series of tracks that an animal had to learn to navigate) or the *Skinner box*, an apparatus containing a lever or buttons that had to be pressed in a particular sequence to produce a reward. These psychological studies, collectively designated *behaviorism*, were based on theories of learning in which conditioning produced and refined behavior patterns. How does this approach differ from the competing ethological perspective?

The emphasis in ethology is on inborn tendencies such as the FAP and instinctual bases for behavior, rather than on conditioning. Additionally, the ethologist studies a broad variety of organisms, often under field conditions, in contradistinction to the narrower laboratory setting of the behaviorist school. The ethologist is also much more interested in evaluating behavior from an evolutionary point of view—selection acting on genetic components of innate behavioral potentials.

**23.3**   How would you say fatigue differs from habituation?

In fatigue, which may involve sensory as well as motor neural components, the functional state of the neural pathway is impaired. In the familiar case of fatigue developed by the continuous contraction of a muscle, the accumulation of lactic acid and the depletion of fuel (glycogen) impair the ability of the muscle to continue to work. The site of these key changes is probably at the motor end plate; other chemical changes may also occur that prevent an impulse from reaching the contractile apparatus. In a sensory circuit, a similar case of fatigue may entail interference at a synaptic junction with the end receptor. Fatigue, then, involves cellular changes that render the nerve, nerve-muscle junction, or the muscle itself partially or totally inoperative.

In the case of habituation, which is a very simple kind of learning, the adjustment comes from action at the level of the central nervous system rather than through changes at the neuromuscular unit. The organism simply ignores repeated stimuli that appear to be trivial. The local sites of activity are fully capable of functioning. As a general rule, habituation lasts for a much longer period of time (up to several days) than the much more brief insensitivity of fatigue. Another feature of habituation is that it may be rapidly overcome by the central nervous system following a situation in which the level of alertness is upgraded. Such a heightened alertness would not alter the cellular events associated with fatigue, although it might elicit some activity from a fatigued neural circuit if the stimulation were markedly increased.

**24.4**    A variety of flying or swimming animals travel long distances to reach specific locations, and many return as well. What mechanisms might enable these migratory creatures to find their way?

Animals that move over long distances, whether as a single life event or as a recurrent regular migratory pattern, incorporate many innate and even learned bits of behavior to accomplish the final goal. The migrant may respond to innate programs in which directions are fixed for particular periods of time. In some cases, the birds fly in one direction for several weeks and then shift to another direction for the balance of the trip. Such a pattern is coordinated with shifting wind patterns that help the birds in their flight. The danger in these relatively rigid behavior patterns is that an unusual alteration in the pattern of prevailing winds might spell disaster for the migrants. Presumably a program of fixed direction may also operate in migratory turtles and some fish.

In some migratory birds a capacity for judging location can substitute for a fixed pattern of flight. The ability to know an exact location, almost as if latitude and longitude were being monitored, is called *map sense.* Evidence for this ability exists but is difficult to study in a laboratory or even a field setting. Homing pigeons, an extremely popular experimental animal, possess a *compass sense*—an ability to distinguish compass direction, so that they behave as if a compass were made available to them. The chief basis for this ability to orient in terms of north and south is the position of the sun. Since the sun changes its position with the time of day, the compass sense must involve an interaction of solar position with an internal clock. If such birds are kept under artificial light conditions and their internal timer altered by a shift in day-night timing, then their sense of direction is also shifted. William T. Keeton showed that when sunshine is not available, a backup system involving an orientation to the earth's magnetic field takes over. He attached magnets to some pigeons and brass bars to others. On sunny days no effect was observed; on cloudy days the birds carrying the magnets were confused and seemed to lose their compass sense, while the brass bar birds continued to function.

Those birds that fly at night probably orient themselves in terms of the stars, like humans on voyages of exploration in ancient times.

**24.5**    From the standpoint of genetics and evolution, what problem does the concept of altruism pose?

In developing his theory of natural selection, Charles Darwin was puzzled by the continued existence of altruistic behavior, since the altruist is likely to die without leaving its genetic imprint upon the next generation. Modern biologists were also puzzled by the persistence of such traits, which seemingly should be selected against. How could a willingness to adopt vulnerable postures for the good of unrelated colleagues be passed on rather than eliminated? It is, of course, true that the altruist enhances the fitness of the entire group by promoting the general welfare, but the altruist should be less likely to reproduce successfully than the more selfish members of the group and eventually should disappear in favor of the genetic complexes of more selfish individuals. The persistence of altruism has generated a number of speculative hypotheses.

**24.6**    The general understanding of natural selection is that the environment challenges individual phenotypes; those that are more adaptive tend to survive and reproduce, while those that are less adaptive produce fewer offspring. Under some circumstances poorly adaptive individuals may produce no offspring at all (see Chap. 12). As a result of this differential reproduction, populations change in time and come to reflect, in their gene distributions, the more successful individuals. Thus, natural selection acts on individuals, but it is the population and the species that evolve.

In the mid-1950s V. C. Wynne-Edwards of the University of Edinburgh advanced the view that selection might operate on entire groups rather than on individuals. Relationships existing within groups might confer advantages or disadvantages that would provide "grist for the evolutionary mill." Wynne-Edwards was particularly impressed with the necessity of limiting the size of natural populations so as to avoid overcrowding and consequent mass starvation. Among the techniques for limiting the numbers of a population to ensure social tranquillity are the altruistic gestures that are displayed by the weak to augment the welfare of the strong. Where dominance hierarchies exist, the less dominant males may step aside and relinquish their reproductive activity in favor of those higher up in the hierarchy who are presumably more fit. Conciliatory gestures by the less aggressive member of a fighting pair might also be regarded as an altruistic action to conserve blood and energy that might be expended in a protracted fight. Can the notion of group selection be used to explain the continued existence of altruism in a group?

The concept that group selection could explain altruism was attractive, but it did not really explain the persistence of altruism. Altruism might very well augment group survivability, but the tendency for altruistic individuals to produce fewer offspring within that group should lead to its dying out in comparison with the selfish individuals whose fitness is augmented by the existence of altruists. Group selection makes clear the advantage of altruism to the population, but it does not explain the reproductive viability of the altruist, which is the heart of the problem.

**24.7** In bees, workers are actually the sisters of the queen, since the queen is produced from the same clutch of fertilized eggs as the large number of workers. The queen is given special treatment (food, nest site, etc.) that results in distinctive royal development. The workers are rendered sterile. When the queen lays her eggs, the young bees that eventually emerge after larval and pupal stages are tended and fed by the queen's worker sisters. Caring for another's brood is certainly a prime example of altruism, as is the sometimes deadly assignment of guarding the hive that some workers assume. What mechanism would you say supports the spread of genes for such altruism in bees? [Hint: The male bees (drones) are haploid; the queen, diploid.]

Since the male is haploid and gives all its genes to a fertilized egg while the female gives a randomized group containing half of her genes (recall meiosis), all sisters have three-fourths of their genes in common. Because they are genetically so close and share more genes with their sisters than with their own potential offspring (who would contain only half of their genes) caring for their sister's offspring is genetically more productive than would be the case in other species. The altruism existing among bees is thus a particularly apt illustration of the role of kin selection and inclusive fitness.

**24.8** Kin selection operates among genetically related individuals. *Reciprocal altruism* involves an altruistic relationship between nonrelatives. In this situation the altruist is not ensuring the continuation of his or her genes. What, then, might be the adaptive advantage of such altruism?

The extending of unselfish behavior here can be understood in an adaptive sense only if there is an expectation of reciprocity, a return of the favor at some future time. Many acts of altruism center on relatively trivial tasks, such as the scratching of one primate's back by another who may more easily perform that task. In tight little bands such altruistic activity is often seen, and the intimacy of the group encourages reciprocity and leads to ready identification of those who accept favors but do not offer their services.

In species in which reciprocal altruism occurs, reciprocation must be assured or else selfish behavior will become a more adaptive lifestyle. The existence of specific altruistic activity may, in and of itself, create models within a group that encourage altruism from others. A genetic component in such sensitivity to model altruism may be operative, but certainly difficult to measure.

# Supplementary Problems

**24.9** Cultural influences on behavior would be strongest among (*a*) amoebas. (*b*) flatworms. (*c*) humans. (*d*) ducks. (*e*) sharks.

**24.10** Long-term evolutionary explanations (adaptation, etc.) for specific behaviors involve (*a*) proximate causes. (*b*) ultimate causes. (*c*) an absence of cause. (*d*) none of these.

**24.11** In the nature versus nurture debate, nature focuses on an individual's (*a*) training. (*b*) tendency to eat unprocessed food. (*c*) genes. (*d*) early influences. (*e*) all of these.

**24.12** A sign stimulus is (*a*) more general than or (*b*) less general than a releaser.

**24.13** Karl von Frisch was the first discoverer of imprinting in geese.
(*a*) True  (*b*) False

**24.14** Feeding of the young in some birds is elicited by releasers within the open mouths of the young.
(*a*) True  (*b*) False

**24.15** FAPs are never found in humans.
(*a*) True  (*b*) False

**24.16** Olfactory imprinting in salmon is involved in the return of these fish to the specific stream from which they migrated to the ocean.
(*a*) True  (*b*) False

**24.17** A pheromone from the queen prevents sexual maturation of worker bees.
(*a*) True  (*b*) False

**24.18** Self-sacrifice by parents for progeny is not really altruism.
(*a*) True  (*b*) False

# Answers

| | | | | | |
|---|---|---|---|---|---|
| **24.9** | (*c*) | **24.13** | (*b*) | **24.17** | (*a*) |
| **24.10** | (*b*) | **24.14** | (*a*) | **24.18** | (*a*) |
| **24.11** | (*c*) | **24.15** | (*b*) | | |
| **24.12** | (*a*) | **24.16** | (*a*) | | |

# Chapter 25

# Evolution: The Process

Evolution, in the broadest sense, is the view that all reality is in a continual state of change. Process is primary, and flux (constant change) is universal. This shifting reality is best understood by studying the forces that underlie change.

**EXAMPLE 1** The face of the earth is gradually changing. New mountain ranges are arising at some sites, while high places are worn away at others. Understanding the forces of erosion enables us to explain the wearing away of older mountain ranges. The exciting theory of continental drift provides insight into the building up of new ranges and, in the movement of tectonic plates, explains much of the turbulence at boundary areas of these geological plates or land masses. (See Chap. 11, *Earth Sciences*, Beiser, Schaum Outline Series.)

As applied to biology, evolution maintains that all the diverse forms of life that are now in existence have come into being through a gradual and continual process of modification of ancestral forms. This process of "descent with modification" does not lead to a finished final product. Evolution modifies *all* living things and will continue to produce change in the future as it has in the present and past. Furthermore, all life shares in the history of evolving, so that protists are as highly evolved as humans, but the specific kinds of modifications are quite different for each ancestral line.

## 25.1 A BRIEF HISTORY OF THE CONCEPT OF CHANGE IN ORGANISMS

No theory better integrates our understanding of the diversity of living organisms, their relationships and degrees of relatedness, and their interaction with a changing environment than evolution does. Though called a *theory*, evolution is accepted by many biologists as a fact. Some controversy continues in the community of biologists as to the precise mechanisms involved in evolutionary change, as well as its tempo, but a virtually complete consensus exists as to its occurrence.

### BEFORE DARWIN

Although organic evolution was not an alien idea among ancient Greek philosophers, it was not a prevalent view. In Plato's idealism each species was fixed and existed as a perfect archetype, or ideal representation. Individuals in nature were imperfect realizations of this archetype. To Aristotle, every species fit into a single hierarchy of increasing complexity. This "scale of nature" was a gross simplification of the nature of life, but it pervaded natural history until the eighteenth century. However, Lucretius, a Roman philosopher who drew upon Greek thought, did suggest that gradual changes in living forms could occur.

Georges-Louis Leclerc de Buffon (1707–1788) was the first of the serious naturalists of the modern era to develop a concept of evolving living forms. He tended to denigrate the classification schemes of naturalists like Carolus Linnaeus (1707–1778) with their emphasis on fixed forms fitting into permanent natural slots. He recognized that species change and utilized findings in comparative anatomy (vestigial organs, etc.) to buttress his views of a fluidity among species. Erasmus Darwin (1731–1802) also proposed that species change, largely by modifications in individuals during their lifetime, which modifications are passed on to their progeny. This view by Charles Darwin's grandfather anticipated the more extensive mechanism for evolution developed by Lamarck.

Georges Cuvier (1769–1832) was a firm believer in the fixity of species. However, his contributions to biological theory and practice were extremely useful to those who did work along evolutionary lines. He virtually founded the science of *paleontology*, the systematic study of fossils, and became skilled in reconstructing whole organisms from fossilized remnants. He developed a comprehensive system for the classification of animals and instituted remarkably thorough studies in comparative anatomy.

His studies of fossils demonstrated that many kinds of animals alive at one time are no longer extant. He devised the theory of *catastrophism* to account for a succession of animal populations. This theory states that a series of catastrophes periodically wiped out most of the forms of life then present and new groups were subsequently formed from the living remnants that were left. This accounts for the variation found in the fossil record according to Cuvier. He did not accept the notion of new species arising after each catastrophe but claimed that these new forms probably existed in some distant part of the world and migrated to the places where their fossils were found.

Jean Baptiste Lamarck (1744–1829) was perhaps the most significant pre-Darwinian contributor to the concept of evolution. Like Charles Darwin, he began his career (in botany) believing in the fixity of species. Later, he switched to zoology and became convinced that all living forms have evolved in a process of diversification. His great work, *Philosophie zoologique* (1809), made landmark contributions to classification as well as providing an impressive list of evidence for an evolutionary process. Most important, it suggested a mechanism for that process and a method (since reevaluated) for the origin of variations in individuals.

Lamarck believed that during the lifetime of any organism those parts that are used tend to develop or enlarge, while parts not challenged by use tend to atrophy. This *use–disuse theory* is illustrated in humans by the large arm muscles formed by blacksmiths or other practitioners of great muscular effort or the withering of limbs of those who do not use them regularly. Lamarck believed that such changes occurring during the lifetime of an individual are then transmitted to the next generation— the *inheritance of acquired characters*. In this fashion the activities of organisms in one generation tend to direct long-term changes in the future. Evolution, in Lamarck's view, is fashioned by biological need and reflects a pragmatic program leading to success in dealing with the challenges of the environment. An often-cited example of Lamarckianism is the long neck of the giraffe, supposedly produced by countless generations of giraffes reaching for the topmost sprigs of leaves on trees to more effectively compete with other herbivores who were confined to more accessible foliage.

## DARWIN AND NATURAL SELECTION

Charles Darwin (1809–1882) is most closely associated with evolution for two significant reasons. First, he amassed such a comprehensive and convincing body of evidence demonstrating organic evolution that it was no longer reasonable for biologists or even open-minded laypeople to dispute the existence of such a process. Second, his explorations of the fauna of South America and Africa during his five-year (1831–1836) voyage as naturalist aboard the HMS *Beagle* provided him with the insights necessary to develop a compelling theory of the mechanism of evolution. That mechanism is known as *natural selection*, and it was first presented at a scientific meeting in 1858. In 1859, his landmark work, *The Origin of Species by Means of Natural Selection*, was published in London. It created a storm of controversy but generated a fervent band of supporters.

The theory of natural selection rests upon three major tenets. First, there is a remarkable *overproduction* of young in each generation—many more than can possibly be supported by the limited resources (food, water, shelter, mates) of the environment. Second, heritable *variations* exist within this too-large population of young. Third, a *competition* for survival occurs in which those variants that are better adapted to a particular environment are successful and continue to produce offspring with their adaptive characteristics. Over time the characteristics which confer adaptiveness, or *fitness*, come to accumulate in the population, while those characteristics which diminish fitness tend to dwindle or die out. It is this last aspect—the greater reproductive success of better-adapted forms—that is properly termed *natural selection*. At one time natural selection was described in terms of a struggle for existence in which the fittest survive. Such a formulation by Darwin's followers tended to portray nature in terms of incessant fighting and bloodshed, but it failed to take into account the significance of cooperative mechanisms in survival. The concept of differential *reproductive* success of variant forms is a more accurate summary. It also stresses that over long periods of time the only criterion for continuing success is reproductive—what fails to reproduce cannot be represented in future generations, no matter how fit. Thus, fitness can be assessed only in retrospect, after reproductive success is considered.

The strongest influence in shaping Darwin's speculations about evolution were the meticulous observations he carried out during the voyage of the *Beagle*. Since he was prone to seasickness, he welcomed the opportunity to remain on land, where he studied both extant forms and the fossils of ancient eras. He was singularly impressed by the varieties of birds and reptiles on the Galápagos Islands and their relationship to similar groups on the Ecuadorian mainland. He also recognized that the presence of marine fossils high up on the Andes mountains was inconsistent with a concept of static species in a constant environment. Opportunities to revisit many sites, especially along the South American coast, enabled him to study plant and animal life over extended periods. He noted that despite great fecundity in nature, little change in numbers of a population from generation to generation occurred; this elimination of nature's bounty helped to shape his recognition of a selection process in nature.

A geologist friend of Darwin named Charles Lyell had written a book (*The Principles of Geology*) whose major thesis was that the natural forces acting on the earth were the same in the distant past as they are at present. This continuity of geological forces produces a constantly changing environment as glaciers advance and retreat, mountain ranges rise and fall, and rivers erode the land through which they flow. If conditions were constantly changing over long spans of time, it was reasonable to suppose that different kinds of life might have existed under these variant conditions. Darwin was able to observe both the geological changes and the fossil evidence of the existence of different life forms at different times and in different environments.

Shortly after his return to England, Darwin came across an essay on population written by a clergyman, Thomas R. Malthus. The thesis of this powerfully drawn work was that the human population tends to increase at a much greater rate than the food supply necessary to sustain the population does. Specifically, Malthus contended that the world's human population tends to increase according to a geometric progression $(2, 4, 8, 16, 32, \ldots, 2^n)$ while available resources increase in arithmetic fashion (addition of a constant increment, e.g., $1, 2, 3, 4, 5, \ldots, n$). Eventually, the ratio of people to food and other resources would reach unmanageable proportions and lead to a chaotic struggle for bare subsistence. Malthus suggested that the existence of pestilence, war, floods, and similar disasters serves to maintain the population at levels commensurate with available resources. For Malthus, the situation was interpreted as a justified intervention of divine providence to maintain a balance by visiting these plagues upon humankind. For Darwin, however, it sowed the seeds for the concept of overproduction of young leading to a struggle for existence.

The possibility of a selection process operating in nature was also suggested by the practices of plant and animal breeders. Darwin was familiar with the achievements of animal husbandry, by which, in a few generations, purposeful alterations in the characteristics of domestic animals could be effected. *Artificial selection* for specific traits is carried out by humans; could not a similar process occur in nature with the gradual accumulation of traits that enhance survival and reproductive success?

## 25.2  THE NOTION OF A GENE POOL: HARDY–WEINBERG EQUILIBRIUM

The totality of the alleles of every gene in a population is the *gene pool* of that population. Each individual carries some of the alleles, but individuals come and go. However, the total gene pool continues as a constant representation of a population. Changes in the specific frequencies of particular alleles constitute the raw material of evolution. At first, modest alteration in allelic frequency does not produce observable changes in a population, but over long periods of time such changes produce marked alterations in the characteristics of a population.

A better understanding of the slow way in which shifts in gene (allele) frequencies may generate the patterns of evolutionary change comes from a study of hypothetical populations in which *no such shifts* in gene frequencies occur. Using the genetic principles of Mendel, G. H. Hardy and W. Weinberg determined that the frequencies of alleles and even the ratios of genotypes tend to remain constant from one generation to the next in sexually reproducing populations under certain conditions. These conditions include

1. A very large population
2. No change in mutation rates
3. Complete randomness in mating so that reproductive success is the same for all allelic combinations
4. No large-scale migrations into or out of the mating pool

In such stable populations, gene frequencies follow simple laws of probability. For example, if allele $A$ has a frequency of $p$ in a population, and allele $B$ has a frequency of $q$, and there are no other alleles for this gene, $p + q = 1$. We saw in Chap. 9 that the probability that two events will occur at the same time is equal to the probability that the first will occur multiplied by the probability that the second will occur. The probability that allele $A$ will occur is equal to its frequency $p$; likewise, the probability that $B$ will occur is $q$. Thus, in a given population the frequency of homozygous $AA$ individuals is equal to the probability that two $A$ alleles will be in the zygote at the same time, which is equal to $p \times p$, or $p^2$. By similar reasoning, the frequency of $BB$ homozygotes is $q^2$. Since there are two ways of forming the heterozygote $AB$ (the $A$ allele from the mother and the $B$ allele from the father, or vice versa), the frequency of $AB$ in the population is $2pq$ (rather than simply $pq$). The sum of all three genotype frequencies $= p^2 + 2pq + q^2 = 1$. Note that this is a binomial expansion of the term $(p + q)^2$. If there were three alleles in the population, the frequencies of each genotype could be determined from the trinomial expansion $(p + q + r)^2$, where $r$ is the frequency of the allele $C$.

**EXAMPLE 2**  In a population in Hardy-Weinberg equilibrium, with only two alleles for a particular gene, if we know that allele $A$ has a frequency $p$ of 0.3, we can find the frequency of allele $B$. Since $p + q = 1$, we know that $q = 1 - p = 0.7$. Furthermore, we can determine the frequencies of the various genotypes, as follows:

$$AA = p^2 = (0.3)(0.3) = 0.09$$
$$AB = 2pq = 2(0.3)(0.7) = 0.42$$
$$BB = q^2 = (0.7)(0.7) = 0.49$$

Since allelic frequencies tend to remain constant from generation to generation, departures from this constancy help to expose selection pressures operating in the population.

## 25.3  NATURAL SELECTION: A MODERN SYNTHESIS

The weakest link in Darwin's theory of natural selection was his ignorance of the mechanism of heredity. Without a sufficient appreciation of the laws of genetics he could not account for the variations that arose as exceptions to the tendency of "like to beget like," even though these variations were essential to his theory.

In 1901 the basic features of Mendelian genetics were discovered. In the next four decades a clarification of classic genetics was paralleled by the development of population genetics as a separate discipline. Among those who provided significant insights into the nature of gene flux within breeding populations were Sewall Wright, Ernst Mayr, Theodor Dobzhansky, and L. C. Dunn. The enrichment of Darwin's theory by the perspectives of population genetics as well as the findings of paleontology and biogeography became what Julian Huxley called a *modern synthesis of evolution*. Essentially this modern synthesis (neo-Darwinism) is a recognition that it is *populations* that undergo the gradual changes that are effected by natural selection acting on individuals.

**EXAMPLE 3**  Sickle-cell anemia originated as a single-gene recessive mutation within an African population. At first it was relatively rare within that population. However, individuals carrying the sickle-cell allele as heterozygotes are relatively immune to malaria because their slightly contorted red blood cells do not accommodate the malarial parasite well; consequently, the frequency of that allele increased in the population to its present count. It is the enhancement of reproductive potential that is crucial to the spread of this or any allele. The serious debility of individuals carrying two copies of the mutant allele serves to limit the spread of the allele, so that negative and positive influences on reproductive potential are involved in evolutionary changes.

## 25.4  PUNCTUATED EQUILIBRIUM

Students of evolution, from Darwin to the present generation, have recognized the comparative rarity of transitional, or linkage, forms in the fossil record between one species and another, or among major groups such as reptiles and mammals. If all populations undergo slow changes in their evolution into new forms, one would expect to find a continuous spectrum of fossil representatives all through the transition process. In 1972, Niles Eldredge of the American Museum of Natural History and Stephen J. Gould of Harvard University proposed the theory of *punctuated equilibrium*, which (among its contributions) explained these gaps in the fossil record. In punctuated equilibria, evolutionary changes occur in irregular jumps—very rapid changes are followed by long periods of relative constancy. Under certain conditions new species form from old, and the major modifications are compressed into several thousand years rather than many millions. Although thousands of years can hardly be considered abrupt, in the time spans associated with evolutionary trends it represents a tiny proportion of the existence of any particular species. Since fossils are found as samples of any era's flora or fauna and are unearthed in a random fashion, the chances of obtaining records of each form for a short period of intense change are unlikely.

The theory of punctuated equilibrium has excited considerable controversy. Its opponents contend that the compressed time scales for change do not radically alter the Darwinian view of gradualism, nor does a demonstrated jump in one lineage prove that similar jumps occur in all lineages. Further, they say, the periods of relative constancy hypothesized by Eldredge and Gould may not actually occur, since profound changes at the molecular and soft tissue levels may be occurring without detection by an examination of the fossil record. It is less easy to dispute the contention of the punctuationalists that the discreteness of present-day forms is due to the fact that transitional states last a relatively short time, so that we observe present species after they have undergone the changes that produce the relatively fixed form.

## 25.5  MOLECULAR BIOLOGY OF EVOLUTIONARY CHANGE

Natural selection acts upon individual *phenotypes*, determining the reproductive success of specific individuals. However, it is the *genotypes* of selectively favored forms that tend to persist, since genotypes are passed on in each generation. The sum total of all genotypes constitutes the gene pool of a population. The gene pool is the entity that actually undergoes evolution and defines, at any particular time, a specific population. The character of DNA and its changes are the ultimate bases for diversity in the living world.

Molecular biology provides the tools for determining the nature (base sequence) of DNA and the proteins it codes for. Mutations, a major source of variation for the evolutionary mill, are for the most part altered base sequences within a DNA strand. The pace of evolution can be gauged by the rate of change of DNA sequences in a specific time frame, while the relatedness of different species can be measured by how closely matched their DNA is. Since proteins are specified by DNA, commonality of protein structure (amino acid sequences) may also be used to determine relationships of descent among groups. It is rather remarkable that the evolutionary perspectives of a man who knew nothing of chromosomes and their chemical units have been confirmed and reinforced by discoveries in molecular biology.

Darwin believed that all living things could trace their ancestry back to a single set of progenitors. The existence of one genetic code in all living things is strongly consistent with this concept of a descent with modification. The basic relatedness of diverse forms is also seen in the similarity of many of their proteins.

**EXAMPLE 4**  Specific proteins such as the $\alpha$ and $\beta$ chains of hemoglobin and cytochrome $c$ have been studied extensively in an effort to provide molecular probes for following evolutionary diversification. The $\beta$ chains of hemoglobin in gorillas and humans are virtually identical in amino acid composition. The monkey $\beta$ chain has fewer than 10 amino acid differences from the human protein, but the $\beta$ chain of jawless fishes differs in more than 100 amino acids from that of the human. These *genetic distances* confirm the relative relatedness of these forms suggested by other data.

## 25.6  CONTROL OF GENE POOLS

The Hardy–Weinberg equilibrium applies to hypothetical populations in which no net change of gene frequencies or genotypic ratios occur. In nature, however, changes in these characteristics do occur. Such changes in a population's alleles are called *microevolution.* A major cause of microevolution is *natural selection*, the Darwinian process of differential reproduction that results in increasing the adaptability of a population. Alterations of the gene pool in a random (nonadaptive) manner may also occur through *genetic drift.* In small populations, loss of particular alleles or genotypes may occur accidentally through a kind of sampling error in which the specific allele or genotype is "overlooked" reproductively.

**EXAMPLE 5**  In a pea patch population containing five TT, twelve Tt, and four tt individual pea plants, a rabbit may enter and by chance alone eat only the five TT plants. Such a loss of the T allele would tend to alter the frequency of that allele for many generations, an example of microevolution. Such accidental elimination of a specific genotype would be highly unlikely in a larger population.

A special case of genetic drift is known as the *founder effect.* If a new and isolated colony is started by a small group atypical of the larger population, the individuals of that colony will resemble the founders rather than the original population. The founder effect has been noted in cases in which high frequencies of genetic disorders exist in small groups tracing their ancestry back to afflicted founders. If an island were inhabited by humans all of whom were blue-eyed, subsequent populations would probably all lack the dominant allele for brown eyes.

*Gene flow*, in which changes in a population's alleles are produced by immigration into or emigration out of a population, is another mode producing microevolution. Such changes tend to be neutral in an adaptive sense, unless migration favors the movement of either less adapted or better-adapted individuals.

A change in the net *mutation rate* is another way in which random changes in gene frequencies occur in a population.

Since complete randomness in mating is one of the prime requisites for stable gene frequencies, a departure from mating randomness may lead to changes in gene frequencies. In *assortive mating*, another source of gene alteration in populations, individuals tend to show preferences in their selection of a mate. In humans, individuals of similar social classes or ethnic identification tend to mate with one another more than with outsiders. Among nonmigratory mammals, mating occurs more readily with near neighbors than with others. Such mating practices can lead to the eventual formation of new varieties within the larger group.

Only natural selection produces a consistent increase in the fitness of populations, as deleterious alleles decrease in frequency and adaptive alleles increase. Since alleles usually operate in coherent groups, the selective process probably influences gene complexes and may be expressed through a variety of interdependent processes such as *coadaptation*—a situation in which several independent components evolve together in the production of a major new structure.

## 25.7  SPECIATION

A *species* consists of all individuals who can share a common gene pool. This means that they may interbreed with one another to produce *fertile* offspring but not with members of another species. The species, as the unit of the various kinds of life that exist in nature, also is defined by gross anatomical features that characterize the specific group and differentiate it from other species.

**EXAMPLE 6**   The horse and the donkey (ass) belong to separate species. They differ in size and other anatomical features. They do not share a common gene pool although they are capable of mating with one another to produce offspring. The cross of a male ass with a female horse yields a *mule*. A male horse crossed with a female ass produces a *hinny*. But the gene pools of each do not blend because both mules and hinnies are sterile and represent a reproductive dead end. Over time then, the horse and the ass maintain separate gene pools, the major criterion of species integrity.

Since species are separate from one another on the basis of their reproductive isolation, the key to maintaining species discreteness lies in mechanisms that produce separate breeding populations. One form of reproductive barrier is found in the *geographical* boundaries that prevent individuals from two populations from reaching one another. Similar inhabitants of two different islands or populations that live on different sides of a high mountain range tend to remain isolated and eventually develop such differences in their gene pools that they are reproductively incompatible.

Where geography does not impose barriers, *biological* mechanisms may arise to maintain separateness between populations. Even in the same locale, two species may remain isolated because of *ecological* requirements. Differences in moisture requirements may separate worms into occupation of different strata underground. Differences in the *time* of reproductive activity may separate two populations that can otherwise interbreed: species that are active during the day have little likelihood of interacting sexually with nocturnal species. The incompatibility of the reproductive organs may also serve as an isolating barrier. All dogs belong to a single species, but a Great Dane cannot readily mate with a Chihuahua. The common gene pool is maintained because each of these breeds can mate with an intermediate variety such as a terrier to produce offspring that share the Great Dane and Chihuahua lineage. However, after long periods of time the varieties of dog existing today may very well evolve into separate species with no gene flow between them.

Mating is often aborted at an early stage because *courtship* rituals are not observed by the outside species. A variety of precise behaviors must precede mating, especially for birds and mammals, and failure to observe the rules of the mating game effectively precludes a sharing of genes. In many cases in which sexual union does occur between separate species, the gametes do not fuse or an embryo fails to develop. Such gametic incompatibility is an effective barrier to a breakdown of species lines. Even when successful mating does occur between members of separate species, the hybrid forms are quite often sterile. The boundaries that preserve the separateness of species thus appear to be quite firm.

If reproductive isolation maintains the segregation of species, then clearly the road to speciation must involve an event that isolates a segment of a species from other populations of that species. As we have seen, in this state of reproductive separation, the isolate group develops mutations and recombinant features that are not shared with the parental species. Given enough separation time, the breakaway population may develop into a new species that can no longer share in a common gene pool. If the isolate group is separated *physically* from the original larger population, this type of speciation is called *allopatric*. However, a subpopulation *within* the parent group may develop characteristics that tend to isolate it from its neighbors, and it goes its merry developmental way to yield ever-increasing uniqueness while dwelling in the same neighborhood. Such a phenomenon is labeled *sympatric* speciation. The smaller the size of the pioneer population, particularly in allopatric patterns, the more likely are the chances that create a separate species.

A particularly rapid type of speciation occurs when some common ancestor reaches an environment in which a great number of distinct opportunities and challenges exist. Such a process is called *adaptive radiation*, the production of many species from a common ancestral form. Island chains such as Hawaii offer examples of many unique species on each island developing from some ancestral migrant. Species developing on one island may colonize a different island and become a distinct species. If this process (generally involving small pioneer groups) continues, a rapidly branching speciation process may occur.

## 25.8  MICROEVOLUTION VERSUS MACROEVOLUTION

*Macroevolution* focuses on the creation of taxonomic groups above the level of species. Although many of the same mechanisms involved in speciation operate in macroevolution, the time spans required are much greater. Much of our knowledge of the broad trends of macroevolution comes from the fossil record. However, changes within a group that lead to less drastic modifications of a population or even the creation of a new species (microevolution) can be studied by measurements of gene frequencies in a population. Patterns of selection that have been discerned include

1. *Stabilizing selection*, in which extremes at either end of a spectrum are disproportionately selected against so that the population, despite continual variation produced in every generation, tends to cluster about its average (mean)
2. *Directional selection*, in which one extreme is favored over its opposite so that the average value slowly moves toward the favored extreme
3. *Diversifying (disruptive) selection*, in which two or more subtypes are favored and the population tends to evolve into several subgroups or new species

Diversifying selection may very well operate in both microevolution and macroevolution, and directional selection is similar to the macroevolutionary process known as *phyletic change*. The basic patterns for the broad changes of macroevolution revealed by the fossil record are

1. Phyletic change (*anagenesis*)—gradual change in a *single* lineage so that eventually descendents are radically different from their ancestors. Anagenesis may be likened to directional selection over long periods of time.
2. *Cladogenesis*—a macroevolutionary trend in which a branching occurs so that one lineage gives rise to two or more lineages. Small populations budding from a lineage may be in a particularly favorable position to produce new groups. Cladogenesis has been emphasized as a major macroevolutionary pattern by Ernst Mayr.
3. *Adaptive radiation*—a relatively sudden formation of many new groups, which are able to move into and exploit new environments. The relatively rapid diversification of early mammals during the extinction of the dinosaurs is a good example of such a major outreach. Adaptive radiation incorporates features of both cladogenesis and anagenesis, since new lineages formed during this volatile evolutionary period may each be undergoing progressive transitions.
4. *Extinction*—more than 99.9 percent of all the species that have ever evolved are no longer present. This loss of diversity is an inexorable feature of the evolution of all kingdoms. A changing environment renders yesterday's fit as today's unfit, doomed to obliteration.

# Solved Problems

**25.1**  Why is it so difficult for many people to accept the notion of evolution?

Evolution, as applied to the biological world, is in conflict with a literal interpretation of the biblical story of creation described in Genesis. In the biblical account each creature was created separately by divine action and supposedly reproduces its own kind with little change through the generations. In our own time, the *scientific creationists* contend that a relative fixity of species is still a correct view and the earth's span should be measured in tens of thousands of years rather than the 4.6 billion years usually accepted by the evolutionary biologists. For those who are oriented to fundamentalist views, evolution becomes a challenge to religious convictions. Most adherents to particular religious faiths have no problem reconciling their religious convictions with an evolutionary perspective, but spiritual concerns did discourage many at an earlier time. Darwin himself was concerned with how his espousal of evolution would affect his wife's pious sensibilities.

A second problem arises from the fact that evolution removes humans from the center of the living world and ends the traditional separation of "lower" animals from "higher" humans. Humans have long been regarded as the goal and ultimate fulfillment of the creative process. That humans are just one of many advanced groups that have arisen from distant ancestors may be a blow to the collective ego of humankind.

An argument has been raised by some nonscientists that a belief in evolution may encourage a dehumanization of shared social values. If we are merely highly evolved animals, then perhaps the sanctity of human life may be ignored in the name of expediency. Evolution may create a more permissive moral climate according to its critics. At the same time, evolution's immersion in continual change presents a challenge to the stability of traditions that check our more aggressive tendencies.

Scientists do not generally accept these criticisms; they feel that the truths of science are not instruments for directing human activity but are worthwhile in and of themselves. The grandeur of life, which should inspire our conduct and concepts, is probably best illuminated by the perspective of evolution.

**25.2**  Evidence for evolution comes from many fields: geology, biogeography, comparative anatomy, embryology, animal husbandry, and molecular biology. How might each of these fields support the theory of evolution?

Among the older lines of evidence, which Darwin drew upon to support his theory, was the *geological* evidence. The slow pace of geological change could never have produced the present diversity of animals and plants within a time span of 6000 years, the age of the earth according to many creationists. But the dating techniques of modern geology confirm earlier realizations that the earth is many billions of years old, a time frame sufficient to encompass the development of modern forms from early primitive ancestors. In addition to fixing a long history for our planet, geology also provides a fossil record in rock that can lead to a reconstruction of the succession of plant and animal forms in time. The oldest strata of rock contain only prokaryotes, whereas more recently formed layers contain the fossils of forms believed to have evolved more recently. The vertebrate record is particularly well documented because of the greater likelihood that their skeletal remains have been preserved.

A second line of evidence that was especially influential in directing Darwin's thinking was *biogeography*. The distribution of plant and animal forms in various regions makes most sense in terms of a descent with modification. Australia, which has been longest isolated from other land masses, has a strikingly unique variety of plant and animal life, with virtually no native placental mammals. If all living things were created at one time, these biogeographical differences would not be likely.

A third line of evidence comes from *comparative anatomy* and consists of two major types. One is the underlying similarity of comparable structures in different species. As an example, the arm of a man, the forelimb of a pig, and the flipper of a whale all contain similar bones and muscles. Since these structures have very different end functions their resemblance is a likely indication that they were separate modifications over time of some common ancestral structure. The term *homology* is applied to such structures, and the evolutionary process producing these discrete forms is called *divergent evolution*. The hand of evolution is also seen in the opposite effect—the development of similar structure and function in very different evolutionary lines. This is called *convergent evolution* and is seen, for example, in the close resemblance between the fins of fish and marine mammals. These *analogous* (versus homologous) structures serve similar functions but have quite different anatomical origins. Another source of evidence for evolution in comparative studies is the existence of *vestigial* organs. These are body parts that were presumably functional at an earlier time in an organism's evolutionary background but have since become mere remnants. Their presence can easily be accounted for by an evolutionary progression involving a gradual loss of function as a lineage is modified. It would be much harder to account for nonfunctional vestiges, such as the human appendix or coccyx (a remnant of tail vertebrae), if each creature were created by divine design.

Still a fourth type of evidence for evolution derives from a comparison of the *embryonic development* of related forms. The early embryonic stages of many organisms are remarkably similar to one another, with divergence appearing at later stages. In the case of humans and chimpanzees a remarkable parallelism exists until an advanced stage of pregnancy. If humans are compared with rabbits, clear differences can be discerned much earlier in embryonic development. These similarities in embryogenesis are regarded as evidence for a common descent, a descent "reincarnated" in the developmental sequences of each organism. At one time the argument for evolution from a consideration of comparative embryology was even more strongly stated by Ernst Haeckel as the biogenetic law: ontogeny recapitulates phylogeny—the development of the individual recalls the development of the group (lineage). Because gill slits are evident at one stage in the embryonic life of all mammals, Haeckel supposed that this represents a fish stage that the mammalian lineage passed through early in its history. While there are suggestions of a recapitulation in each animal's embryonic unfolding, current thinking regards Haeckel's view as extreme.

A fifth line of evidence is the *artificial selection* practiced by plant and animal breeders. The fact that their stocks could be modified in a directive manner by the breeders' selection from variations that were spontaneous and undirected made a strong impression upon Darwin.

*Molecular biology* has supported the evolutionary perspective by demonstrating the universal sharing of the same genetic code by all organisms. Similarities in DNA are used to confirm evolutionary relationships by establishing genetic distance. Immunological compatibilities are also used to support degrees of relatedness. Antigenic similarities of corresponding proteins certainly fit an evolutionary pattern. These similarities can be determined by provoking an antibody response in a test animal by a protein from an organism A. If a corresponding protein from an organism B induces a similar response, then A and B bear a close relationship of descent.

Other recent evidence includes the actual formation of new species of *Drosophila*, the creation of a separate species of hare, and the alteration in antibiotic sensitivity of bacteria subjected to an environment containing antibiotics.

**25.3**   In its time, Lamarck's explanation of evolution as the result of inheritance of acquired characteristics was widely accepted and was even invoked by Darwin in some instances. Why is it generally discounted today?

Studies of genetics have shown that although the genes can be altered by certain mutagens, they are generally stable entities, passed unchanged from generation to generation. Since they are the unit of inheritance, their relative immutability is in direct conflict with Lamarck's explanation. Since the gene for neck size, for example, does not change in response to the stretching of a giraffe's neck, any acquired characteristic of a longer neck is not passed on to the next generation. Instead, natural selection chooses those animals *born* with alleles for longer necks, resulting in an increase in the frequency of these alleles in the population.

**25.4**   Suppose that out of a sample population of 650 rabbits researchers find 39 that express a recessive allele *s* for short ears. Assuming there are only two alleles for ear length in the population, how many rabbits are homozygous for long ears (*SS*) and how many are heterozygous (*Ss*) if the population is in Hardy–Weinberg equilibrium?

Since there are 39 rabbits with the *ss* genotype, the frequency $q^2$ for this homozygote is

$$q^2 = \tfrac{39}{650} = 0.06 \quad \text{and} \quad q = \sqrt{0.06} = 0.24$$

Since $p = 1 - q$ = frequency of the *S* allele for long ears,

$$p = 1 - 0.24 = 0.76 \quad \text{and} \quad p^2 = (0.76)^2 = 0.58$$

Since the frequency $p^2$ of the *SS* homozygote is 0.58, the number of homozygous long-eared rabbits is

$$(0.58)(650) = 377$$

The frequency of the heterozygotes is given by

$$2pq = 2(0.76)(0.24) = 0.36$$

and the number of heterozygous *Ss* rabbits is

$$(0.36)(650) = 234$$

**25.5**   The peppered moth (*Biston betularia*) is an inhabitant of England and parts of continental Europe. In the early part of the nineteenth century almost all these moths were light-colored with irregular splotches of dark pigmentation. This pigmentation pattern afforded a high degree of camouflage against the relatively light lichen-covered tree trunks in which they lived. However, by the late 1800s a population of dark-colored moths with increased melanin had risen dramatically, especially in industrial areas such as those that surrounded Manchester. How would you explain this shift in coloration?

H. B. D. Kettlewell (Oxford), working with E. B. Ford and Nicolaas Tinbergen, has shown that the dark-colored moths enjoy a greater degree of camouflage against the trunks of trees that have become darkened in polluted environments. The industrial revolution, then, prompted a microevolutionary change in coloration involving an alteration in the frequency of a single allele. Kettlewell counted the number of moths eaten by birds in rural and in industrial settings and confirmed that dark-colored moths are eaten in greater numbers in country areas whereas light-colored moths are captured in greater numbers in industrial areas. Kettlewell actually identified natural selection as it was occurring and further analyzed the mechanism (predatory birds) for the selective process. Since a change in allelic frequency in a population represents evolutionary change, *industrial melanism* has since been eagerly studied by a generation of biologists.

**25.6**   What is the basic difference between microevolution and macroevolution?

Microevolution involves the small, usually gradual changes in allelic frequency that may eventually lead to the formation of new species. Macroevolution involves the origin of new major trends in evolution and covers much longer periods of time. At least three aspects of macroevolution may be identified:

1.   Formation of new and radically different structures such as the vertebrate eye or the wing of a bird
2.   Adaptive radiation of such broad groups as the flowering plants or the eukaryotes
3.   Major extinctions such as those of the dinosaurs in the Cretaceous Period

The smaller changes of microevolution, though different from the greater disruptions and alterations of macroevolution, probably illustrate basic ways in which organisms interact with their environments to yield change at all levels. The broad changes that species undergo are known as a *phylogeny*. It is the creation of new *taxa* (groups within a hierarchical classification order) and the elimination of older groups that macroevolution focuses upon.

**25.7**   Natural selection does not always push a population toward a single allele for a particular gene. Frequently, several alleles within a population, each with a characteristic phenotype, are maintained in a stable ratio. This stable phenotypic diversity is referred to as *balanced polymorphism*. One mechanism believed to promote balanced polymorphism is the phenomenon known as *hybrid vigor*, or *heterosis*, in which heterozygotes demonstrate greater hardiness and viability than homozygotes for the same gene. It has long been known that hybrids are often more vigorous than the offspring of purebred strains. Puppies with mixed parentage are usually less highstrung than inbred puppies are. In the United States, development of hybrid corn markedly improved the yield and quality of the corn crop and catapulted the plant geneticist Henry Wallace, a prime mover in the hybridization experiments, to the vice presidency of the nation. Using sickle-cell anemia as an example, show how hybrid vigor might work to maintain polymorphism in a population.

The reasons for hybrid vigor are not always known but are probably related to reduced likelihood of homozygosity for deleterious recessive genes—hybrids are heterozygous for a large proportion of their gene pairs. Regions of the world where malaria is still a problem, such as Africa, also support an unusually high frequency of the allele for sickle-cell anemia, in addition to the non-sickle-cell alleles. In such populations homozygotes, both for the sickle-cell allele and for the normal alleles, are at a selective disadvantage. People with normal alleles are subject to the ravages of malaria; people homozygous for the sickle-cell allele suffer the debilitations of anemia. Those who are heterozygous for the sickle-cell allele, however, do not express the gene enough to be debilitated by it, but have their hemoglobin sufficiently affected to make them poor hosts for the malarial parasite. Thus the heterozygote serves as conservator of both types of alleles, protecting them from the selection pressures that eliminate them in the homozygous state, and maintaining a balanced polymorphism in the population.

# Supplementary Problems

**25.8** Charles Darwin, like his grandfather Erasmus, was a lifelong adherent of the theory of evolution.
(*a*) True   (*b*) False

**25.9** Darwin was encouraged by his many findings of transitional fossil forms.
(*a*) True   (*b*) False

**25.10** The existence of rudimentary limb structures in snakes constitutes an argument for special creation.
(*a*) True   (*b*) False

**25.11** Convergent evolution involves the creation of similar end structures from different ancestral structures in unrelated organisms occupying similar environments.
(*a*) True   (*b*) False

**25.12** Convergent structures are described as homologues.
(*a*) True   (*b*) False

**25.13** Social Darwinism, in stressing the importance of survival of the fittest in a social context, tended to condemn social welfare programs.
(*a*) True   (*b*) False

**25.14** The high incidence of dwarfism in the Amish of Pennsylvania probably illustrates a founder effect.
(*a*) True   (*b*) False

**25.15** The frequency of the sickle-cell allele in black Americans is 0.05. The frequency of the full-blown disease, then, is 0.0025 (0.25 percent).
(*a*) True   (*b*) False

**25.16** The frequency of heterozygotes in the previous problem is 0.095 (9.5 percent).
(*a*) True   (*b*) False

**25.17** Both Lamarck and Wallace cited similar mechanisms for evolution.
(*a*) True   (*b*) False

**25.18** All dogs   (*a*) have the same number of chromosomes.   (*b*) belong to the same species.   (*c*) demonstrate a menstrual cycle.   (*d*) both *a* and *b*.   (*e*) both *a* and *c*.

# Answers

| | | | | | |
|---|---|---|---|---|---|
| **25.8** (*b*) | **25.11** (*a*) | | **25.14** (*a*) | **25.17** (*b*) |
| **25.9** (*b*) | **25.12** (*b*); they are analogous structures | | **25.15** (*a*) | **25.18** (*d*) |
| **25.10** (*b*) | **25.13** (*a*) | | **25.16** (*a*) | |

# Chapter 26

## Ecology

The word *ecology* is derived from the Greek root *oikos*, which means "household," and ecology is indeed a study of the household of life. It is a *holistic* (broad-based and integrative) approach to understanding living things *in context* as they relate both to their physical environment (*abiotic* aspects) and to each other (*biotic* aspects). It is these *interactions* of living things that provide the raw material for ecological studies. Ecologists also tend to support appropriate responsible social actions to ensure that life in its diversity and richness of connections will continue.

**EXAMPLE 1** (1) The Ecological Society of America has set up an office in Washington, D.C., to provide information that may affect legislative decisions on the environment. (2) Professor E. O. Wilson of Harvard University is leading a campaign to save the tropical rain forests of the world. (3) Ecologists often serve as expert witnesses in legal procedures dealing with the environment.

### 26.1 ANATOMY OF AN ECOSYSTEM

The ecological unit is the *ecosystem*, which is a group of diverse interacting populations found within the regional limits of a neighborhood. This neighborhood (*habitat*) might be as small as a local pond or as broad as the vast Sahara Desert. Various interacting populations within an ecosystem make up the community, the living components of the ecosystem. Some ecologists focus almost exclusively on the living organisms of an ecosystem, while others study the way in which the physical characteristics of the neighborhood constrain and regulate the ecosystem.

Although ecosystems vary from tidal pools and barrier reefs to arid stretches of scrub grass, they all share features that have been discovered as ecology, once an almost completely descriptive branch of biology, moved in an increasingly experimental direction. These features are

1. Energy flow
2. Nutrient cycling
3. Regulation of population size (numbers of individuals)

#### ENERGY FLOW

Energy moves through the community of an ecosystem in a single direction by means of a food chain (web) in which there are the eaters, the eaten, and a combination of both. Populations are assigned an "occupational" role in an ecosystem in terms of their relation to the overall flow of this energy (food) in the food chain:

*Producers*: the first group in a food chain, usually consisting of green plants, which convert some of the energy of the sun (through photosynthesis) into organic molecules they use and store in their tissues.
*Consumers*: animals that feed on green plants and each other. *Primary* consumers are *herbivores*, which subsist on the primary producing plants. *Secondary* consumers feed on primary consumers, while *tertiary*, *quaternary*, etc., consumers are further along the chain.
*Decomposers*: bacteria, fungi, plants, or animals that feed on dead organisms and release the bound organic material of the organisms to the food chain.

**EXAMPLE 2** A deer dying in a meadow may be picked apart by scavenger species such as vultures or crows. Material not eaten undergoes decay by bacteria and fungi, so that inaccessible portions of the corpse not consumed are made available to other organisms in the community.

A *niche* is an occupational or functional slot within an ecosystem that is generally filled by a particular species. Since energy flow is so vital to the maintenance of an ecosystem, the niche is usually categorized in terms of a relation to the food chain. Thus, every ecosystem has a niche for a primary consumer, usually a herbivore that feeds on a primary producer and is, in turn, eaten by a secondary consumer. Niches, which have been likened to an organism's caloric profession, are often defined in more specific ways than the broad category of producer, consumer, etc.

## NUTRIENT CYCLING

As energy in a food chain is passed along from one link to another, its useful capacity for work is diminished in accordance with the second law of thermodynamics (see Chap. 2). Heat is given off with every transformation. At the end of the food chain little or no free energy is left, so that recycling is not possible. On the other hand, matter is not lost as it passes from one component of the food chain to another—the passage of organic molecules and their elemental units along the food chain may be described in terms of a cycle. Generally, ecologists follow specific atoms through the cycle, e.g., carbon (C), nitrogen (N), and sulfur (S), and chart their fates as they pass through the food chain, into the environment, and back again into the community. The nitrogen and carbon cycles are particularly well documented.

We studied the nitrogen cycle in Chap. 13. As may be seen in Fig. 26.1, the carbon cycle is powered alternately by the reduction (through photosynthesis) and the oxidation of carbon. Carbon enters the cycle as atmospheric $CO_2$, which is converted by plants to food during photosynthesis. Ultimately, this carbon is converted back to $CO_2$ through either respiration or combustion, and the cycle begins again.

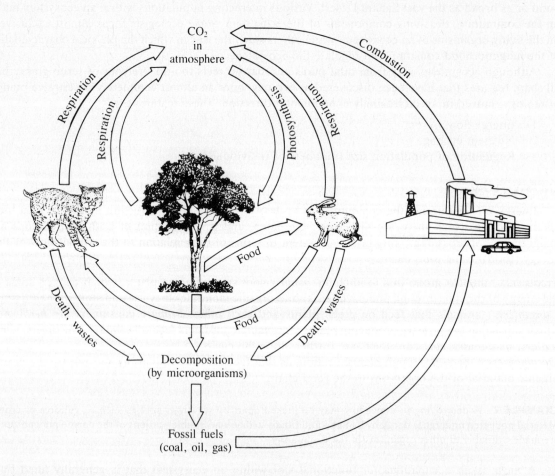

**Fig. 26.1**

## REGULATION OF POPULATION SIZE

As Malthus and Darwin well knew, natural populations tend to increase exponentially, rather than incrementally by addition of a constant amount. Because of their high reproductive potential, such populations *tend* to double, then double again, and so on. A constant rate (reproductive potential) produces dramatic increases over time because as the population increases, the rate is being multiplied by an ever-increasing base value (the number of individuals in the population). The situation is summed up in the differential equation $dN/dt = rN$, where $N$ is the number of individuals, $t$ is time, and $r$ is the intrinsic rate of increase. The left side of the equation, the change in number of individuals over time, is the rate of growth. This rate is equal to the intrinsic rate of increase $r$ multiplied by the numbers present, since each individual shares in the tendency to increase in numbers. Such a situation produces the typical exponential growth curve seen in Fig. 26.2. Growth increases slowly at first and then rapidly as the steady rate is applied to increasing numbers of individuals. The slope may even approach a vertical line. But such a depiction is only hypothetical, since a population that increases in size at such a great rate will be subject to limiting constraints. Its birthrate will soon be matched or even overtaken by an increase in its death rate, which will tend to maintain the population at a steady number.

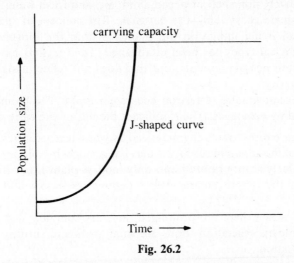

**Fig. 26.2**

Stable population size is usually characterized by a modification of the previous equation as follows: $dN/dt = rN[(K - N)/K]$, where $K$ is the *carrying capacity*, or the maximum number of individuals a habitat can support. Here the rate of growth is modified by a factor $(K - N)/K$. As seen in Fig. 26.3, an early period of slow growth, the *lag phase*, is followed by a very marked period of growth,

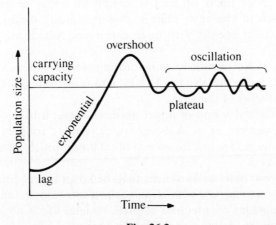

**Fig. 26.3**

the *exponential (log) phase*. This then gives way to a long-term period of stability of numbers, the *plateau phase*. Some variation may occur in this third phase, including overshoot, oscillation about a mean, stability at a lower number of individuals, or even a crash in the population leading to extinction.

Mechanisms for maintaining uniform population size are of two types:

1. *Density-dependent mechanisms*, which are sensitive to the size of the population and increase in effectiveness as population size increases
2. *Density-independent mechanisms*, which operate independently of population size and exert similar destructive effects on crowded or sparse populations, i.e., weather, climate, natural catastrophe

The control of population size is one of several areas in which evolution and ecology coincide. Field experimentation in a variety of habitats has enabled ecologists to sharpen our understanding of the mechanisms of natural selection.

## 26.2 TYPES OF ECOSYSTEMS

In one sense, the relatively thin shell of ocean, land, air, and fresh water that makes up the arena for all life on earth is a single ecosystem—the *biosphere*. The concept of *spaceship earth* emphasizes the interactive unity of all living things on the planet. Within the biosphere an older traditional subdivision of major ecological types has been established. *Land regions* have been more intensively probed than *marine* or *fresh aquatic* habitats, but this does not reflect assigned importance or even intrinsic interest or complexity.

A *biome* is any of several unique terrestrial ecosystem types. The biomes constitute the largest community units classified by ecologists. The significant biomes of the earth are as follows:

1. *Tropical rain forest*: dense forest tracts characterized by warm temperatures and very heavy rainfalls. Trees are abundant, but the apparent fertility is deceiving, since the soils are actually of poor quality.
2. *Desert*: areas of extremely scanty rainfall and only modest plant life. Although the sandy Sahara is the best known of the deserts, many desert regions are rocky and different from popular conceptions.
3. *Chaparral*: regions in which there are prolonged, hot, dry summers and temperate rainy winters and in which the dominant vegetation forms are small trees and shrubs. The animals are usually small with bland coloration.
4. *Savannas*: grassland regions in the tropics characterized by light and seasonal rain.

**EXAMPLE 3** The grasslands of Africa, which begin below the desert, are dominated by deep-rooted grasses with very few patches of shrubs and trees. In this ideal grazing region a rich diversity of large mammals (giraffes, zebras, wildebeests, etc.) dominates the ecosystem.

5. *Temperate grasslands*: large tracts of land in temperate zones characterized by limited water availability during much of the year; clumps of scrub grass, shrubs, and some annual plants predominate. Small rodents coexist with large carnivores, which are dependent on the smaller mammals.
6. *Taiga*: northern forests thick with massive cone-bearing evergreen trees. Animal life includes smaller animals, such as hares, mice, shrews, and lynxes, and larger ones, such as bears, elks, deer, and moose. Snow is present most of the year.
7. *Tundra*: a modified grassland region of upper northern areas; it is so cold that a permanent layer of frozen undersoil (*permafrost*) exists. A short growing season during the northern summer provides sustenance for shrubs and rushes and for animal life (fauna), which includes multitudinous insects, birds, lemmings, and foxes.
8. *Temperate deciduous forest*: rich stands of trees that shed their leaves during the cold season, bushes and shrubs, and grasses interspersed with *cryptogamic plants* (mosses and liverworts). Cold winters alternate with warm summers of adequate rainfall. Animal life is abundant, ranging from mice, chipmunks, and raccoons to wolves and mountain lions.

Most of our planet's surface consists of water. The marine (saltwater) environment makes up about 70 percent of that surface. Both freshwater and marine environments possess a rich array of community life and significantly affect economic aspects of human societies.

From an ecological perspective, the oceans can be divided into a *neritic* region above the continental shelf and the *oceanic depths* beyond the relatively shallow shelf (see Fig. 26.4). The portion of the neritic region just offshore is named the *littoral zone*. Because of its currents and complete penetration by the sun due to its shallowness, it is particularly rich in plant and animal life. Shoreward of the littoral zone, an intertidal zone is periodically covered with water at high tide and exposed at low tide. The ocean depths are divided into a *pelagic zone*, rich in plankton, and the even deeper *abyssal* zone.

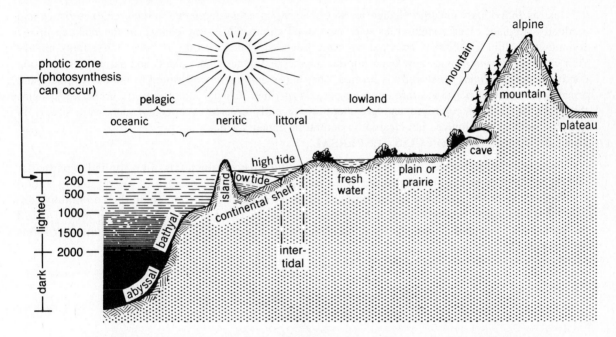

Fig. 26.4 Some common ecological environments. (*From Storer et al., General Zoology.*)

## 26.3  STABILITY AND ECOLOGICAL SUCCESSION

Some ecologists have described ecosystems as supraorganisms possessing inherent properties of growth, metabolism, periodicity in activities, and eventually death. Since ecosystems are far more open and less firmly bounded than individual organisms, such a view may be an oversimplification of the actual case, but it is useful in classifying some aspects of ecosystem function.

The interdependent relationships obtaining within a community are described as the *web of life*, reflecting the complexity of interactions that occur.

**EXAMPLE 4**  The food chains that exist within communities are actually complex networks in which many different species may play the role of producer or consumer or decomposer. In a forest in the United States, up to 60 different species of birds might feed upon many hundreds of species of insects. All these birds are part of a single *trophic level*—they get their food in the same way and bear a similar nutritional relationship to other members of the food chain.

Stability of ecosystems was once equated with complexity, particularly the complex interactions occurring within a food web. More recent field studies, however, show that some simple ecosystems may possess considerable durability; however, if only one population exists at a particular level, the loss of that population may doom the entire ecosystem.

Although ecosystems demonstrate some flexibility and tend to maintain their integrity, they may be irreparably harmed by

1. Sudden shifts in the environment (temperature change, drought, flooding) that destroy a significant portion of the community
2. Uncontrolled increase in the numbers of particular populations due to failure of the mechanisms for population control
3. Loss of key minerals or other nutrients in the ecosystem
4. Human interference, which may lead to destruction of habitats, an overkilling of specific species, or pollution with toxic materials that cannot be handled within the ecosystem

Just as individuals undergo change as they mature, so ecosystems evolve new characteristics and gradually supplant older communities with new populations. This slow change in the makeup of the community within a habitat is called *succession*. Succession often occurs as older inhabitants modify their environments to provide new opportunities for the next generation's plants and animals. Succession continues until a *climax community* is formed, one that is extremely well suited to the environment and remains essentially unchanged through long periods of time. Primary succession occurs in ecosystems that have never been colonized and are therefore recently created. Secondary succession occurs in systems previously colonized but cleared by natural or human agents.

## 26.4 BIOMASS AND SPECIES DISPERSAL

*Biomass* refers to the weight of living organisms within an ecosystem. It is often applied to particular trophic levels to provide an insight into what transpires during the passage of energy along a food chain. Because of the continual loss of biomass in proceeding along a food chain the community can, in terms of biomass, be considered a *pyramid* (see Fig. 26.5). In a food (energy) pyramid a broad producer base is topped by ever-diminishing populations of consumers. The terminal consumer forms the apex of the pyramid.

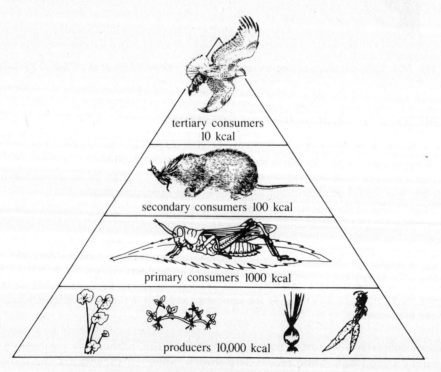

tertiary consumers
10 kcal

secondary consumers 100 kcal

primary consumers 1000 kcal

producers 10,000 kcal

**Fig. 26.5** Energy transfer in a typical food pyramid, starting with 1 million kcal of sunlight. (*Partially from Storer et al., General Zoology.*)

The *carrying capacity* of a habitat refers to the upper limit of life-support capabilities of that habitat. It is usually expressed in terms of numbers of individuals that can survive within a stable community. For animals the carrying capacity is generally a function of available food resources, while for plants it may be mineral nutrients, $CO_2$ levels, or the availability of sunlight. When the carrying capacities are relatively high, population densities tend to be large. When the carrying capacities are low, populations tend to be sparse. A particularly high carrying capacity for many species of both plants and animals is characteristic of a tropical rain forest.

Population density is a quantitative feature of ecosystems; its qualitative aspect is the dispersal of individuals in space. Eugene P. Odum of the University of Georgia has cited three broad distribution patterns:

1. *Random* distribution, in which individuals are scattered without pattern throughout the habitat
2. *Uniform* distribution, in which regular patterns of dispersal occur, such as flowers in a flower bed
3. *Clumping*, in which irregular groupings are discerned, such as flocks of wild birds

Patterns of dispersal are influenced by degrees of socialization in a population, the nature of the terrain and the arrangement of its plant life, interactions with other species, availability of resources, and so forth. *Dispersive* factors tend to scatter members of a population, whereas *cohesive* factors tend to bring individuals together.

By the time of World War I ecologists had demonstrated that because of competition no two species can occupy the same niche for very long—the *niche rule*. In the 1930s in an elaborate series of experiments with *Paramecium* a Russian biologist, G. F. Gause, expanded this rule by showing that in competing for a scarce resource one species tends to drive out a competing species. This *competitive exclusion principle*, or Gause's principle, has been extensively confirmed in a variety of laboratory experiments and has stressed the role of competition in determining species survival within the ecosystem. However, by the 1980s ecologists became aware of the possibility of species occupying the same niche and surviving together in nature. At the present time, competitive exclusion is believed to operate but is still being evaluated.

## 26.5  UPSETTING THE STABILITY OF AN ECOSYSTEM

Ecosystems possess resiliency and resist a variety of perturbations in a manner suggesting the operation of homeostatic control mechanisms. Slow changes over time tend to lead through a series of successions to the relatively long-term stability of a climax community. Environmental catastrophes can plunge even a stable climax community into chaos, but many less dramatic disturbances can be borne. Younger communities, which are far more active and productive than more mature systems, are less able to resist environmental insult.

### SIMPLIFICATION AND SHRINKAGE OF NICHES

A great diversity of species and a complex and intricate pattern of interactions among the populations of a community were once thought to confer greater stability on an ecosystem. Computer simulations of hypothetical ecosystems as well as the poor resistance of species-rich tropical rain forests to physical disruption have challenged this traditional view. Simpler ecosystems with fewer branches in their functional webs (food chains, nutrient cycles) demonstrate a greater apparent degradation to environmental assaults, but they tend to recover rather rapidly and assume new stable arrangements. Complex ecosystems are not markedly disrupted at first by dislocations in the physical environment, but long-term reverberations are set up that may eventually lead to permanent disruption.

A clearer positive effect of the diversity of the physical environment on community stability is observed. Where several influences converge to influence temperature, rainfall, wind velocities and direction, etc., complex ecosystems seem better able to adjust to temporary dislocations.

Human intervention generally simplifies all aspects of an ecosystem. Where a variety of species at separate trophic levels exist in nature, the introduction of agriculture or the setting up of a village reduces these species to one or even none.

**EXAMPLE 5** As a result of monoculture organization, many farmers clear a field and plant a single crop such as a particular strain of corn. Should disease strike that corn crop the entire "artificial" ecosystem would crash, since the corn is the primary producer here. The use of highly inbred strains of particular staples increases the vulnerability of these crops to natural enemies and represents a source of great vulnerability for both the ecosystem and the farmer.

## POLLUTION

*Pollution* is generally defined as the introduction of harmful materials to an ecosystem. Although pollution has been regarded as a human activity in which plastics, synthetic toxins, nondecomposing chemicals, etc., are brought into the flow channels of the ecosystem, it may also involve natural processes that produce materials that cause ecosystems to "belch," "vomit," or even die. Volcanoes and forest fires spew noxious ash and other atmospheric pollutants that can seriously damage or even destroy ecosystems.

One problem with pollutants, especially applicable to toxic organic substances, is that they become more concentrated as they move along a food chain. A lake may contain only moderate levels of the pesticide DDT, but the tiny invertebrates in that lake may concentrate the pesticide 100 times; still greater concentrations are accumulated and stored in the fatty tissue of the fish that feed on the small invertebrates. By the time the DDT reaches the birds that feed on the fish, its levels are many thousands of times greater than in the lake.

Pollution of ecosystems by elemental metals such as lead and mercury has been well documented. These substances have been particularly devastating to higher-level predators such as the tertiary consumers in a food chain. These include not only large wild carnivores but human beings as well.

Lead belongs to the class of heavy metals that tend to interact with and precipitate protein. Tissues like the brain and cells like erythrocytes (red blood cells) are particularly sensitive to the effects of lead. Among the symptoms of lead poisoning are weakness and muscle tremors, interference with thought processes, impaired transport of oxygen by erythrocytes, and nerve destruction.

Lead moves into communities through its presence in gasoline. Combustion of leaded gasoline tends to spew compounds of lead into the atmosphere. Most cars manufactured today are required to use lead-free gasoline, and the atmospheric contamination by lead has considerably abated. At one time lead-based paints were used extensively, but are no longer manufactured. Children chewing on paint in old apartments and homes may contract lead poisoning.

**EXAMPLE 6** Some historians believe that Rome fell as a result of the poisoning of its elite by lead. This intriguing theory is based on the fact that only the wealthy upper class could afford the elaborate plumbing that brought water directly to the home or the pewter flasks in which wine could be stored. Both the plumbing and the pewter storage vessels contained large amounts of lead. Over long periods of time the wealthy gentry from whom the leadership was garnered may have gradually lost their initiative, intellectual ability, and even physical fitness to the demon lead. Many of the effects of lead poisoning are manifested over long time periods and may cause only subtle degeneration initially.

Mercury has also drawn the interest of historians. Mercury salts used by hatters in France to soften their felts were believed to have caused some nerve damage. Many hatters supposedly were irritable and developed symptoms of nervous disorder as a result of slow mercury poisoning. The image of the mad hatter, a character in *Alice in Wonderland*, springs from that historical perception. Some skeptics have challlenged the authenticity of such reports in recent years. However, mercury salts from industrial wastes in Minamata Bay (Japan) have been shown to have killed many citizens who ate contaminated shellfish.

Pollution generally involves the introduction of some harmful substance to an ecosystem, causing disturbances within that ecosystem. *Eutrophication* involves too much of a *good* thing. An overabundance of nutrients is provided in the waters of a river or lake, stimulating overgrowth of phytoplankton (floating microscopic plants) or algae. This floral population soon reaches a density at which vital gases and nutrients are used up, and the overgrown "blooms" produce toxins and die as an unpleasant rotting mass.

Eutrophication, although rooted in the concept of excessive nutrient levels, may also be caused by a rise in water temperature, usually due to the dumping of hot effluents by factories. The increased temperature may speed up activity within the algal community and produce overgrowth. The end result is the same as that brought on by an excess of nutrients.

Eutrophication may also involve the overgrowth of larger plants such as weeds or water lilies. In lakes used for recreational purposes such overgrowth of weeds may cause economic hardship to those dependent on tourists or summer residents for their livelihood.

# Solved Problems

**26.1** What are the strengths and weaknesses of an ecological perspective in understanding life?

The great strength of ecology is that it deals with living things in terms of their natural surroundings rather than in artificial isolation. Other branches of biology focus upon particular aspects of organisms and may thereby lose the reality of connections and interrelationships that are vital to understanding life in its entirety. Ecology, by its very nature of focusing on the context within which organisms develop and function, brings together a variety of disciplines. Evolution, which provides a central structure in the study of biology, can be fully appreciated only from the ecological vantage point of populations interacting with their environment. Anatomy, systematics, biochemistry, and even molecular biology become more fully focused when utilized in analyses of ecosystems. On a more practical level, the study of existing ecosystems has vitalized the conservation movement and enhanced the active effort to maintain ecosystems or even individual species of plants or animals threatened with extinction.

A possible weakness of ecology stems from the broad spectrum of its concerns—it tends to be diffuse. Individual ecologists do focus on particular aspects of communities or the physical environment, but the novice might find the field daunting by its very breadth. In its descriptive phase, that breadth was more easily handled. The recent emphasis on experimentation tends to encourage a specialization within the field, but experimentation introduces another problem. With experimentation, controlled conditions may not be readily obtained in a field situation. Ecosystems can be studied from a passive approach, in which the experimenter merely observes such phenomena as energy flow, diversity of species, etc., but more ambitious attempts to establish an ecological science require interventions that may be difficult to isolate and control. Some success has been achieved, particularly where clear boundaries to a habitat exist, such as on islands.

**26.2** The biomass of primary consumers in a food chain is considerably less than the biomass of (primary) producers. Why?

The energy trapped within the plants of the first trophic level cannot be transferred intact to the next trophic level because of the waste and increase in entropy associated with each transfer of energy. In addition, the plant utilizes a great deal of the energy it obtains from the sun to maintain itself and to build up its own vital structures during its lifetime. Certainly no more than 10-15 percent of the calories stored by the plants pass into the herbivores that constitute the next trophic level.

**26.3** Why might primary producers also be referred to as *transducers*?

The photosynthetic plants, be they oceanic algae, grass in grasslands, or corn in a cornfield, convert the *radiant* energy from the sun into the *chemical* energy stored in the bonds of sugars, fats, and proteins. It is these foodstuffs that are utilized by the plant during its lifetime and then passed on to higher links of the food chain with the death of the plant or its shedding of expendable parts such as fruits. These plants can thus be aptly called *transducers*, since they are not generating energy but instead are transforming that energy and then transferring it to other trophic levels of their ecosystem.

**26.4**   What would you say are some density-dependent mechanisms that might come into play to control population when overcrowding develops?

As a population increases in density, available food resources may be used up, increasing the death rate. Terminal consumers are most vulnerable to growing shortages of food. Other material resources, such as sites for shelter, breeding territories, building materials, etc., are also critical in maintaining populations within reasonable limits.

Under crowded conditions intraspecific competition and aggression increase. This may lead to emigration of less hardy individuals out of the habitat. Stress imposed by crowded conditions has, in some mammal populations, led to infanticide, malfunction of the endocrine system (particularly the adrenal glands), sterility, and even an increased degree of homosexual activity—all these factors tending to lower a population's growth rate.

As a population increases in density the spread of contagious diseases rises. This may be due to increased contacts between individuals, which facilitates the spread of bacterial infections, as well as to lowered resistance because of the crowded conditions.

In populations hunted by large carnivores, a predator-prey relationship exists. An increase in the numbers of the prey population makes it easier for the predator population to capture prey. As a result the size of the predator population, with an increase in its food supply, increases and this tends to check any further increase in the numbers of the prey species.

Homeostasis at the level of population size is largely a matter of density-dependent factors acting as self-regulatory mechanisms. Natural catastrophes (density-independent) limit population size but not in any direct relation to density, since sparsely populated groups are destroyed as readily as crowded aggregations by earthquakes, fire, volcanic eruptions, etc.

**26.5**   We have seen that in the absence of limiting factors, population growth can be expressed by the equation $dN/dt = rN$, which yields an exponential growth curve. The introduction of the carrying capacity factor $(K - N)/K$ changes this curve into the sigmoid curve characteristic of actual populations. Compare what happens mathematically as $N$ rises with what is happening in the population.

When $N$ is very small, $(K - N)/K$ is almost equal to 1 and the equation $dN/dt = rN(K - N)/K$ can basically be expressed as $dN/dt = rN$. Therefore, for low $N$, there are few density-dependent factors operating to limit growth and the actual growth curve resembles the exponential curve. However, as population size increases, the term $(K - N)/K$ gets smaller and the slope decreases. This corresponds to increasing growth-limiting factors. At very high population densities, with a high $N$, the term $(K - N)/K$ approaches zero and the slope becomes flat, as is seen when populations are in equilibrium.

**26.6**   The reproductive strategy of populations can be described in terms of two theoretical types. The *r-selected strategy* is essentially prodigal in approach. Species selected for this high-stakes reproductive approach tend to have large numbers of small young that mature rapidly with little or no involvement of parents in either care or training. Reproduction occurs only once or a very few times, prompting Helena Curtis to label this style the "big bang." In contrast, the strategy of *K-selected* species involves the production of few young of relatively large size, which mature very slowly. Parental care is intensive and may involve extensive social development to ensure that the offspring survive their young, vulnerable period and prepare for their adult responsibilities. Reproductive activity may not begin until the individual attains maturity, usually after a rather prolonged period of development. Once adulthood is attained however, reproductive capability extends over long periods of time. The two strategies do not always exist in a clear-cut and discrete dichotomy. Species frequently fall between these two extremes. How would you say species of each type respond, as a population, to environmental stress? What do their growth curves look like?

In r-selected populations a great increase in numbers is possible over short periods of time. Although individuals may be vulnerable to environmental insults, the population as a whole can survive because even a small surviving remnant can provide a nucleus for rapid restoration of the original population density. Growth curves for species of this sort resemble the exponential curve, with rapid increases in numbers followed by sharp declines. Among the species that are plainly $K$-selected, the individuals are more durable, but the population is more liable to destruction from environmental perturbations because of the relatively low reproductive potential at any point. These populations, which are stable and tend to maintain their numbers around a particular level, are represented by a sigmoid curve.

**26.7**   Organisms in an ecosystem interact in many different ways. One very common relationship is that of predator to prey. In almost all cases, the predator seeks out prey for nutrient need alone. Occasionally, individuals may kill for sport, as in the playing with and slaying of moles by cats. In humans, hunting for sport is quite common; in the case of rare and highly prized groups (alligators, elephants) hunting has driven whole species to near extinction.

Another type of relationship involves long-term, intimate association between two different species: *symbiosis*. Often, one of the organisms lives within or upon the other. Three types of symbiotic relationship exist, differentiated by how the relationship affects both organisms. What would you say is the nature of these three types?

1.   *Commensalism*: a relationship in which one of the partners benefits while the other is neither harmed nor benefited. A good example is the remora, a fish that attaches itself to the underside of sharks by means of a sucker device atop its head. The shark, because of its constant activity and sloppy table manners, provides both transportation and "crumbs" for the remora. The remora does not advance the interests of the shark.
2.   *Mutualism*: an association in which both members benefit. A lichen consists of a fungus tightly intertwined with a green alga. Lichens are particularly prominent in barren areas such as deserts and frozen wastelands. The fungus provides water and a tight hold on the sandy or rocky substrate, while the alga provides food through its photosynthetic capacity.
3.   *Parasitism*: a widespread nutritional strategy in which one member of the couple harms the other. The exploiting member is the *parasite*, the other member the *host*. Parasitism generally involves a very intimate association of parasite and host. It has been estimated that there are more parasitic than free-living groups. All bacterial diseases involve a parasitic infestation by pathogenic microorganisms. Many species of flatworms (*Platyhelminthes*) and roundworms (*Nemathelminthes*) are parasites in vertebrates ranging from frogs to humans.

**26.8**   *Laterization* refers to the process by which the supposedly rich soils of tropical rain forests convert to a hard crusty layer when their indigenous plants are cut down. The *laterite* that forms cannot support the growth of domestic (or any other) plants. At one time the jungles in broad areas of Brazil were slashed in the expectation that highly fertile soils lay under the thick underbrush. The soils proved to be relatively infertile and, in combination with laterization, thwarted agricultural efforts. Given the rich proliferation of plant life, how would you explain the soils' infertility? What effects would slashing have on the soils?

The seeming fertility of jungle land actually lies in the surface layer of undergrowth. When that is cut away, the underlying soil, which has been rapidly depleted of its nutrients by its teeming floral populations, proves to be a poor basis for agriculture. At the same time, mineral nutrients tend to wash away because of the heavy rains that reach the soil and carry soluble elements and mineral compounds away from the land. When these areas are cleared not only is an infertile crust produced, but rainfall is markedly diminished since much of the water in the atmosphere was previously produced from the once-lush jungle growth itself. Alternatives to the stripping of jungle rain forest are currently being investigated.

**26.9**  Life may abound in the relatively shallow waters of the *littoral zone*, but how can life exist in the deeper waters of the pelagic zone and even the abyssal depths where light cannot penetrate?

The largely animal populations of these deeper waters may be far removed from the layers of water into which light reaches, but they may retain functional links to their biological cohorts above them. As plants and animals die, the products of their dissolution settle to the lower depths and are utilized by deep-dwelling organisms. Such settling is enhanced by the occurrence of vertical currents caused by temperature differences in different regions of the ocean.

Not only do the remains of shallow-dwelling organisms filter down, but minerals and organic matter from the floor of the ocean may be brought up by strong upwellings, which further supports life in the oceanic depths. Additionally, some communities exist in fissures on the ocean floor. Gases such as hydrogen sulfide provide a chemosynthetic substitute for photosynthesis, and a variety of bizarre creatures are sustained by the oxidation of hydrogen sulfide and other energy-rich gases.

**26.10**  Explain how DDT can be found in only low concentration in the general environment yet occur at lethal levels in the inhabitants of the same area.

Like many toxic substances, DDT tends to accumulate in fat tissue. Although it may initially be dispersed throughout the environment, the producers remove it from the general environment and concentrate it. Since they are selectively eaten by the primary consumers, these consumers are thus eating higher concentrations than are found in the general environment. They too tend to accumulate the toxins, and when eaten, thus pass on even higher concentrations. Therefore, each trophic level exhibits toxic concentrations greater than those of the level below it, with the highest levels showing concentrations much greater than those at the bottom of the food chain.

**26.11**  Air pollution has proved to be an intractable and pervasive threat to the environment. What are the most serious pollutants of the atmosphere, and how do they affect our well-being?

Carbon monoxide, which is a product of incomplete combustion from automobile engines and industrial smoke, may compete with oxygen for binding sites on hemoglobin. In a closed space, this gas is capable of killing. Its toxicity at lower levels may lead to headaches, dizziness, and blurred vision.

The effects of another product of combustion, carbon dioxide, are just becoming evident. Carbon dioxide is being produced in such high quantities that plants, which normally regulate its levels, are unable to cope, and it is beginning to build up in the atmosphere. Although $CO_2$ is not poisonous at current levels, the buildup is trapping heat and causing a gradual increase in temperatures at the earth's surface—the *greenhouse effect*—with many possible consequences, including melting of the polar ice caps.

A variety of oxides of nitrogen make up another class of toxic pollutants. The most serious of these is $NO_2$. If the concentration of $NO_2$ at the tip of a burning cigarette were present in an entire room, it would be lethal to anyone in the room.

The oxides of sulfur are responsible for respiratory distress, chest pains, and excessive tearing. In the presence of water, $SO_2$ and $SO_3$ become $H_2SO_3$ and $H_2SO_4$—powerful corrosive acids. These acids are also responsible for the eroding of statues and art treasures in polluted areas. Acid rain containing these and similar corrosives threatens the forests and other plant life of the temperate zones.

Particulate matter may also cause discomfort but is not life-threatening. Far more serious is the action of sunlight on gaseous pollutants, producing such highly toxic chemical complexes as peroxyacetylnitrate (PAN). The resulting photochemical smog can be lethal for people with respiratory problems.

# Supplementary Problems

**26.12**  At each trophic level of a food chain (pyramid) the energy not used or passed along is given off as (*a*) matter.  (*b*) free energy.  (*c*) heat.  (*d*) water.  (*e*) none of these.

**26.13** The total weight of Inuits (Eskimos) subsisting on polar bears would be   (*a*) greater than   (*b*) the same as   (*c*) less than   (*d*) unrelated to the total weight of polar bears.

**26.14** Materials tend to become more concentrated as they move along a food chain.
(*a*) True   (*b*) False

**26.15** Ozone ($O_3$) is an irritating toxin, but its presence in the atmosphere serves to filter out harmful ultraviolet rays.
(*a*) True   (*b*) False

**26.16** Changes in the makeup of a community from pioneer forms to the mature species of a climax community are described as a *succession*.
(*a*) True   (*b*) False

**26.17** Permafrost (frozen layer of subsoil) is a characteristic of a grasslands biome.
(*a*) True   (*b*) False

**26.18** A niche is an actual physical site in a habitat.
(*a*) True   (*b*) False

**26.19** All deserts are hot.
(*a*) True   (*b*) False

**26.20** Which is larger   (*a*) the increase in numbers of a population in the lag phase or   (*b*) the increase in numbers during the log (exponential) phase?

**26.21** Which is greater   (*a*) DDT levels in the algae of a lake or   (*b*) DDT levels in birds that eat the fish of that same lake?

**26.22** Which is greater   (*a*) the degree of parental care for mayfly babies or   (*b*) the degree of parental care for human babies?

# Answers

| | | | |
|---|---|---|---|
| **26.12** (*c*) | **26.16** (*a*) | **26.20** (*b*) | |
| **26.13** (*c*) | **26.17** (*b*); tundra | **26.21** (*b*) | |
| **26.14** (*a*) | **26.18** (*b*) | **26.22** (*b*) | |
| **26.15** (*a*) | **26.19** (*b*) | | |

# Chapter 27

## Origin of Life

The *cosmos*, which comprises all the material of reality, may have originated from 10 to 20 billion years ago. Our particular portion of the cosmos is the universe known as the *Milky Way*. A *universe* is an island comprising millions of stars, although the term *universe* is sometimes loosely applied to the entire cosmos. The sun is a medium-sized star lying about two-thirds of the way from the center of the Milky Way. The sun and its planetary satellites make up the *solar system*. (One alternative theory is that the cosmos always existed much as it does today.) The prevalent view is that the cosmos *began* with a massive explosion of tightly condensed matter many billions of years ago—the *big bang* theory. Remnants of that long-ago explosion can be studied with powerful telescopes that can pick up light originating many billions of years ago.

Our solar system probably began as a swirling cloud of gases that eventually condensed into the sun and the planets. The early earth started out as gaseous, but after a while a core of heavy metals, such as nickel and lead, formed. Overlying this core is a relatively thick *mantle* and a relatively thin *crust* forming the surface of the earth. One theory holds that the earth was originally cold but heated up under forces of compression in the settling and synthesis of core materials. Radioactivity also produced a great deal of heat. After about 750 million years the earth cooled and the present crust developed. At this time we live on a relatively cool earth.

The universe we inhabit is not unique and is similar to other kinds of island universes. Nor is the sun a special kind of star. Its location is not unusual, and in size it is medium to small. The planet Earth is larger than Mercury but much smaller than Jupiter or Saturn. In sum, life has arisen under circumstances and in a milieu that fall within a middle range of properties. Conditions on earth, however, are ideal for the development of life as we know it. It is conceivable that such conditions exist on planets of other solar systems we cannot easily observe.

Scientific theories about the origin of life on earth require that the earth be billions of years old. There is evidence to support this assumption. One line of evidence comes from observations of other universes and the measurement of the atmospheres of our sister planets. Further evidence is contained in the kinds and proportions of radioactive materials found throughout the universe, but especially on the earth. For example, Lord Rutherford devised a technique, known as a *radioactive clock*, that gave an age for rocks on the earth's crust of at least 2 billion years. More recent work with the measurement of two isotopes of lead ($^{206}$Pb and $^{207}$Pb) yields a minimum value of 3.35 billion years. All these lines of evidence are based on a constant rate of decay of one radioactive element into the next in a radioactive series. Support for the antiquity of the earth is also drawn from the oceans. If the salinity of the oceans today is divided by the corrected rate of salt deposition by the rivers of the earth, one arrives at a figure suggesting that the oceans are at least several billion years old. Thus, a tentative figure of billions of years is clearly justified.

Two separate views exist regarding the origin of life. The *creationist* view, largely inspired by the original narrative in Genesis, maintains that the earth is no more than 10,000 years old, that each species was *created* separately during a short burst of divine activity some 6000 years ago, and that each species tends to maintain its unique and discrete character through time. *Scientific creationism*, a recent remodeling of this perspective by some conservative geologists and engineers, inspired several unsuccessful battles by fundamentalists to alter curricula in U.S. schools to include a creationist alternative in biology classes where evolution is taught.

An alternative view is that life emerges as a selected point along a continuous spectrum of increasingly complex arrangements of matter. When matter becomes sufficiently complex, we encounter the characteristics associated with life. This is a mechanistic view, but there is room for epiphenomena such as love, conscience, morality, etc., in such advanced forms as humans. Biologists support a natural origin for life.

## 27.1  THE OPARIN HYPOTHESIS

The mechanistic view of life suggests that the complex reactions of living things can best be explained by the properties of their component parts, and that an orderly progression of cause and effect brought about an emergence of life from aggregations of simple inorganic materials into ever more complex organic macromolecules. A clear and rigorous explanation of how this evolution of life from the abiotic realm of chemistry and physics could have come about was presented by A. I. Oparin to his Russian colleagues in 1924. In 1936 his views received worldwide attention.

The Oparin hypothesis starts with the origin of the earth, about 4.6 billion years ago. The early atmosphere was almost certainly a reducing one, possibly with large amounts of methane ($CH_4$), steam ($H_2O$), ammonia ($NH_3$), and some hydrogen ($H_2$). Such an atmosphere would promote chemical synthesis. As the earth cooled, much of the steam condensed to form the primitive seas. Turbulence in the atmosphere during that cooling period produced violent lightning and thunderstorms. Along with the heat rising from the interior of the earth and the ultraviolet rays of the sun, these bursts of energy produced a variety of simple organic substances in the atmosphere, and these substances soon collected in the early seas. Since (1) no living things were then present to break down these organic materials and (2) the reducing atmosphere promoted an increasing synthesis of energy-rich molecules, the seas incorporated these molecules until the seas took on the characteristics of a hot, dilute soup (a metaphor provided by J. B. S. Haldane). The seas were constantly recharged with fresh organic material because a cooling earth produced torrential lightning storms over many thousands of years.

The next stage was extremely crucial to Oparin's hypothesis. The organic material of the seas, becoming increasingly concentrated, accreted into larger molecules of spatial, or structural, complexity—colloids with special properties of electric charge, adsorptive powers, translational movement, and even the ability to divide after reaching a certain size. Oparin called these specific colloids of great organizational complexity *coacervates*. They tended to be shaped into droplets by surrounding "cages" of highly ordered water molecules. There thus existed a very clear line of demarcation between the molecules of the coacervate and the surrounding water. The absorptive properties of the droplet caused it to grow, and eventually an actual membrane may have formed at the coacervate-water boundary, increasing the selective permeability of the droplet.

Much of Oparin's experimental work involved an exploration of the properties of coacervates and their possible role in the evolution of living cells. He believed that at an early stage in the development of living material, amino acids were incorporated into proteins. Since proteins can serve as catalysts, their formation provided the means for an ordering of chemical reactions—the arising of a controlled metabolism. Oparin did not, of course, deal with the reproduction of these complex organizations of organic molecules, because the role of polynucleotides was then unknown. Clearly though, the formation of such information-carrying molecules is crucial to a theory of a gradual evolution of life from simpler abiotic *systems*.

Stanley Miller provided experimental support for Oparin's belief that the conditions and simple inorganic molecules present during the earth's early history could combine to create the complex organic molecules of living organisms. Miller, a student of the Nobel laureate Harold Urey (University of Chicago), set up a Tesla coil that discharged electric bolts into a closed system containing methane, ammonia, water vapor, and some hydrogen gas. The results of this energetic stimulation of an atmosphere resembling that of the early earth were spectacular. A variety of organic molecules were generated, including ketones, aldehydes, and acids, but most important of all—amino acids. Since proteins are vital to both the structure and the function of living cells, the creation of amino acids under conditions that were believed to prevail upon the early earth supported the Oparin hypothesis.

Geologists have more recently revised their estimate of the makeup of the atmosphere during the beginnings of our planet. It is now believed that carbon monoxide, nitrogen, and carbon dioxide were significant constituents of that atmosphere. The Miller-Urey experiments of the 1950s were repeated using the revised atmosphere, and similar yields of organic molecules were achieved. This supported the earlier theories of a primordial transformation of inorganic molecules into the organic building blocks of life.

Sidney Fox (University of Miami) showed that ultraviolet light can induce the condensation of amino acids to dipeptides and, later, that under conditions of *moderate dry* heat, amino acids can be polymerized to *proteinoids*, short polypeptides containing up to 18 amino acids. These proteinoids show a nonrandom arrangement of the amino acids, an advance over random accretions. Particularly exciting was his finding that polyphosphoric acid increases the yields of these polymers, a result suggestive of the present role of ATP in protein synthesis. The proteinoids produced by Fox generally assume a specific spherical shape. These tiny spheres (*microspheres*) show some of the properties of living cells, but they are a long way from a true living structure.

The seminal work of Fox has been extended by Cyril Ponnamperuma, a chemist at the Ames Research Center in California. In 1964, he showed that during the thermal polymerization of amino acids, small amounts of guanine form; he thus linked nucleotide synthesis to the synthesis of polypeptides. Later, he reported that adenine and ribose are products of long-term treatment of reducing atmosphere gases with electric current.

Not all biologists believe that the first living forms were produced in primitive oceans. Some theorize that early life began in the hot and extremely thick atmosphere of a long-ago time. The basis for that belief comes from the tendency of polymers to dissociate back into their constituent monomers when water is plentiful and heat and other forms of energy abound. Under such conditions hydrolysis rather than condensation would be encouraged.

Others have stressed the role of wet soils (J. B. S. Haldane) and clays (Bernal) as stabilizing media for coacervates that were first formed in the turbulent seas. There is, then, a difficulty with the view that life could have arisen in stormy seas, where maintaining both structural and functional integrity would have been difficult.

The inevitable question raised by formulations such as Oparin's is why life doesn't continue to evolve from abiotic sources. The *spontaneous generation* of life had long ago been disproved by Louis Pasteur and, more recently, by investigators of microbial systems, and its impossibility is a cornerstone of an evaluation of the distinctiveness of life. But conditions on the earth many billions of years ago were quite different from what they are today. The reducing atmosphere then promoted complexity. Today's atmosphere, an oxidizing one, tends to degrade large molecules and complex structures that are not stable. This oxidizing atmosphere promotes a drift toward simplicity rather than complexity.

Another significant factor is the presence of living forms now universally distributed in the environment. These organisms gobble up any available energy-rich structures in their never-ending quest for food. The chances of developing complex systems are quite unlikely now that living organisms cover the globe.

## 27.2  HETEROTROPH TO AUTOTROPH

Oparin realized that early living entities dwelt in an environment rich in energy-yielding organic molecules that could be absorbed as food. This ingestion of preformed organic fuels represents a *heterotrophic* habit. However, in local regions intense competition for vital substances among expanding populations might lead to critical shortages. Let us designate one such depleted nutrient as *A*. Under such circumstances, if a mutant appeared that could synthesize *A* from nutrient *B*, it would tend to survive while its maladapted competitors would die for want of enough *A*. As *B* became depleted, an organism that could synthesize *B* from *C* would demonstrate greater survival ability. In this way, organisms would tend to evolve complex enzyme systems enabling them to synthesize their requisite materials from simpler substances—a nutritional habit called *autotrophy*. According to N. H. Horowitz, organisms in areas where nutrients were scarce soon evolved long chains of enzyme-catalyzed reaction sequences which afforded them freedom from dependency on the materials of the dilute soup. The evolution of autotrophy as a significant advance in the early evolution of life has been coupled to the earlier contributions of Oparin and is known as the *Oparin–Horowitz theory*.

## 27.3  ORIGIN OF CELLS

Complex coacervate droplets maintain their structure within an amorphous (unstructured) liquid medium. Further, an exchange of materials with that environment occurs across the limiting boundary of the coacervate. Although this boundary seems to be made of oriented water molecules and other simple inorganic materials, its properties approach the permeability characteristics of cells and may have been the forerunner of early prokaryotic cells. A growing complexity of organic materials within the coavervate was dependent on the "foreign policy" of the droplet, as dictated by the outer membrane. In turn, the membrane could become increasingly complex as materials brought into the cell were carried to the surface.

Although the evolution of the first cells is crucial to the establishment of a mechanistic hypothesis for the origin of life, a great deal of speculation regarding the transition from prokaryotic to eukaryotic cells also intrigues the imagination of many biologists.

# Solved Problems

**27.1**  Are all possibilities for the origin of life on earth covered by the two alternatives of creationism and the slow evolution of complexity?

No, there are other possibilities. Some biologists, both past and present, prefer to believe that life always existed somewhere in the universe and that living material reached the earth through a long journey of spores or seeds of some kind coming from a distant portion of the universe (J. B. S. Haldane). This view does not acknowledge the need to postulate an origin, and it may be associated with the belief that the cosmos itself had no beginning but always was.

A variation of the mechanistic view of Oparin is that life did indeed evolve from simple chemical substances but that the steps involved were not always ordinary chemical reactions. A key improbable event may have occurred that was crucial to eventual development of living material. However, this perspective does not depart from a natural explanation since even improbable chemical events are bound to happen, given enough time and materials.

**27.2**  Why are scientists opposed to a mandated balanced treatment in which creationism is presented along with the theory of evolution in science classes?

A science class should be guided by the processes that are valid for scientific investigation. Creationism, even as scientific creationism, is not an alternative scientific theory that is arrived at by examination of evidence. Instead, in most unscientific fashion, it is motivated by an *a priori* commitment to the authority of the Bible and its various interpretations. It operates in a realm not guided by the ground rules of science, so scientists cannot accept it as legitimate material for scientific evaluation. This does not negate the possible truth of the operation of a supernatural creator; it merely focuses on the *appropriateness* of such exploration within the framework of science.

Another problem is the emotional atmosphere that may be generated by a consideration of creationism in the classroom. In debates with scientists, creationists often do well, but their success may spring from appeals to what is morally correct and what one *ought* to think. These considerations may blunt the searing dissection of reality that science engages in. Scientific knowledge must not be fettered with prior constraints rooted in particular moralities or religious interpretations. While there should be room for a consideration of ethical positions and even theological perspectives, that room is not the one in which science classes are being conducted. A confusion of the different roles of science and religion in uncovering the truth of reality could easily occur in the minds of the young should creationism be introduced within the boundaries of a biological discussion.

In a more partisan fashion, many biologists resent the tactics that have been used by the proponents of creationism. Most of the new breed of scientific creationists are geologists, engineers, and physicists.

Virtually none are biologists. They tend to distort the necessary tentativeness of all theories in science; in terms of evolution they characterize the process as merely a theory and hence not valid. Some scientists feel that such characterizations are not fair arguments. Again, most scientific creationists come from a fundamentalist background, so that their ideas spring from a commitment to a literal interpretation of the Bible rather than from analysis of the available evidence. On the positive side, the intellectual and legal challenge by the creationists has forced biologists to reappraise the concept of evolution, the separate force of natural selection, and the various patterns that may occur as organisms change in a shifting environment.

**27.3**  In order for organic molecules to form from the chemicals of the early earth's atmosphere, high sources of energy would have been necessary. What are some possibilities for these early energy sources?

Electric energy would have been abundant because of the prevalence of thunderstorms during the earth's early history. The numerous volcanoes provided a ready source of thermal energy. Radioactive energy was also abundant. Finally, because an ozone layer had not developed at that point, ultraviolet radiation would have been ubiquitous.

**27.4**  Given the characteristics of coacervates, etc., what were the earliest cells probably like?

First, it should be pointed out that there are vast differences between coacervates, microspherules, and other complex colloidal aggregates on the one hand and living cells on the other. Oparin showed that a coacervate could selectively incorporate material into its colloidal interior and even divide after reaching a certain size, but this is still a magnitude below the capabilities of even the most primitive of living cells. These cells represent a big jump from complex colloids.

The earliest cells were probably like the bacteria of today—a limiting membrane, covered with a cell wall, enclosing a "cytoplasm" with very few organelles. They were anaerobic in their metabolic pattern and heterotrophic in food habit. Slowly the heterotrophic habit may have been modified with the development of autotrophic capability: some of these early cells (*chemosynthesizers*) could build energy-rich organic molecules using energy obtained from prying apart the bonds of simple inorganic molecules. But others (*photosynthesizers*), eventually far more widespread, could tap the energy of the sun and produce energy-rich organic molecules from compounds of hydrogen ($H_2S$, $H_2O$) and $CO_2$.

**27.5**  Are viruses alive?

In terms of the definition of life given in Chap. 3, viruses are not alive. They are complex associations of two macromolecules—protein and nucleic acid—but they are neither self-regulating nor capable of metabolism. Perhaps most crucial is their inability to reproduce independently. They may be crystallized and kept in an inert state in test tubes for long periods of time. Their talent lies in their ability to seduce living cells into manufacturing new viral material following the injection of viral nucleic acid into those cells. The genetic message of the virus literally captures the protein-synthesizing machinery of the cell, which then carries out the bidding of the viral information tape. The cell also produces the nucleic acids required for viral replication.

At one time viruses were felt to be links in the stepwise increasing complexity of macromolecules on their journey toward becoming full-fledged cells. The more likely explanation offered by many virologists at this time is that viruses are degenerate products of more complex forms, even of once-living cells.

Forms resembling living organisms exist that are even simpler than viruses. In 1971 a scientist at the U.S. Department of Agriculture showed that infections of potato plants are caused by very small bits of circular RNA lacking a protein coat. These tiny lengths of naked RNA, soon found to cause a variety of diseases in flowering plants, were called *viroids* by their discoverer, T. O. Diener. Although viroids cause disease, they do not destroy the cells they parasitize. The mechanisms whereby they enter the cell and take over part of its polynucleotide synthesizing machinery are not fully known.

Several neurological diseases, including *scrapie* in sheep, arise from an infestation of central nervous system cells by a self-replicating protein called a *prion*. Prions have been isolated from diseased tissues, but their mode of action and reproductive strategy are unknown.

# Supplementary Problems

**27.6**   Recent astronomical findings suggest that the earth's early atmosphere may have contained   (*a*) $CO_2$. (*b*) CO.   (*c*) nitrogen.   (*d*) all of these.   (*e*) none of these.

**27.7**   Coacervates   (*a*) have a stable structure.   (*b*) show selective permeability.   (*c*) are complex colloids. (*d*) are capable of dividing.   (*e*) all of these.

**27.8**   The major difference between todays' atmosphere and that of the early earth is that today's atmosphere contains   (*a*) $CO_2$.   (*b*) nitrogen.   (*c*) oxygen.   (*d*) helium.   (*e*) none of these.

**27.9**   Most biologists believe that the earliest organisms were   (*a*) heterotrophs.   (*b*) autotrophs.   (*c*) eukaryotes.   (*d*) all of these.   (*e*) none of these.

**27.10**   Most geologists agree that the earth is approximately 4.6 billion years old.
(*a*) True   (*b*) False

**27.11**   Most living organisms exist within the mantle of the earth.
(*a*) True   (*b*) False

**27.12**   Nucleotide bases may have arisen from the condensation of HCN.
(*a*) True   (*b*) False

**27.13**   Louis Pasteur demonstrated that under conditions prevalent on today's earth the spontaneous generation of living organisms does not occur.
(*a*) True   (*b*) False

**27.14**   Sidney Fox showed that the earliest organisms were probably heterotrophs.
(*a*) True   (*b*) False

**27.15**   The oldest known fossils are approximately 3.5 billion years old.
(*a*) True   (*b*) False

**27.16**   A universality of the genetic code suggests that all organisms have had a common origin.
(*a*) True   (*b*) False

**27.17**   In the mid-1930s Wendell Stanley isolated and crystallized tobacco mosaic viruses. Injection of the crystals into tobacco plants resulted in an active infestation of tobacco mosaic disease. This suggests that viruses are not   (*a*) organic in nature.   (*b*) alive.   (*c*) the agent of tobacco mosaic disease.   (*d*) protein in nature.   (*e*) none of these.

**27.18**   The protein coat of a virus is known as a   (*a*) capsid.   (*b*) cell wall.   (*c*) cell membrane.   (*d*) centrosome.   (*e*) lysogenic particle.

# Answers

| | | | | | | | |
|---|---|---|---|---|---|---|---|
| **27.6** | (*d*) | **27.10** | (*a*) | **27.13** | (*a*) | **27.16** | (*a*) |
| **27.7** | (*e*) | **27.11** | (*b*); crust | **27.14** | (*b*); Horowitz | **27.17** | (*b*) |
| **27.8** | (*c*) | **27.12** | (*a*) | **27.15** | (*a*) | **27.18** | (*a*) |
| **27.9** | (*a*) | | | | | | |

# Chapter 28

# The Kingdom Monera

The kingdom *Monera* comprises the prokaryotic, single-celled organisms. They are all bacteria possessing ribosomes and a naked, circular strand of DNA that serves as a chromosome, but they generally lack membrane-enclosed organelles such as mitochondria, lysosomes, peroxisomes, an endoplasmic reticulum, and a true nucleus. They divide by binary fission rather than mitosis but may undergo genetic recombination. Their fossils have been found in rock strata that are 3.5 billion years old. The Monera are divided into two broad subkingdoms: *Archaebacteria* and *Eubacteria*. The eubacteria are both more common and more recently evolved.

## 28.1 ARCHAEBACTERIA AND EUBACTERIA

As their name implies (*archae* means "ancient"), the archaebacteria are probably the oldest of living cells. Their cell walls lack a substance, peptidoglycan, found in the walls of all eubacteria. The lipids in their plasma membrane are branched, differing not only from those of other bacteria but from those of all other organisms. This unusual lipid makeup is probably related to the extreme environments to which they have adapted. The phototrophic forms use the pigment *bacteriorhodopsin* instead of *bacteriochlorophyll* (chlorophyll *a*) used by eubacteria.

At the molecular level, their transfer RNA and ribosomal RNA possess unique nucleotide sequences found nowhere else, although structurally more similar to those of eukaryotes than to those of eubacteria. On the basis of these fundamental molecular differences as well as the chemical differences of their primitive organelles, some microbiologists have suggested that the archaebacteria should be placed in a separate kingdom.

All the archaebacteria live in extreme environments where other kinds of bacteria could not survive. It has been suggested that these ancient monerans evolved at a time when those extreme environments were actually typical of conditions throughout the biosphere and that they are merely adhering to the patterns of their early origins. One group, the *methanogens*, lives in bogs and swamplands, where they produce methane through pathways of anaerobic chemosynthesis. These pathways would have been adaptive in the reducing atmosphere of early earth. The *halophilic* (salt-loving) archaebacteria are found in regions of high salinity. The Dead Sea in Israel abounds with these forms, which use the high salt gradients of their environment to generate high-energy intermediates such as ATP or similar trinucleotides. A *thermoacidophilic* group has carved out a niche in hot springs and volcanic vents. The sulfurous vents of Yellowstone National Park are especially rich in these forms, which thrive under conditions of high heat and low pH.

The eubacteria are extremely wide ranging in their characteristics, and their classification is still quite fluid. Among the recognized major groups are the purple and green bacteria; each of these is photosynthetic, but they differ from other bacteria in the pathways involved in the process and in the use of $H_2S$ as a source of reducing equivalents rather than water. The cyanobacteria carry on photosynthesis in a manner similar to that of the higher plants. A variety of heterotrophic forms are also part of the kingdom. Those bacteria that stain positively with Gram stain are placed in a separate group, while the corkscrew-shaped *spirochetes*, which include the pathogens responsible for syphilis, are characterized in part by their negative reaction to Gram staining. Some eubacteria, unlike archaebacteria, contain a flagellum, although it is different in structure from flagella seen in eukaryotic cells.

### BACTERIA AND BLUE-GREEN ALGAE

At one time the monerans were subdivided into bacteria and blue-green algae. The blue-green algae are now considered to be bacteria and are classified as a division, *Cyanobacteria*, within Eubacteria.

The basis for the earlier designation of the cyanobacteria as algae was the similarity of their photosynthetic process to that of algae and even to the metaphyta (many-celled plants). They split water to produce oxygen and contain chlorophyll *a*. The cyanophyta of the ocean are a significant source of atmospheric oxygen. Other characteristics, however, particularly those discovered by molecular probes, have indicated that the cyanobacteria belong with the eubacteria.

Because they synthesize their own food, the cyanobacteria are widespread and may have made up pioneer populations in difficult terrains in many parts of the primitive world. Cyanobacterial colonies exist in many colors besides the blue or green tints usually associated with them. Their blooms may give large bodies of water a characteristic color.

## BACTERIAL MIDGETS

At one time the *rickettsiae*, which are even smaller than the minute bacterial cells, were placed in a separate category. Current opinion among microbiologists is that they are bacteria, although they are not much larger than a virus. They do possess cell structure, and they synthesize their proteins. *Rocky Mountain fever* is a disease caused by a rickettsia carried by a tick.

The *chlamydiae* are another group of minute bacteria that are also cellular and, hence, classified as bacteria. Like the rickettsiae they contain both DNA and RNA (a complete system for encoding, transcribing, and translating molecular information), are capable of reproducing independently, demonstrate organellar organization at least at the ribosome level, and synthesize the coupled enzyme systems that permit the establishment of metabolic pathways. Some forms of chlamydiae cause chronic infections of the human female reproductive tract.

The smallest of the bacteria are the *mycoplasmas*, a group unique among monerans in their lack of a cell wall.

## BACTERIAL MORPHOLOGY

Bacteria are generally one of three shapes—round, rodlike, or spiraled. The *cocci* are perfectly round or else egg-shaped. Many of the eubacteria have this shape, including the pneumococcus, which causes bacterial pneumonia, and *Streptococcus*, responsible for "strep" throat and the complication of rheumatic fever. The rodlike bacteria are known as *bacilli*. Bacteria existing as short, helical strands are designated *spirilla*. The causative agent of syphilis, *Treponema pallidum*, is a delicate spiraled moneran belonging to the spirochetes.

Cell division patterns differ for each of these morphological groups. In pneumococcus, the two daughter cells remain together as a diplococcus pair; in other cocci a long strand of cells may be produced as a result of many divisions. Bacilli and spirilla often break apart after each division.

## 28.2  THE SIGNIFICANCE AND ORIGIN OF ORGANELLES

As far back as the beginning of the twentieth century biologists noted a similarity between various membrane-enclosed organelles and specific bacteria. Particularly striking is the resemblance between the chloroplast and the chlorophyll-laden cyanobacteria. Many biologists also noted a similarity between mitochondria and other free-living bacteria. The fact that both chloroplasts and mitochondria contain their own DNA and may replicate independently of the rest of the cell seems to support the view that these and other organelles were independent bacteria that somehow invaded early cells and set up a permanent housekeeping arrangement within these cells. The invaders are supposedly symbionts that profited from the protected environment of the host cell and donated their capabilities and talents to the cell. Thus, chloroplasts may have been cyanobacteria whose early lodging within the cell conferred a photosynthetic capacity. Other moneran cells, particularly those of extremely small size, may have been responsible for still other organelles characteristic of the eukaryotic cell.

Lynn Margulis of Boston University has amassed an impressive array of evidence for this theory of the origin of organelles, the *endosymbiotic hypothesis*. It has been accepted by many cell biologists and has stimulated a variety of experimental approaches to either confirm or disprove the position.

## 28.3  BACTERIA, THE ECOSPHERE, AND HUMAN INTERACTIONS

Bacteria are the major decomposers of most ecosystems. They not only break down dead remnants of larger organisms but also liberate the constituent molecules and atoms for use by other members of the community.

**EXAMPLE 1**   A dead elephant in a grassland area of Africa is a repository of carbon, nitrogen, phosphorus, and other elements. Bacterial decay not only breaks up the large segments of flesh and bone but also returns the carbon, nitrogen, and phosphorus as available $CO_2$, $NH_3$, $N_2$, and compounds of phosphorus. The action of bacteria (with some help from fungi) is a major feature of the carbon, nitrogen, and phosphorus cycles in nature.

In earlier chapters we dealt with the role of certain bacteria, some living in root nodules and some free-living in the soil, in the fixation of gaseous nitrogen from the atmosphere. Without this process the vital fertilizers of the soil would be used up and normal growth of plants stunted or completely prevented.

The photosynthetic capacity of most of the members of the cyanobacteria fits them admirably for the role of primary producers in fresh water but especially in the seas. They are probably the major movers in the oxygen revolution that brought aerobiosis to the biosphere some 2.8 billion years ago.

The various types of fermentation carried out by bacteria are particularly useful to humans. Alcohol, acetic acid (vinegar), and acetone are just some of the products produced by bacteria. Fibers are often refined, leather tanned, and textiles prepared by the use of bacteria. Bacteria also provide us with cheese and yogurt. The bacterium *Escherichia coli* is the principal living tool of molecular biology. Genetic engineering has also enabled scientists to insert human genes into bacteria. These bacteria then multiply, cloning the inserted gene, to provide large amounts of such vital proteins as insulin, interferon, and growth hormone.

Although most bacteria enhance the quality of life in the ecosystem and within human communities, they also present a negative aspect as the causal agent of many diseases. These diseases range from leprosy and tuberculosis to typhoid fever and bacterial pneumonia. At the same time, bacteria have been sources of many antibiotics used in the fight against bacterial infestations.

In the latter part of the nineteenth century, Louis Pasteur, a French scientist, advanced the idea that bacteria cause many diseases—*the germ (bacterial) theory of disease.* This significant insight was rejected at first, but Robert Koch, a German physician, recognized its importance. He investigated a number of diseases he believed were caused by bacteria, including anthrax and, later, tuberculosis. In developing his investigations he devised a set of criteria for proving that a specific bacterium (or other microorganism) is the causal agent of a specific disease. This set of standards has become known as *Koch's postulates.* As set down by Koch:

1.   The putative microorganism must always be present in hosts suffering from the disease.
2.   The microorganism must be isolated from the sick host and grown under laboratory conditions (in vitro) as a pure, uncontaminated culture.
3.   Samples of the pure culture injected into healthy hosts that are vulnerable to the disease must produce the appropriate infection.
4.   These hosts, now carrying the disease, must yield the microorganism in question to a new set of pure cultures. The pure cultures obtained from these secondary hosts must be similar to the original pure cultures when closely compared.

Through the careful application of this protocol, microbiologists have established a long list of human diseases caused by bacteria.

The pathological changes associated with bacterial invasion may be caused by a direct interference with tissue function, the production of *exotoxins* (toxins released by the bacteria during their lifetime), or the production of *endotoxins* (toxins released after the death and breakdown of the invaders). In some cases, the harm produced by the invader is due to an overextended immune response which may produce shock or set off an autoimmune response.

# Solved Problems

**28.1**  List the differences between archaebacteria and eubacteria.

|                           | Archaebacteria           | Eubacteria               |
|---------------------------|--------------------------|--------------------------|
| Cell wall                 | Lacks peptidoglycan      | Contains peptidoglycan   |
| Plasma membrane lipid     | Uniquely branched        | Straight-chained         |
| Photosynthetic pigment    | Bacteriorhodopsin        | Bacteriochlorophyll      |
| Ribosomes                 | Similar to eukaryotes    | Unlike eukaryotes        |
| Flagellum                 | Absent                   | Present (often)          |
| Habitat                   | Generally extreme        | Generally moderate       |

**28.2**  *Spores* are usually found among the bacilli. A spore arises within the single bacterial cell as an *endospore*, which is enclosed by a highly durable, impenetrable spore coat. The endospore has enough material from the larger cell to ensure its survival and capability to develop into a new bacterial cell. What function would you say spores serve?

The endospore is hardier than the parent cell and, so, represents a strategy for survival of the cell line under adverse conditions. Spores can lie dormant for many years and are highly resistant to heat and cold, chemical insults, and desiccation (drying). Spore formation clearly raises the reproductive potential of a bacterial species. In terms of the challenge to human populations, spore formation enhances the infectivity, virulence, and parasitic persistence of bacteria. Even treatments such as steam exposure or strong disinfectant may fail to eradicate spores.

**28.3**  What facts of cellular structure and function would you say support the endosymbiotic theory of Lynn Margulis?

One line of evidence derives from the prokaryotic nature of many cellular organelles. They contain their own genetic blueprint, at least in the case of basal bodies, mitochondria, and chloroplasts, which are also roughly the size of bacteria, and they contain ribosomes which participate in organellar protein synthesis. The absence of a nuclear membrane is also consistent with prokaryotic characteristics, as is the naked DNA and absence of nucleosomes.

Another line of evidence derives from present instances of bacteria that have invaded eukaryotic cells and "made the decision" to remain there permanently. Margulis has described many examples, but the most dramatic is cellulose-digesting bacteria within the protozoan that dwells in the digestive tract of termites that subsist on wood.

In the case of the chloroplast, similarities between the chlorophylls of this organelle and the chlorophyll of cyanobacteria support the idea that cyanobacteria are the progenitors of the chloroplast. In mitochondria, the interior membrane that overlies the matrix is similar in both structure and function to the bacterial plasma membrane.

Although the endosymbiont hypothesis is both attractive and fruitful in ordering experimental protocols, it is by no means universally accepted. Certainly the formation of the nuclear membrane and even the endoplasmic reticulum cannot easily be accounted for by an invasion of bacteria. Also, while many bacteria have flagella as their motile apparatus, the structure of these flagella, in almost every case, is completely different from that of their eukaryotic counterparts: the basal body is much smaller, the typical eukaryotic 9 + 2 arrangement of microtubules is absent, and, most important, the flagellum rotates rather than beats.

**28.4**  Koch's postulates for proving that a particular microorganism is responsible for a given disease might seem elaborate and redundant. Justify each of the four steps in the process.

As prescribed in step 1, if there are any cases of a particular disease for which the suspected microbe is absent, the microbe clearly cannot be the agent of infection. In isolating and identifying the suspected organism, step 2 ensures that this microorganism is the sole variable in inducing infection in the test animal used in the next step. Step 3 establishes the empirical link between the microorganism and the disease, and step 4 double-checks to make sure another organism was somehow not introduced during the test. This emphasis on identification of the microorganism at each stage of the process is analogous to the situation in forensics in which a piece of evidence must remain in custody in an *unbroken* chain from the crime scene to the courtroom.

# Supplementary Problems

**28.5**  All monerans  (*a*) are bacteria.  (*b*) lack an endoplasmic reticulum.  (*c*) contain DNA and RNA.  (*d*) demonstrate a long circular strand of DNA not found enclosed in a nuclear membrane.  (*e*) all of these.

**28.6**  Antibiotics block metabolic reactions in bacteria so that they cease growing and fail to multiply. Which of the following is not an antibiotic?  (*a*) penicillin.  (*b*) streptomycin.  (*c*) automycin. (*d*) cytochrome c.  (*e*) tetramycin.

**28.7**  Many antibiotics are derived from bacteria themselves, particularly from the *Actinomycetes* group, which form branching chains as they grow. Penicillin, however, is derived from a  (*a*) mold.  (*b*) roundworm.  (*c*) protozoan.  (*d*) flowering plant.  (*e*) none of these.

**28.8**  Bacteria may be  (*a*) free-living organisms.  (*b*) saprophytes.  (*c*) parasites.  (*d*) nitrogen fixers.  (*e*) all of these.

**28.9**  Some bacteria were recently found to accumulate about 20 granules of iron ore ($Fe_3O_4$) within their cells. This allows them to orient to  (*a*) gravity.  (*b*) a magnetic field.  (*c*) flowing water.  (*d*) all of these.  (*e*) none of these.

**28.10**  Which are larger  (*a*) mycoplasmas or  (*b*) cyanobacteria?

**28.11**  Which are more complex  (*a*) viruses or  (*b*) bacteria?

**28.12**  Which have greater reproductive potential  (*a*) bacteria or  (*b*) humans?

**28.13**  Obligate anaerobes are killed by molecular oxygen.
(*a*) True  (*b*) False

**28.14**  Facultative anaerobes may live with or without oxygen.
(*a*) True  (*b*) False

**28.15**  The methanogens are among the most recently evolved of monerans.
(*a*) True  (*b*) False

**28.16**  The cytoplasm of all monerans is extensively compartmentalized by membrane.
(*a*) True  (*b*) False

**28.17**  If the generation time of a bacterium is 20 minutes, then in 2 h there will be 64 bacteria.
(*a*) True  (*b*) False

**28.18**  Enteric bacteria (those found in the gut) such as *E. coli* may be used as an indicator of fecal pollution in swimming pools or lakes.
(*a*) True  (*b*) False

**28.19** All bacteria are heterotrophs.
(*a*) True  (*b*) False

**28.20** Pathogenic bacteria which kill their hosts are, in a sense, less successful parasites than those which cause nonfatal disease.
(*a*) True  (*b*) False

# Answers

| | | | |
|---|---|---|---|
| **28.5** (*e*) | **28.9** (*b*) | **28.13** (*a*) | **28.17** (*a*) |
| **28.6** (*d*) | **28.10** (*b*) | **28.14** (*a*) | **28.18** (*a*) |
| **28.7** (*a*) | **28.11** (*b*) | **28.15** (*b*) | **28.19** (*b*) |
| **28.8** (*e*) | **28.12** (*a*) | **28.16** (*b*) | **28.20** (*a*) |

# Chapter 29

## The Kingdom Protista

The kingdom *Protista* includes all the eukaryotic unicellular species. Some of these organisms are animal-like (*protozoans*), others resemble plants (*algal protists*), and still others demonstrate the characteristics of fungi. Some taxonomists include colonial forms and simple multicellular forms in this kingdom because they are more closely related to protists than to the other three multicellular kingdoms. In that case the kingdom has been given the name *Protoctista*. We will use the more common designation *Protista* here.

Divisions within the kingdom are not always based on evolutionary descent but, rather, may be more practically rooted in functional characteristics. As with the monerans, taxonomy is in a state of flux, and different classification schemes are found in a variety of biology texts.

The protists evolved about 1.6 billion years ago. They are extremely complex; their cells show even more diversity than is found among the cells of the multicellular kingdoms. Their phylogeny is equally complex and is not yet completely understood. It is believed that they gave rise to the fungi, higher plants, and multicellular animals, although from forms considerably different from today's protistan representatives.

### 29.1 PROTOZOANS

The protozoans are heterotrophic organisms found in every major habitat. Some are free-living, whereas others exist as parasites within the bodies of animals. Protozoans also demonstrate the symbiotic lifestyles of commensalism and mutualism. The parasitic protozoans cause some of the most widespread and debilitating of human ills.

Generally, reproduction is asexual, but complex sexual patterns also occur. Protozoa as a division has been divided into five major phyla. Some protozoologists recognize six phyla.

#### MASTIGOPHORA

The Mastigophora (zooflagellates) are probably the most primitive of protozoans. They are distinguished by one or more flagella, which are their principal organelles of motility. These flagella, unlike the rotating flagella seen in bacteria, have the characteristic 9 + 2 microtubular substructure found in the cilia and flagella of multicellular animals. The Mastigophora usually have no cell wall, making undulation of the cell body possible in some species. Some biologists feel that the Mastigophora are the descendents of ancestral forms that gave rise to multicellular animals.

Of the almost 2600 species in existence most are parasitic. *Trypanosomes* are flagellates that cause African sleeping sickness. *Trichonymph* is a genus of Mastigophora capable of digesting cellulose; it lives within the gut of the termite.

#### SARCODINA

This phylum comprises amoebas and their shelled relatives. No cell wall exists outside the cell membrane, but several groups possess a shell that affords considerable protection against abrasive materials in the immediate environment.

The *Foraminifera* are among the most numerous of all protozoans, with almost 40,000 species extant and extinct. They are amoebas with snail-like shells, which are trundled about as the animal moves by amoeboid movement from place to place. The shells often bear beautiful designs and are attractively colored. The Foraminifera live in the sea and extract calcium carbonate from their environment to produce their gaudy shells. The white cliffs of Dover are a formation built up from the ocean bottom by billions of accumulating Foraminifera shells—a process taking eons.

Another shelled group, the Radiolarian, probably goes back to an even earlier (Precambrian) era in the earth's history. Their shells are rich in compounds of silicon, leading to more durable rock formations than those of the Foraminifera. The very deep ooze of some ocean bottoms is composed of radiolarian shells, since silicates are more resistant to pressure and low pH than many other compounds.

### SPOROZOA

The sporozoans are all parasites and possess virtually no means of locomotion. They often undergo complex life cycles as they move from one host to another. Quite often the specialized structures that enable free-living forms to move about, capture prey, and explore new habitats are considerably reduced or absent in the parasite. Locomotor structures may disappear during the course of the parasite's evolution, but reproductive structures become more prominent. Prodigious numbers of gametes, spores, or other reproductive cells are usually produced, especially in metazoan parasites. In some parasites special holdfast devices develop, which prevent the host from "washing out" the parasite. Mechanisms to prevent dissolution of enteric (intestinal) parasites by the host's digestive juices include the weaving of a slimy resistant coat by the parasite or the formation of resistant spores. This theme will be explored further in Chap. 32.

Malaria, caused by the sporozoan parasite *Plasmodium*, is probably the most pervasive and widespread of all infectious diseases. It may eventually kill or cripple up to half the human population, principally in tropical regions. While plasmodia are the actual infectious agents of the disease, the mosquito is the vector that passes the protozoan from one host to another.

**EXAMPLE 1**   The female *Anopheles* mosquito is the primary host in the complex life cycle of the protozoan responsible for malaria (see Fig. 29.1). When the mosquito bites an infected human, it releases an anticoagulant and sucks up *gametocytes* of the *Plasmodium*. In the digestive tract of the mosquito the gametocytes develop into male and female gametes. Fertilization occurs within the stomach, and the zygote burrows into the wall. Following zygotic meiosis a series of mitoses occurs, resulting in the formation of many *sporozoites*, which migrate to the

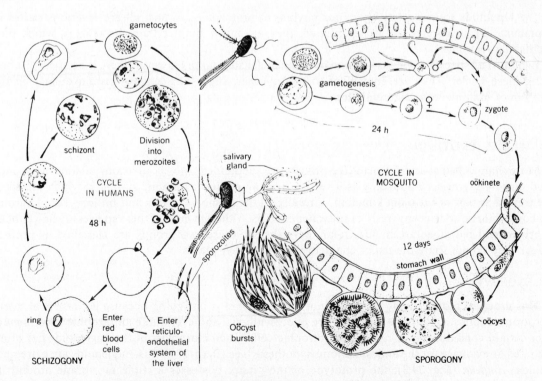

**Fig. 29.1** The life cycle of *Plasmodium vivax*. (*From Storer et al.*)

insect's salivary glands. When a new human victim is bitten the sporozoites enter the bloodstream and migrate to the liver, where they multiply asexually. A new stage, the *merozoite*, is formed in the liver. The merozoite is transported to red blood cells, where they reproduce and rupture the red cells, causing chills and fever. These symptoms are the result of toxins released when the red cells rupture. Some merozoites develop into gametocytes.

In this case the female of the species is deadlier that the male, since only the female feeds on blood.

Mosquito control is one arm of a campaign to flight malaria; other measures include ameliorating drugs such as quinine and, most necessary but elusive of all strategies, development of a vaccine. While malaria has generated great concern because of the havoc it creates in human populations, it also infects other primates, many rodents, and even representatives of birds and reptiles. Several species of *Plasmodium* cause malaria. The most virulent in human populations is *P. falciparum*. Some success in fighting these protozoan strains was achieved in the late 1940s, but resistance soon developed on the part of both the protozoans and their mosquito hosts, a development that discouraged many leaders of international programs.

## CILIATA

The ciliates are regarded as the most highly evolved of all protozoans. The complicated organellar structure encountered in this phylum is unmatched by that of other protozoans. More than 7000 species are found in ponds and lakes as well as in the sea. They are, for the most part, free-living. The prototype and best known of the ciliates belongs to the genus *Paramecium* (see Fig. 29.2*a*). As their phylum name implies, the ciliates are covered with short hairlike cilia, which show a 9 + 2 microtubular infrastructure. Almost all ciliates possess a *micronucleus*, which is the repository for genetic information that will be passed along, and a *macronucleus*, which contains multiple copies of the genetic material found in the micronucleus. The exchange of genetic material between two paramecia, a process known as *conjugation*, involves the micronuclei (see Fig. 29.2*b*).

## THE OPALINIDA

The Opalinida form a relatively minor phylum of protozoans. Almost all are enteric parasites in nonmammalian vertebrates. Although ciliated, they are much simpler than the ciliates, which they superficially resemble.

In almost all species of Opalinida, two or more nuclei are found. However, there is no specialization among these nuclei. They are equivalent to one another, suggesting a more primitive stage in the evolution of multinucleate differentiation.

## 29.2  ALGAL PROTISTS

More than 25,000 species of plantlike protists are subdivided into six divisions of what were once called *algae*—a term used now only as a vestige from an earlier taxonomic system in which the algae were treated as part of the plant kingdom. Virtually all the members of this half-billion-year-old group are photosynthetic and occupy fresh- or saltwater habitats. Although most of these species are unicellular, the brown and red algae are multicellular; nevertheless, these two groups are classified as protists because they do not share the major characteristics of higher plants.

## EUGLENOPHYTA

This group of monocellular protists is not numerous but is intriguing because it consists of many heterotrophic forms. Only some members are photosynthetic, and even these autotrophic species may, under certain conditions, give rise to heterotrophic cells. The chloroplast of the euglenoids has often been cited as evidence for the endosymbiotic hypothesis (see Chap. 28). Most euglenoids are extremely complex. *Euglena* (Fig. 29.3), the prototype of the group, possesses a triple membrane around its chloroplast, contains both chlorophyll *a* and chlorophyll *b*, and actively swims about by means of a

(a)

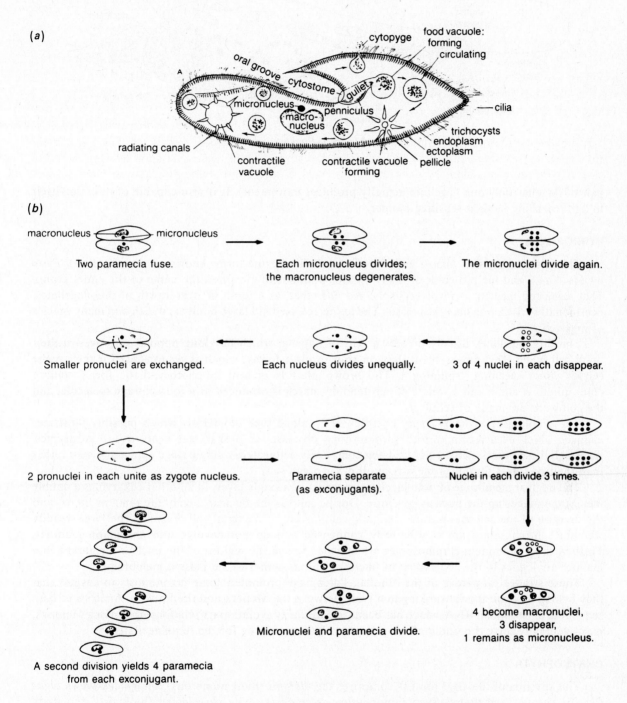

(b)

macronucleus ——— micronucleus

Two paramecia fuse.

Each micronucleus divides;
the macronucleus degenerates.

The micronuclei divide again.

Smaller pronuclei are exchanged.

Each nucleus divides unequally.

3 of 4 nuclei in each disappear.

2 pronuclei in each unite as zygote nucleus.

Paramecia separate
(as exconjugants).

Nuclei in each divide 3 times.

A second division yields 4 paramecia
from each exconjugant.

Micronuclei and paramecia divide.

4 become macronuclei,
3 disappear,
1 remains as micronucleus.

Fig. 29.2 The paramecium: (a) structure; (b) conjugation. (*From Storer et al.*)

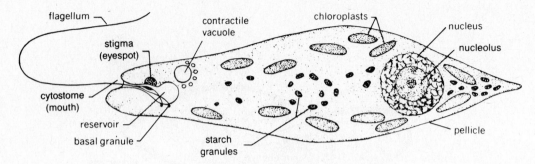

**Fig. 29.3** *Euglena.*

pair of flagella (only one flagellum actually produces movement). It is also capable of orienting itself to light by using its light-sensitive *eyespot.*

## PYRROPHYTA

This division consists almost entirely of unicellular marine forms known as *dinoflagellates. Pyrro* means "fire," and the red color of many species undoubtedly inspired the name of the entire group. That same red pigment is involved in the *red tides* that, as a result of overgrowth of dinoflagellates, occasionally arise along many seacoasts. The toxins released kill large numbers of fish and many species of invertebrates.

The dinoflagellates usually possess a pair of flagella, which lie along opposing grooves in their thick cell walls. This arrangement imparts spin in most forms, which is responsible for their name (Greek *dinos* meaning "spinning"). The brown plastids present in photosynthetic forms contain chlorophylls *a* and *c* and a variety of carotenoids; starch is produced as a food storage molecule, and the cell walls are made of cellulose.

The dinoflagellates have a very complex and unique type of meiosis which possibly illustrates changes which occurred in various chromosomal processes as prokaryotes evolved into eukaryotes. Although these processes have not yet been completely delineated, a single stage appears to exist rather than the two stages found for meiosis in all other organisms.

The nuclear membrane of dinoflagellates consists of a single layer; in all other eukaryotes a double membrane makes up the nuclear envelope. During mitosis the nuclear membrane remains intact, and the division of the cell that occurs is slightly reminiscent of bacterial cell division. The large, readily stainable chromosomes are continuously condensed and do not devolve into chromatin granules. Further, these prominent chromosomes are attached to specific regions of the nuclear membrane in a manner analogous to the attachment of bacterial chromosomes to the plasma membrane.

These strange properties of the dinoflagellates have prompted some taxonomists to suggest that they belong to some special subkingdom lying between the Monera and the Protista. Analysis of base sequences of ribosomal RNA, which has been used to clarify evolutionary relationships among Metozoa, may eventually solve the riddle of the proper evolutionary niche for the dinoflagellates.

## CHRYSOPHYTA

This division of the algal protists comprises the *diatoms* (most numerous), the *golden-brown algae* (less numerous), and the relatively minor *yellow-green algae.* Some *phycologists* (biologists who study algae) place diatoms in a separate phylum—*Bacillariophyta.* All members of this group contain chlorophylls *a* and *c* in their plastids, produce a yellow-brown carotenoid (*fucoxanthin*) that gives the cells their characteristic color, and store their food as fats, oils, and a unique polysaccharide called *laminarin.* The walls contain hydrated silica instead of cellulose. Diatoms are encased in a double shell, each half of which fits together like the top and bottom of a biscuit tin. Pores in the elaborately etched glassy shells afford communication between the interior and the environment.

## CHLOROPHYTA

*Chlorophyta* is an extremely diverse group of more than 7000 known species. Their chloroplasts contain chlorophylls *a* and *b* as well as the carotenoids usually found in higher plants. The Chlorophyta probably are ancestral to the plant kingdom. Most are freshwater forms, but they also exist on land and in the sea. They are the algal component of the *lichens*, a tight mutualistic union with fungi found in cold climates. *Chlamydomonas* is a typical chlorophyte (see Fig. 29.4). Complex sexual and asexual reproductive stages characterize the group. Both unicellular and multicellular forms exist.

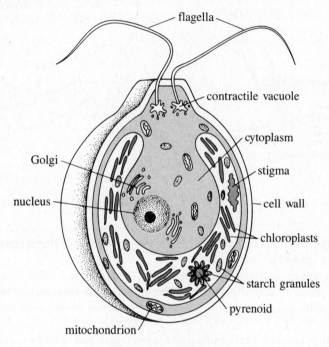

**Fig. 29.4** Chlamydomonas.

## PHAEOPHYTA

The *brown algae* are mostly multicellular and make up the seaweed population of temperate and cold ocean environments. Like Chrysophyta, they have chlorophylls *a* and *c* and the carotenoid fucoxanthin. They also store their calories as oils and the polysaccharide laminarin. In the form of giant kelps, they may reach a length of more than 50 m. An alternation of generations characterizes their reproductive pattern.

## RHODOPHYTA

The *red algae* consist mainly of seaweeds. They contain chlorophyll *a*, but never chlorophyll *b* or *c*. Their red color is due to the presence of an accessory pigment, *phycoerythrin*, belonging to the *phycobilin* group. All Rhodophyta are multicellular and reproduce sexually. Although few species have been studied in detail, alternation of generations has been commonly observed.

## 29.3 FUNGILIKE PROTISTS

Protists in this group consist of two heterotrophic groups of slime molds. The *Myxomycota* are *plasmodial slime molds* (see Fig. 29.5*b*). They are highly pigmented, amoeboid cells that alternate between a "multicellular" aggregate and individual cells. The aggregate stage is called a *plasmodium*.

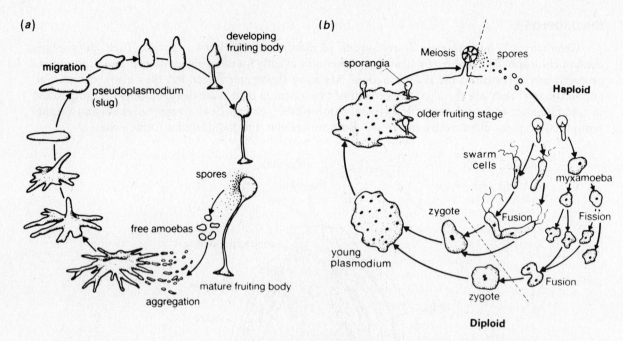

**Fig. 29.5** Slime molds. (*a*) Life cycle of a cellular slime mold; (*b*) life cycle of a plasmodial slime mold.

It consists of a large mass of cytoplasm within which are many nuclei, so that it is not really multicellular. Such an arrangement is termed a *coenocyte*.

The *Acrasiomycota* are *cellular slime molds* (see Fig. 29.5*a*). They differ from the Myxomycota in that their aggregation phase is actually multicellular rather than coenocytic. When food is in short supply, the individual cells aggregate but individual membranes persist and each cell can be distinguished.

The *Oomycota*, which include water molds and some rusts and mildews, are superficially similar to fungi. Because their cell walls are made up of cellulose rather than chitin, they are grouped within the protists. The presence of flagella in this group also distinguishes them from the true fungi, as do the dominance of a diploid phase in their life cycle and their production of ova. The potato blight that wrecked the farm economy of Ireland in 1848 was caused by an oomycete.

# Solved Problems

**29.1**   Criticize the statement that the cells of Protista are simpler than those of the higher plants or animals.

Actually, protistan cells are far more complex than their counterparts among the *Metaphyta* (multicelled plants) and *Metazoa* (multicelled animals). This situation should not be unexpected, since the single cell among Protista must carry out all the functions of an independent organism. Cells in plants and animals need not be involved in all the functional properties of the entire organism, since cells can specialize.

The most complex of all cells are found among the protozoans. Specialized structures exist for ingestion and digestion, motility, water balance, and reproduction. The ciliates probably demonstrate the most complex organelles, some of which are unique to that group.

The many variations of cell division found among the Protista illustrate the complexity of protistan cells. Mitosis usually occurs, but it varies greatly in a way that is not paralleled in any other kingdom.

Even the replication of chromosomes shows unique patterns in some algal protists. All protists are capable of asexual reproduction, but most show sexual patterns as well. Exchanges of genetic material are common, often involving patterns of meiosis similar to meiosis in multicellular organisms.

**29.2**    Despite the great heterogeneity of representative protistans, are there features that all members share?

Common traits exist in specific structures (organelles), particular habitat requirements, and metabolic patterns. All protistans are made up of eukaryotic cells. They almost all possess flagella (or cilia) of a 9 + 2 microtubular pattern during some portion of their life cycle, although a few use ameboid movement instead. None can survive in completely dry environments except as inactive, resistant *cysts*. All manner of moist environments teem with active protistan organisms.

In algae and protozoans an aerobic metabolic apparatus exists, including both a Krebs cycle and pigments and enzymes of the electron transport chain. All protists evidence patterns of asexual reproduction, but most also carry out sexual recombinations.

**29.3**    Is it correct to say that ciliates like *Paramecium* are more evolved or advanced than the amoeba (Sarcodina)?

Not exactly. Both types of organisms have been evolving toward greater adaptiveness for the same length of time and have had equal opportunities to adjust to the pressures of natural selection. Each of these organisms is quite successful in maintaining a foothold within its respective ecosystem.

However, the ciliated paramecium is more specialized and possesses a greater number of organelles than the amoeba. Its permanent organellar digestive tract has a grooved mouth lined with cilia, with the ability to produce food vacuoles at its terminus, and even an anal pore into which to expel undigested material. The amoeba makes no provision for sexual recombination; the paramecium has added conjugation and its consequent recombination to its bag of genetic tricks. The presence of a micronucleus and a macronucleus is further evidence of increased specialization. Clearly, the cilia of paramecium afford a more diverse and rapid means of locomotion than amoeboid motion. Both organisms are capable of avoidance behavior (moving away from noxious materials in the environment), but the paramecium responds both more quickly and in a more precise manner.

Perhaps the most dramatic difference between the two forms is the elaborate system of *trichocysts*, which may discharge long, spearlike threads into the environment. These threads play a minor role in defense and may serve to anchor the paramecium to its substratum.

**29.4**    Why are reproductive structures so well developed in parasitic forms in comparison with the mechanisms for motility or sensing environmental alterations?

A successful parasite must live in or on a host, usually in a particular region of its victim. Once established in such a neighborhood, the parasite does not need to wander or constantly cope with rivals. Instead, it must resist the defenses of the host, hold tight, and flourish with the assistance of prepared materials derived from the host. Intestinal parasites, such as flatworms and roundworms, often lose their digestive capacity, since they subsist on the already-digested foodstuffs of the host. At the same time, motility is not useful in a circumstance in which holding on is the key to survival. Thus, in the course of adapting to the parasitic way of life, many regressive changes occur in the body (or cytoplasm) of the parasite.

On the other hand, given the restrictive habitat of the parasite, sexual activity is severely constrained. For many parasites, the solitary mode of existence precludes the selection of a mate. Limitations also exist for the dissemination of zygotes once fertilization has occurred. This may explain, in part, the high frequency of *hermaphroditism* (possession of male and female sex organs by an individual animal) among parasites that have only limited access to sexual partners. Perhaps most striking among parasites is the heavy commitment to sexual activity. This may involve large reproductive structures, production of great numbers of eggs, and the existence of intermediate hosts in which alternate forms of the parasite may develop during a complex life cycle. Evolution toward a heightened reproductive capacity characterizes those successful parasites which must contend with an environment which limits their reproductive opportunities.

**29.5**   The volvocine series consists of a group of related chlorophytes. Some species of this series are made up of single cells; others, of many cells. The individual cells of both forms are strikingly similar to one another. The unicellular prototype of this series is *Chlamydomonas* (see Fig. 29.4), which is remarkably like the cells of the more advanced multicellular forms of the series.

In *Pandorina*, aggregations of from 4 to 32 individuals embedded in a gelatinous matrix make up the spherical colonial unit. All cells of the colony reproduce either sexually or asexually in a synchronous manner. In sexual reproduction, female gametes tend to be larger, but there is little other evidence for specialization of form or function.

*Pleodorina* consists of colonies with many more cells than *Pandorina*. In addition, a group of four or more smaller cells performs vegetative functions only, while the larger cells of the colony are both reproductive and vegetative. A clear-cut specialization of both cell size and function thus exists.

The most complex representative of the volvocine series is *Volvox*. Here we encounter a large spherical colony consisting of thousands of cells. As in the less complex members of the series, the flagella of each cell are oriented toward the surface, so that the entire colony demonstrates motility. The cells of the colony are connected by thin cytoplasmic strands that allow communication. Relatively few of the many cells are reproductive, and the mode of reproduction is *oogamous* (production of eggs).

Each of the members of the volvocine series is a separate extant genus. Other genera which have been included in that series are *Gonium*, a primitive colonial form in which flagellated gametes are all similar in structure and function (*isogamy*), and *Eudorina*, in which a fair degree of specialization distinguishes the cells of different portions of the colony.

Why do you suppose the volvocine series is considered so important?

The various genera of the volvocine series range from single-celled to multicelled animals, and the multicellular forms show the rudiments of specialization and integration. Specialization in *Volvox* is so advanced that some taxonomists consider it to be a multicellular plant. Because of these features, many algologists feel that the series illustrates the steps involved in the evolution of multicellular plants from single-celled organisms. This is not to say that the series contains the actual links in this evolution; its diverse members merely provide examples of what the actual intermediate forms may have been like.

**29.6**   The Rhodophyta grow at deeper levels of the ocean than other kinds of algae. Why? (*Hint:* Blue light penetrates water more effectively than red and yellow light waves do.)

The accessory pigments of red algae, particularly their phycobilins (phycoerythrin), tend to absorb blue light. (Their red color is due to the fact that they absorb the blues but reflect the reds.) Since blue light, with its higher-energy, short wavelengths, tends to penetrate to greater depths than reds and yellows, the red algae can carry out photosynthesis at great depths. In some warm ocean regions, red algae grow to depths of almost 300 m. Representatives of the phylum may vary in pigmentation and flourish at various depths. It is little wonder then that the Rhodophyta are among the most numerous of the seaweeds.

**29.7**   Why are the two kinds of slime molds and the Oomycota usually classified as protists although they so closely resemble the fungi in their physical appearance?

Most taxonomists believe that the resemblance of these three protists to the fungi is due to convergence in evolutionary development, much like the wings of an insect (epithelial) and the wings of a bird (forelimb modification). A better and more fundamental basis for the assignment of an evolutionary relationship is the nature of the cell wall. In almost all fungi the cell wall is made up of *chitin*, an aminopolysaccharide. In the three groups of fungilike protists the cell walls are usually made up of cellulose. Except for yeast, all fungi are clearly multicellular. The slime molds are not definitively multicellular. The plasmodial slime molds are coenocytic with no clear cellular partitions. In the case of the cellular slime molds (*Acrasiomycota*), the cells tend to maintain their individual identities, even while they are aggregated, and the cells are always haploid, which is quite different from the alternation of diploid stage with haploid stage found in most fungi. In Oomycota prominent flagellated zoospores occur during the life cycle. In fungi, flagellated cells are absent.

No final disposition is really possible in terms of classification, but the bulk of the significant evidence seems to justify a protistan assignment for these three funguslike forms.

# Supplementary Problems

**29.8**   Match the individual organisms in column **A** with their appropriate phyla in column **B**.

|  | A |  | B |
|---|---|---|---|
| 1. | Primitive flagellates | (*a*) | Sporozoa |
| 2. | *Amoeba proteus* | (*b*) | Sarcodina |
| 3. | *Plasmodium falciparum* | (*c*) | Ciliata |
| 4. | *Paramecium aureus* | (*d*) | Mastigophora |
| 5. | Foraminifera |  |  |
| 6. | Trypanosomes |  |  |

**29.9**   Most protozoans are photosynthetic.
(*a*) True  (*b*) False

**29.10**   A 9 + 2 arrangement of microtubules is found only in the cilia of bacteria.
(*a*) True  (*b*) False

**29.11**   The male *Anopheles* mosquito synthesizes an anticoagulant to keep the blood of its prey flowing during feeding.
(*a*) True  (*b*) False

**29.12**   Malaria is caused by only one species of *Plasmodium.*
(*a*) True  (*b*) False

**29.13**   Most of the Opalinida are parasites living in the intestine of lower vertebrates.
(*a*) True  (*b*) False

**29.14**   Conjugation is a type of asexual reproduction.
(*a*) True  (*b*) False

**29.15**   *Euglena* is always heterotrophic.
(*a*) True  (*b*) False

**29.16**   Dinoflagellates are always motile.
(*a*) True  (*b*) False

**29.17**   Fucoxanthin is found in diatoms and brown algae.
(*a*) True  (*b*) False

**29.18**   Most chlorophytes (green algae) reproduce by means of flagellated gametes. A well-known exception is
(*a*) *Gonium.*  (*b*) *Chlamydomonas.*  (*c*) *Spirogyra.*  (*d*) *Volvox.*  (*e*) yeast.

**29.19**   All photosynthetic Protista possess  (*a*) laminarin.  (*b*) chlorophyll *a.*  (*c*) fucoxanthin.  (*d*) chlorophyll *d.*
(*e*) all of these.

**29.20** Protoplasmic streaming occurs within the aggregate phase of plasmodial slime molds. This helps in (*a*) distribution of materials. (*b*) locomotion. (*c*) scaring predators. (*d*) all of these.

# Answers

| | | | |
|---|---|---|---|
| **29.8** | 1—(*d*); 2—(*b*); 3—(*a*); 4—(*c*); 5—(*b*); 6—(*d*) | **29.15** | (*b*): *sometimes* heterotrophic |
| **29.9** | (*b*) | **29.16** | (*a*) |
| **29.10** | (*b*) | **29.17** | (*a*) |
| **29.11** | (*b*) | **29.18** | (*c*) |
| **29.12** | (*b*) | **29.19** | (*b*) |
| **29.13** | (*a*) | **29.20** | (*a*) |
| **29.14** | (*b*) | | |

# Chapter 30

# The Kingdom Fungi

Fungi are placed in a separate kingdom on the basis of several distinct characteristics. They are all eukaryotic, heterotrophic, and, except for yeasts, multicellular (or multinucleate). They obtain their food by absorption rather than by ingestion. They secrete their digestive enzymes outside their bodies and then absorb the products of digestion produced outside. Most fungi possess cell walls made of *chitin*, an amino-containing polysaccharide mentioned in Chap. 29. They all lack flagella and are restricted in terms of motility.

Yeasts are unicellular fungi believed to have developed from multicellular ancestors. Molds and mushrooms are other examples of fungi. The fungi are at least 400 million years old.

## 30.1 BASIC STRUCTURE OF FUNGI

Fungi consist of a tangled mass of multibranched threads called *hyphae* (see Fig. 30.1). These threads, or filaments, are only incompletely subdivided into separate cells by walls (*septa*), which are scattered throughout the hyphal network at right angles to the long axis of the hyphae. In most fungi the septa are porous and permit cytoplasmic flow from one "cell" to another. In other groups the nuclei are scattered through a continuous mass of cytoplasm—a *coenocytic* structure. The entire filamentous mass is called a *mycelium*. Within the rapidly growing mycelium of parasitic fungi, specialized hyphae called *haustoria* often appear. In fungi that parasitize plants, these short processes penetrate the plant cells and rapidly absorb whatever nutrients are present there.

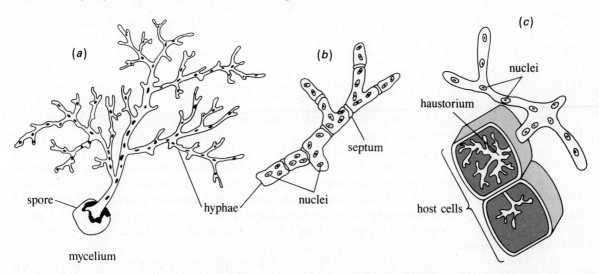

**Fig. 30.1** The major features of fungal structure: (*a*) mycelium; (*b*) septate hyphae with multinucleate cells; (*c*) coencytic hyphae with haustorium projecting into host cell.

## 30.2 DIVISIONS OF FUNGI

The more than 100,000 recognized species of fungi fall into four basic divisions. These divisions are roughly comparable to the phyla, which are the major groups within the animal kingdom.

The *Zygomycota* (conjugation fungi) occupy a terrestrial habitat, invading the soil or decaying organic matter. They usually produce asexual *spores* at the tips of specialized hyphae (*sporangiophores*) that extend into the air; these spores are carried by the wind to new territory.

**EXAMPLE 1**   *Rhizopus* is a mold that grows on bread. Three varieties of hyphae exist in *Rhizopus* (see Fig. 30.2). The *stolon* is a relatively thick filament that grows along the *substratum* (surface on which the mold grows) in a fairly straight pattern. *Rhizoids* are thin, highly branched hairs that penetrate into the interior of the bread and anchor the mycelium. The rhizoids, as a result of both their thinness and great surface area, are particularly effective in absorbing soluble materials formed along the surface. The third type of hypha is the sporangiophore, bearing a sporangium at its end. Each sporangium produces many thousands of *spores*, asexual cells that give rise to new mycelia. So many spores are produced within each sporangium that an entire slice of bread may be covered with new mycelial growth in a matter of hours.

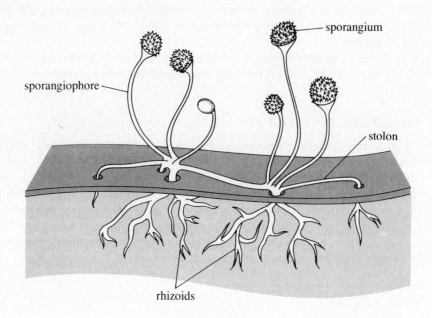

**Fig. 30.2**

The division is named for the tough resistant *zygospores* that are formed when haploid gametes fuse, usually under adverse conditions. Then, under more comfortable circumstances, the diploid zygospore, which is essentially a zygote with a tough coat, undergoes meiosis, and one or more of the meiotic products give rise to a new haploid mycelium. We will study this in more detail in the next section.

The *Ascomycota* (sac fungi) include yeasts, some mildews, the *ergot* (which infects rye), and *Penicillium.* The very complex and often colorful *cup fungi* that decorate the ground of many rain forests are also part of this group. Separate cells with perforated septa are common. The name of the group comes from the presence of a reproductive sac called an *ascus*, which is formed during the sexual cycle of all its members.

The *Basidiomycota* (club fungi) include the very conspicuous mushrooms and toadstools and a variety of puffballs. Here, too, the hyphae are compartmentalized by septa.

In most representatives an extensive underground hyphal mass sporadically sends up vertical hyphal fruiting bodies in which the spores are formed. The reproductive structures found in the fruiting bodies are club-shaped, which gives the group its name.

A separate catchall group known as the *Deuteromycota* comprises all the forms in which a sexual cycle has not been discovered.

The lichens are also often separated from other fungi. These intimate mutualistic associations of an algal or cyanobacterial autotroph with a fungus are of great ecological significance. Generally, in temperate or frigid zones the fungal component is an ascomycote; in warm climates the fungus may be a basidiomycote.

## 30.3  REPRODUCTIVE STRATEGIES OF THE FUNGI

Most fungi are haploid through much of their life cycle. When conditions for growth are adequate, reproduction is asexual. Haploid spores form in the sporangia of the sporangiophores. Since these hyphae are haploid, no meiotic process is necessary to produce the many spores that form within the sporangia. When the sporangia burst, the spores are borne by air currents, water, or animals to near or distant sites where suitable substrata may provide opportunities for new mycelial growth.

Sexual reproduction usually occurs when food supplies are low or optimal conditions of moisture and temperature do not exist. In the Zygomycota specialized hyphae from two mycelia of different mating types (+ and −) grow toward each other, forming a *conjugation bridge* (see Fig. 30.3). Each tip separates as a gamete-producing cell (*gametangium*) but remains attached to its parent hypha. The gametes fuse to form a zygote (the zygospore) in the middle of the conjugation bridge. The zygospore is surrounded by a thick, spine-covered coat. It usually separates from each of the parent mycelia and may lie dormant for a full season. When reawakened, the zygote undergoes meiosis and one of the haploid cells forms a short hypha, which quickly gives rise to a sporangium at its tip. The spores formed within the sporangium give rise to new mycelia.

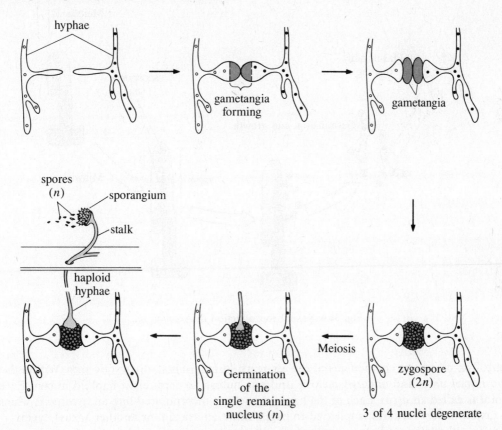

**Fig. 30.3** Conjugation in *Rhizopus*.

In Ascomycota, asexual reproduction occurs through the creation of spores called *conidia*, which essentially bud off from *conidiophore* hyphae of the parent plant and form new mycelia on germination. In sexual reproduction (see Fig. 30.4), a plus (female) and a minus (male) strain each form a bulbous, multinucleate body (called an *ascogonium* in the plus strain and an *antheridium* in the minus strain). A cross bridge then forms from the ascogonium to the antheridium. The bridge allows migration of male nuclei into the female organ, which then develops hyphae that are dikaryotic (containing two different nuclei). These hyphae intertwine with haploid hyphae of both parents to form a cuplike

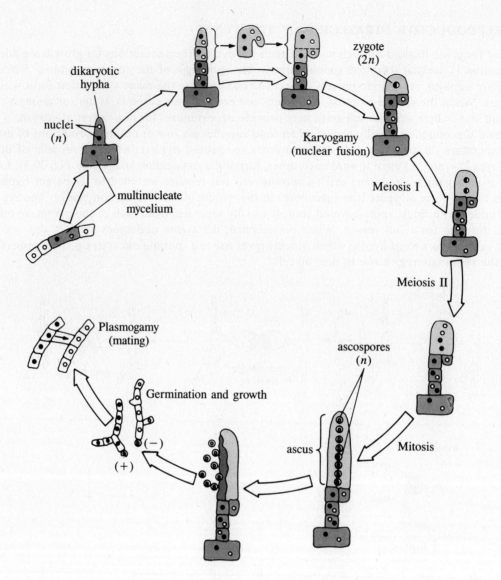

**Fig. 30.4** Sexual reproduction in *Ascomycota*.

*ascocarp.* The dikaryotic components of the ascocarp form terminal, dikaryotic sacs. Within these sacs, the two nuclei unite and undergo meiosis and one mitosis to form eight haploid nuclei. Each sac at this point is called an *ascus.* Each of the haploid nuclei is incorporated into an *ascospore,* which, when the ascus ultimately ruptures is released to begin either an asexual or another sexual cycle.

Yeasts, although unicellular, are grouped with the Ascomycota. Rather than formation of conidia, their asexual reproduction involves *budding,* in which a new, smaller cell pinches off from the single parent cell. A single yeast cell may also fuse with another of opposite mating strain to form a diploid cell. The fused nuclei then undergo meiosis usually followed by a mitotic division in which eight ascospores are produced. This structure in yeasts is equivalent to the ascus with its eight ascospores formed during the sexual phase of other Ascomycota.

In mushrooms and other Basidiomycota there is no separate asexual phase in the reproductive cycle (Fig. 30.5). The dense underground mycelial layer gives rise to the compact and dense reproductive *fruiting body,* which we identify as the mushroom, toadstool, or puffball. Club-shaped *basidia* line the *gills,* or membranous partitions of the fruiting body, and within the basidium a fusion of haploid cells

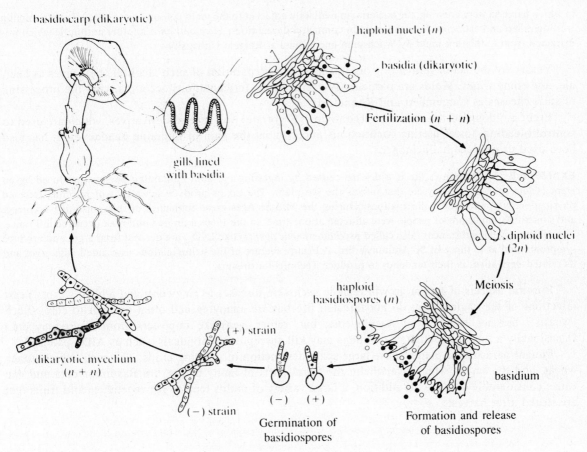

**Fig. 30.5** The life cycle of a mushroom.

produces a zygote. Shortly after fertilization, meiosis occurs, and each meiotic product becomes a *basidiospore*, which leaves through a special extension of the basidium to form new mycelia.

The Deuteromycota do not demonstrate a sexual phase in their reproductive cycle. They are classified in a single group, but probably they are quite heterogeneous.

In lichens, a fungal spore containing an algal cell embedded in a tangle of hyphae breaks away from the main mass of the lichen. It germinates to form a new spreading mass of intimately connected algal and fungal cells. Sexual reproduction is unknown in the lichens.

## 30.4  FUNGI AS FRIENDS AND FOES

Along with bacteria, fungi are the most effective decomposers within ecosystems. They not only attack dead material but also break down feces and processed material that could clog vast tracts of the terrestrial environment if not broken down.

Many fungi maintain tight mutualistic relationships with plants. They invade the roots and send hyphae into the soil, substantially increasing the absorption of both water and minerals by the plant. The fungus profits from the carbohydrate and other nutrients provided by the plants. These associations of fungi and root cortex are named *mycorrhizae*; their extensive distribution significantly increases the density of plant growth in most terrestrial ecosystems.

A variety of fungi produce antibiotics that can be extracted and used to fight bacterial infections.

**EXAMPLE 2**   *Penicillium chrysogenum* is one of several species of fungus that synthesize and secrete the antibiotic *penicillin.* Alexander Fleming, in 1928, first noted that when a *Penicillium* mold accidentally invaded a petri dish in which bacteria were growing, the bacteria immediately adjacent to the mold colonies were killed. Later penicillin was identified and extracted from the mold in commercial quantities. *Streptomycin* is another antibiotic which was extracted from a different mold by Wachsman and Schatz at Rutgers University.

Yeasts are the major fermenting organisms in the preparation of such alcoholic beverages as beer, ale, and some wines. Molds are particularly useful in the fermentations accompanying the processing of such cheeses as Camembert and Roquefort.

Ergot mold is also the source of ergonovine (Ergotrate) and several derivatives, which are used to control bleeding, foster uterine contractions and combat the pain of migraine headaches. It has also been used to produce hallucinogens.

**EXAMPLE 3**   St. Anthony's fire is a disease caused by ingesting rye flour containing poisons produced by an ergot (*Claviceps*), an ascomycete that infects the rye plant. The toxins produce sensations of burning, frenzied physical activity, and wild hallucinations. During the Middle Ages great epidemics of this poisoning occurred, and sometimes thousands of people were afflicted at one time. In the 1960s a limited outbreak occurred in France.

*Hallucinogenic* substances (also called *psychotomimetic agents*) like $LSD_{25}$ are derived from ergot and produce symptoms similar to those of St. Anthony's fire. A bizarre feature of the hallucinations associated with ergot and its related derivatives is their tendency to produce Technicolor dreams.

Harmful aspects of fungal growth include such skin diseases as ringworm and athlete's foot. Yeast infections of the vaginal tract do not threaten life but are annoying and often difficult to cure. (Such "yeasts" are not members of Ascomycota, but, rather, yeastlike imposters from Deuteromycota.) Occasionally a fungal infection of the lungs may kill susceptible individuals, such as AIDS patients.

Fungal parasites of plants cause large-scale destruction of growing crops. The *rusts*, which attack wheat, and the smuts, which overwhelm the floral parts of many plants, are Basidiomycota and can cause considerable damage. In addition, a large variety of molds tend to rot vegetables and fruits that are stored after harvest.

# Solved Problems

**30.1**   Many fungi resemble protists or even plants. What is the justification for placing fungi in a separate kingdom?

At one time the fungal forms that taxonomists have currently placed in a separate kingdom were grouped as a division of the plant kingdom, the *true fungi*. However, fundamental differences distinguish the fungi from both plants and protists. All fungi are heterotrophs, which obtain their food by absorption from outside. Although many fungi do store some cellulose in their cell walls, the predominant cell wall constituent is almost always chitin, clearly distinguishing fungi from the plants. Unlike many plants and some colonial protists, the fungi do not have flagellated zoospores, nor are the vegetative cells of the mycelium motile. The rapidly spreading growth of the mycelium perhaps compensates for the lack of a locomotor capacity.

Although most of the members of the kingsom are multinucleate, the partitions (*septa*) dividing the cytoplasmic compartments are usually imperfect, so that a continuity of cytoplasm exists. This permits a rapid transport of materials from regions where food is being absorbed to the hyphal termini, where rapid growth takes place.

**30.2**   Some taxonomists have classified the water molds (Oomycota) as fungi. In this book we have placed them among the Protista. Why?

Perhaps the strongest argument for excluding the water molds from the kingdom Fungi is the absence of chitin in their cell walls. Chitin is found in all other fungi. Atypical, too, is the dominance of the diploid phase of their life cycle. Almost all fungi are predominantly haploid.

No fungal group contains flagellated zoospores, while the water molds do present this feature. They also produce eggs, which is not a characteristic of fungi. Also, septal compartments are most inconspicuous in their mycelia.

For these reasons, the water molds, which have done such serious damage to potato crops and vineyards, are best regarded as protists. Their multicellular (actually, *coenocytic*) character, however, is an argument against a protistan assignment.

**30.3**   State whether the following structures are haploid, diploid, or dikaryotic: mycelium, hypha, gametangium, zygospore, sporangium, basidium, gills of mushroom, stolon, sporangiophore, conidium, ascocarp (excluding sac), ascocarp sac.

| | |
|---|---|
| Mycelium | Haploid |
| Hypha | Haploid |
| Gametangium | Haploid |
| Zygospore | Diploid |
| Sporangium | Haploid |
| Basidium | Diploid |
| Gills of mushroom | Dikaryotic |
| Stolon | Haploid |
| Sporangiophore | Haploid |
| Conidium | Haploid |
| Ascocarp (excluding sac) | Haploid and dikaryotic |
| Ascocarp sac | Diploid |

**30.4**   Which probably contributes more to the evolutionary survival of Ascomycota—the conidia or the ascospores?

Ascospores, though haploid, are the result of sexual union between the hypha of two different sexual strains; the conidia are formed asexually. Because of the genetic recombination inherent in sexual reproduction, the ascospores impart greater genetic diversity to Ascomycota and, therefore, greater adaptability (and survivability) in the face of environmental stress.

**30.5**   In the *dikaryotic* condition two unfused nuclei lie within the same cell. This is a typical condition encountered in the reproductive tissue of many fungi just before *syngamy* (union of nuclei or union of cells). If the two unfused nuclei are genetically different, the cell is heterokaryotic. What term would be used if a single haploid nucleus divided to form two identical, unfused nuclei?

The resultant cell would be said to be *homokaryotic.*

**30.6**   Suggest some ways fungal diseases might be fought.

Some fungi, such as *wheat rust*, have two intermediate hosts. When one of the hosts is economically important and the second is not, destruction of the second host may be used to halt the spread of the fungus. Complete destruction of all the members of a particular species is necessary for this technique to be effective, but this is extremely difficult. In the case of wheat rust, destruction of the barberry plant is called for.

Using sterilized soil and disinfecting seeds before planting may stem the spread of fungal disease. In cases in which insects are known to carry fungal diseases, all-out war against the insect vector with insecticides or biological agents is effective.

Dusting with copper salts or other fungicides is an effective technique to halt the spread of infestation in crops. In fungal infections in humans, such potent preparations as *griseofulvin* have eliminated serious conditions. Since fungi prefer warm, moist conditions, maintaining a dry environment at low temperatures could inhibit fungal growth.

Plant geneticists have developed resistant plant strains, but fungi have also altered their genetic makeup and may, in the course of evolution, regain the upper hand.

**30.7**  Most fungal spores, whether sexual or asexual in derivation, are disseminated by wind. Most algal zoospores are disseminated by water. Why should this be the case?

    Fungal spores are not flagellated and have no means of locomotion. Their passive character requires relatively rapid air currents for their dispersal. The zoospores of algae are usually flagellated. They may move quite effectively and rapidly through water. Also, most algae dwell within lakes, ponds, or oceans, whereas the common fungi are generally terrestrial.

# Supplementary Problems

**30.8**  A major difference between algae and most fungi is that the former possess   (*a*) chloroplasts.   (*b*) cellulose cell walls.   (*c*) pseudopodia.   (*d*) both *a* and *b*.   (*e*) haustoria.

**30.9**  The above-ground portion of the Basidiomycota (mushrooms) is   (*a*) diploid.   (*b*) autotrophic.   (*c*) a sexual structure.   (*d*) both *a* and *b*.   (*e*) both *a* and *c*.

**30.10**  Mycelia consist of a tangle of   (*a*) hyphae.   (*b*) pseudopods.   (*c*) cell walls.   (*d*) sporangia.   (*e*) none of these.

**30.11**  Chitin structurally resembles cellulose but substitutes an *N*-containing group for one —OH group.
(*a*) True   (*b*) False

**30.12**  Mycelia generally tend to grow slowly.
(*a*) True   (*b*) False

**30.13**  Fungal spores, whether sexual or asexual, generally are surrounded by a tough coat and are distinctly colored.
(*a*) True   (*b*) False

**30.14**  LSD tends to antagonize the action of serotonin. This may explain the psychedelic influence of this ergot derivative, since serotonin is a known neurotransmitter.
(*a*) True   (*b*) False

**30.15**  The asexual spores known as *conidia* are extremely large.
(*a*) True   (*b*) False

**30.16**  Unleavened bread (matzoh) is made without   (*a*) flour.   (*b*) water.   (*c*) salt.   (*d*) yeast.   (*e*) mushrooms.

# Answers

| | | | |
|---|---|---|---|
| **30.8** | (*d*) | **30.13** | (*a*) |
| **30.9** | (*e*) | **30.14** | (*a*) |
| **30.10** | (*a*) | **30.15** | (*b*); their name comes from a Greek root meaning "dust" |
| **30.11** | (*a*) | **30.16** | (*d*) |
| **30.12** | (*b*) | | |

# Chapter 31

# The Kingdom Plantae

Plants are multicellular, autotrophic organisms that have successfully invaded terrestrial environments. They probably arose from the algal division Chlorophyta. Opportunities for active photosynthesis were considerably enhanced when plants established themselves on land.

**EXAMPLE 1** Light is more available on land, where it is filtered only through the atmosphere, than it is in the often churning and murky waters of lakes and seas. Another vital ingredient for photosynthesis, $CO_2$, is also much more abundant and readily absorbable on land than in the water. Virgin invasions of land were probably highly successful for these reasons, especially at a time when few enemies of the green plants had gained a foothold in terrestrial habitats.

Transition stages from water to land probably occurred at or near the shoreline. There evaporation concentrated minerals, leaving rich deposits. On the shores of rivers, *alluvial* (riverborne) deposits also provided a rich matrix for the development of new plant forms.

Life on land required new styles of reproductive activity. Films of water allowed some plants to cling to the old aquatic reproductive strategies, but as plants migrated to drier regions, these strategies had to be modified (see Chap. 11).

Maintenance of the plant on land also required protection against desiccation.

**EXAMPLE 2** The outer cell layer (epidermis) of all plant bodies is completely covered by a layer of waxy material, *cutin*. Although cutin is secreted by individual epidermal cells, the cuticle is not subdivided but exists as a continuous waxy layer. The water-resistant cuticle is generally thicker in exposed areas, such as the upper surface of a leaf, than in less exposed areas, such as the lower surface of the leaf.

More than 400 million years ago the ancestral forms of today's land plants began to spread through the terrestrial environment. As they moved into major land areas, they evolved adaptations that facilitated their survival. A major division into two separate lineages occurred early in the colonization of the land. One group was the *bryophytes*, and the other, far more numerous in the modern era (Cenozoic), the *tracheophytes* (vascular plants).

## 31.1  BRYOPHYTES AND THE CHALLENGE OF A TERRESTRIAL ENVIRONMENT

Bryophytes consist of three extant groups: *mosses*, *liverworts*, and *hornworts*. About two-thirds of all species of bryophytes are mosses. Although the bryophytes have developed some protective structures (see Fig. 31.1), they are not eminently suited to a terrestrial existence. They require moist environments, particularly for their reproductive cycles. Although they do have processes that resemble the roots of higher plants, and even demonstrate chlorophyll-laden scales, they are clearly only a stage in the evolution of the higher land plants. The bryophytes are much more common in warm climates than in cold, or even temperate, regions.

The mosses are both more numerous and more visible than other bryophytes (*bryo* means "moss"). Lacking the internal support structures of the higher plants, mosses tend to spread extensively but grow very close to the ground. Like other bryophytes, they show a dominant gametophyte (haploid) generation and a dependent sporophyte (diploid) generation (see Fig. 31.1). Most mosses are *dioecious* (have separate sexes) but some are *monoecious* (have both sexes in the same *thallus* or plant body).

The liverworts are named for their flat, lobular appearance, resembling the lobes of the liver, *wort* means "herb" or "plant." *Marchantia* is a liverwort that shows clearly the partial adaptation to terrestrial life of the bryophytes (see Chap. 11, especially Fig 11.7).

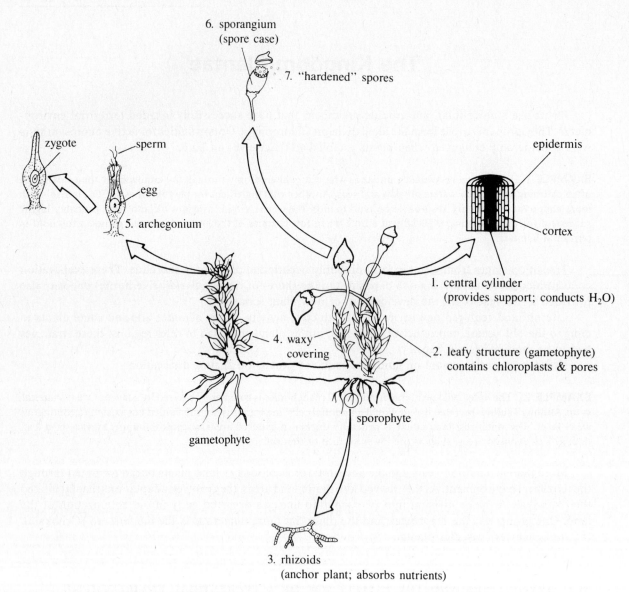

**Fig. 31.1** Some adaptations of bryophytes to terrestrial environments.

**EXAMPLE 3**  *Rhizoids*, single hairlike cells, extend from the ribbon-shaped thallus of *Marchantia* into the ground, where they absorb water. Motile sperm that must swim through a layer of water to the egg are produced, so moisture is a necessary ingredient in the fertilization process. Both sperm and egg are produced within special protective receptacles, a major advance over the algae. The sperm are formed within *antheridia*, and the eggs are produced in *archegonia*.

The *hornworts* are a minor group of bryophytes, possessing some features of higher plants. They get their name from the hornlike sporophyte that grows up from the flattened gametophyte.

## 31.2  THE VASCULAR  PLANTS

Vascular plants differ from the bryophytes in their greater adaptiveness to land environments. Unlike the thallus of a moss or a liverwort, the body of a vascular plant is divided into separate parts,

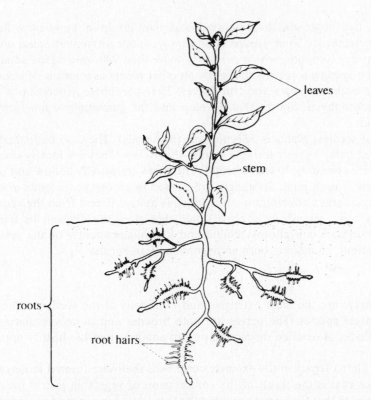

**Fig. 31.2** The basic differentiation of tissues in vascular plants.

or organs, which show specialization of both structure and function: *roots*, *stems*, and *leaves* (Fig. 31.2). In the seed-bearing plants, *cones* or *flowers* comprise special reproductive structures.

**EXAMPLE 4** The roots are an extensive system of ramified fibers that anchor the plant firmly in the ground. The slender *hairs* that extend from the ends of younger roots absorb water and minerals and bring these vital materials to the rest of the plant. This specialized root system supplies the above-ground structures of the plant with water to carry out photosynthesis. The stem affords an opportunity for the plant to grow upward, sometimes, as in the case of trees, to great heights. The vascular tissue within the stems conducts vital raw materials from roots to peripheral areas and brings synthesized food down from the leaves; it also provides the support that enables vascular plants to stand tall. The *leaves* of tracheophytes are ideally suited for receiving maximum exposure to the sunshine necessary to their role as primary photosynthetic agents for the entire plant.

Possible transition forms in the evolution of vascular plants from multicellular algae belong to the division *Psilophyta*. These plants were once abundant on the earth but now occur in only a few genera. They possess only some of the features of the vascular plants, but a primitive vascular system can be discerned. Some botanists regard the Psilophyta as forms that gave rise to other vascular types, but others consider them to be degenerate descendents of the more advanced types, such as ferns.

**EXAMPLE 5** *Psilotum*, an extant tropical genus, is a relatively simple form, which may have descended from the fossil psilophytes which died out about 350 million years ago. It has no true roots; its lower portion is a *rhizome*, a horizontal stem that lies along the ground and extends tiny rhizoids into the soil. What might be viewed as pseudo leaves are pairs of scales along the shoot. However, the presence of vascular bundles marks the genus as a vascular plant.

## CLUB MOSSES AND HORSETAILS

The club mosses, or lycopods, belong to the division *Lycophyta*. They share a peculiar characteristic with the Psilophyta—the gametophyte generation is nonphotosynthetic and must rely on symbiotic fungi for nourishment. Lycophyta was the dominant land plant group in the swamps that covered the earth about 300 million years ago. Some groups of ancient lycopods formed tall trees; nothing but their fossils remains. Present-day lycopods, which consist of more than 900 species, are small. The tropical representatives are often *epiphytes*—plants that grow on other plants as a means of structural support.

The lycopods have both true roots and true leaves. The sporophyte generation is dominant, the usual situation for tracheophytes. Spores that develop into the gametophyte are formed on special leaves called *sporophylls*.

Another division of seedless plants is *Sphenophyta* (horsetails). They too flourished in an ancient time but are represented today by only a single genus—*Equisetum*. They are mostly small, herbaceous (nonwoody) plants rarely exceeding 70 cm in height. The stems are usually hollow and jointed. Leaves typically form in whorls at each joint. Sporangia are carried in groups at the ends of a central stem. Such groups of sporangia form a *strobilus*, or *cone*, which is quite different from the sporophylls of the club mosses. It is this cone, resembling a horse's tail, that gives the division its name. The small gametophytes of the horsetails are photosynthetic and live independently of the sporophyte. Each gametophyte is monoecious, containing both archegonia and antheridia.

## FERNS

The ferns (*Pterophyta*) are the most extensive and numerous of the seedless plants, comprising more than 12,000 separate species. The leaves are both broader and more vascularized than in the lycophytes or sphenophytes. A detailed discussion of fern anatomy and life history appears in Chaps. 11 and 13.

The seedless plants left a legacy in the extensive coal beds that were formed largely more than 300 million years ago. Since coal is the result of the compression of vegetable matter beneath the earth, the great forests that grew at that time have provided the fossil fuel for our modern industrial furnaces. The time span during which coal was formed from those great tracts of seedless plant forests is labeled by geologists the *Carboniferous Period*.

## 31.3  SEED PLANTS

The development of a seed represented the height of adaptiveness to a terrestrial environment in the plant kingdom. In all seed plants not only is the sporophyte dominant, but the gametophyte is reduced to a dependent structure retained within the archegonium of the sporophyte. Further, the flagellated sperm of the lower plants is supplanted by a process of pollination. In pollination, the plant achieves an independence from water as a vehicle of fertilization. The zygote and developing embryo, which will become a new sporophyte, are also liberated from dependence on water, since the seed houses the embryo in a tough coat and provides other sources of protection for it.

**EXAMPLE 6**   The seed contains a partially developed sporophyte that has been arrested in its development. It is surrounded by stored food material and protected from a variety of environmental stresses by a tough integument (skin). The seed and its embryonic sporophyte can remain dormant for quite some time and then germinate (resume growth) when conditions are appropriate. Such properties in a reproductive structure considerably enhance the possibility for survival in the sometimes harsh terrestrial habitat.

The seed probably arose independently at different times in the evolution of plants. Some fossil "ferns" have even been found with structures that resemble the seeds of the *spermatophytes* (seed-bearing plants). Present-day seed plants are classified as *gymnosperms* and *angiosperms*.

## GYMNOSPERMS

The *gymnosperms* (from the Greek for "naked seeds") consist of four divisions. The earliest gymnosperms probably arose a little less than 400 million years ago. The four divisions of gymnosperms are quite distinct from one another, and some botanists regard their inclusion in a single group as artificial. The extinct seed ferns (*pteridosperms*) have also been considered a fifth division of this catchall taxon. Many taxonomists consider each of these divisions to constitute a class: *Cycadophyta, Ginkgophyta, Gnetophyta, and Coniferophyta.*

The Cycadophyta were particularly abundant during the age of the dinosaurs (Mesozoic Era). They may have arisen from the ancient seed ferns. The cycads that are present today bear a striking resemblance to the palms; their common name is *sago palm.* They generally grow in warm climates.

The Ginkgophyta survive today as a single species, the *ginkgo*, or maidenhair tree. It is very common as an urban ornamental tree because it is exceptionally resistant to air pollution (a great asset in large cities where both air and water may be quite polluted). Ginkgos are dioecious. Unfortunately, the female ginkgo emits an overpowering foul odor in the spring, which is related to its reproductive cycle. This characteristic strongly militates against its selection for a small household garden patch.

The Gnetophyta include three diverse groups of tropical or desert plants. *Gnetum* exists as a thick, upright vine; *Ephedra* and *Welwitschia* are desert shrubs.

The *Coniferophyta* (conifers) are the most conspicuous of the gymnosperms, especially in colder northern climates. Although particularly prominent during the Mesozoic Era, conifers continue to dominate many forests.

**EXAMPLE 7**   On the west coast of the United States are a group of trees known as *redwoods.* They may reach a height of 100 m, and their trunks may be 4 m in diameter. Some redwoods are reputed to be 1500 years old. The redwoods are gymnosperms belonging to the genus *Sequoia.* The giant sequoias are another species of this genus. Pines, spruces, firs, and hemlocks are among the familiar species. In almost all cases the leaves are considerably reduced in breadth to form needles or flattened scales.

The most prominent feature of the conifers is the reproductive structure known as the *cone.* Female cones tend to be larger than male cones. The pine, which is a typical conifer, usually has both male and female cones on the same tree. The life cycle of the pine is similar in many respects to the life cycles of the angiosperms, which were discussed in Chap. 11.

**EXAMPLE 8**   Each scale of a pine cone is a sporophyll that gives rise to a sporangium (also called a *nucellus* in the female). The sporangium of the male cone meiotically produces many haploid *microspores*, each of which becomes a hard-coated, winged pollen grain. Inside these pollen grains, the haploid nucleus divides mitotically, producing four cells. These four haploid cells constitute all there is of the male gametophyte (called the *microgametophyte*), and even two of these cells degenerate. The two remaining cells are called the *degenerative cell* nucleus and the *tube cell* (nucleus). Their functions will be discussed in a moment.

Each sporangium of the female cone is surrounded by an integument with a *micropyle* (a small opening) at one end (see Fig. 31.3). Inside the sporangium, four haploid *megaspores* are produced, three of which degenerate.

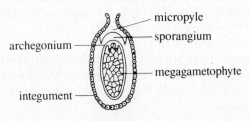

**Fig. 31.3**

Through repeated mitotic divisions inside the sporangium, the fourth megaspore creates the female gametophyte, a haploid, multicellular mass called a *megagametophyte.* This female gametophyte then produces several archegonia at its micropylar end. Each of these develops an egg cell. The combined integument, sporangium, and megagemetophyte are called an *ovule.*

When a pollen grain enters the female micropyle and contacts the sporangium, it develops a pollen tube, which penetrates the sporangium. The two pollen nuclei enter the tube, and the generative nucleus divides. (The tube cell is associated with growth of the pollen tube and does not divide.) One of the resulting daughter nuclei divides again, producing two sperm cells. When the pollen tube enters an archegonium, one of these sperm cells fertilizes the waiting egg. The zygote develops into an embryo inside the megagametophyte, which is still in the sporangium. Eventually, the ovule is released as a seed (composed of a seed coat derived from the integument, stored food from the megagametophyte, and the embryo). The endosperm, or stored food, in the seed is formed by fusion of the second sperm cell with a diploid endosperm mother cell.

## ANGIOSPERMS

The flowering plants (angiosperms), whose seeds are contained within a "vase" (the ovary), are the most successful and widespread of all plants. They arose more than 100 million years ago in the Cretaceous Period of the Mesozoic Era and became the dominant land plants of the Cenozoic, or modern, Era. Whereas the reproductive structures of the gymnosperms are cones which bear their ovules, or future seeds, on exposed scales of the cone, the reproductive structures of the angiosperms are flowers which completely enclose the future seeds. The longitudinal subsections of the pistil (the innermost structure of the flower containing the female reproductive structures) are called *carpels*. The carpels may have arisen as an infolding of sporophylls to make up the ovary, the style, and the stigma (see Chap. 11).

Close to 175,000 species of flowering plants exist. Those that are pollinated by insects tend to have large, magnificently colored petals and are often pleasantly scented. They have been extraordinarily successful in every possible terrestrial habitat and in lakes and ponds, but they are quite meagerly represented in the seas. Although almost all angiosperms are autotrophs, a few lack chlorophyll and survive as saprophytes. Some have become parasites.

Angiosperms that endure through only a single growing season are known as *annuals*. Those that grow through two growing seasons, such as the *carrot*, are dubbed *biennials*. *Perennials* last for long periods of time.

## 31.4  ECONOMIC IMPORTANCE OF PLANTS

Plants are a seemingly endless source of benefit to humans. The 250,000 species of angiosperms, in their roots, stems, leaves, and even flowers, provide food for almost all the animal world.

**EXAMPLE 9**  Among the *dicots* are the *Rosaceae*: roses, strawberries, cherries, and almonds. They include some of the most common fruits and ornamentals. Alfalfa, beans, clovers, peas, etc., make up the *Leguminosae* family. The bacterial nitrogen fixation that occurs in the roots of many of these plants marks them as vital to agriculture. This family also includes many of the vegetables and stock feeds that are a mainstay of the agriculture of developed nations. Citrus fruits are part of the *Rutaceae*. *Cruciferae* include mustard, cabbage, cauliflower, and broccoli. Gourds, pumpkins, squashes, cucumbers, and the citron belong to the *Cucurbitaceae*.

The most important *monocots* are the grasses, *Graminae*, a family with over 5000 species. This family includes bamboo, wheat, rice, maize, barley, and timothy. It is the single most important group of food plants. Other monocots include pineapples (*Bromeliaceae*) and date palms (*Palmales*).

Many of the flowering food plants are *herbs*, which are relatively small, soft-stemmed plants. Most of the herbs are annuals, but they are also found as biennials and perennials.

Many medicines are derived from plants. For example, quinine comes from the bark of the cinchona; digitalis comes from the foxglove.

Although synthetics have made great inroads in the manufacture of fabrics, natural fibers such as flax (from the family *Linaceae*) and cotton are still commonly used for clothing, drapes, upholstery, etc.

The decorative uses of plants range from vast landscaping designs, involving great tracts of trees, flowers, and shrubs, to a simple flower arrangement in a vase.

Wood is used in furniture making and construction, and use of wood as an alternative source of energy has seen a resurgence in recent decades in the United States in response to unstable oil prices. It is the only source of energy in many countries (a situation that has led to widespread deforestation, which in turn has precipitated devastating erosion and flooding). Coal itself is a plant product—albeit from ancient trees and more primitive tracheophytes.

Thousands of industrial chemicals are derived from the plant world—plastics, turpentine, tannin, and rubber, to name just a few. (Although it is now generally produced synthetically, rubber was originally made from the latex of the rubber tree.)

## 31.5  THE GREEN REVOLUTION

Toward the end of World War II Norman Borlaug, supported by the Rockefeller Foundation, began a series of genetic studies that led to the development of high-yield dwarf varieties of various grass grains. Eventually his work was expanded under the aegis of the Food and Agricultural Organization of the United Nations. Special varieties of wheat and rice were sent to those developing countries that were threatened by famine. The intense cultivation of those strains of agricultural staples dramatically increased food production in those regions. So successful was this program that it was named the *green revolution*. Food production in parts of Asia quadrupled following the green revolution.

Along with the breeding of strains that possessed large grains was the need to support growth with heavy applications of fertilizer. These fertilizers are generally synthetics derived from the processing of petroleum. Pesticides were also used in large amounts. Unless the farmers used great amounts of DDT and similar chlorinated hydrocarbon pesticides, the crops could be destroyed by droves of insects either during growth or after harvesting and storage in bins or silos. The green revolution has involved a total mobilization of machinery, chemicals, and human resources in achieving success. (See Prob. 31.7 for the problems the green revolution wrought.)

Hybridization studies of wheat, rice, and corn continue in such places as the International Rice Research Institute in the Philippines and the International Maize and Wheat Improvement Center in Mexico.

# Solved Problems

**31.1**  What is the evidence that all plants arose from the Chlorophyta (green algae)?

All plants share with the chlorophytes the presence of both chlorophyll *a* and chlorophyll *b*. They also all contain $\beta$-carotene as an accessory photosynthetic pigment. In almost all cases, starch is the principal storage carbohydrate. Unlike the fungi, chlorophytes and plants have cell walls made of cellulose, although other substances may be present in the cell walls of some plants.

Supplementing the biochemical similarities is the common pattern of cytokinesis. A cell plate is formed at the center of the dividing cell and ultimately divides the cytoplasm into two compartments. A fundamental characteristic such as this is not likely to arise independently in unrelated lineages. Also, the chloroplasts of both commonly demonstrate a stacked grana structure.

**31.2**  In the transition to a terrestrial habitat all plants developed along their surfaces a waxy cuticle, which is impervious to water. How then can plants exchange water vapor and other gases with the environment?

The surfaces of all plants are dotted with pores that open to the ambient (surrounding) environment and thus permit gas exchange. In the bryophytes these pores are always open, posing a danger of desiccation. It is this feature of bryophytes, together with the requirement that their flagellated sperm swim to the egg in the archegonium, that mandates that they grow only in moist environments.

In the tracheophytes the pores, called *stomata*, are surrounded by guard cells that periodically open and close. At a time when photosynthetic activity is high, the guard cells become taut as a result of the accumulation of sugars and the stomata remain open. This permits $CO_2$ and other gases to enter while oxygen (in excess of the plant's needs) may diffuse outwards. At night, when little photosynthetic activity occurs, the guard cells become flaccid and the stomatal slit is closed. This flexibility of open and closed stomata is one important factor in the ability of ferns and the seed plants to adapt to dry terrestrial habitats.

**31.3** Since bryophytes lack many of the features of the tracheophytes (vascular plants), how have they managed to survive on dry land?

The bryophytes can live only in relatively moist regions and have not been able to invade dry areas of the land. They are usually quite short, so that the risk of drying out in the air is reduced. Like other plants, they possess a water-insoluble cuticle that diminishes water loss. However, the pores that are scattered over the surface are open and provide a means for the absorption of water from the damp ambient air. In many mosses, the dominant gametophyte is spongelike in its ability to both absorb and hold on to water.

Although bryophytes lack true roots, stems, and leaves, they do have rudimentary structures that carry out some of the functions of these organs (which are found in the tracheophytes). Rhizoids are hairlike, single-cell processes that anchor the gametophyte to the soil. They may also play a minor role in water absorption.

Leaflike scales in the mosses and the lobes in liverworts contain chlorophyll and are major photosynthetic sites, thus providing some of the independence seen in higher plants. The thick body (thallus) of the bryophyte provides support for the plant and may also represent sites for food storage. The sporophyte usually shows a tendency to grow upward, but since it remains a dependent structure upon or within the gametophyte, it is not threatened by an independent need for food, water, and a sheltered growth site.

**31.4** Contrast ferns and gymnosperms in terms of adaptation to a terrestrial environment.

Ferns and gymnosperms are each members of *Tracheophyta*. They are vascular plants with the capacity to transport water, minerals, and organic substances to and from roots, stems, and leaves. This, of course, is true only for the fern sporophyte. The free-living gametophyte is as limited in its terrestrial adaptiveness as a bryophyte would be. In the gymnosperms no free-living gametophyte exists. Instead the gametophyte remains protected within the sporophyte. The formation of a winged seed, with its tough outer coat and internal supply of food, provides the gymnosperm with another advantage over the ferns in adapting to a terrestrial existence. Among the conifers, the formation of needlelike leaves prevents water loss, a distinct advantage in dry habitats.

**31.5** How do angiosperms differ from gymnosperms?

First, the seeds of the angiosperms, a more recently evolved and highly successful division of the tracheophytes, are enclosed within a protective chamber, the *ovary*. A ripened ovary containing seeds is called a *fruit*. The fruit not only encloses the seeds but may also aid in seed dispersal. The fruit is either carried or eaten by animals that migrate to distant sites, carrying the seeds with them; the seeds are ultimately dropped or eliminated from the animals' digestive tracts.

The seed also forms a little differently in angiosperms. One sperm nucleus from the pollen tube unites with an egg nucleus to produce the zygote. A second sperm nucleus unites with two haploid nuclei in the gametophyte (embryo sac) to produce the triploid endosperm, an important source of food within the seed for such seeds as corn.

In gymnosperms, pollination (transfer of pollen to female reproductive structure) can be only wind-borne. In flowering plants, pollen may be transferred by wind or by animals. Although insects have been stressed as prime agents for pollination, recent studies suggest that mice and other small mammals may also play a role, particularly in tropical plants.

A major internal modification of the angiosperms is the development of specialized xylem cells, the *vessels* and *fibers*, in addition to tracheids. The vessels are particularly significant because they are large-bore columnar cells that anastomose (join) end to end. When their inner cellular contents degenerate, they collectively form long tubes that greatly facilitate the passage of water in the plant. Fibers, on the other hand, function solely to provide support. In conifers the single xylem elements, the tracheids, represent a more primitive condition.

**31.6** In perfect flowers (flowers containing in one plant both stamen and pistil) self-pollination may be prevented by the superior height of the pistil or the ripening of pistil and stamen at different times. Why do you suppose such inhibitory mechanisms have evolved?

Such adaptations may prevent the loss of variability that self-fertilization would produce.

**31.7**   What would you say are some of the weaknesses of the green revolution?

To start with, the wide-scale sowing of a single strain of rice or wheat increased the vulnerability of the crop to particular plant diseases and pests. The special dwarf varieties of rice and wheat were often planted by thousands of farmers in broad regions of Asia and Africa. Should the particular strain go under, a whole nation could lose its entire staple crop.

An early element of possible disappointment was the fact that the enlarged grains of the dwarf strains were richer in carbohydrate but were not commensurately higher in protein. Yet, in most impoverished areas, protein deprivation is the great challenge to survival. Such diseases as kwashiorkor are caused primarily by protein deficiency rather than a general caloric deprivation. Later work did produce higher protein yields; particularly important was the increase in such essential amino acids as lysine.

The success of the green revolution depended on the mechanization of agriculture and the large-scale preparation of synthetic fertilizers and pesticides. Such mechanization required that the developing nations remain dependent on the largesse and industrial capability of the developed nations. Continued dependency of "have-not" on "have" nations sometimes produced tensions that threatened a necessarily cooperative relationship.

In the 1970s, when the price of oil sharply increased, the accessibility of petrochemically produced fertilizers was curtailed. Although many of the leading oil producers were politically allied with countries dependent on the green revolution, no effective adjustments were made and fertilizers became a bottleneck item for effective growing programs.

Another significant weakness of the green revolution became apparent with the growing realization that such pesticides as DDT are dangerous and are spreading rapidly through the biosphere. The specific effects on humans are not completely understood, but the damage to the ecosystem is quite apparent. Several species of birds, such as the osprey, are threatened with extinction because of the effects of DDT on the formation of their eggshells. The eggs crack when brooded, a result of an interference with calcium metabolism in the reproductive tract. In the United States, DDT has been banned for internal sale, but it is still manufactured for export.

Although the green revolution has been successful in markedly increasing grain yields, it can be only a temporary solution unless population size is regulated so that an appropriate balance exists between food supplies and the generation of more hungry mouths. Greater stability of ruling governments is a prerequisite for such population control programs.

# Supplementary Problems

**31.8**   A substance that impregnates the walls of tracheophyte cells to provide a necessary stiffness for structural support is   (a) tannin.   (b) chitin.   (c) lignin.   (d) nectar.   (e) none of these.

**31.9**   Both Chlorophyta and plants share which of the following characteristics?   (a) cell plates.   (b) grana in chloroplasts.   (c) starch as a storage fuel.   (d) cellulose in cell walls.   (e) all of these.

**31.10**   The archegonia are least prominent in   (a) mosses.   (b) liverworts.   (c) ferns.   (d) conifers.   (e) angiosperms.

**31.11**   Motile, flagellated sperm are found in   (a) ferns.   (b) bryophytes.   (c) conifers.   (d) both a and b.   (e) both b and c.

**31.12**   The gemmae found within the gemmae cups of *Marchantia* are actually   (a) male gametes.   (b) female gametes.   (c) asexual spores.   (d) pollen grains.   (e) reduced gametophytes.

**31.13**   Rhizoids contain tracheids.
(a) True   (b) False

**31.14** *Heterospory* refers to the production (by a plant or algal species) of two different classes of spore, one of which gives rise to the male gametophyte, the other the female.
(*a*) True   (*b*) False

**31.15** Each sorus of a fern contains many sporangia.
(*a*) True   (*b*) False

**31.16** The ginkgos are monoecious (hermaphroditic).
(*a*) True   (*b*) False

**31.17** Pine is a softwood, but maple and oak are hardwoods.
(*a*) True   (*b*) False

**31.18** Vessels and fibers are xylem elements found in the gymnosperms.
(*a*) True   (*b*) False

**31.19** Monocots have parallel venation in their leaves, while dicots have netted venation.
(*a*) True   (*b*) False

**31.20** Secondary woody growth in angiosperms is usually found in dicots.
(*a*) True   (*b*) False

**31.21** The characteristic spotted pigment patterns in the petals of some flowers may serve to attract insects and even help them to navigate toward reproductive structures.
(*a*) True   (*b*) False

**31.22** Both rice and wheat are dicots.
(*a*) True   (*b*) False

**31.23** Norman Borlaug invented DDT.
(*a*) True   (*b*) False

**31.24** As we go from bryophytes through ferns and gymnosperms to angiosperms, an increasing dominance of the sporophyte generation can be noted.
(*a*) True   (*b*) False

**31.25** Xylem cells perform their water-carrying function best after they die.
(*a*) True   (*b*) False

**31.26** Because they possess vessels in their woody stems and show a marked reduction in the female gametophyte the gnetophytes (a group of gymnosperms) are regarded as a possible link to the angiosperms.
(*a*) True   (*b*) False

# Answers

| | | | |
|---|---|---|---|
| **31.8** | (*c*) | **31.18** | (*b*); evolved in angiosperms |
| **31.9** | (*e*) | **31.19** | (*a*) |
| **31.10** | (*e*) | **31.20** | (*a*) |
| **31.11** | (*d*) | **31.21** | (*a*) |
| **31.12** | (*c*) | **31.22** | (*b*); both are monocot grasses |
| **31.13** | (*b*); they are nonvascular | **31.23** | (*b*) |
| **31.14** | (*a*) | **31.24** | (*a*) |
| **31.15** | (*a*) | **31.25** | (*a*) |
| **31.16** | (*b*) | **31.26** | (*a*) |
| **31.17** | (*a*) | | |

# Chapter 32

# The Kingdom Animalia

Animals are multicellular, eukaryotic organisms that are characterized by their nutritional habits—they eat (ingest) other living organisms. Many animals prey on other animals and are known as *carnivores*. Others ingest plants and are classified as *herbivores*.

**EXAMPLE 1**  A lioness stalks a herd of impala and kills one of the slower members of the herd. She shares the meal with other members of her group (pride). In this case the lioness is the *predator* (hunter), and the impala is the *prey* (the hunted). The remains may be picked on by such *scavengers* as vultures (see Chap. 26). The impala graze on the grass, so they are herbivores.

Humans belong to a subphylum of the phylum Chordata known as *Vertebrata*. The vertebrates, or backboned animals, make up approximately 5 percent of the animal kingdom, but they figure prominently in the lives of humans. All other animals are classified as *invertebrates*.

## 32.1  SUBKINGDOM PARAZOA: THE SPONGES

The animal kingdom is usually subdivided into groupings designed to reflect the evolutionary relationships of major lineages (see Fig. 32.1). The *sponges* are generally placed in a separate subkingdom

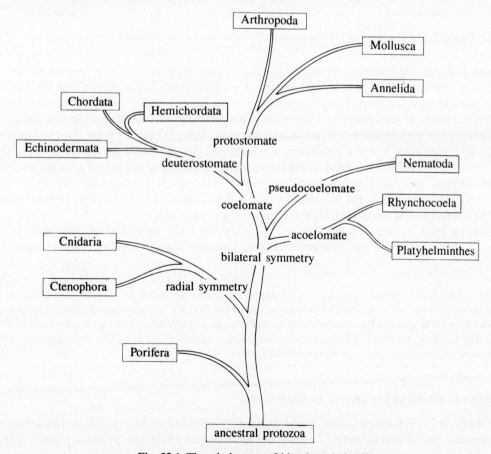

**Fig. 32.1** The phylogeny of kingdom Animalia.

405

known as *Parazoa*, while all other animals, supposedly derived from a separate protistan modification, form the subkingdom *Eumetazoa*.

The sponges are sessile (permanently attached, hence nonmotile), sacklike organisms belonging to the phylum *Porifera*. The name of the phylum is derived from the anatomy of the sponge (see Fig. 32.2). Many pores dot the surface of the sponge body. Water is sucked in through these pores, moves through the interior cavity (*spongocoel*) within the body of the sponge, and exits through an *excurrent vent (osculum)*. Food particles in the water are filtered out by special cells, known as *collar cells*, or *choanocytes*.

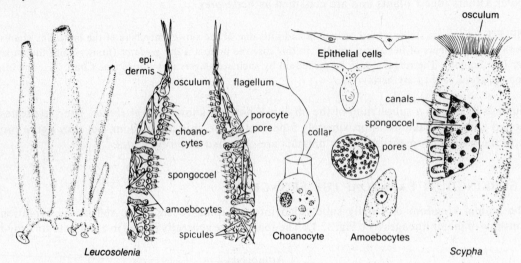

**Fig. 32.2** The structure of simple sponges. (*From Storer et al., adapted from Hyman.*)

Although sponges are classified as multicellular animals, they demonstrate less integration and specialization of function than other animal groups. They lack tissue organization, and their cells are the primary units of structure and function. The body is two-layered—an outer epidermis and an inner sheet consisting mainly of choanocytes. A gelatinous compartment, the *mesohyl*, separates these layers.

Most sponges are marine, but a few species inhabit fresh water. There are more than 10,000 species of sponges which vary greatly in size and shape. Some are brightly colored and enhance the appearance of marine bottoms. Aside from their use as effective cleansing swabs, sponges also produce antibiotics and a chemical that, when modified, is used in cancer chemotherapy. Sponges may be disaggregated by passing them through a mesh; the individual cells are then capable of reassociating to form an intact sponge. This is evidence of the autonomy the individual cells possess.

Four classes exist within phylum Porifera. These classes are separated on the basis of such characteristics as the nature of the inorganic *spicules* (spinous support structures) deposited in the mesohyl.

**EXAMPLE 2**   The class *Calcarea* possesses a skeleton consisting of spicules of calcium carbonate. The *Demospongiae*, which make up more than 90 percent of all sponge species, usually contain silicon compounds in their spicules. Often these spicules also combine with *spongin* (tough collagen fibers). The glass sponges (*Hexactinellida*) are named for their six-rayed, siliceous (containing-silica), glasslike spicules. The *Sclarospongiae* contain siliceous spicules above a dense network of calcium carbonate.

## 32.2   RADIATA: CNIDARIA AND CTENOPHORA

Above the level of the sponges, animals can be broadly divided into two classifications on the basis of body symmetry: radial and bilateral. The radially symmetrical phyla, the *Radiata*, possess a central mouth around which the rest of the organs are radially arranged.

The Radiata consist of two phyla: *Cnidaria* (coelenterates) and *Ctenophora* (comb jellies). Each of these phyla consists of individuals with two layers of cells, *radial symmetry*, and an inner *gastrovascular cavity* (*coelenteron*). The gastrovascular cavity serves as a digestive structure. Food is taken in through the mouth, and what is undigested passes out again through the mouth. Whereas sponges are *filter feeders*, the coelenterates and comb jellies can ingest large food particles and pass the digested nutrients throughout the body.

The Radiata display a *tissue level of organization*. Specialized nerve cells make up a nerve net, which may permeate the entire body of the organism. Polarity across nerve junctions, or synapses, is not apparent, since impulses appear to travel in either direction across the synapse. *Epitheliomuscular cells* make up a contractile tissue, which is actually derived from the outer (ectodermal) layer of cells.

## CNIDARIA

A striking feature of the cnidarians, giving the phylum its name, are the *cnidocytes*. These specialized cells in the tentacles and parts of the main body wall contain a coiled tube, the *nematocyst*, compressed within a capsule (see Fig. 32.3). When appropriately stimulated, the cnidocyte "pops" the nematocyst, which may sting and lasso prospective prey. The nematocyst may also function as an anchoring device.

The almost 10,000 species of Cnidaria are grouped into three major classes and one minor class: *Hydrozoa* (e.g., hydras), *Scyphozoa* (jellyfish), and *Anthozoa* (corals and sea anemones) are the major classes, the *Cubozoa* (sea wasps) the minor class. In most hydrozoan life cycles a clear-cut *polyp* and

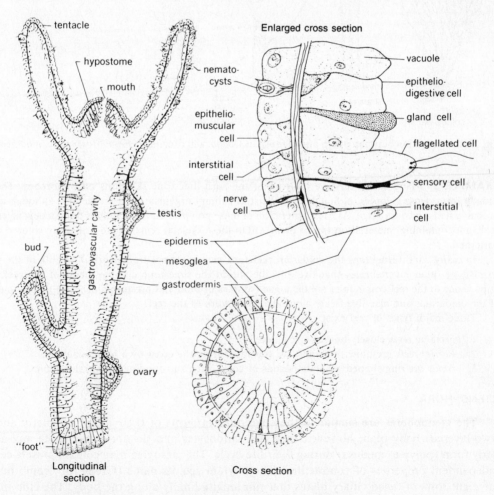

**Fig. 32.3** The body plan of *Hydra*, a cnidarian. (*From Storer et al.*)

*medusa* stage alternate (see Fig. 32.4). In the scyphozoans, the medusa stage clearly predominates, the *mesoglea* (matrix between ectoderm and endoderm) is thick and occasionally cellular, and all live in the sea. The anthozoans ("flower animals"), the largest and most varied of the three classes of cnidarians, completely lack a medusa stage and consist of large, fleshy cylindrical bodies. They are responsible for the coral reefs found in warm ocean waters.

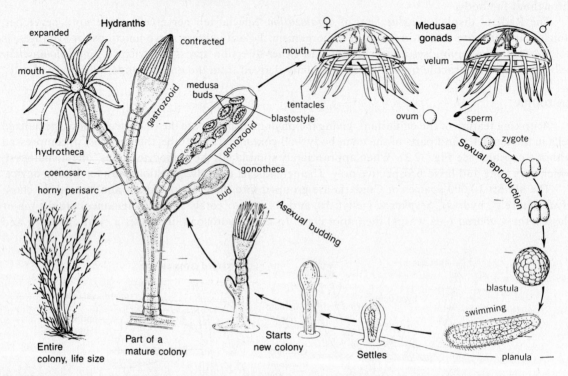

**Fig. 32.4** Alternation between polyp and medusa in *Obelia*, a hydrozoan. (*From Storer et al., modified from Wolcott.*)

**EXAMPLE 3** Many coral reefs are built up of the hard limestone skeletons of anthozoans. These cnidarians usually form dense colonies, although a few exist as solitary organisms. All are polyps, although ancestral forms probably had a medusa stage. As each generation of these polyps dies, other polyps build a new layer of calcareous (calcium-containing) material on top of them. Often these skeletal remains are brilliantly colored and intricately patterned.

In many coral formations the cnidarian skeletons merely form a framework. The bulk of the reef formation consists of other invertebrates that live and die within the limestone compartments of the reef. Of particular importance to the reef community are the algae (sources of food and producers of food) that permeate the tissues of the cnidarians and also live freely along the latticework of the reef.

Three major types of reef exist:

1. *Fringing reefs* closely border a coastline.
2. *Barrier reefs* are quite thick and are separated from the coast by a broad channel.
3. *Atolls* are ring-shaped separate islands of coral, each surrounding a central lagoon.

## CTENOPHORA

The ctenophores are similar to the cnidarians in terms of their radial symmetry and *diploblastic* (two-layered) body plan; however, they lack cnidocytes and do not demonstrate any alternation of body form (polyp to medusa) during their life cycle. The mesoglea may contain muscle cells that show independent properties of contractility. The phylum gets its name (Greek for "comb bearers") from the eight rows of fused ciliary plates that run longitudinally along the body. The cilia in these plates (combs) beat synchronously to propel the animal in the water. Usually these *sea walnuts* (another common name) are passive passengers of ocean tides and stronger currents. Many of the species are luminescent, and all capture their food with the aid of sticky cells lining the tentacles.

## 32.3  BILATERIA: DEUTEROSTOMES AND PROTOSTOMES

All the remaining members of the animal kingdom probably arose from a common ancestral form that first demonstrated bilateral symmetry. This type of symmetry involves a similarity of morphology (shape and form) on each side of a longitudinal axis. These *Bilateria* usually have a *head*, or *anterior* end, and a tail, or *posterior* end. The head end, which first enters a new environment, shows an increasing complexity of sensory structures in the more recently evolved groups. All the bilateral forms are *triploblastic*, possessing a third primary germ layer known as *mesoderm* (see Chap. 11).

All the Eumetazoa above the level of Radiata are differentiated in terms of the presence or absence of a true coelom.

**EXAMPLE 4**   The flatworms (*Platyhelminthes*) and proboscis worms (*Rhynchocoela*) have no coelom whatsoever. The digestive cavity, lined by the inner germ layer (endoderm), is the only body space. These two groups are known as *acoelomates*. The *pseudocoelomates* have an unlined cavity between endoderm and mesoderm. These include such phyla as *Rotifera* and *Nematoda* (roundworms). The *coelomates* possess a true coelom—a fluid-filled cavity within the mesoderm that is completely lined with epithelium. All the more complex animals beyond the roundworm level of organization possess a coelom.

The coelomate *Bilateria* soon diverged into two major branches in the animal kingdom, based on differences in the developing embryo (see Fig. 32.1). In the *deuterostomes* (Greek for "second mouth"), the blastopore forms the anus, and the mouth is formed secondarily. In the *protostomes*, the blastopore forms the mouth, and the anus is formed secondarily. Deuterostomes also demonstrate radial, indeterminate cleavage, whereas protostomes tend toward spiral, determinate cleavage. The major phyla showing deuterosotome developmental patterns are Echinodermata and Chordata; the major protostome phyla are Platyhelminthes, Nematoda, Mollusca, Annelida, and Arthropoda.

Deuterostomes and protostomes also differ in the origin of the mesoderm and in the formation of the coelom cavity. In deuterostomes, the mesoderm arises from endoderm in the archenteron; in protostomes, the mesoderm arises from a single cell determined early in development. Further, in deuterostomes, the coelom is formed from an outpocketing of the gut. As this bulge of mesodermal cells expands into the blastocoel and pinches off, it forms a separate "inner tube" of mesoderm, the hollow core of which is the coelomic cavity. In protostomes, the single mesodermal cell gives rise to a solid ring of cells around the gut. An internal split in this solid mass produces the coelom. These two methods of producing the coelom end in the same result.

The coelom endows an organism with several benefits. In separating the gut and other internal organs from the body wall, it permits independent movement of these two sectors of the body. Thus, peristaltic movements interior to the coelom need not interfere with the locomotor contractions of muscles in the body wall. The fluid of the coelom may also serve as a hydrostatic skeleton against which muscles in the body wall can contract to produce movement. Finally, the membranes of the coelom, such as the *peritoneum*, provide double-layered sheets of support for internal organs.

## 32.4  ACOELOMATES

### PLATYHELMINTHES

The platyhelminths are the simplest of the triploblastic animals, but they exemplify an *organ level* of complexity compared with the tissue level shown by cnidarians and ctenophores. Their ancestral origin is not clear. The almost 15,000 species are usually assigned to four separate classes.

*Turbellaria* is the only class of free-living flatworms. The name comes from the turbulence created in their water habitat by their vigorously beating cilia. They feed on small, live organisms or may tackle larger dead animals. The digestive cavity is branched and extends throughout the body of the flatworm. Considerable storage of food occurs, so turbellarians can survive long periods of starvation. Circulation of gases and solutes is extremely limited.

**EXAMPLE 5**   The planaria is a typical turbellarian. It lives in fresh water, although most of its "classmates" are marine. Since no circulatory system exists, planarias are relatively long and flat. This morphology allows most cells of the flatworm to maintain contact with the surroundings so that gas and solute exchanges can occur readily. Planarias move by both beating of cilia and undulation of the body to produce a gliding motion.

The three other classes of platyhelminths are all parasitic. The *Monogenea* tend to have simple stages in their life cycle. They are hermaphrodites, but only rarely do they fertilize themselves. The *Trematoda* (flukes) are parasites of a high order of complexity. They may have as many as four intermediate hosts, usually at least one of which is a mollusk (see Fig. 32.5.)

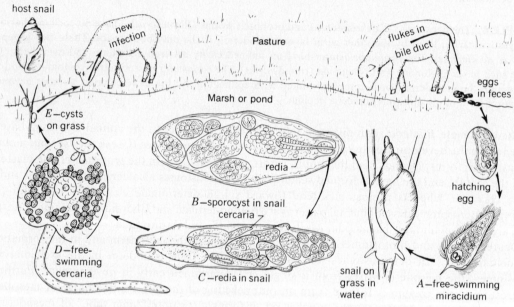

**Fig. 32.5** The life cycle of the sheep liver fluke, *Fasciola hepatica.*

**EXAMPLE 6**   The Chinese liver fluke, *Clonorchis sinensis*, is a major parasite in many Asian countries. Adult worms live in the liver and biliary canals of humans and other mammalian hosts. The many eggs pass out of the host and must be ingested by a specific freshwater snail. After undergoing changes in form, the fluke, as an active swimming *cercaria*, burrows into a freshwater fish or crustacean, where it exists as a *metacercaria*. This second intermediate host, when eaten by a large carnivore, passes encysted metacercaria into the carnivore. The metacercaria become adult worms, thus completing the cycle.

More than 200 million Asians suffer from a blood fluke called a *schistosome*. The resulting parasitic disease (schistosomiasis) involves only two hosts, humans and snails. The effects of this parasitic disease are chronic fatigue, lowered resistance to other diseases, and a general malaise.

Tapeworms (*Cestoda*) are the fourth class of flatworm. They consist of an anterior head (*scolex*), which contains hooks or similar mooring devices to grip the interior of the intestine, where the worm passes most of its adult life. Behind the head lie repeating segments known as *proglottids*. Each proglottid contains male and female reproductive structures and little else, since the tapeworm has no digestive tract and little in the way of a nervous system. Tapeworms, long ribbons of proglottids, are eminently suited to a parasitic existence.

Members of phylum Rhynchocoela (the nemerteans) resemble the platyhelminths, but they possess an extensible *proboscis* (nose), which is used to capture prey, hence their name "proboscis worms."

In some cases the proboscis has a stylet (small lance) at its tip, which acts as a needle for the injection of toxins. They are more complex than flatworms in that they possess an anus and a moderately well-developed circulatory system.

## 32.5  PSEUDOCOELOMATES: ROTIFERA AND NEMATODA

The *Rotifera* and *Nematoda* possess a pseudocoelom. Rotifers are tiny and, despite their multicellularity, may be smaller than some large amoebas. Their name ("wheel carrier") comes from the ridge of beating cilia (*corona*) that surrounds their mouth. Rotifers are of particular interest to embryologists because all adults have a fixed number of cells, resulting from a constant number of zygotic cell divisions. They also demonstrate many varieties of parthenogenesis in which unfertilized, but diploid, eggs routinely produce a new generation of young. In some cases populations contain only females. Other groups produce haploid eggs that develop parthenogenetically into haploid males, a process resembling the situation in ants and bees. When sperm and egg unite to form a zygote, the fertilized egg is usually thick-shelled and highly resistant to adverse environmental conditions. Rotifers living in drainage pipes or mudholes may, when their environment dries out, change to a desiccated, long-enduring, inactive state and resume activity even years later when water becomes available.

Several other phyla of tiny pseudocoelomates exist. They include the hermaphroditic *Gastrotricha* and the spiny *Kinorhyncha*, which are virtually trapped within an external layer of cuticle.

*Nematodes* (roundworms) are the most numerous of all phyla in terms of actual numbers of individuals. Most of these worms, with their characteristic tapered ends, are free-living; however, such diseases as *hookworm*, *trichinosis*, and *elephantiasis* arise from parasitic roundworms.

## 32.6  PROTOSTOME COELOMATES

*Annelida*, *Mollusca*, and *Arthropoda* are the three major protostome phyla that probably share a common evolutionary divergence. Arthropods and most of the mollusks have an open circulatory system in which large spaces (*hemocoels*) stand between arteries and veins. Annelids possess a closed circulatory system in which enclosed vessels (arteries, capillaries, and veins) ramify throughout the transport system. All three phyla are coelomates.

### ANNELIDA

The phylum Annelida (Latin for "ringed") comprises worms whose bodies are divided into segments (*metameres*). This segmentation is clear externally but also exists internally in the form of membranes (*septa*) that subdivide the interior of the worm. Metamerism also exists in arthropods, believed to be directly related to the annelids, and in chordates as well.

**EXAMPLE 7**  In the earthworm, a typical terrestrial annelid, the entire body is subdivided into segments that are generally similar to one another. Each segment contains four pairs of external bristles, which are used to anchor and move the worm. The segmentation is expressed internally in the existence of nephridial (excretory) tubes and excretory pores in almost all segments. However, the digestive tract is not segmented.

Approximately 12,000 species of annelids dwell in terrestrial, marine, and freshwater habitats. Three classes exist:

- *Polychaeta* ("numerous bristles") are the bristle worms, usually found at the seashore. Separate sexes generally exist. Pairs of fleshy paddles (*parapodia*) jut out on each segment, supplying locomotion and respiratory surfaces.
- *Oligochaeta* ("few bristles") are predominantly terrestrial, but a few marine and freshwater forms exist. They are hermaphroditic, and generally a pair of worms copulate to produce a double fertilization event. They live in moist environments, since gas exchange across their

body surface requires that the gases be dissolved. Earthworms (*Lumbricus*) are a principal example of the class.

- *Hirudinea* are the leeches—relatively small, flattened, tapered worms with a sucker at each end. Many are parasitic on invertebrates or vertebrates. They attach to live organisms with the front sucker and ingest blood and other bodily fluids. They are hermaphroditic.

## MOLLUSCA

The phylum *Mollusca* ("soft-bodied") consists of seven different classes. A tremendous diversity of form and lifestyle characterizes these classes; most are shelled, but some have lost the calcareous shell in the course of their evolution. All classes except the Bivalvia (clams and oysters) possess an extensible, raspy tongue called a *radula*. Rows of strong teeth along this usually flattened structure serve as a hacksaw when the radula is extended and pulled back over a food-laden area. Since no other phylum possesses a radula, it clearly serves to establish a tie of common descent among the mollusks. We will examine four of the seven extant classes.

The *Polyplacophora*, named for the many (actually eight) shells borne upon the dorsal surface, are represented by the *chitons*. Although chitons are a relatively small group of molluscans, they bear a marked resemblance to the hypothesized ancestral form of all mollusks. The three basic regions found in almost all mollusks are the flat, highly muscular *foot*, the shell, and the *mantle*, which encloses the internal organs dorsally and lies alongside the foot (see Fig. 32.6). In shelled mollusks the mantle secretes the shell. The mantle cavity, which lies between the mantle and the foot, contains the *gills*, which are usually leaflike flaps extending from the foot.

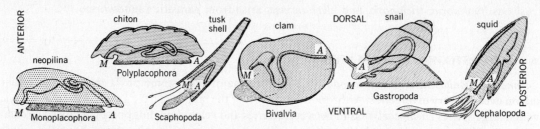

**Fig. 32.6** The body plan of various mollusks. *M* = mouth, *A* = anus. (*From Storer et al.*)

The *Gastropoda* are the snails, periwinkles, conchs, and slugs (nonshelled gastropods). They are the largest and most diverse group of mollusks, with more than 75,000 species. The shells are often quite complex and even ornate, and the entire animal can usually retreat within the shell for protection.

The *Bivalvia* include the mussels, scallops, clams, and oysters, which are major sources of fine seafood. They consist of two complementary shells, which are joined along the middle of the dorsal line of the body. Powerful muscles (*adductors*) pull the shells tightly together. When these muscles relax, the wedge-shaped foot can be extruded and the bivalve may move about or grasp the substratum. Almost all bivalves are filter feeders, and they use the enlarged gills for both respiration and ingestion. The sexes are separate; eggs are usually fertilized and may even develop through early stages within the tissues of the female.

The *Cephalopoda* are perhaps the most specialized and advanced of the mollusks. In almost all cases they lack a complete shell.

**EXAMPLE 8**   The chambered nautilus is the only extant cephalopod with a full shell. It resembles the *squids* and *octopuses* in basic body plan, but it occupies a very ornate striped shell. Most of the shell serves only to keep the entire animal afloat, and some zoologists feel that this is the route taken by other lineages of cephalopods in the evolutionary loss of the shell.

The cephalopods have a highly developed nervous system and are capable, especially the squids, of very rapid locomotion. They are usually carnivores and possess many adaptations for hunting, including camera-type eyes capable of very fine resolution. They are unique among mollusks in having a closed circulatory system. Some giant squids weigh 500 kg.

## ARTHROPODA

In terms of numbers, diversity of species, and distribution in the biosphere, the *Arthropoda* are the most successful of all animal phyla. The five classes of arthropods ("jointed feet") comprise well over a million species, and many new species continue to be discovered. The body of an arthropod is usually covered by a thick, protective *exoskeleton*, the *cuticle*, which contains chitin. The body is usually segmented, as seen in annelids, with the segments giving rise to various types of appendages.

**EXAMPLE 9**  Among crustaceans and insects the most anterior appendages become *antennae* (flexible sensory hairs) followed by laterally moving jaws (mandibles). These groups are classified as *mandibulates*. In arachnids and horseshoe crabs no antennae or mandibles develop. Instead, the most anterior pair of appendages form fleshy pincers known as *chelicerae*. These arthropods are known as *chelicerates*; they have four pairs of legs.

Most arthropods have powerful *compound eyes*, which are composed of many tightly arrayed units known as *ommatidia*. Each unit has its own lens and retinal screen for the formation of an image. The coelom of arthropods is considerably reduced. This is also true for mollusks; however, most zoologists believe that arthropods are more closely related to the annelid worms than to mollusks. There are five major classes of arthropods:

- *Arachnida* includes spiders, mites, ticks, scorpions, and daddy longlegs (see Fig. 32.7). All have four pairs of legs. Since they are largely terrestrial, they respire by means of tracheae (see Chap. 19) or delicate internal leaflike plates known as *book lungs*. Some poison their prey. The spiders spin webs to ensnare unwary victims. Web building is species-specific and apparently controlled genetically.

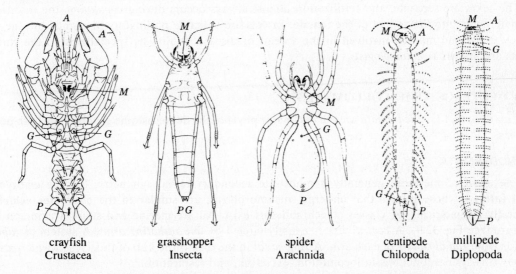

| crayfish | grasshopper | spider | centipede | millipede |
| Crustacea | Insecta | Arachnida | Chilopoda | Diplopoda |

**Fig. 32.7** Representatives of the five major classes of arthropods. (*From Storer et al.*)

- *Crustacea* primarily marine or aquatic forms, utilizing gills for respiration. It includes lobsters, shrimps, crabs, and barnacles; all have two pairs of antennae (sensory probes) on their heads. The second pair of antennae may serve in defense or even to obtain food in some crustaceans. Large crustaceans such as lobsters and shrimp are known as *decapods*. Smaller crustaceans make up the *isopods*, and the very tiny *copepods* are found as part of the plankton communities of oceans and lakes.

- The *Diplopoda* consists largely of the clearly segmented *millipedes*. They are herbivores and are almost entirely terrestrial. Each body segment has two pairs of legs, thus providing a great deal of support.
- The class *Chilopoda* contains the *centipedes*. They are all carnivores and are particularly adept at rapid running. They are, however, highly vulnerable to drying out. Both chilopods and diplopods are part of the superclass *Myriopoda*.
- The class *Insecta* has more species than all other animal organisms combined. Though rare in the sea, they are the most successful invaders of the terrestrial environment. They also abound in aquatic regions and in the air. The more than 25 orders of insects probably owe their adaptiveness to relatively small size, the existence of wings, a very acute compound eye, the existence of metamorphosis, and in some cases an elaborate social system.

The body of most insects is patently divided, like Gaul, into three parts—a *head*, *thorax*, and *abdomen* (see Fig. 32.8). The head is particularly complex, consisting of many individual parts that are elegantly integrated into sensory and mechanical devices.

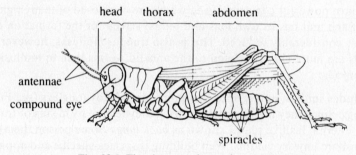

Fig. 32.8 The three sections of a grasshopper.

The sexes are separate, and fertilization almost always occurs during *copulation*, the insertion of the male penis into the vagina of the female. Insects are valuable pollinators of plants, but they also destroy crops and may be involved in the spread of disease. Biologists specializing in the study of insects are known as *entomologists*.

## 32.7 DEUTEROSTOME COELOMATES

*Chordata* and *Echinodermata* are the two major phyla of the deuterostome group. They both possess coeloms.

### ECHINODERMATA

The adult echinoderms generally demonstrate secondary radial symmetry, but the embryos, or larval forms in those groups that undergo metamorphosis, are similar to the chordates, which are bilaterally symmetrical. Six classes of echinoderms exist. All are marine and share a common basic structure (see Fig. 32.9). A *central disk* is usually circled by five *radiating arms*. A system of *tube feet* extends from *radial canals* in each arm. A *ring canal* in the disk is the hub of this closed *water-vascular system*. This system functions in locomotion, excretion, and respiration.

### CHORDATA

The nonvertebrate Chordata consists of two subphyla: the *Urochordata* (tunicates), which inhabit a *benthic* (ocean bottom) habitat, and the *Cephalochordata* (e.g., lancelet). They share, with the vertebrates, the three basic characteristics of all chordates:

1. A *notochord*, a flexible but sturdy rod of fibrous tissue just ventral to the neural tube
2. A hollow dorsal *nerve cord* derived from ectoderm
3. Pharyngeal *gill slits*

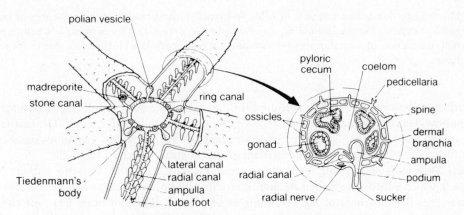

**Fig. 32.9** The structure of an echinoderm.

The specific features of the subphylum Vertebrata are a *skeletal backbone* and *cranium*, which encloses the spinal cord and brain; a high degree of *cephalization* (the specialization of the anterior end of the nervous system into a complex brain with associated specialized sense organs); and segmentation of the muscles of the trunk into *somites* at some time in the course of development.

Of the seven living classes of vertebrates, three are fishes. The remaining four classes are grouped as *tetrapods* (Greek for "four feet") because they usually possess two pairs of limbs.

The *Agnatha* are jawless fishes that were once numerous but now consist only of *hagfish* and *lampreys*. A large proportion of present-day agnathans prey on other fish by attaching to their outer surface and boring through to the internal organs with their jagged tongue. The *Chondrichthyes*, like the agnathans, have skeletons made of cartilage. Both groups probably evolved from bony ancestors. Extant chondrichthyes include sharks, rays, and skates. Though composed of cartilage, the jaws of the shark are a fearsome sight. The *Osteichthyes* are the bony fish, which are found in both fresh and salt water.

**EXAMPLE 10**   The seven species of salmon are among the most popular food and sport fishes, with the large *sockeye salmon* being the most extensively sought. Salmon are usually born in fresh water, with parents preferring a fast-moving stream. Fertilization during *spawning* (reproductive activity) occurs externally, with a male placing its sperm over eggs the female has deposited in "nests" in the shallow stream bottom. After a varying period of time the young salmon swim to the ocean, with only a few surviving the perilous trip. They usually return to the same stream to spawn after spending a few years or less in the ocean. Many species die shortly after spawning.

*Amphibia*, the frogs, toads, and salamanders, are only partially successful as terrestrial inhabitants. Many amphibians (a word meaning "both lives," i.e., water and land) are dependent on water for their sexual activity. In frogs, the fishlike *tadpole* (larval form) undergoes metamorphosis to become a lung-bearing adult land dweller. Many amphibians, however, respire principally through a moist surface membrane rather than through their lungs.

The *Reptilia* all have scales on a rough, waterproof integument that shields against drying out. Other modifications for successful terrestrial life include functional lungs; an enclosed (*cleidoic*) egg in which the embryo can absorb yolk and develop within a protective bath of amniotic fluid, well protected by a leathery shell; internal fertilization; and a variety of behavioral patterns that enable the reptile to survive extremes of temperature and shortages of food.

During the Mesozoic Era, which ended about 65 million years ago, reptiles were the dominant land forms. Many of these reptiles were dinosaurs (Greek for "terrible lizard") of rather awesome proportions. The more than 20 orders of reptiles that lived long ago have dwindled to a mere four or five orders today—lizards, snakes, alligators and crocodiles, and turtles and tortoises.

*Aves*, the birds, have been aptly described as "flying lizards." In addition to the fact that they probably descended from dinosaurs, they possess several features in common with reptiles.

**EXAMPLE 11**   A scaly skin is characteristic of birds and reptiles, most readily seen on the legs of birds, which are not covered by feathers. The telolecithal eggs are very similar, except that birds produce a brittle, rather than a leathery, shell. Production of a similar nitrogen end product (uric acid) is a biochemical clue to the relationship of the two classes.

The birds have become exquisitely modified for flight. The skeleton is quite light, and many of the bones are hollow. Heavy reptilian teeth are replaced by a tough, but lighter, beak. The feathers, which probably evolved from reptilian scales, provide an ideal surface for flight. The four-chambered heart and the endothermic lifestyle support the extreme physical effort required for flight.

The class *Mammalia* shares with the insects an extreme mastery of the terrestrial habitat. Mammals have hair (fur), which affords an efficient mechanism for insulation, especially when *piloerection* (erection of hairs) increases the thickness of the insulating layer. The mammalian heart has four chambers. A diaphragm is present (see Chap. 19) to enhance the efficiency of the lungs. Endothermy is always maintained. The salient characteristic of all mammals is the ability of the mother to produce nourishing milk for the young in mammary glands. The very primitive mammals known as *monotremes* do not have nipples or breasts, but they do produce milk, which runs down the ventral hairy surface and is lapped up by the young. Most mammals possess very complex brains and show highly sophisticated behavior patterns, including play. There are three subclasses of Mammalia: *Protheria, Metatheria*, and *Eutheria*.

The monotremes (Protheria), like the duck-billed platypus and the spiny anteaters (*echidnas*) living in Australia, are rather avian in character. They lay eggs, lack teeth, and possess a beak. Their young hatch from eggs at a comparatively early stage and cling to the hair of the mother.

The mammalian subclass *Metatheria* consists of the *marsupials*, or pouched animals. They are the dominant mammals of Australia, but in the United States they are represented largely by the opossum. They are mammalian in their possession of teeth rather than a beak, and they give birth to live young. But the young are born at an early stage in their development and complete their fetal life within a pouch on the underside of the female body. Each of the young fastens on to a teat, and feeding goes on continuously.

The *Eutheria* are the "true," or placental, mammals, the most familiar mammalian representative of the northern hemisphere. They are distinguished, especially in larger organisms, by a protracted period of pregnancy during which the young reach a more advanced state of development than is the case for monotremes and marsupials. A placenta forms within the uterus, which is a major metabolic depot for exchange of food, oxygen, and waste between mother and fetus.

Approximately 12 major orders of placental mammals exist, some of which have secondarily returned to oceans and lakes.

**EXAMPLE 12**   Whales, dolphins, and porpoises make up the order *Cetacea*. Their bodies are streamlined for swimming in the oceans, and a thick layer of blubber (fat) protects them against the cold waters. Their forelimbs function like oars, and their hind limbs have been lost completely. Like other Eutheria they give birth to their young and nurse them through their early development.

*Carnivora* is a particularly varied mammalian order, ranging from dogs, wolves, and large cats to otters and seals. Almost all carnivores are efficient predators, exemplifying Tennyson's notion of a "nature red in tooth and claw." *Rodentia*, the order that includes squirrels, beavers, rats, and mice, is another widely dispersed group. Rats, in particular, are generally present in all urban areas and share the spoils created by humankind. Bats are found in the order *Chiroptera*. Much of the folklore dealing with the harm that they cause is generally false. Sheep and cattle, with an even number of toes on each foot, belong to the *Artiodactyla*, while horses, zebras, and rhinoceroses are part of the *Perissodactyla*, with an odd number of toes. Both these herbivorous orders contain species that live in close relationships with humans.

# Solved Problems

**32.1**  Most of the distinguishing features of the more advanced animals can be understood in terms of their nutritional mode. Explain.

Animals take in live or dead organisms. They must seek out, possibly kill, and finally ingest their prey. To accomplish this nutritional task they must be alert to the presence of prey or edible plant materials. Those animals that feed on live animals must be able to track down their victims and vanquish them. This suggests a need for highly developed sense organs and a neuromuscular apparatus for rapid and precise movement.

Most animals have complex sense organs such as eyes and ears to receive information about what is available. Since animals have radiated to all kinds of environments, such organs also allow adaptation to environmental challenges.

Sensory stimuli reaching the brain and nerve tube are processed rapidly, and precise motor responses occur, usually very rapidly. Darting, striking, and leaping movements attest to the level of complexity of muscular responses in the vertebrates. In most animal lineages an evolutionary trend toward increased neural complexity and enhanced motor capabilities occurs. Cephalization is part of this process of increased neural development. Since most animals above the level of coelenterates are bilaterally symmetrical, cephalization can occur more readily (a longitudinal axis possesses a front and rear end). Teeth and strong jaws are also evident as means to subdue prey.

**32.2**  Which invertebrate groups have sponges given rise to?

None. The sponges are an offshoot of the major invertebrate line. No evolutionary divergence occurred beyond the separation of sponges into four classes. The ability of sponges to reassociate after their cells have been separated suggests that little specialization has occurred beyond the differentiation of cells into functional specialties.

Sponges may have arisen from a protozoan, the *choanoflagellate*. This protozoan possesses a collar surrounding its flagella that is similar in structure to the choanocytes of the sponges. The other invertebrates probably arose from a very different protistan ancestor. Sponges, then, arose from a unique forebear and have produced no evolutionary neotypes, thereby justifying their assignment to a separate subkingdom, the Parazoa.

**32.3**  What possible processes might have led to the production of multicellular animals like sponges and the Eumetazoa from the Protista?

*Aggregation* of single cells into a multicellular mass could produce a multicellular organism. Cellular slime molds demonstrate this pattern, and so do a number of protozoans. Additionally, aggregation of cells into sheets or ridges often occurs during embryogenesis in invertebrates. Most biologists, however, do not accept aggregation as the origin of multicellularity, because almost all animals begin life as a single cell that divides rather than as an association of independent cells.

A second theory is the *colonial hypothesis*—the concept that animals arose from colonial protists. The type of colony most often cited as a forerunner of the animal kingdom is *Volvox*. Other colonial protists include the choanoflagellates and several species of amoeba. It is of considerable interest that some botanists recently have claimed that *Volvox* is a multicellular plant.

A third view is the *syncytial hypothesis*. As its name implies, this concept stresses the possibility of a protistan organism consisting of a large mass of cytoplasm with scattered nuclei (a syncytium) becoming compartmentalized by the development of septa. This mass, now clearly cellular, could become ever larger, more complex, and more perfectly integrated.

There is no way of determining which of these routes actually occurred in the evolution of animals. Since 95 percent of all organisms that once existed have probably become extinct, it may not ever be possible to identify the specific steps leading to multicellularity.

**32.4**   Contrast the feeding behavior of sponges and cnidarians.

Sponges are filter feeders. Water enters the sponge through the pores that open to the exterior. Currents are set into motion by the flagella of the choanocytes, or collar cells, which line the inner layer of the body. In the simplest arrangement, or *ascon* body type, there is a thin body wall, and the inner cavity (spongocoel) is lined by a single layer of choanocytes. In the more complex arrangements a thick wall contains outpocketings which are lined with many choanocytes, a strategy which increases the surface exposed to the currents. Food particles are filtered out by the collar, and trapped food particles are engulfed by the base of the collar cell; the water coming through the pores exits through the osculum atop the body. Special cells that move through the middle area of the body, the *amoebocytes*, may also ingest food particles. It is of interest that the currents that bring food into the sponge move in a single direction through the body. This parallels the one-way digestive tract associated with more advanced metazoans.

In hydra and other cnidarians a gastrovascular cavity possesses a single opening, the mouth; this is the more primitive two-way digestive tract, in which food enters and undigested materials leave through the same opening. The extendable tentacles that surround the mouth play a key role in capturing prey and carrying it to the mouth. The tentacles are richly endowed with various types of nematocysts (produced in cnidocytes), which may lasso and poison the prey. In hydra, the presence of prey induces rapid feeding movements by the tentacles and even the body stalk. These movements may be induced by specific molecules contained in the potential food organisms, which dissolve in the water surrounding the hydra. Glutathione, a tripeptide present and released by most injured animals, elicits a marked feeding reaction.

Digestion in cnidarians occurs within the gastrovascular cavity by enzymes liberated by gland cells within the inner cell layer. This is a distinct advance over the intracellular digestion of sponges, since cnidarians can ingest large particles and even whole organisms. Once breakdown of the larger mass occurs, nutritive amoeboid cells engulf the smaller particles and the final stages of digestion are completed intracellularly.

**32.5**   The cnidarians and ctenophores are radially symmetrical while later animal phyla primarily demonstrate bilateral symmetry. What are some possible advantages of radial symmetry?

A radially symmetrical organism can strike or move in any direction from its unaroused state. The tentacles of a cnidarian reach out in all directions, since no left-right restrictions are imposed on its body. In a hydra, for example, defense structures (nematocysts) exist uniformly along the entire circumference, so that the defenses cannot be easily breached. This capability is particularly important in sessile, floating, and slow-moving forms, which may be more vulnerable to attack. The radial arrangement also increases the likelihood of snaring food, a benefit that is important, since the radially symmetrical animals either are attached or move slowly and therefore are basically dependent on food's coming to them.

Polyps are generally less motile than medusae, but the hydra is an example of a polyp that does move around, even showing the capability of turning somersaults with its tentacles.

**32.6**   What are the major differences between flatworms (platyhelminths) and roundworms (nematodes)?

The flatworms lack a coelom, while the nematodes possess a pseudocoelom. The flatworms arose much earlier than the roundworms, and their digestive tube is a two-way conduit, similar to the situation in the cnidarians. On the other hand the nematodes possess both a mouth and an anus, so that their digestive tube is a one-way system, with greater specialization of mouth and anal regions possible. Because they possess a body cavity (the pseudocoel), nematodes are not solid, whereas the flatworms are. Also, the nematodes have a tough *cuticle*, which is protective and especially adaptive in parasitic roundworms; flatworms do not have such a cuticle. Motile flatworms, such as planaria, move largely by beating their surface cilia. Nematodes move only by a whiplike movement of the whole body produced by the contraction of longitudinal muscles.

Most nematodes are dioecious (having separate sexes), but the flatworms are usually hermaphroditic. A greater complexity of the nervous system also characterizes the nematodes.

**32.7**   How does the segmentation of an annelid worm differ from the repeating units of the body of a tapeworm?

In a tapeworm, each of the segments is virtually identical, differing mainly in size. Each segment, or *proglottid*, is derived from the *scolex* (head) of the tapeworm. The older proglottids are pushed back by new proglottids formed at the margin of the scolex and the most recent segment. This formation process in tapeworms constitutes a type of asexual reproduction. The proglottids are actually buds. When the mature proglottids, swollen with self-fertilized eggs, arrive at the end of the tapeworm, they break off, and surviving eggs eventually reach a new host.

In annelids, segmentation is part of a unitary body plan, and the segments are not by themselves reproductively competent. Such structures as nephridia (excretory organs) and setae (bristles) do recur, but many other structures extend through the entire body of the worm and do not exist as separate structures in each segment. Some examples are the circulatory system and the reproductive system. Also, the digestive tract is not segmented but exists as a continuous, one-way tube with varying specialization occurring along its length. Segmentation in Annelida then is an advancement that facilitates development. It continues as a *morphogenetic* (showing a hereditary pattern of shape) feature in Arthropoda, as well.

**32.8**  The *trochophore larva* is a characteristic developmental stage of annelids, mollusks, and a few minor phyla. It is a ciliated, free-swimming form with the shape of an inverted top. Why do you suppose the trochophore larva is considered a valuable taxonomic tool?

Quite often adult organisms are very different from ancestral forms, thus obscuring their actual natural history. An examination of the early developmental stages, such as the trochophore larva, may provide significant clues to that history and show relationships between phyla that in the adult specimens appear unrelated. Similar insights may be derived from the transient presence of nonfunctional gill slits in the embryos of reptiles, birds, and mammals.

**32.9**  Mollusks generally carry a shell, which is secreted by the mantle. Yet some mollusks, such as most cephalopods, appear to have evolved toward loss of a shell. What would you say are the advantages and disadvantages of the shell?

Slow-moving animals like clams and snails are highly vulnerable to predators. The shell is a protective device to guard against enemies. In addition to serving as a coat of armor, the shell may demonstrate elaborate patterns and color schemes that camouflage the mollusk. In bivalves such as scallops, rapid opening and closing of the shell actually provides a means of movement.

A disadvantage of the shell is that it is extra baggage that slows the organism down. The extra amount of energy required to carry the shell around, as in the case of land snails, must also count as a negative factor. However, since many mollusks are not capable of rapid locomotion even in the absence of a shell, on balance the shell confers greater advantages than disadvantages.

In the cephalopod, like the squid, an increasing reliance on rapid movement, both for hunting and for evading enemies, occurs. In these cephalopods the shell is usually absent except, in some cases, as a remnant. A siphon within the body uses a water jet to propel the animal through its marine environment. In addition, the squid ejects an inky material from its anus (forced by a funneled structure into the water) that serves as camouflage.

**32.10**  The Arthropoda (and especially Insecta) are regarded by most biologists as the most successful of all animal groups. Why do you suppose this is?

Quantitatively, those groups that have the greatest number of species and the largest number of individuals are more successful in a survival sense than groups with fewer species and numbers. Almost 80 percent of the animal species now alive are arthropods, with almost 1 million species of insects alone. Only the nematodes rival arthropods in numbers of living individuals.

Second, a great diversity of species invading all possible habitats signals a successful evolutionary development. The arthropods have indeed spread to almost every kind of microenvironment and demonstrate a global distribution. Insects and arachnids probably outdo the mammals as conquerors of land. The chitinous, jointed exoskeleton of the arthropods provides maximum protection against both desiccation and predators, while allowing considerable freedom of movement. Reproductive patterns, including the development of an extendable penis and internal fertilization, also provide for life in dry habitats.

A third indication of biological success is the way in which arthropods, particularly insects, have coevolved with plants and other animals to ensure survival by mutualistic interactions. The pollination of flowers by insects is one of many examples of such adaptive exchanges. Perhaps no other group has achieved a greater exploitation of specific human-created microenvironments. Cockroaches and bedbugs, for example, have taken full advantage of the riches and pleasures of civilization. Bees provide a service and are cultivated by beekeepers, and locusts (order *Orthoptera*) and beetles (order *Coleoptera*) have exploited field crops, with devastating effects.

**32.11**  What are the possible disadvantages of a chitinous exoskeleton (cuticle)?

Every structure produced by an organism entails a price in energy and material resources allotted to that structure. However, the advantages gained from that protective exoskeleton apparently justify the small expenditure of metabolic resources.

A second problem might be some curtailment in flexibility. Sufficient numbers of joints compensate for the stiffness of the skeleton between the joints. Border areas between head and thorax, between thorax and abdomen, and at the point of attachment of the legs are deeply jointed to provide a high degree of bending.

The greatest disadvantage of the skeleton stems from the necessity for growth of the arthropod within the containing walls of the chitinous coat of armor. In most arthropods a series of *molts* (sheddings) occurs in which the stiff outer cuticle bursts and is cast off; the softer inner cuticle is largely resorbed. The resultant naked and vulnerable individual then proceeds to secrete a new cuticle around itself. These molts enable the arthropod to grow while maintaining a recurrently refashioned cuticle. Molting is integrated into the specific patterns of complete or partial metamorphosis found in each species. Metamorphosis in insects is particularly striking.

**32.12**  What very dramatic example of convergent evolution is found in cephalopods, most arthropods, and vertebrates?

All three groups have evolved a very complex and efficient eye, a sensory organ that is highly sensitive to light and affords imaging of the *ambient* (surrounding) environment. In representatives like humans (vertebrate) and the octopus (cephalopod) the eyes are remarkably alike and are called *camera eyes* because of their similarity to a camera. A tiny *pupil* admits light to a chamber which contains a focusing lens. An image is then cast on a light-sensitive plate—the retina. This eye structure, though remarkably similar in the two major groups, arose independently in the course of evolution. The arthropods have a compound eye. In all three cases, the independently derived eyes function in a similar and highly efficient manner.

**32.13**  In both vertebrates and arthropods, the jaw is a jointed, movable apparatus surrounding the mouth. It is used for both capturing and beginning the process of breaking up prey or large food particles. Jaws may snap shut with considerable force, so that they may also serve a defense function. How does the vertebrate jaw differ from the jaw of many arthropods?

The jaws of vertebrates derive from bones of the skeleton located in the gill region. The jaws found in many arthropods are modified anterior appendages. Vertebrate jaws are articulated (connected) so that they move up and down. The arthropod jaw moves laterally.

**32.14**  What is the function of each of the following structures in the *cleidoic* (amniotic) egg of reptiles and birds: yolk sac, allantois, amnion, chorion, and shell?

*Yolk sac*:  a food supply for the developing embryo, especially rich in high-caloric fats
*Allantois*:  a chamber for gas exchange and the elimination of nitrogenous wastes
*Amnion*:  a compartment acting as a watery cushion within which the embryo may develop with a minimum of mechanical shock and an ample supply of water
*Chorion*:  an outer membrane surrounding and protecting the embryo and its vital sacs
*Shell*:  a leathery or brittle outer egg cover affording protection and free access for gas exchange

**32.15** What are the major differences between mammals and reptiles?

Most of the differences center on the increased adaptation of mammals to a terrestrial habitat. Mammals are endothermic, whereas reptiles only partially regulate their body temperatures. This permits mammals to maintain a high and relatively constant degree of metabolic activity independent of environmental temperature fluctuations. The scaly skin of reptiles is replaced by a hair-covered, smooth skin in mammals. Hair, or fur, provides a more efficient insulation system.

The ventricle of the heart in mammals is completely partitioned, so that the heart consists of four complete chambers. The reptilian ventricle is only partly partitioned. The mammalian heart is thus far more efficient in its separation of oxygenated (arterial) and deoxygenated (venous) blood, a useful feature in organisms with a high metabolic rate. Mammalian brain development is far more advanced than that of reptiles, affording mammals a more elaborate set of neural responses such as feeling, association, and fine motor skills. The legs are more directly below the body, which makes running easier. Mammals have a muscular diaphragm, which greatly enhances the efficiency of air flow in the lungs, another requisite for active organisms.

Perhaps the key difference lies in the reproductive patterns of mammals. Not only is fertilization accomplished internally, but the embryo and later fetus develop within the uterus of the female. In almost all mammals a placenta provides a metabolic exchange structure between mother and fetus. Later, following *parturition* (birth), the young are nourished with milk, which is formed within the mother's mammary glands and expelled through the *teats* to the suckling young.

# Supplementary Problems

**32.16** An alternative to a broad animal classification scheme based primarily on the dichotomy of protostomes and deuterostomes is one rooted in the absence or presence of particular kinds of   (*a*) antennae.   (*b*) coeloms.   (*c*) dermal coverings.   (*d*) metamorphoses.   (*e*) blood pigments.

**32.17** Most animals are   (*a*) diploid.   (*b*) haploid.   (*c*) heterotrophic.   (*d*) both *a* and *c*.   (*e*) both *b* and *c*.

**32.18** The body sac of a sponge is   (*a*) porous.   (*b*) watertight.   (*c*) made of chitin.   (*d*) both *a* and *c*.   (*e*) both *b* and *c*.

**32.19** Select the matching term from column **B** that best fits the descriptive phrase in column **A**.

| A | B |
|---|---|
| 1. Porifera and Cnidaria | (*a*) Diplobastic |
| 2. A distinguishing feature of flatworms | (*b*) Triploblastic |
| 3. Arthropoda and Annelida but not Chordata | (*c*) Protostomes |
| 4. Echinodermata but not Mollusca | (*d*) Deuterostomes |
| 5. Annelida but not Nematoda | (*e*) True coelom |
| 6. A cavity surrounded by mesoderm | |

**32.20** Which have a greater number of species?   (*a*) cnidarians.   (*b*) mollusks.

**32.21** Which has a greater shell size?   (*a*) Gastropoda.   (*b*) Cephalopoda.

**32.22** Which is larger?   (*a*) larva after second molt.   (*b*) the original larva.

**32.23** Which have more legs?   (*a*) insects.   (*b*) spiders.

**32.24** Which have more pairs of wings?   (*a*) Lepidoptera (butterflies).   (*b*) Diptera (flies).

**32.25** Which has more arms? (a) a starfish. (b) an octopus.

**32.26** Which is longer? (a) the gestation (pregnancy) period in elephants. (b) the gestation period in humans.

**32.27** Fish never possess lungs.
(a) True  (b) False

**32.28** Bony fish (teleosts) are always larger than the chondrichthyes (cartilaginous fish).
(a) True  (b) False

**32.29** The lobe-finned fishes, which possess lungs and gills, may have given rise to terrestrial vertebrates.
(a) True  (b) False

**32.30** Frogs are capable of internal fertilization.
(a) True  (b) False

**32.31** After the mass extinctions of approximately 65–70 million years ago, mammals began their expansion.
(a) True  (b) False

**32.32** Aquatic mammals do not produce milk.
(a) True  (b) False

# Answers

| | | | |
|---|---|---|---|
| **32.16** (b) | **32.21** (a) | **32.25** (b) | **32.29** (a) |
| **32.17** (d) | **32.22** (a) | **32.26** (a) | **32.30** (b) |
| **32.18** (a) | **32.23** (b) | **32.27** (b) | **32.31** (a) |
| **32.19** 1—(a); 2—(b); 3—(c); 4—(d); 5—(e); 6—(e) | **32.24** (a) | **32.28** (b) | **32.32** (b) |
| **32.20** (b) | | | |

# Chapter 33

# The Primates

The immediate ancestors of present-day primates were probably a group of relatively nondescript (in human terms) insectivores that looked like the tree shrews. The primates ("of first rank") were named by humans, who naturally regarded their order of mammals as the most important. If mice or rats were doing the taxonomy, then rodents would probably be called "primates."

The primates appear to have evolved, at least at first, for an arboreal existence. The digits are relatively unspecialized and best serve for grasping vines or branches of trees.

**EXAMPLE 1**  Many mammals, such as the whale, have modified the original five-digit arrangement at the end of the forelimb into specialized structures, such as the flipper. In the horse the five digits have become a single-digit hoof. (Actually the hoof is the nail of that one digit.) In primates the digits have all remained and continue the jointed-arrangement of the primitive reptilian structure.

Changes in vision, modification of the pelvis for an upright stance, behavioral variations, and, eventually, great expansion of the higher brain centers also occurred during primate evolution. In the more advanced apes and in humans a tendency to descend from the trees and to assume a terrestrial existence characterizes phylogenetic development.

## 33.1  PRIMATE LINEAGES

Primates came into existence more than 65 million years ago (see Fig. 33.1). The present-day order Primates (class Mammalia) consists of two suborders: the *prosimians* (lemurs, tarsiers, etc.) and the *anthropoids* (monkeys, apes, and humans). Early in the evolution of the primates the prosimians dominated, but now they are fewer and confined to only a few regions such as Madagascar. Ancient prosimian stock probably gave rise to both modern prosimians and the anthropoids.

There are six families of prosimians, although some zoologists place the family *Tupaiidae* (tree shrews) in the order *Insectivora* because of their primitive nature. The best-known modern prosimians are lemurs, tarsiers, and lorises (including the bush babies). The prosimians have characteristics that are part way between those of insectivores and monkeys.

**EXAMPLE 2**  The tarsier is about the size of a large rat and covers ground much as a kangaroo does. It can turn its head almost completely around, so that it can view terrain directly behind itself. In terms of its relative brain size and the shape of its nose, it resembles the monkeys.

With the separation of the prosimian line, the remaining phylogenetic trunk (*Anthropoidea*) consists of the lineages for monkeys, apes, and hominids (humans). Monkeys first arose about 50 million years ago. First, the new world monkeys branched from the ancient primate line. Later, the old world monkeys evolved as a separate lineage.

**EXAMPLE 3**  The new world monkeys are native to South America. All tend to live in trees but only roughly half have prehensile tails. Most lack the opposable thumb, which is a hallmark of later primates. They belong to the superfamily *Ceboidae*. The nostrils are wide and flare to the side, giving the nose a flattened appearance, hence the former designation *platyrrhine* ("flat nose"). The capuchin monkeys used by organ grinders are an example of the new world monkeys.

The old world monkeys are members of the superfamily *Cercopithecoidea*. Their snout is often long and narrow, and the nostrils are close together, a feature that led to their being named at one time the *catarrhines*. The tails are usually short and are never used for swinging from branches. Their thumbs are opposable, and they have only two premolars rather than the usual three. They may have given rise to some extinct groups of apes, but otherwise they have no clear descendent lineage. Baboons are an example of this group.

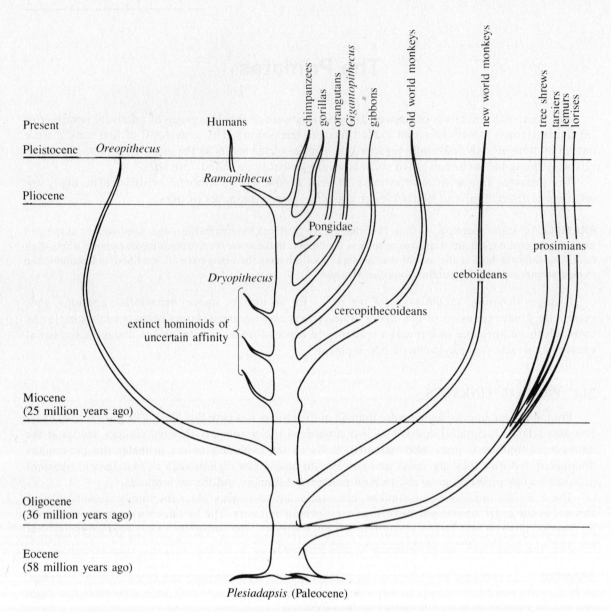

**Fig. 33.1** Primate evolution.

The lineage remaining after the separation of the monkeys is called the *hominoid line*. The paleoanatomist George Gaylord Simpson suggested grouping this line in the superfamily *Hominoidea*. This large grouping includes (1) the lesser apes (*Hylobatidae*), (2) the great apes (*Pongidae*), and (3) humans (*Hominidae*). Some taxonomists no longer recognize a separation of great apes from lesser apes. The *gibbon* is an example of a lesser ape. The great apes consist of the orangutan (*Pongo*), chimpanzee (*Pan*), and gorilla (*Gorilla*). Two species of chimpanzees exist, but both the gorilla and the orangutan consist of but a single species. Modern humans also constitute a single species (*Homo sapiens*).

The earliest fossils of primitive apes are approximately 35 million years old. They are named *Aegyptopithecus*, the "dawn ape." Since the lineage they established was the hominoid line, these primitive apes were the common ancestor of both modern apes and humans. The divergence between apes and humans was once thought to have occurred more than 10 million years ago. More recent theories suggest that divergence occurred about 6 million years ago.

**EXAMPLE 4**  It is common practice to lump apes together as a single group quite separate from humans. This tends to be justified by an analysis of morphological characteristics. However, a cladistic (evolutionary branching) view of the evolution of advanced primates suggests that the gibbon diverged from the hominoid line almost 12 million years ago and the orangutan separated from that line close to 10 million years ago. The evidence for this divergence comes from an analysis of the differences in protein structure among present apes. This evidence is not accepted universally as a measure of the time of separation of the two groups. If one accepts such "dating," however, then humans are closer to gorillas and chimpanzees than these two groups of apes are to the gibbon or orangutan. This is a minority position at this time, but support for this cladistic analysis is growing.

The establishment and tracing of the hominoid lineage is extremely difficult, but not because of a lack of evidence. Rather, an embarrassment of riches in terms of fossil finds complicates the picture and fuels the controversy that exists among rival paleoanthropologists (scientists who study humans through an analysis of fossils). At one time a fossil known as *Proconsul* supposedly represented a common ancestor for apes and humans. The current consensus is that *Proconsul* more properly belongs exclusively to the ape lineage.

## 33.2  PRIMATE CHARACTERISTICS

The initial expansion of the first prosimian primates was into the trees. Subsequent modifications of these ancient prosimian ancestors tended to fit their descendents for arboreal life.

The very flexible shoulder joints of most monkeys and apes permit them to swing from one branch to another. This "moving pendulum" type of locomotion was termed *brachiation* (Latin *brachium*, "arm") by Sir Arthur Keith, who recognized its great efficiency in a jungle setting. Other modifications include the "migration" of the eyes toward the center of the face. This enables the primate to superimpose the images from both eyes, an arrangement that creates the depth perception so vital to those who live in high places. An intense and prolonged period of care for the few or even single young produced is necessary to ensure survival in the perilous perches within the branches of a tree. Even those primates who, like humans, have descended from the trees continue to lavish great care upon their young.

Most primates can grasp objects with a powerful and limber big toe, but this ability has been lost in bipedal, upright primates like humans. However, almost all primates, especially those most recently evolved, have hands with an opposable thumb, which permits not only grasping but a variety of actions involving the manipulation of the environment. This ability, coupled with a marked development of the *neocortex* (cerebrum), probably provides apes, especially humans, with their greatest talent for survival. In upright primates the freeing of the hands for activities ranging from toolmaking to lovemaking has provided fertile ground for complex group interactions and eventually the creation of a culture. These social characteristics are fostered by both the large size and the many convolutions of the brains of the hominoids. In hominids a particular expansion of the capacity for association, speech, and fine motor movement has occurred.

## 33.3  HUMANKIND AND ITS DESTINY

A great number of fossil apelike creatures were found in Africa and later in Europe and Asia as well.

They flourished for many millions of years, and they or their descendents are believed to have evolved into modern apes and humans. This group, the *dryopithecines* ("tree apes"), eventually diverged into several different genera. One of the divergences, *Ramapithecus*, lived about 10 million years ago. Because of some of its hominidlike characteristics, many anthropologists believe that it represents a specific ancestor of humans. Others dispute this assertion and claim that the final divergence between humans and apes occurred only 5 million years ago, so that *Ramapithecus* may be ancestral to both apes and humans.

**EXAMPLE 5**  *Ramapithecus* possessed a smaller but much wider dental arch than the apes do. This approximates the shape of the human dental arch. In humans the hands do much of the breaking and shredding of tough foods, while in most apes the teeth are virtually the only agent for these tasks. The wearing-down patterns of the teeth of *Ramapithecus* individuals suggest that they too used their hands in addition to their teeth for shredding food.

The earliest humans, small and quite different from modern humans, were the *australopithecines* ("southern apes"), so named because they were first discovered in a quarry in South Africa in 1924. They were identified by Raymond Dart. A particularly interesting aspect of the australopithecines is that they were clearly upright walkers, but their brain was not particularly large.

**EXAMPLE 6**   Early australopithecines probably lived about 3.5 million years ago. One of the best known of these fossil australopithecines is *Lucy*, a fairly complete skeleton found in the Afar region of Ethiopia and named by the paleoanthropologist Donald Johanson in 1974. Johanson felt that this remarkable specimen deserves a separate species designation, *Australopithecus afarensis*. Others feel that Lucy (and later-discovered members of that family) are merely variants of the more common *Australopithecus africanus*.

More recent australopithecines have been found that are far brawnier and more robust than the early *A. africanus* (or *A. afarensis*). They have been named *A. robustus* and *A. boisei*. Most paleoan-thropologists believe that the more graceful and smaller australopithecines (*gracile* forms) eventually gave rise to *Homo* and modern man. One scenario is that *A. afarensis* gave rise to *Homo habilis* and then *H. habilis* led to modern humans. With a brain capacity of 450 mL, these australopithecines were more like modern chimpanzees (about 400 mL cranial capacity) than modern humans (about 1400 mL).

The first fossil representative of the human genus *Homo* was *H. habilis*, with a brain capacity of about 650 mL. *H. habilis* ("handy man") was found with simple stone tools in the Olduvai Gorge region of South Africa. It probably arose about 2 million years ago and coexisted with the aus-tralopithecines.

Much closer to modern humans are the various representatives of *H. erectus*, which first evolved more than 1.5 million years ago. These "erect humans" may have lived until about 250,000 years ago.

*H. erectus* is the first example of a true human in a modern sense. *H. erectus* was much larger than earlier forms, approaching the dimensions of our own species. The cranial capacity of about 1000 mL is not too far from that of modern humans either.

The skeletal remains suggest that *H. erectus* walked with a stride similar to our own. A great deal of walking must have occurred because, unlike the australopithecines, which were found exclusively in Africa, *H. erectus* lived in Europe and Asia as well. Possibly related to the migration to colder northern regions is the finding that some groups of *H. erectus* (in China and Spain) had tamed fire and were using it for heating the caves in which they dwelt and hunting. Fire in hunting may have been used to drive large animals into swamps or other inhospitable areas where they could more readily be killed by bands of human hunters.

One striking finding among *H. erectus* remains is the presence of high-quality stone tools. Stones were not merely cut along a single plane but were fashioned into multifaceted *hand axes* with a separate handle. The more elaborate stone tools of *H. erectus* are sometimes described by anthropologists as in the *Acheulian* style. Since the tribes of *H. erectus* were mainly big-game hunters, the tools may have been mainly used for stripping meat from carcasses.

*H. erectus* probably overlapped both *H. habilis* and *H. sapiens*; the latter ("wise human") of course is our own species. *Java Ape Man*, discovered by Eugene Dubois in Java in 1891, and, later, *Peking Man* from central China, are now classified as *H. erectus*. Initially they were designated *Pithecanthropus erectus*. They are among the first group of fossils to be found in Europe and Asia.

Modern humans are represented by fossil fragments found in Swanscombe, England, and in the caves of the Neander region in Germany. The Neanderthal specimens represent a group, *H. sapiens neanderthalensis*, that lived from about 200,000 years ago to only 35,000 years ago. Characterized by a massive jaw, large brow ridges, and a sharply receding forehead, these modern humans are sometimes described as ignorant brutes, but they actually developed a rather sophisticated culture with burial rituals and other complex social practices. They made flint balls, perforating tools, primitive knives, and mineral pigments. They may actually have been absorbed into our own contemporary group, sometimes designated as *H. sapiens* or, alternatively, may have died out. The *Cro-Magnons*, part of our own group of contemporary humans, first arose about 35,000 years ago in western Europe and disappeared about 10,000 years ago. They were physically indistinguishable from modern races of humans, and their cultural level was equal to that of contemporary groups of humans. Brain size and facial features were equivalent to those of present-day humans. Particularly striking are the magnificent cave drawings they created, which are marked by both a high degree of technical proficiency and the possibility that they used rituals of sympathetic magic (see Chap. 1).

Evolution in humans of the present day is characterized by a self-conscious and self-directing set of influences. A complex symbolic culture permits modern civilizations to direct the course of hominid evolution through the manipulation of the environment, the control of reproduction, the overcoming of disease, and even the engineering of genes to produce preordered outcomes. Some of the unfortunate consequences of that civilization also include unintentional threats to the survival of other species and, in the case of nuclear weapons, a challenge to human welfare and survival as well.

## 33.4 MISCONCEPTIONS OF THE EVOLUTIONARY RELATIONSHIP BETWEEN HUMANS AND APES

It is inaccurate to assume that humans developed from a creature similar to or identical with present-day apes. Actually, both humans and apes developed from a common ancestor that was probably not at all like present-day apes in terms of specific features. Both humans and apes have diverged and have gone their separate adaptive ways for about 8 million years. Many antievolutionists assume that evolution advances the notion that an ape much like the contemporary gorilla gave rise to a human in a relatively short span of time (thousands of years). These views prevent an open consideration of the value of evolutionary theory in explaining diversity among all organisms.

Closely related to this perspective is the equally false idea that evolution always proceeds in straight-line fashion (orthogenic evolution) from one ancestral form directly through a series of descendent forms to some highly adaptive and relatively permanent organism. The process of evolution goes on continually, and the lineage develops, in most cases, like a bush rather than a tree. There are no highly deserving special forms that are the fulfilment of a single lineage; instead, continual branching occurs and only rarely does a series of forms form a single unswerving lineage. In the case of humans, the hominid lineage produced a number of different genera and species, which may have coexisted with one another for a considerable period. At the same time various groups of apes carried on their evolutionary lines to produce contemporary apes; other apes became extinct. Many of these extinct forms were branches of an ape lineage and not directly ancestral to present apes. The important lesson is that once the long-ago divergence occurred between the hominid (human) and pongid (ape) trunks in primate evolution, a separate selective process operated in each of these major groups.

A third misconception is that all the traits associated with hominids either arose at one time or at least began their development together. This is clearly not the case. Such traits as an erect posture appear to have settled in long before other distinctive hominid characteristics appeared.

A fourth belief long held by proponents of the creationist view is that acceptance of the evolutionary descent of humans is incompatible with either a belief in the Judeo-Christian deity or a commitment to religion in general. This is not true, and its falsity is demonstrated by the active participation of religious figures such as Father Teilhard de Chardin in evolutionary theory and fieldwork. Although some fundamentalist sects that accept uncritically a literal interpretation of scripture may encounter conflict with an evolutionary framework, the bulk of religious adherents can resolve their intellectual and spiritual commitments.

## Solved Problems

**33.1**  What characteristics of primates place them in the subphylum Vertebrata?

All vertebrates possess a vertebral column surrounding the spinal cord and a true complex brain surrounded by a cranium. This situation applies to all extant primates. Vertebrates also possess a postanal tail. All primates either possess such a tail [which actually serves as a grasping (prehensile) structure in some prosimians and monkeys] or show evidence of descent from ancestors who did show such a tail and may still have a vestige of that tail in the form of a *coccyx*. The ventral heart is a general feature that primates share with the other vertebrates, but it is not exclusive to this subphylum.

**33.2**   What would you say has been the relative significance of the "smell" brain versus the "vision" brain in primate evolution?

Many of the salient features of primate evolution are related to adaptation to an arboreal life. These include the flexibility of the arm sockets, the prehensile tail, the opposable thumb, which permits grasping and manipulation of features of the environment, the reduction of the long snout, and the conversion of specialized claws to flattened nails, which are more closely molded to the long and delicate digits.

Since vision is so much more important to a life in the trees than smell is, one trend that occurs in primate evolution is an increasing dominance and development of the vision areas of the brain (largely centered in the occipital region in the back of the cerebrum) as opposed to the smell centers (the rhinencephalon), which are located in the anterior part of the cerebrum and project into the snout.

The most primitive of the living primates are the lemurs, usually not much larger than a good-sized rat. They are largely arboreal, but they possess a smell brain, which suggests little divergence from their terrestrial forebears, the insectivores. On the other hand, the arboreal, but more recently evolved, tarsier not only has a vision brain, but has eyes located in a more medial position than those of the lemurs. This repositioning permits an overlapping of visual fields and affords a capacity to assess depth. The importance of vision for a prosimian like the tarsier also shows in a peculiar ability captured in the following lines:

> The Tarsier, weird little beast
> Can't swivel his eyes in the least
> But when sitting at rest
> With his tummy due west
> He can screw his head round to face east.

Only some monkeys and apes are arboreal, but all have a vision brain. The great apes and certainly humans have evolved toward a terrestrial existence, but the brain remains essentially a vision brain—an example of the lag that may occur between modification in lifestyle and an attendant evolutionary modification in morphology. In almost all primates the trend toward an enhanced vision brain is associated with the reduction of the snout. The snout serves as a probe to extend the olfactory senses, so this evolutionary tendency of the primates accords with prediction.

**33.3**   What are the major differences between apes and humans?

1. The earliest difference, clearly encountered even in early forms of australopithecines, is an upright posture and an efficient bipedalism (walking on the hind legs) in humans. Associated with the erect posture is the tendency of the skull to sit roughly centered atop the vertebral column. In apes the column attaches to the rear of the skull. In upright human forms, this repositioning of the skull provides greater support for the cranium and encourages a straight vertical position for the entire body.

2. In apes, the big toe of the foot is long and opposable to the other toes. In all members of *Homo* the toe is smaller and is in line with the other toes. Although the human foot is no longer readily prehensile, the five aligned toes provide a better base support for the upright animal.

3. Both the jaws and the teeth are considerably reduced in humans. This makes the snout narrower. The lips are extremely motile and more readily serve such social functions as speaking and kissing.

4. The brow ridges of the ape are considerably reduced in humans. Instead a smooth, high forehead exists and is associated with extra cranial room for the forebrain.

5. Perhaps the most significant characteristic that separates apes and humans is the great enlargement of the brain that occurred in later hominids. That is best measured by comparing the cranial capacities

of representatives of the two lineages:

| Species | Cranial Capacity, mL |
|---|---|
| Chimpanzee | 400 |
| Gorilla | 550 |
| *Australopithecus africanus* | 500 |
| *Homo habilis* | 650 |
| *Homo erectus* | 1000 |
| *Homo sapiens* | 1400 |

6. Closely associated with the increased brain size, especially in *H. sapiens*, was the creation of a complex symbolic culture. Among the characteristics of that culture was an elaborate toolmaking ability.

**33.4** Compare *H. sapiens neanderthalensis* with Cro-Magnon (*H. sapiens sapiens*).

Both groups are classified as modern humans, but Cro-Magnon is considered closer to modern forms, probably having developed into various European populations of today. Neanderthals first appeared about 300,000 years ago and seem to have died out about 35,000 years ago. Cro-Magnon lived from about 30,000 years ago until about 10,000 years ago.

Neanderthal's face was rugged, with relatively large brow ridges and a very prominent jaw. Cro-Magnon was taller and less burly than Neanderthal and had a straight face, a modern projecting nose, no brow ridges, and a definite chin. Because of their sloping foreheads, Neanderthals are often erroneously thought to have been less intelligent than present-day humans. Their brains may have been slightly larger. Both groups had significantly larger brains than *H. erectus*.

The elements of a relatively sophisticated civilization are apparent in both groups, but Cro-Magnon was more advanced. The Neanderthals made flint balls, perforating tools, primitive knives, and mineral dyes. Evidence suggests that they buried their dead in ceremonial fashion. Cro-Magnons used both stone and bone tools and are well known for their cave paintings.

**33.5** The best-known creations of Cro-Magnon civilization are the highly skilled cave drawings found in Lascaux (France) and Altamira (Spain). These often depict wounded animals and are found deep within the caves. Of what significance might the location and subject matter of these drawings be?

Some anthropologists suggest that because the drawings are found far back in the caves, they might have been part of some magic rituals. Their depiction of wounded animals may signify a form of sympathetic magic in which successful hunting was being sought. These drawings may demonstrate not only a creative drive but also a developing sense of religion.

**33.6** The continuing evolution of modern humans is related to their establishment of a symbolic culture. What is the difference between a symbol and a sign?

A sign is more direct and less abstract than a symbol. A sign is a display or action that indicates a desired response or communicates a specific feeling. Thus, an arrow pointing in a particular direction is a sign that directly points to the specific action intended. In the same way, a snarl or an upraised fist indicates hostility. The sign, in a sense, contains the message.

A symbol is far more subtle and does not intrinsically demonstrate the desire of the symbol creator. A symbol is an abstract formulation that stands for something. It is more flexible than a sign. The best example of a symbolic structure is language. Written or spoken words represent concepts that are not inherent parts of the words themselves. One can say "I love you," and regardless of facial expression those words symbolically convey care and attachment.

One advantage of a *symbolic culture*, a complex of ideas and actions that are built upon symbols, is that it can be stored and passed along from one generation to another. Books may contain concepts, beliefs, and recipes for manufacture that could not be embodied within the biological framework of DNA and enzymes. A symbolic culture gives humans a tremendous power over the environment. It provides a function for postreproductive individuals whose prolonged lives can be devoted to training of the young. It underlies the development of computers and the industrial and medical instruments that enable evolution to be manipulated to serve human goals. Most important of all, a symbolic culture means that each generation can lay its hands on all past accomplishments and add new insights to those of the past.

# Supplementary Problems

**33.7**    Primates possess   (*a*) a coelom.   (*b*) three primary germ layers.   (*c*) an opposable thumb.   (*d*) all of these.   (*e*) none of these.

**33.8**    All primates probably evolved from primitive   (*a*) prosimians.   (*b*) annelids.   (*c*) old world monkeys.   (*d*) new world monkeys.   (*e*) *Ramapithecus.*

**33.9**    Many primates lack   (*a*) a forebrain.   (*b*) a small intestine.   (*c*) hair.   (*d*) a tail.   (*e*) enucleated red blood cells.

**33.10**   Primates probably arose about   (*a*) 1 billion years ago.   (*b*) 200 million years ago.   (*c*) 60 million years ago.   (*d*) 2 million years ago.   (*e*) 100,000 years ago.

**33.11**   A baboon is an example of   (*a*) a primate.   (*b*) a monkey.   (*c*) an ape.   (*d*) both *a* and *b*.   (*e*) both *a* and *c*.

**33.12**   Acute vision is most important for animals living   (*a*) on the ground.   (*b*) in water.   (*c*) in trees.   (*d*) in close-knit societies.   (*e*) none of these.

**33.13**   The ape most distant from the human lineages is   (*a*) the gibbon.   (*b*) the orangutan.   (*c*) the chimpanzee.   (*d*) the gorilla.   (*e*) they are all equally distant.

**33.14**   Brow ridges are never present in apes.
(*a*) True   (*b*) False

**33.15**   Most advanced primates are solitary.
(*a*) True   (*b*) False

**33.16**   Dryopithecines were ancestral to apes and humans.
(*a*) True   (*b*) False

**33.17**   Lucy was discovered by Louis Leakey.
(*a*) True   (*b*) False

**33.18**   Australopithecines were larger than modern humans.
(*a*) True   (*b*) False

**33.19**   *Homo erectus* used fire and hand tools.
(*a*) True   (*b*) False

**33.20**   Neanderthals were members of *Homo erectus*.
(*a*) True   (*b*) False

**33.21**   All humans are probably descended from   (*a*) African stock.   (*b*) a North American australopithecine.   (*c*) Eskimos of the Arctic.   (*d*) Australian ancestors.   (*e*) none of these.

**33.22** The significant instrument for the passage of knowledge from one human generation to the next is (*a*) the genome.  (*b*) blood.  (*c*) language.  (*d*) planting of trees.  (*e*) none of these.

# Answers

| | | | | | | | |
|---|---|---|---|---|---|---|---|
| **33.7** | (*d*) | **33.11** | (*d*) | **33.15** | (*b*) | **33.19** | (*a*) |
| **33.8** | (*a*) | **33.12** | (*c*) | **33.16** | (*a*) | **33.20** | (*b*) |
| **33.9** | (*d*) | **33.13** | (*a*) | **33.17** | (*b*) | **33.21** | (*a*) |
| **33.10** | (*c*) | **33.14** | (*b*) | **33.18** | (*b*) | **33.22** | (*c*) |

# Subject Index

# Name Index

NOTES

NOTES

NOTES

480